Lecture Notes in Civil Engineering 438

Lecture Notes in Civil Engineering (LNCE) publishes the latest developments in Civil Engineering—quickly, informally and in top quality. Though original research reported in proceedings and post-proceedings represents the core of LNCE, edited volumes of exceptionally high quality and interest may also be considered for publication. Volumes published in LNCE embrace all aspects and subfields of, as well as new challenges in, Civil Engineering. Topics in the series include:

- Construction and Structural Mechanics
- Building Materials
- Concrete, Steel and Timber Structures
- Geotechnical Engineering
- Earthquake Engineering
- Coastal Engineering
- Ocean and Offshore Engineering; Ships and Floating Structures
- Hydraulics, Hydrology and Water Resources Engineering
- Environmental Engineering and Sustainability
- Structural Health and Monitoring
- Surveying and Geographical Information Systems
- Indoor Environments
- Transportation and Traffic
- Risk Analysis
- Safety and Security

To submit a proposal or request further information, please contact the appropriate Springer Editor:

– Pierpaolo Riva at pierpaolo.riva@springer.com (Europe and Americas);
– Swati Meherishi at swati.meherishi@springer.com (Asia—except China, Australia, and New Zealand);
– Wayne Hu at wayne.hu@springer.com (China).

All books in the series now indexed by Scopus and EI Compendex database!

Zinoviy Blikharskyy · Piotr Koszelnik ·
Lech Lichołai · Piotr Nazarko · Dušan Katunský
Editors

Proceedings of CEE 2023

Civil and Environmental Engineering and Architecture

Springer

Editors
Zinoviy Blikharskyy
Lviv Polytechnic National University
Lviv, Ukraine

Piotr Koszelnik
Rzeszów University of Technology
Rzeszów, Poland

Lech Lichołai
Faculty of Civil and Environment
Engineering and Architecture
Rzeszów University of Technology
Rzeszów, Poland

Piotr Nazarko
Faculty of Civil and Environment
Engineering and Architecture
Rzeszów University of Technology
Rzeszów, Poland

Dušan Katunský
Faculty of Civil Engineering
Technical University of Košice
Košice, Slovakia

ISSN 2366-2557 ISSN 2366-2565 (electronic)
Lecture Notes in Civil Engineering
ISBN 978-3-031-44954-3 ISBN 978-3-031-44955-0 (eBook)
https://doi.org/10.1007/978-3-031-44955-0

This Springer imprint is published by the registered company Springer Nature Switzerland AG
The registered company address is: Gewerbestrasse 11, 6330 Cham, Switzerland

Paper in this product is recyclable.

Preface

Building constructions, construction technologies, renewable energy sources, heat and gas supply, ventilation and water supply of the buildings are the areas where extreme progress has been made in recent years.

After many successful international conferences in the field of construction and environmental engineering (CEE) «LVIV—KOŠICE—RZESZÓW» Current Issues of Civil and Environmental Engineering that took place over the past 20 years in Rzeszow, Poland (1995, 2000, 2004, 2009, 2015), Košice, Slovakia (2002, 2005, 2011, 2017), Lviv, Ukraine (2001, 2003, 2007, 2013, 2019). Rzeszow University of Technology has organized and held the XVIII International Scientific Conference «LVIV—KOŠICE—RZESZÓW» on September 06–08, 2023, in Rzeszow.

The *Lecture Notes in Civil Engineering* (LNCE) contain the latest advances, innovations, and applications in the field effective methods of calculation, resource-saving technologies and progressive materials in civil and environmental engineering, as presented by leading international researchers and engineers at the XVII International Scientific Conference «LVIV—KOŠICE—RZESZÓW» Current Issues of Civil and Environmental Engineering «LVIV—KOŠICE—RZESZÓW», held in Rzeszow, Poland on September 06–08, 2023. It covers highly diverse topics, including structural shaping and optimization, aspects of structural behaviour and modelling of advanced methods of analysis, experimental tests and numerical simulations, design codes, in particular Eurocodes and other national and regional limit state codes, highway and bridges engineering, modern architectural and structural solutions, innovative materials and products, durability and maintenance, fabrication and erection, sustainability in construction, renewable energy sources, heat, gas and water supply, ventilation and air-conditioning, ecological and energy-saving technology, modern technology in water purification and treatment, protection of water ecosystems. The contributions, which were selected by means of a rigorous international peer-review process, highlight numerous exciting ideas that will spur novel research directions and foster multidisciplinary collaboration among different specialists.

In particular, this book introduces a number of new advances and includes interdisciplinary research, experimental and theoretical studies that advance achievements in civil engineering and environmental engineering. These advances have significant implications in a number of research areas such as sustainable construction and infrastructures, waste and wastewater management, enhanced sustainability, renewable and clean energy, sustainable materials and industrial ecology, and building automation.

We express our sincere gratitude to the members of the Organizing Committee, to the members of the Scientific Committee, and in particular to all authors and participants for their substantial and valuable contributions.

We express our sincere gratitude to Dr. Dieter Merkle, the Vice President of Applied Sciences and Mr. Pierpaolo Riva, the Editor of Engineering and Applied Sciences at Springer for their great support in preparing the final book.

We hope that the reader will find the book inspirational for further research and professional life Rzeszow, Poland.

Lviv, Ukraine Zinoviy Blikharskyy
Rzeszów, Poland Piotr Koszelnik
Košice, Slovakia Dušan Katunský
Rzeszów, Poland Lech Lichołai
Rzeszów, Poland Piotr Nazarko
September 2023

About This Book

This Lecture Notes in Civil Engineering (LNCE) have as its principal aim the fostering of professional exchanges between scientists and practitioners who are interested in the field of construction and environmental engineering. Topics include structural shaping and optimization, aspects of structural behaviour and modelling of advanced methods of analysis, experimental tests and numerical simulations, design codes, in particular Eurocodes and other national and regional limit state codes, highway and bridges engineering, modern architectural and structural solutions, innovative materials and products, durability and maintenance, fabrication and erection, sustainability in construction, renewable energy sources, heat, gas and water supply, ventilation and air-conditioning, ecological and energy-saving technology, modern technology in water purification and treatment, protection of water ecosystems.

The journal includes high-quality research articles accepted for publication on the basis of thorough peer reviews. Each scientific work of the conference was reviewed by at least two members of the scientific committee. Their efforts have contributed to the high quality of final works, and therefore their reviewing activities are acknowledged and appreciated.

Zinoviy Blikharskyy
Piotr Koszelnik
Dušan Katunský
Lech Lichołai
Piotr Nazarko

Editorial Board/Scientific Commitee

Contents

About the Editors

Prof. Zinoviy Blikharskyy professor of civil engineering, Director of the Institute of Building and Environmental Engineering of the "Lviv Polytechnic" National University, Lviv, Ukraine. He has more than 35 years of teaching, research and administrative experience in higher education institutions and universities in Ukraine and abroad. He has received engineering and construction education in the field of Industrial and Civil Engineering, as well as both a Ph.D. and Dr. Sc. degree in the field of building constructions. He is the winner of the State Prize for "Restoration of the Palace of Bondinelli", which is an award of the President of Ukraine. His scientific interests are corrosion of reinforced concrete structures taking into account load action; features of restoration of bearing capacity, operational suitability, strengthening of reinforced concrete structures under the action of loading; use of non-metal reinforcement for strengthening of normal and inclined sections of reinforced concrete structures; reconstruction and restoration of buildings and constructions of historical and modern building. He has published more than 150 scientific papers, supervised ten candidates' theses and was a co-founder of six international conferences "Current issues of Civil and Environmental Engineering and Architecture" (2007–2017 years). Also he is an active member of the Academy of Construction of Ukraine, editor-in-chief of the magazine "Theory and Practice of Construction" since 2007.

Prof. Piotr Koszelnik Associate professor of Environmental Engineering at the Faculty of Civil and Environmental Engineering, Rzeszów University of Technology, Poland, and since 2010 the Head of the Department of Environmental Engineering and Chemistry. He serves as the Dean of the Faculty and a member of the Committee on Environmental Engineering of the Polish Academy of Science. Main academic interests include environmental chemistry especially water chemistry, carbon and nitrogen cycling, stable isotopes, eutrophication, micropollutants in water, man-made lakes. After obtaining the habilitation, he conducted the project "Analysis of nitrate sources in the surface waters of the San River catchment based on nitrogen and oxygen isotopic ratios". These studies were thematically related to the implementation of the EU Nitrates Directive, under which it is necessary to identify sources and then to limit the runoff of nitrates from agricultural sources to waters. Moreover he was one of the investigators for the research project "Emission of greenhouse gases from dam reservoirs". As part of the project, study was conducted on the production of methane and carbon dioxide in the bottom sediments of six dam reservoirs located in south-eastern Poland and emission of these gases into the atmosphere. He was also the manager of the project "Study on resilience to degradation and the possibility of reclamation of small water retention reservoirs with the use of calcium preparations". He is the author of more than 100 scientific papers.

Prof. Lech Lichołai professor of civil engineering, Dean of the Faculty of Civil and Environmental Engineering and Architecture, Head of the Department of Building Engineering. He scientifically deals with issues related to material and technological solutions in buildings, utilization of renewable energy in buildings, passive solar heating systems, sustainable building. The Department of Building Engineering has external isothermal chambers for testing passive solar heating systems. Various configurations of thermal-storage walls operating in real conditions of the external climate are being tested. Among other things, the accumulation part of the partitions is equipped with phase change materials. The thermal efficiency of such walls is determined, which can be used in building envelope elements. He is a member of various societies and scientific organizations; member of the Section of Building Physics in the Committee of Civil and Water Engineering of the Polish Academy of Sciences, member of the Polish Solar Energy Association, member of the Science Committee of the Polish Association of Construction Engineers and Technicians. He is the organizer and chairman of the Organizing Committee of the Conference Sustainable Development, Architecture—Building Construction—Environmental Engineering and Protection, Innovative energy-efficient technologies, utilization of renewable energy sources, which is organized regularly at Lake Solina. At the Rzeszów University of Technology, he conducts classes in the following thematic areas: Building engineering, Energy efficiency in architecture and urban planning, Energy-saving technologies, Sustainable development in building. He is a member of the Senate of the Rzeszów University of Technology.

Prof. Piotr Nazarko Associate professor at the Faculty of Civil and Environmental Engineering and Architecture, Rzeszow University of Technology, Poland, and he serves there as Vice-Dean for Development. He is a graduate of civil engineering and pursued doctoral studies in the field of mechanical engineering. His main academic interests include structural mechanics and non-destructive damage detection, with a particular emphasis on utilizing soft computing methods. At his university, he is recognized as an expert in the field of BIM technology and is responsible for organizing postgraduate studies in this area. He is an enthusiast for implementing active learning methods into the education process and recently took on the responsibility of leading the university's Centre for Teaching Excellence. He is also an Associate Editor of the *Journal of Civil Engineering, Environment, and Architecture* (JCEEA), published by the Publishing Office of Rzeszow University of Technology.

Prof. Dušan Katunský is a full-time professor at the Technical University of Košice, which is the second largest university in Slovakia. Professor Katunský is a recognized researcher in building physics and dean of the Faculty of Civil Engineering at the Technical University of Kosice. He was appointed as a university professor by the President of the Slovak Republic in January 2008. Prior to this he was the head of the Institute of Architectural Engineering and Institute of Building Technology. His research focuses on building physics, including heat and moisture analysis, aerodynamics, hydrodynamics, acoustics and daylight lighting. He is the author of several well-cited and recognized scientific papers in indexed international journals. He is also the author and co-author of several book publications intended for students and university teachers. He is the author of several monographs that were published in foreign languages. He is a reviewer for publishers such as Elsevier, Thompson Reuters. He acts as the editor in chief for University's Selected Scientific Papers—*Journal of Civil Engineering*, which has been issued for 17 years. He is a qualified expert in construction and has completed several expert assessments for practice, courts, the police and the prosecutors office.

An Influence of Roof Lattice Girder Inclination on the Work of Structural Systems Supporting Shed Roof Sheeting

Jacek Abramczyk[1](✉) and Katarzyna Chrzanowska[2]

[1] Rzeszow University of Technology, Al. Powstańców Warszawy 12, Rzeszów 35-959, Poland
jacabram@prz.edu.pl

[2] Doctoral School of the Rzeszow University of Technology, Al. Powstańców Warszawy 12, Room V229, Rzeszów 35-959, Poland

Abstract. The article presents the results of the research related to an influence of changes in the shape of flat tubular frames, constituting structural systems of sheds, on their strength work and stability. The shape changes are made as a result of inclining the roof lattice girders of these frames to a horizontal plane. An initial rectangular frame system, called the basic frame configuration, is optimized first. The optimal base configuration is then converted to a few derivative rectangular trapezoid configurations. The purpose of creating diversified frame configurations is that they can be arranged along the length of a building to obtain subsequent diversified transverse supports for spatial shell roofs. Each examined configuration consists of two vertical columns and a horizontal or oblique lattice girder loaded uniformly with the successive folds of the roof sheeting along the length of its top chord. Each sheeting is composed of thin thin-walled folded sheets connected to each other with their longitudinal edges. The area of the knowledge under consideration is one of several uniform and coherent research areas analysed so far. The previously considered area has been related to the influence of frame column inclination on the stability and strength performance of flat frame systems. The results of the conducted analysis are to be the basis for the development of a uniform comprehensive method for parametric shaping of unconventional building forms covered with metal folded sheeting supported by effective flat frame structural systems characterized by diversified shapes and arranged along the length of the designed building sheds.

Keywords: Flat structural systems · Oblique rigid tubular frames · Transformed cladding made of folded sheets · Unconventional forms of buildings

1 Introduction

The geometric and mechanical properties of nominally flat thin-walled steel sheets folded in one direction allow for their specific torsional and transverse flexural deformations. Thus, it is possible to give different curvatures to their transverse shapes by varying the shape and mutual position of the directrices supporting the transformed roof sheeting due

Z. Blikharskyy et al. (Eds.): CEE 2023, LNCE 438, pp. 1–12, 2024.
https://doi.org/10.1007/978-3-031-44955-0_1

to its low torsional and transverse bending stiffness [1]. Subsequent flat frame systems with variously sloped roof directrices are arranged transversely to the directions of all roof folds. Differentiation of the mutual position or shapes of the roof directrices arranged along the roof (building) length forces an unconventional folded shell form of each roof (building) [2] (Fig. 1).

(a) (b)

Fig. 1. Roof sheeting composed of repetitive folded shell units and its structural system: **a** external view, **b** internal view

Flat transverse frame systems are the most commonly used simple structural systems leading to different mutual positions of roof directrices supporting folded spatial coverings of a medium span [3]. As a result of using different inclinations of the subsequent roof directrices to the horizontal plane, unconventional forms of the roof coverings can be achieved. This action leads to diversified innovative general forms of buildings roofed with transformed corrugated sheeting [1]. The flat transverse systems with diversified inclination of their subsequent directrices are arranged successively along the length of a building to attract its general form.

As a result of the inclination of the frame roof directrices belonging to the same structural system, their forms change from rectangular to trapezoidal-rectangular, which makes the changes in the direction of not only the external load applied to the directrices, but also the characteristics of the load transferred from the roof girder to the frame columns. Therefore, not only quantitative but also qualitative changes in the strength work and stability of the above-mentioned flat systems are expected [4]. If the subsequent folds of roof sheeting are supported by the roof directrices with additional individual plates constituting the tables fixed to the directrices, the load carried and transferred by them is treated as distributed uniformly along the directrix length, which distinguishes the considered covers from other ones used in the construction industry.

2 Critical Analysis of the Present Knowledge

Several original methods for shaping innovative general forms of buildings and their rational structural systems are presented in the monograph [5]. A number of rational structural systems intended for various unconventional forms of buildings is dealt with by Eekhout [6] and Motro [7] within the structural morphology [8]. Each method of shaping spatial building forms and their rational structures depends to a large extent on the material from which the structures and their coating are made. Smart materials for construction members are presented by Abbas et al. [9].

The generally known geometric and mechanical properties of the nominally flat thin-walled folded metal sheets significantly limit a freedom to shape free forms of façades and roofs with the folded sheeting and frameworks supporting the sheeting [10]. To diversify the general forms of buildings, sheeting made up of the folded sheets is transformed into spatial shell forms that are shallow parabolic-hyperbolic folded roof shells called hypars [11, 12]. The directrices supporting these shells are most often rectilinear and skew and their edges are closed spatial quadrangles.

The important limitations concerning creation of the only shallow and little-diversified hypars are exceeded by Reichhart [1]. The Reichhart's method uses an appropriate technique for arranging the nominally flat folded sheets along the respectively designed directrices so that the sheets successively built into the roof or elevation sheeting have a freedom of their transverse and longitudinal deformations assuring their effective mechanical work under a characteristic load. In addition, the length of their transverse edges has to be equal to the length of the relevant sections of the precisely designed directrices. In other words, the spacing of these tables located along each directrix must be adjusted to the changes in the fold's width of each sheeting resulting from the shape transformation of this sheeting. The Reichhart's method makes it possible to impose many desired forms for the roof or facade sheeting by means of the appropriately developed shape, length and mutual position of the subsequent directrices included in the building structural system designed.

The changes in the shape and relative position of the subsequent directrices cause respective changes in the shell form and transformation of each sheeting's fold. In addition, there are observed important changes in the lengths of almost all supporting lines of each transformed shell sheeting at the same time. Thus, a rational design of the forms is difficult [13]. On the basis of the Reichart's method, it is possible to develop a method for shaping plane transverse structural systems containing mutually skew directrices forcing various attractive novel forms of the designed folded roof and elevation sheeting.

A significant modification of the Reichhart's method has been introduced by Abramczyk using the lines of striction of ruled surfaces to model the contraction appearing in each folded sheet subjected to torsional or bending-torsional transformation [14]. This approach requires designing several properly correlated directrices to enforce the required shape transformations of the entire folded sheeting. The achieved forms and the performance of their structural systems show the decisive influence of the mutual distance and inclination of all skew directrices belonging to the designed structural system on the form, effort and stability of these systems and the shell thin-walled sheeting supported by these systems.

Several diversified methods for geometrical shaping unconventional general forms of buildings covered with transformed folded sheeting are presented in the monograph by Abramczyk and Chrzanowska [14]. These methods are based on the diversified shape, length and mutual position (inclination and distance) of a few directrices supporting the transformed shells. The results of the preliminary studies related to computer simulations of the static-strength work of a few selected types of structural systems and transformed folded shells supported by these systems are also presented by Abramczyk [15].

In the case of buildings of the medium span, the simplest structural systems consist of a number of flat transverse frames spaced regularly along the length of each building [4]. Each frame of these systems is usually composed of two vertical columns and one flat lattice girder. The upper chord of the girder is the directrix of the designed shell roof [16]. Diversified inclinations of the subsequent directrices force atypical shell forms of roofs made up of the transformed sheeting [2].

The article presents issues related to shaping the flat frames of structural systems [17], where the subsequent frames positioned along the length of a designed building have various forms [18, 19]. The main attention is paid to some geometric, strength and stability properties of the systems supporting the transformed folded shell roofs [20]. A few methods for shaping metal trusses are presented by Ram and Gupta [21] and Somma and Vit [22]. Wardenier et al. [23] pay special attention to static-strength shaping and stability of truss joints. Methods for shaping stabile plane structural systems are presented by Qu et al. [24] and Ziemian et al. [25].

The results of the research related to static-strength work of some spatial and planar structural systems are precisely presented by Zhang et al. [19], and Martin and Purkiss [26]. The structural performance of tubular bar structures are described comprehensively by Marshal [27]. Checking the strength work and stability of all individual elements and their joints as well as the entire structure is related to some significant factors used in shaping rational structural systems [25, 28]. The local stability of all members is also to be considered and ensured by adopting the appropriate bar slenderness [29].

The loads acting on building roofs or facades can take various directions, including vertical, horizontal and normal to the wall or roof surface. All permissible loads and their combinations are taken into account in each designing process of buildings and their structural systems [30, 31]. However, the scope of each innovative research may be divided into a number of different fields. Therefore the influence of the changes in the shape of the examined flat bar structural systems on the change in their strength and stability must also be divided into several areas due to the complexity, size and variety of the problems requiring consideration and solution.

One of these areas is considered in the present article. It is related to the number of different types of loads and their combinations and the number of different configurations of the simulated planar structural systems. In addition, the way of modifying the shapes of the tested transverse systems is limited to the changes in the inclination of the examined lattice girders to the horizontal. This action allows for a simple and clear description of the observed changes and relations. The results of these observations and the analysis of the invented relations are to be implemented into the future method for geometric, strength and stability shaping of various unconventional folded building forms supported by effective flat bar frame systems. The frames have to be characterized by some specific diversified and optimized mechanical properties so that their arrangement along the length of the designed buildings ensures the expected novel and rational different general building free forms.

3 The Aim

The main objective is to present the results of the analysis concerning the impact of changes in inclination of the lattice girders belonging to flat frame structural systems supporting thin-walled flat folded sheeting on the strength work and stability of these systems. This analysis involves an optimization of the cross-sections of the bar elements belonging to the simulated frames in terms of the effort of these elements and the overall stability of the entire frames working under a few selected types of the loads characteristic for sheds. The analysis that has been conducted is one of several important coherent steps of the research leading to a comprehensive method for parametric shaping of sheds characterized by innovative, unconventional general forms and covered with nominally flat thin-walled folded sheets transformed elastically into shell forms.

4 Methodology

The analyzed transverse structural systems are flat rigid tubular frames, each of which consists of a horizontal or oblique lattice girder connected by its upper and lower chords to vertical single-branch columns. The top and bottom chords are parallel and joined to each other by V-braces, Fig. 2a, b. These diagonals are inclined to both chords at the angle of 45 degrees. The columns are fixed in a foundation. All bars of one element type, for example all diagonals, have the same cross-sections.

In the first step of the research, an initial rectangular rigid frame Kbo was adopted and then optimized to the effective rectangular basic configuration Kb, Fig. 2a. This optimal Kb was obtained as a result of cyclic selections and changes in the cross-sections (assigned to each type of the elements) achieved by means of the "trial and error" method and the optimizing condition regarding the highest possible effort of its elements and the sufficient capacity to maintain the entire frame stability. The height H of each system is 12 m, the width $L = 16$ m, the height of the girder $h = 1/8\ L = 2$ m, the spacing of diagonals $k = 3$ m, Fig. 2, and the spacing of the successive transverse systems $b = 6$ m.

The cross-sections calculated for the basic rectangular frame Kb were used in the initial derived configurations Kpoj ($j = 1$ to 3) in the process of shaping the optimized trapezoidal configurations Kpj derived from Kbo, Fig. 2.b. Each of Kpoj was built on the basis of Kb by inclining the lattice girder to a horizontal, where the value of this inclination nb is an independent variable affecting the stability and strength of Kpoj. The inclination is defined as a ratio of the ridge d by the distance L between the side columns. The ridge is measured between the levels of the higher and smaller column heads, Fig. 2.b. The tests were carried out on the basic and several derivative frames for which d takes one of the values belonging to the set {0, 1.5, 3.0, 4.5 m}.

In the research carried out, the main step are computer simulations of the models built in the Robot program [32]. The considered configurations of flat frames are perfect systems that do not take the initial geometrical imperfections related to the accuracy of the joint execution and element location into account. In addition, the non-linear analysis has been used in the calculations to capture the increments of bar bending stiffness caused by their longitudinal forces and transversal stiffness of the whole frame [33]. There are used the subspace iteration method in modal analysis and the incremental Finite Element

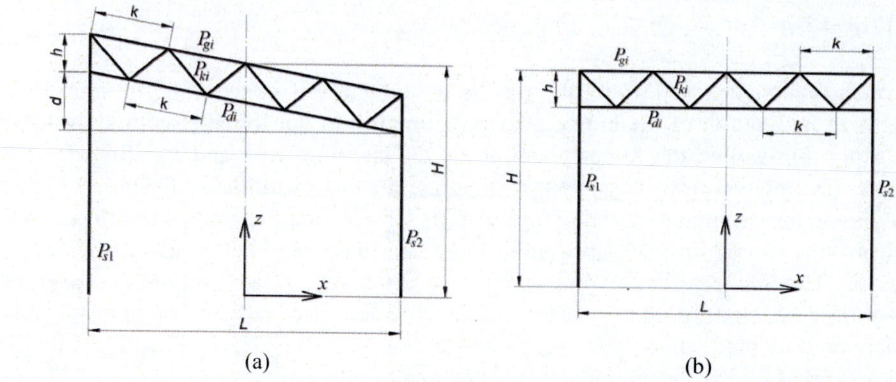

Fig. 2. The diagram presenting: **a** a rectangular, **b** a rectangular trapezial flat frame system consisting of one-branch columns P_{s1} and P_{s2}, top and bottom chords P_{gi} and P_{di}, and diagonal braces P_{ki}

Method in the performed nonlinear calculations to model and compute the geometric and mechanical characteristics of the optimal cross-sections of the examined transverse frame elements. Small displacements and small deformations are assumed. Straight segments divided regularly into finite elements having their length not exceeding 1 m are idealized FEM models of all frame elements.

The performed parameterization made concerned only the shape of the frames. However, parameterization of loads and mechanical properties of the frames is to be considered in the future. The calculation accuracy of the geometry and positions of all bar axes and joints is 1 mm.

For the purposes of the research, the followings are assumed: (1) the vertical characteristic roof load directed down to the frame base is 3 kN/m^2, from which it is possible to calculate the uniformly distributed load acting along the upper chord of the girder q = b·3 kN/m^2 = 18 kN/m, Fig. 3a, (2) the load uniformly distributed on the surface of an oblique shell roof and directed normally to the surface is $r = 6$ m·1.5 kN/m^2 = 9 kN/m^2, Fig. 3b, (3) the vertical asymmetrical characteristic roof load directed downwards and equal to 3 kN/m^2 and 1.5 kN/m^2, from which a uniformly distributed load acting along one half of the top chord can be calculated as q = b·3 kN/m^2 = 18 kN/m and a uniformly distributed load acting along the other half of the same top chord can be achieved as $q/2$ = b·1.5 kN/m^2 = 9 kN/m, Fig. 3c, (4) the another vertical asymmetrical characteristic roof load directed downwards to the base equal to 3 and 1.5 kN/m^2, from which a uniformly distributed load acting along one half of the top chord can be calculated as $q/2$ = b·1.5 kN/m^2 = 9 kN/m, and a uniformly distributed load acting along the other half of the same top chord can be obtained as q = b·3 kN/m^2 = 18 kN/m as the result of changing the positions of the loads defined in the previous point (3) and acting on the upper chord, Fig. 3d.

The self-weight of each frame system is calculated using the internal computer program algorithm. It is not expected to use load and material reduction factors that are adopted in the calculations based on the standards because of the cognitive nature of the research.

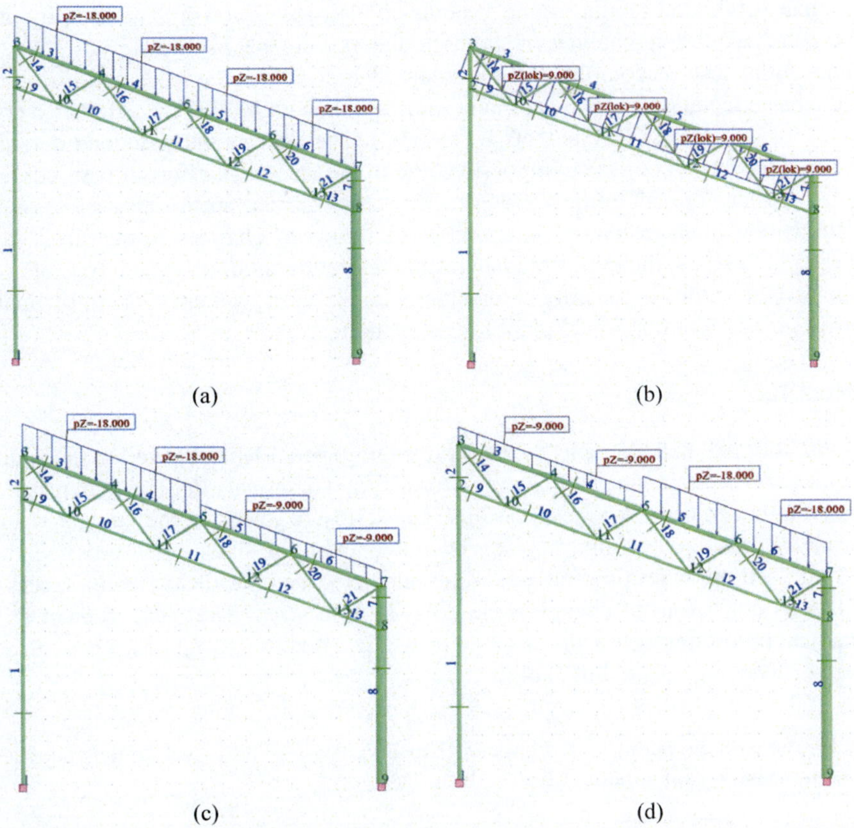

Fig. 3. Four types of the load combinations used for optimizing the considered transverse frame systems, **a** the downward vertical load (corresponding to the live load), **b** the normal load (corresponding to the wind load), **c** the first asymmetrical vertical load (corresponding to the live load), **d** the second asymmetrical vertical load (corresponding to the live load)

For each tested derivative frame configuration, four different combinations of loads were adopted, see Fig. 3a, b, c, d, and the cross-sections of each of both columns are diversified and optimized separately due to the significant difference in their lengths. For the rectangular basic configuration, considering both configurations from Fig. 3c and d does not make sense because of the same results. Therefore, identical cross-sections of P_{s1} and P_{s2} columns in the rectangular base configuration Kb are assumed. Simultaneous action of the self-weight of the bars of all frame systems and the respective characteristic load is adopted.

For the case of the strength work of the tested frames made of S235 steel, the stresses forced by the compressive and tensile forces resulting from the above-mentioned loads combinations were analysed first. Local stability of all bars is maintained by adopting an appropriate cross-section class for each element type. The overall stability of the whole flat frames is defined by the critical load factor, which in the performed theoretical analysis should be greater or equal to 1.0. All considered issues are related to the influence of the shape of the flat tubular frame systems on their strength work and stability in the elastic range, excluding analysis of statics and displacements of joints.

The initial tests accomplishing the adopted first concept on the optimization of only the basic rectangular frame configuration and then exploiting the calculated cross-sections for the analogous elements belonging to the subsequently defined derivative configurations turned out to be impossible due to the very high effort of their columns, especially. Therefore, an optimization processes was carried out for all derivative configurations, and then instead of examining the trends of changes in the stress levels appearing in the bars, the trends of changes in the section module of each type of cross-section and the values of the critical load factor φ_{cr} (caused by the change of inclination of lattice girders to a horizontal plane) were examined.

5 Results

The most strengthened elements of the basic configuration Kb subjected to the adopted load types are the upper chord P_g, lower chord P_d and diagonals P_k. The total resistance of the columns P_s cannot be used because the ability to maintain the general stability of the entire rectangular frame has a decisive impact on the size of their cross-sections. The optimized cross-sections of Kb were calculated as a result of many successive calculations consisting in changing the cross-section sizes. The basic geometric and mechanical properties of the girder's elements calculated as a result of the optimization process of Kb are presented in Table 1.

Table 1. The basic properties of the girder's elements calculated as a result of the optimization process of the basic configuration Kb

Element	Cross-section (mm × mm)	D/d	σ_c (MPa)	σ_c (MPa)	
P_s	177.8 × 4	44	207	– 98	
P_d	114.3 × 4	29	108	– 236	
P_g	219.1 × 4.5	49	238	– 130	
P_k	114.3 × 3	38	157	– 232	
f_{cr}					1.04

The characteristics of the cross-sections presented in Table 1 were then used to create the initial derivative configurations Kpoj ($j = 1$–3). The initial derivative configurations are the basis for the calculations related to the optimization process of the derivative configurations Kpj ($j = 1$–3). The obtained results are given in Table 2.

6 Discussion

After the analysis of the strength work and stability of the examined frames, section modulus were calculated for the cross-sections of the most strengthened elements that are the columns and bottom chords of the examined configurations Kb and *K*pj. Two

Table 2. The compressive σ_c/tensile σ_r stresses calculated for the most stressed bars belonging to each type of the derivative configurations Kpj ($j = 1$–3)

Configuration	Kp1	Kp2	Kp3	
Element	Cross-section (mm × mm)	Cross-section (mm × mm)	Cross-section (mm × mm)	
P_{s1}	139.7 × 4	168.3 × 4	177.8 x 4.5	
	355.6 × 5.6	457 × 6.3	457 x 6.3	
P_d	114.3 × 4.5	127 × 4.5	159 x 4	
P_g	244,5 × 4	219.1 × 5	244.5 x 4.5	
P_k	101.6 × 4	101.6 × 4	101.6 x 4	
σ_c/σ_c (MPa)				
P_{s1}	228/ – 122	237/ – 124	232/ – 136	
P_{s2}	217/ – 234	222/ – 230	218/ – 222	
P_d	125/ – 221	146/ – 218	165/ – 225	
P_g	228/ – 131	231/ – 136	225/ – 138	
P_k	156/ – 223	164/ – 229	198/ – 232	
f_{cr}	7.07	5.81	5.76	1.04

diagrams presented in Fig. 4a, b were built on the basis of these section modulus, where three continuous lines represent three relationships between the values of the above-mentioned section modulus of the P_{s1} and P_{s2} columns and the bottom chord P_d, and the inclination nd of the lattice girders used. The dotted lines: Tend P_{s1}, Tend P_{s2} and Tend P_d illustrate three general trends of changes in the size of the cross-sections of the above elements caused by the inclination of the considered frame girders to a horizontal plane. Absolute values of the calculated section modules are measured along the ordinate axis, Fig. 4a. The relative increments of these modules are measured along the ordinate axis in Fig. 4b. The girder inclination is measured along the axes of abscissa.

The analysis of the values given in the above tables and the properties of the lines from Fig. 4a, b show: (1) the strong significant influence of the inclination of the examined lattice girders on the strength work of their lower columns—the P_{s2} line, (2) the less significant influence of the inclination of the lattice girders on the strength work of the bottom chord—the P_d line, (3) the insignificant influence of the lattice girder inclination on the strength work of the higher column—the P_{s1} line.

An important tendency to qualitative changes in the mechanical properties of the analyzed frame configurations is shown in Fig. 5. Changes in the values of the critical load factor φ_{cr}, caused by the change in the lattice girder slope, start with the value 1.04 and increase very fast. This value obtained during the optimizing process of the cross-sections of the Kb's elements limits the effort of the compressed columns of the rectangular configuration to $\sigma_c = 207$ MPa. This value makes it impossible to exploit the total bearing capacity of the columns. Therefore, the overall stability of the frame plays an important role in the optimization of the Kb basic configuration. In the case of the

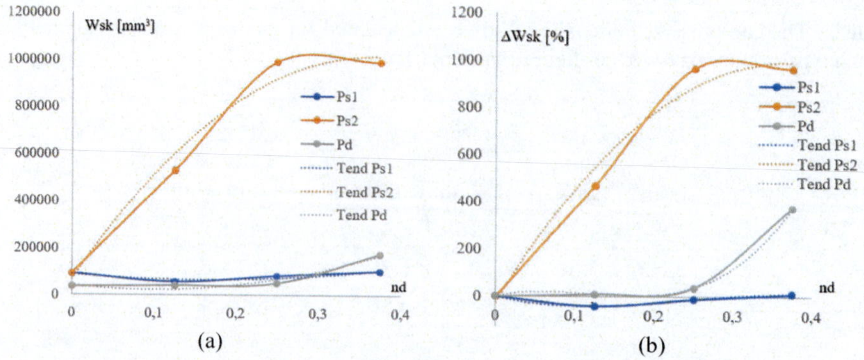

Fig. 4. Three relations occurring between the values of the examined section modules Wsk calculated for the P_{s1}, P_{s2} columns and the bottom chord P_d, and the inclination nd of the lattice girders: (a) the absolute values of Wsk, (b) the relative increments of Wsk

examined derivative configurations Kpj, the decisive role in shaping the cross-sections of all elements is played by their effort.

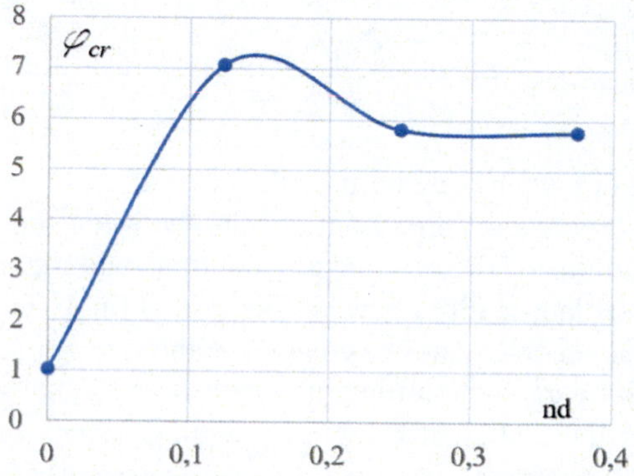

Fig. 5. The dependence of the critical load factor φ_{cr} from the inclination nd of the girders

The frame girder slope causes the radical increase in f_{cr}. Next, the values stabilize themselves around the value 6, and do not change significantly with the increase of the inclination. Thus, a significantly greater ability to maintain overall stability is observed for the derivative configurations than for the rectangular configuration due to the special nature of the loads transmitted from the horizontal or inclined top chords to the vertical columns of different lengths.

7 Conclusions

Inclination of the lattice girders of the examined flat rigid frames to a horizontal plane causes a radical and very significant increase in the effort of their columns and bottom chords as well as the critical loads of the entire frames subjected to the arbitrary types of loads. The optimization processes performed for the successively increased discrete values of the inclinations of the frame girders lead to the important changes in the values of the section modulus of: (a) the P_{s1} columns: by − 40% for nd = 0.125, by − 11% for nd = 0.25 and by 11.5% for nd = 0.375, (b) the P_{s2} columns: by 471% for nd = 0.125, by 968% for nd = 0.25 and 0.375, and (c) the P_d girder bottom chord by 11% for nd = 0.125, by 39% for nd = 0.25 and by 376.5% for nd = 0.375, in relation to the analogous elements of the optimized basic rectangular frame. This increase is especially large for the P_{s2} poles. The increase in the inclination of the columns has a minor effect on the changes in the size of the optimal cross-sections of the bars belonging to the other elements of the examined frame types.

The mechanical changes of the above frames caused by the changes in the inclination of their girders are not only quantitative but also qualitative because the critical loads and the frame's ability to lose overall stability are the limitations determining the minimum size of the column cross-sections of the rectangular configuration. In the case of the examined non-rectangular frames, however, the decisive condition limiting the minimum size of the cross-sections of all their elements is the maximum allowable effort resulting from the yielding of the S235 steel from which the frame bars are made.

References

1. Reichhart A (2002) Geometrical and structural shaping building shells made up of transformed flat folded sheets. Publishing House of Rzeszow University of Technology, Rzeszów
2. Abramczyk J (2017) Shell free forms of buildings roofed with transformed corrugated sheeting. Publishing House of Rzeszow University of Technology, Rzeszów
3. Biegus A (2003) Stalowe budynki halowe (In Polish). Arkady, Warszawa
4. Łubiński M, Żółtowski W (2004) Konstrukcje metalowe cz. 2. Obiekty budowlane (In Polish). Arkady, Warszawa
5. Abel JF, Mungan I (2011) Fifty years of progress for shell and spatial structures. International Association for Shell and Spatial Structures, Madrid
6. Eekhout M (ed) (2000) Form as a bridge between architectural, structural and industrial design. In: 4th international colloquium on structural morphology. IASS: spatial lattice and tension structures, Delft, Netherlands, pp 1000–1010
7. Motro R (2009) An anthology of structural morphology. Universite Montpellier, Word Scientific Publishing Company, Montpellier
8. Li Q, Borgart A, Wu Y (2016) How to understand 'structural morphology'? J Int Assoc Shell Spatial Struct 57(2):145–158
9. Abbas I, Marin M, Saeed TA (2020) GL model on thermo-elastic interaction in a poroelastic material using finite element method. Symmetry 12(3):488
10. Reichhart A (1995) Corrugated deformed steel sheets as material for shells. In: International conference on lightweight structures in civil engineering. Micro-Publisher-C-P, Warsaw, pp 185–197
11. Bryan ER, Davis JM (1982) Manual of stressed skin diaphragm design. Granada Publishing Ltd, London
12. Gergely P, Banavalkar PV, Parker JE (1971) The analysis and behaviour of thin-steel hyperbolic paraboloid shells. In: A research project sponsored by the America iron and steel institute. Technical Report No. 338. Cornell University, Ithaca, NY

13. Abramczyk J (2012) Multi-segment shell structures. In: XIII international conference on lightweight structures in civil engineering (Polish chapter). Micro-Publisher-C-P, Warsaw, pp 38–47
14. Abramczyk J, Chrzanowska K (2022) Complex building forms roofed with transformed shell units. Materials 15(24):8942
15. Abramczyk J (2016) Integrated building forms covered with effectively transformed folded sheets. Procedia Eng 8:1545–1550
16. Nageim H (2016) Steel structures: practical design studies, 4th edn. Boca Raton, CRC Press
17. Lam D, Ang TC, Chiew S-P (2013) Structural Steelwork: design to limit state theory, 4th edn. Taylor & Francis Group, CRC Press
18. Gutkowski W, Bauer J, Gierliński KM, Obrębski J (1975) Obliczenia statyczne płaskich, prętowych przekryć strukturalnych, vol 3. Prace COBPKM Mostostal, pp 1–5
19. Zhang H, Shayan S, Rasmussen KJR, Ellingwood BR (2016) System-based design of planar steel frames, I: reliability framework. J Constr Steel Res 123:135–143
20. Gomes RA, de Oliveira LA, Francisco MB, Gomes GF (2023) Tubular auxetic structures: a review. Thin-Walled Struct 188:110850
21. Ram S, Gupta S (2019) Principles of structural design, wood, steel, and concrete, 3rd edn. Taylor & Francis Group, CRC Press
22. Somma G, Vit M (2023) Experimental investigation of flare groove weld stiffness in lattice girder beams by means of push-out tests. J Constr Steel Res 202:107780
23. Wardenier J, Packer JA, Zhao X-L, van der Vegte GJ (2010) Hollow sections in structural applications. Bouwen met Staal, Zoetermeer, The Netherlands
24. Qu H, Huo J, Li A, Jiang Y, Liu Y (2017) Experimental study on impact behaviour of steel plane tubular frames. Thin-Walled Struct 111:210–223
25. Ziemian RD, McGuire W, Deierlein GG (1992) Inelastic limit states design. Part I: planar frame studies. J Struct Eng (ASCE) 118(9):2532–2549
26. Martin L, Purkiss J (2007) Structural design of steelwork to EN 1993 and EN 1994, 3rd edn. Elsevier, CRC Press
27. Marshal PW (1992) Design of welded tubular structures. Elsevier, Amsterdam
28. Jarmai K, Farkas J (1995) Stability constraints in the optimum design of tubular members. In: Stability of steel structures, international colloquium, European session. Technical University of Budapest, Budapest, pp 11/187–11/194
29. Hillebrand M, Schmidt M, Wieneke K, Hegger J (2021) Investigations on interface shear fatigue of semi-precast slabs with lattice girders. Appl Sci 11(23):1196
30. Kurobane Y, Packer JA, Wardenier J, Yeomans N (2004) Design guide for structural hollow section column connections. Verlag TUV Rheinland GmbH, Koln
31. Packer JA, Henderson JE (1997) Hollow structural section connections and truss—a design guide, 2nd edn. Canadian Institute of Steel Construction, Toronto
32. Autodesk (2023) Robot. Retrieved from https://www.autodesk.com/products/robot-structural-analysis
33. Abramczyk J (2021) Folded sheets as a universal material for shaping transformed shell roofs. Materials 14(13):3582

Behaviour of Steel Lattice Tower with Partially Preloaded Bolted Splices

Edyta Bernatowska, Andrzej Wojnar, and Lucjan Ślęczka(✉)

Rzeszów University of Technology, Rzeszów, Poland
sleczka@prz.edu.pl

Abstract. In steel lattice towers bearing type bolted connections, category A according to EN 1993-1-8, are most often used. Clearance between the bolt and the hole diameters may cause slip of the joined elements, which can influence the structural behaviour of the tower. Usually, such connections are considered as loosely bolted, and slip can occur immediately after loading the joint with a small amount of longitudinal force. The paper presents the structural response of a steel tower with a height of 40.5 m, located in Poland, in which Category A bolted connections were taken into account, but with the possibility of the occurrence of friction forces in contact members, resulting from the tightening of the bolts by the hand or portable cordless impact wrench to a certain level of preload. It is shown how bolt tightening causing partial preload of bolts, being only a fraction of bolt preload applied in the case of Category B or C joints, changes the structural response of the whole tower, with particular emphasis on its deflection in serviceability limit state. Analysis shows, that tightening of the bolts to approximately 60% of the full preload radically changes the structural response of the whole tower compared to loosely tightened connections.

Keywords: Steel lattice tower · Structural response · Category A connections · Effect of joint slip · Partial preload of bolts

1 Introduction

Steel lattice towers are an economical solution of structures used for transmission lines and telecommunication purposes. Very often, they are constructed of open galvanised sections joined by bolted lap splices. The use of slip-resistant bolted connections of category B or C according to EN 1993-1-8 [1] makes high demands on the preparation of contact surfaces and on preloading of bolting assemblies. The nominal minimum preload force obtained by controlled tightening should reach in each bolt value equal to 70% of the ultimate resistance of the bolt. Therefore, in practise, bearing type bolted connections, category A according to EN 1993-1-8, are most often used, which may adversely affect the durability of the structure under variable loads and the overall rigidity of the structure. They are only slightly tightened, and a clearance between bolt diameter (d) and bolt hole diameter (d_0) may cause slip of the joined elements when they are loaded.

Z. Blikharskyy et al. (Eds.): CEE 2023, LNCE 438, pp. 13–21, 2024.
https://doi.org/10.1007/978-3-031-44955-0_2

Such a slip in the bolted joints can influence the serviceability limit state of the towers. It is reported that during the full-scale transmission lattice tower test, the experimentally obtained displacements can be up to three times greater than the numerical ones [2, 3]. To reduce the effect of joint clearance on tower deflections, it is recommended [4] that for applications such as towers and masts, the nominal clearance for normal round holes shall be reduced by 0.5 mm. In such a case for bolts M16-M24 the hole diameter should be 1.5 mm greater than diameter of bolts instead of 2.0 mm, which does not fully eliminate the phenomenon of slip, but only reduce its influence.

The problem of a precise description of the slip has been of interest for years. Models of the behaviour of a bolted joint subject to slipping are formulated by many researchers [3, 5–7], and applied in the analysis of structures [2, 8, 9]. Experimental and numerical studies of the slip factor that influences the shear load-slip behaviour of bolted steel connections are also presented [10, 11]. All this effort helps to provide a more accurate description of the behaviour of the structure in which there are bolted connections.

Existing analyses usually distinguish only Category A connections (with loose tightened bolts) or Categories B and C (with bolts tightened to obtain the full preload force $F_{p,C}$), while currently the most common condition is to use Category A connections, with bolts class 8.8, that are tightened relatively hard, but are not preloaded. Manual tightening of the bolts with the force of the worker's arm [12] results in a tightening torque of approximately 40% of the tightening torque required to preload the M16 bolt class 8.8 and approximately 25% for the M20 bolt class 8.8. The more frequent use of portable cordless impact wrenches increases the value of the tightening torque used in the assembly of category A connections even more. This induces friction force in the connection, proportional to the value of the tightening torque, and thus reduces the resulting slippage.

The paper presents the structural response of a steel tower with a height of 40.5 m, located in Poland, in which Category A bolted connections with M16 and M20 bolts with class 8.8 were taken into account, but with the possibility of inducing friction forces in the contact members, resulting from the partial preload of the bolts. It was shown how the level of bolt tightening changes the structural behaviour of the whole tower, with particular emphasis on its serviceability limit state.

2 Description of the Considered Steel Lattice Tower

The self-supporting tower with height $h = 40.5$ m, used for telecommunication needs as a cellular base station, was chosen for analysis. The tower shaft has a triangular plan, with a base width equal to 4.3 m (Fig. 1). All members are made of hot rolled steel angles, with steel grade S235, and hot-dip galvanised for corrosion protection. Cross-bracing is used as primary spacing patterns on the adjacent faces of a tower for ease of connections and to reduce the number of bolt holes in a section. The tower shaft was vertically subdivided into seven assembly sections, designated as S1 ÷ S7.

All bracing and secondary bracing members are connected to the legs with one bolt only (Fig. 2a), using the gusset plates. Splice joints in leg members, linking assembly sections, are double lap connections, with two shear planes (Fig. 2b). Grade 8.8 bolts are used, and all connections are Category A according to EN 1993-1-8 [1].

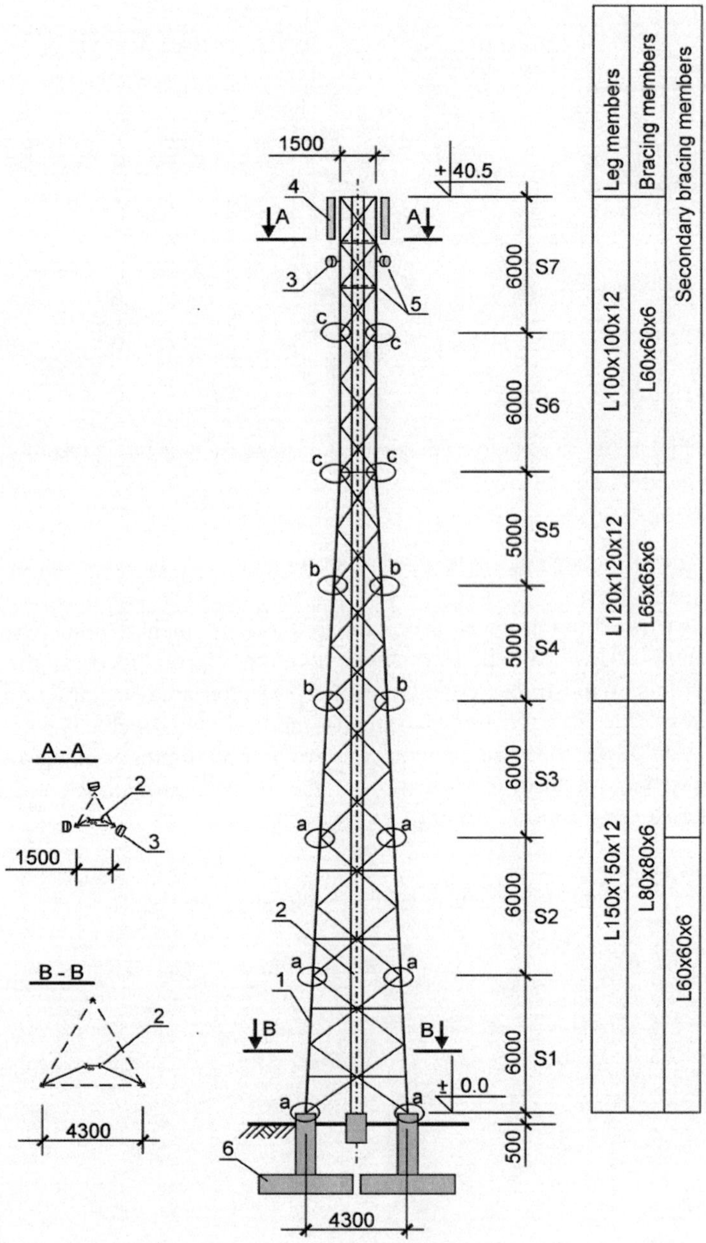

Fig. 1. Shape and dimension of the steel lattice tower analysed: 1—tower shaft, 2—ladder with cables, 3—microwave antennas, 4—panel antennas, 5—working platforms, 6—foundations

The tower is equipped with a ladder installed within the shaft, with a guide for power cables, six panel antennas mounted at a height of 39.5 m, three microwave antennas with

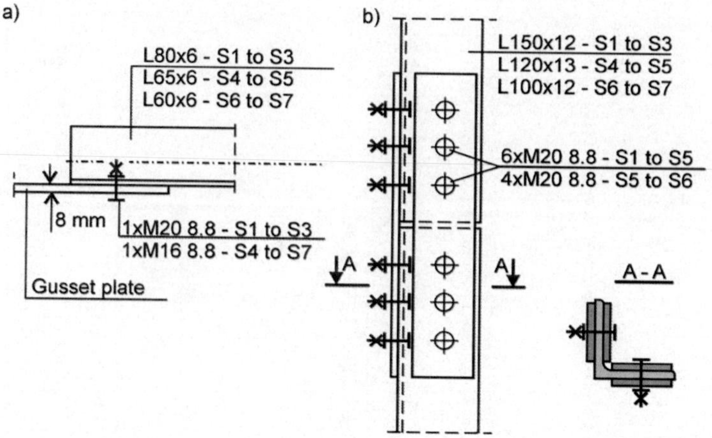

Fig. 2. Bolted joints; **a** connections of the bracing members with the legs, **b** splice joints in the leg members

diameter 0.6 m, suspended at the level of + 38.0 m, and two working platforms (+36.5 and + 38.5 m).

The loads applied to the tower are considered as permanent actions (structure self-weight, antennas and working platforms weights), the imposed loads on the platforms, and wind actions. Wind forces come from the projected areas of structural elements and ancillaries, from the fundamental value of the basic wind velocity $v_{b,0} = 22$ m/s, and the location of the tower in category II terrain. The summary of the resultant the equivalent gust wind load in the direction of the wind on structural components and linear ancillaries is presented in Table 1.

Table 1. Resultant wind forces for individual panels

Designation of sections	Resultant force $F_{T,W(z)}$ [kN]
S1	9.81
S2	10.92
S3	10.80
S4	8.37
S5	8.59
S6	9.94
S7	11.05

Wind forces acting on the antennas were also applied at heights of 38.0 and 39.5 m, with values of 3.97 and 3.91 kN. Ice loading in this analysis is not considered.

3 Modelling of Bolted Joints

In order to capture the effect of bolted joint slip on the tower behaviour, all bolted joints were treated as nonlinear springs, with appropriate force-displacement characteristics.

Throughout the height of the leg members, the following connection types are distinguished: 'a' between the S1-S2, S2-S3 and between the foundation and the bottom section S1; 'b' between the segments S3-S4 and S4-S5, and also 'c'—between the segments S5-S6 and S6-S7 (Fig. 1). Bolted joints connecting the bracing members with the legs were divided into two groups, depending on the diameter of the bolt used (Fig. 2b). Group 'd' includes joints of bracing members located within segments S1, S2 and S3; group 'e' includes joints in segments S4 ÷ S7.

The idealized load-deformation characteristic for the joints was applied taking into account the stiffness of the joint components and the slip between the connecting elements. The parameters of the load-deformation joint model were defined based on the configuration and geometry, which depend on the number of bolts, the number of friction surfaces, and the preload level value of the bolts. The slip range considered was equal to the double clearance between the bolt and the bolt hole diameters $c = \pm (d_0 - d) = \pm 1.5$ mm in case of connections of bracing with legs (single joint), and its doubled value $\pm 2c = \pm 3.0$ mm in case of double lap connections of leg members.

The multilinear force-displacement model presented in Fig. 3 was applied for 'd' and 'e' bracing elements joints. It is characterized of assumption that bolt is centred in the hole (possible slip can occur in case of both compression and tension displacements). The range of displacement in each direction is equal to $c = \pm 1.5$ mm. The level of friction force F_s at which a slip occurs is calculated according to EN 1993-1-8 [1], as shown in Eq. (1):

$$F_s = k_s \, n \, \mu \, k \, F_{p,C} \tag{1}$$

where $k_s = 1.0$, n is the number of the friction surfaces, μ is the slip factor and k is a factor considering the level ($k = 0$; 0.2; 0.4; 0.6; 0.8 and 1.0) of the preload force $F_{p,C}$, defined in Eq. (2).

$$F_{p,C} = 0,\ 7 f_{ub} A_s \tag{2}$$

f_{ub} is the ultimate tensile strength for the bolts ($f_{ub} = 800$ N/mm^2) and A_s is the tensile stress area of the bolt. Friction factor $\mu = 0.3$ was assumed as for hot dip galvanized surfaces.

The stiffness k_j of the B-C region (bearing stage) was calculated according to the component method [1]. The stiffness of the 0-A elastic region should be assumed as rigid according to EN 1993-1-8 (with stiffness $k_{0-A} = \propto$). In order to facilitate calculations in FE programme, the stiffness of this branch was assumed to be $k_{0-A} = 10$ kJ. Similarly, the zero stiffness of the A-B branch (slip stage) was taken into account in the modelling, by giving it a slight slope $k_{A-B} = 0.1$ kJ (Fig. 3b). During structural design, the failure criteria of the joints were checked, so the plastic region (failure stage) of the force-displacement characteristic, which occurs after reaching the shear or bearing resistance of the bolts, was not modelled. Parameters that describe the force-elongation characteristics of the joints considered, used in the analysis, for different levels of bolt preload, are listed in Table 2.

The adopted model of connection behaviour is an idealisation of the model developed by Ungkurapinan et al. [5], which is now widely used [2, 6, 7, 9].

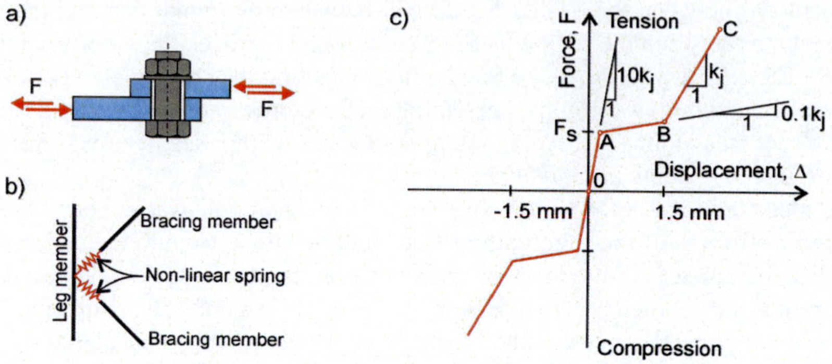

Fig. 3. Bracing member joints; **a** centred position of the bolt, **b** modelling joint as non-linear spring, **c** the force-displacement characteristic

Table 2. Load-deformation joints parameters

Joint type	Stiffness k_j (N/mm)	Friction force F_s [kN], depending on the bolt preloading level				
		$0.2\ F_{p,C}$	$0.4\ F_{p,C}$	$0.6\ F_{p,C}$	$0.8\ F_{p,C}$	$1.0\ F_{p,C}$
a	171360	98.8	197.6	296.3	395.1	493.9
b	156870	98.8	197.6	296.3	395.1	493.9
c	94710	65.9	131.7	197.6	263.4	329.3
d	9870	8.2	16.5	24.7	33.0	41.2
e	7770	5.3	10.6	15.8	21.1	26.4

For splice joints in leg members (joints ‘a’, ‘b’ and ‘c’, Fig. 1), a similar model of joint behaviour was applied; see Fig. 4. In this case, a shifted position of the bolts was used, to influence the assembly conditions. During the erection stage, each sections of the tower is assembled on the ground, then lifted as a unit and bolted to the previously installed segment. In such a case, the weight of the segment affects the small compression force and bearing of the bolts, and then they are tightened. Therefore, later in the working life stage, slippage in such joints is possible only in the case of a tensile force and their range is equal to 2·2c = 6 mm. Stiffness k_j of B-C region (bearing stage) was calculated according to the component method [1]; see Table 2.

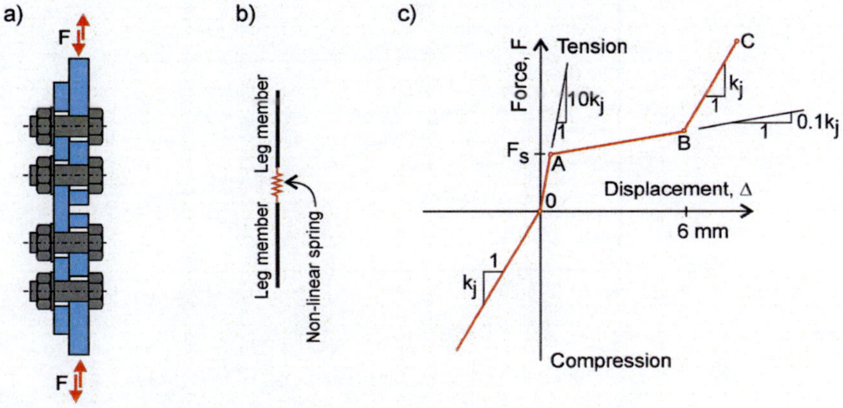

Fig. 4. Splice joints in leg members; **a** initial position of bolts and elements, **b** modelling joint as non-linear spring, **c** the force-elongation characteristic

4 Global Analysis and Results of Structural Response

The response of the whole structure was determined using an elastic first-order analysis, without taking into account global or local imperfections. In the numerical model, individual members were represented by beam elements with proper geometrical properties of cross sections and spatial orientation. Bolted joints were treated as nonlinear springs with zero length, with appropriate force-displacement characteristics (Figs. 3 and 4).

Three wind directions were considered. Basic loading combinations include permanent actions and wind loading, with partial factors for the actions appropriate for ultimate limit state or serviceability limit state. Structural analysis was performed in Autodesk Robot Structural Analysis Professional 2022 software [13]. All members of the tower were checked to fulfil the ULS provisions of Eurocode 3.

The analysis showed that the tower response is vulnerable to the level of bolt preload. Maximum displacement at the top of the structure, caused due to the combined effect of self-weight and wind loads, decreases as the level of bolt preload force is increased. The results obtained in the case of wind load perpendicular to a tower face is presented in Fig. 5. A very similar course of this dependence was obtained in the case of other wind load directions. Similarly, an increase in the bolt preload level in applied Category A connections reduces the angle of rotation of the tower at the level of + 38.0 m, in the place where the microwave antennas are mounted. The course of this change for the wind load perpendicular to a tower face is shown in Fig. 6.

The change in internal forces in the leg and the bracing members caused by the change in the force-elongation characteristics of the bolted connections was practically insignificant.

A change in the dynamic characteristics of the tower, i.e. the fundamental flexural frequency of tower shaft, was also observed (Fig. 7), showing positive influence of bolts tighter tightening. All of these results are based on a single tower analysis, but can be generalised to the need to indicate the level of bolt tightening in category A joints used in towers, which will result in a more rigid shaft structure. Tightening the bolts to 60% of

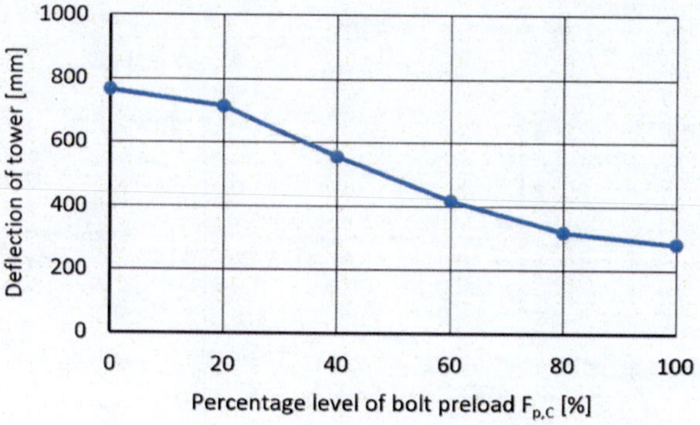

Fig. 5. Tower deflection depending on the level of bolt preload

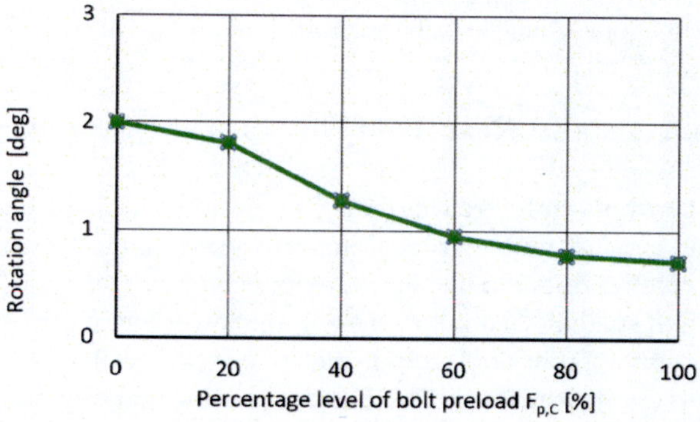

Fig. 6. Rotation angle of the mounting point of microwave antennas depending on the level of bolt preload

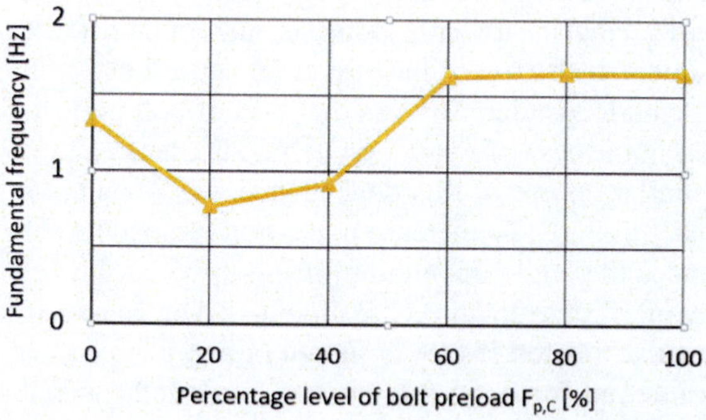

Fig. 7. The fundamental flexural frequency of the tower shaft depending on the level of bolt preload

total preload allows obtaining a structure similar in its structural response to the structure with category C connections.

5 Conclusions

The bolt slip process has random character and depends on the magnitude of the load, the clearance between the bolt hole and the bolt diameters, the manufacturing tolerances, the treatment of the contact surfaces and the tightening of the bolts. Due to the complexity of the description of this phenomenon, simplified models are used.

In this paper, the structural behaviour of the steel lattice tower has been investigated using a non-linear model of a bolted joint with slip, which is influenced by the level of tightening of the bolts. The considered range of bolt tightening varied from zero (loose bolts, ideal Category A according to EN 1993-1-8) to the full preload level $F_{p,C}$ required in Category C joints. The analyses performed confirmed the well-known rule that the harder the bolt is tightened, the stiffer the joints, and thus the whole structure.

Tightening the bolts to 60% of total preload has been shown to reduce tower deflection by approximately 50% compared to tower deflection with loose bolts. A favourable reduction in the angle of rotation of the tower top was also observed at this level of the bolt preload, as well as increasing the fundamental flexural frequency of tower shaft.

Such category A connections, but with a slightly greater tightening of the bolts than assumed by manual tightening by the worker's arm, have a very positive effect on the global behaviour of tower structures, and their execution is still much easier than category C connections.

References

1. EN 1993–1–8 (2005) Eurocode 3: Design of steel structures—Part 1–8: Design of joints. CEN, Brussels
2. Zhan Y, Li B, Wu Z, Cunningham LS, Wu G, Yang Y (2020) Modeling of galvanized lattice steel structures incorporating the effect of joint slip. J Constr Steel Res 173:106252
3. Jiang WQ, Wang ZQ, McClure G, Wang GL, Geng JD (2011) Accurate modeling of joint effects in lattice transmission towers. Eng Struct 33(5):1817–1827
4. EN 1090–2 (2018) Execution of steel structures and aluminium structures—Part 2: Technical requirements for steel structures. CEN, Brussels
5. Ungkurapinan N, Chandrakeerthy SDS, Rajapakse RKND, Yue SB (2003) Joint slip in steel electric transmission towers. Eng Struct 25(6):779–788
6. Gan Y, Deng H, Chao L (2021) Simplified joint-slippage model of bolted joint in lattice transmission tower. Structures 32:1192–1206
7. An L, Jiong W, Wenqiang J (2019) Experimental and numerical study of the axial stiffness of bolted joints in steel lattice transmission tower legs. Eng Struct 187:490–503
8. Kitipornchai S, Al-Bermani FGA, Peyrot AH (1994) Effect of bolt slippage on ultimate behavior of lattice structures. J Struct Eng 120(8):2281–2287
9. Mähr T, Espinha M, Ziegler R (2017) Global structural modelling of space frame structures with consideration of joint behaviour: an integral analysis approach. ce/papers 1(2–3):590–598
10. Cruz A, Simões R, Alves R (2012) Slip factor in slip resistant joints with high strength steel. J Constr Steel Res 70:280–288
11. Lacey AW, Chen W, Hao H, Bi K (2019) Experimental and numerical study of the slip factor for G350-steel bolted connections. J Constr Steel Res 158:576–590
12. P-212 (2009) Joints in steel construction—simple connections. The Steel Construction Institute and the British Constructional Steelwork Association
13. The Autodesk. Retrieved from https://www.autodesk.com. Accessed on 12 May 2023

Digital Image Correlation Pattern for Concrete Characteristics—Optimal Speckle

Yaroslav Blikharskyy(✉), Nadiia Kopiika, Roman Khmil, and Zinoviy Blikharskyy

Lviv Polytechnic National University, Lviv 79013, Ukraine
yaroslav.z.blikharskyy@lpnu.ua

Abstract. Digital Image Correlation (DIC) is an innovative effective method for assessment of stress-strain state of the structural element. The Digital Image Correlation technique enables to study cracking and fracture mechanisms in constructions, in order to obtain the full field of deformations and stresses and to evaluate the load-bearing capacity, reliability and residual resource. Measurement of deformations according to the DIC method is performed with the use of a stochastic contrast speckle pattern, which has to be previously applied to the studied surface. Good pattern determines high confidence of experimental results and low noise level and has to meet the following requirements: high contrast, the consistent speckle sizes and degree of filling of the surface, isotropy and randomness. This paper presents results of experimental research of prism concrete samples with different parameters of speckle pattern. The comparative analysis of experimental results enabled to determine the most optimal speckle sizes, pattern randomness and coverage.

Keywords: Digital image correlation · Optimal speckle pattern · Speckle parameters · Concrete structures · Experiment

1 Introduction

Recent times, ensuring the reliable exploitation of reinforced concrete (RC) structures and prolongation of their service life have become widely discussed issues [1–5]. Therefore, additional attention should be paid to various monitoring techniques, especially the innovative contactless methods, which give wider range of opportunities, than the traditional ones. Among these methods the digital image correlation (DIC) should be highlighted, which has reached the leading positions in engineering scientific community [6].

Digital image correlation could be defined as contactless technique, based on analysis of studied physical surface images. Digital image correlation has proved to be the effective approach for reliable experimental investigation of specific material properties on macro-and micro-levels [7]. For instance, this method can be effectively used to monitor crack propagation and record the full strain field of the concrete surface [8]. The

Z. Blikharskyy et al. (Eds.): CEE 2023, LNCE 438, pp. 22–31, 2024.
https://doi.org/10.1007/978-3-031-44955-0_3

recent research has proved, that the DIC technology is suitable for the assessment of crack expansion in high-strength concrete, providing reliable experimental results [9].

The specific and unique feature of the digital image correlation method is the possibility to monitor the structural behavior and changes of stress-strain field up to the destruction stage. Thus, Yuan et al. [10] successfully used this technique to indicate the maximum strain locations at the destruction stage, which proved that DIC can be adopted to forecast the structural failure. This study also demonstrates, that digital image correlation technique is effective for evaluation of the variation of specimen surface strain distributions on the microstructural level [10].

In addition, the damage level as the integral indicator of structural reliability can be indicated on the basis of energy ration index and the development of damage zones obtained with DIC technology [11].

In the number of research works the DIC results were compared with the data, obtained with other innovative methods. For instance, comparative analysis of digital image correlation (DIC) and acoustic emission (AE) techniques enabled to ensure accuracy of the microscopic failure mechanism of Steel-Fiber-Reinforced Concrete [12]. High coherence between the AE and DIC results was also noted by Sharma et al. from stress-strain field and cracking process in RC beams subjected to four-point bending [13]. Comparison of numerical simulation results and experimental digital image correlation data of strain field within the surface of fiber concrete elements has also confirmed the applicability of DIC [14].

Among the specific features of the DIC is the possibility to measure the displacement over the entire image, unlike to the limitations of standard sensors that only track the deformation at a specific point [15]. The possibility to monitor the cracking process within the whole surface remotely can be used to identify the limit state of the element and danger of its destruction. Thus, this method is highly effective for practical engineering tasks of technical inspections, ensuring high accuracy in prediction of structural residual service resource [16].

As an example, could be considered the study of Dizaji et al., describing full-field three-dimensional digital image correlation (3D-DIC) deformation measurements performed on the Mecklenburg Bridge, which enabled to thoroughly assess the bridge behavior and effectiveness of strengthening measurements [17].

DIC includes following stages: creation of stochastic contrast speckle pattern on the studied surface, recording images of studied physical object in digital form, and further analysis [7, 18]. The general principle of image processing is the comparison of the reference image, recorded before the application of the load with deformed images, recorded at different loading levels [19]. Images analysis is conducted with the use of numerical methods and specialized software, for example Vic-2D/3D, Solid Mechanics DIC, StrainMaster, GOM Correlate Pro, ARAMIS, etc.

The main principles of image processing were thoroughly analyzed and described in number of sources [18, 20, 21]. The first stage includes the application of a stochastic contrast dot pattern on the studied surface. This pattern was recorded in the digital form as the reference image, which could be divided into subsets of adjacent numbers. Subset usually has the square form of 10–50 pixels dimensions.

After that the position of each subset is tracked during the deformation process. Each subset of pixels can be represented as the matrix of natural numbers [7]. For instance, white pixels correspond to value "0", while black ones, —"100" and grey pixels correspond to certain number in the range (0…100). Each subset includes several points with variations of the gray level (I), therefore, the set of different correlation areas id formed [7].

In each subset the center point (A) is considered as the reference point. At different loading stages its position (A') is tracked, which enables to record the deformations of the surface. Digital images of the surface are recorded at different loading stages, represented in the form of matrixes and processed on the personal computer.

The quality and accuracy of the experimental data are usually determined by the parameters of the speckle pattern. In general, the requirements for the pattern include high contrast, the degree of filling of the surface by 50–70%, isotropy and randomness [22]. Also, it is important to use the optimal size of dots, which is usually within 0.5 and 2.5 mm and depends on the specifics of each particular case (for crack monitoring, —smaller dots are required, for full-scale structures, —greater values). It is important to identify the most optimal size of dots in the pattern for each particular case.

2 Aim

The aim of the study is to study the parameters of speckle pattern and identify the most optimal size of dots for digital image correlation measurements. For this purpose, the experimental investigation of concrete samples with different speckle patterns was conducted and experimental results were thoroughly compared and analyzed.

3 Method of Experimental Research of Concrete Samples

The experience of long-lasting research of concrete and reinforced concrete (RC) structures with the use of various experimental and theoretical methods has shown, that in most cases the limit stresses are reached in localized areas. In general, the concentration of stresses and deformations is localized in cracking and non-homogeneous zones, which can lead to the development of irreversible processes. This can result in complex stress-strain state of the structural element, which is complicated to study with the use of traditional methods. Combination of standard loading methods with speckle metrology technologies enables to monitor strains in localized volume of the material. In particular, the digital image correlation technology is based on the comparison of speckle patterns intensity distributions of optically rough surfaces. Combination of the intensities of the correlation peaks with the corresponding calculation algorithms at the subpixel level enables estimation of displacement and deformation fields without restrictions.

For experimental research were manufactured 3 concrete samples,—concrete prisms with dimensions of 600 × 150 × 150 mm. Images of the deformed surface were recorded with the use of monochrome camera Flir Grasshopper 3 with lense Computar F25/2.8.

The digital image correlation method for monitoring of strains and deformations requires the creation of speckle pattern on the investigated surface. For this purpose, the concrete surface was previously polished to the maximum smoothing the surface using a grinder and a grinding stone. After that, the surface was cleaned with a solvent.

The next stage consisted in printing the speckle pattern on paper using a laser printer and application it to the concrete surface. After that, the upper surface of the paper was thoroughly impregnated with acetone and the upper surface of the paper was smoothed out on the concrete. The whole procedure must be done quickly, otherwise acetone will begin to dissolve the speckles and they will become blurred on the surface of the concrete. The obtained speckle pattern is shown on Fig. 1.

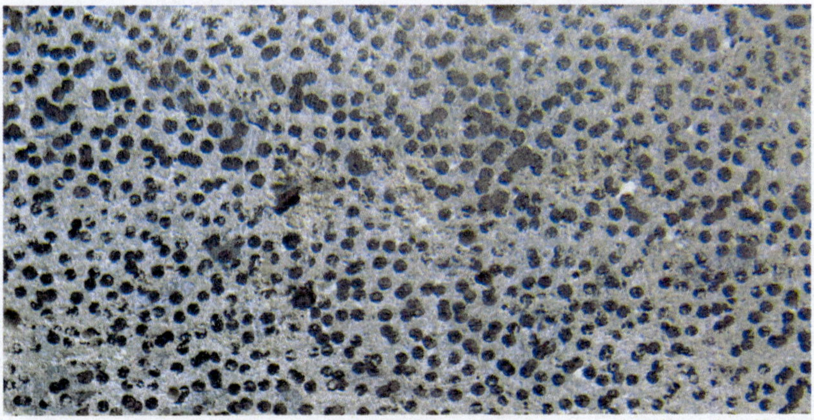

Fig. 1. The speckle pattern on the investigated concrete surface

In general, three samples were prepared. The speckle patterns were generated using a speckle pattern generator from the official web-site of the software developer (Speckle Pattern Fundamentals) with 50% density and 75% variation. Dot sizes were 1, 1.5 and 2 mm.

The load on the concrete samples was applied with the hydraulic press PG-250. Strains were recorded every 250 ms. Loading values were recorded with the use of third camera with *t* a speed of 50 frames/sec. Micro-indicator with an accuracy of 0.001 mm was additionally installed on the prism to control and check results, recorded with DIC approach. The measurement base of the indicator was 200 mm (see Fig. 2).

The general view of the experimental setup is shown on Fig. 3.

Images for DIC approach were recorded by 2 cameras at a speed of 250 ms (1/4 s). 2 LED lamps were used for lighting.

4 Results of Experimental Research of Concrete Samples

For formulation of complete σ-ε diagram, the video of loading process was divided into photos at 4 frames/second. This enabled to synchronize the load with the deformations obtained by the DIC approach and obtained using micro-indicators of 0.001 mm accuracy and 200 mm measurement base.

Fig. 2. The use of micro-indicator and DIC during the experiment

Fig. 3. The general view of the experimental setup

The recorded images were processed with the use of VIC-2D software for each of concrete samples. The area of interest (AOI) was manually identified (see Fig. 4) and the optimal subset size was chosen (15 pixels), which enabled to divide the studied surface into subsets with unique variation of the gray level (see Fig. 5).

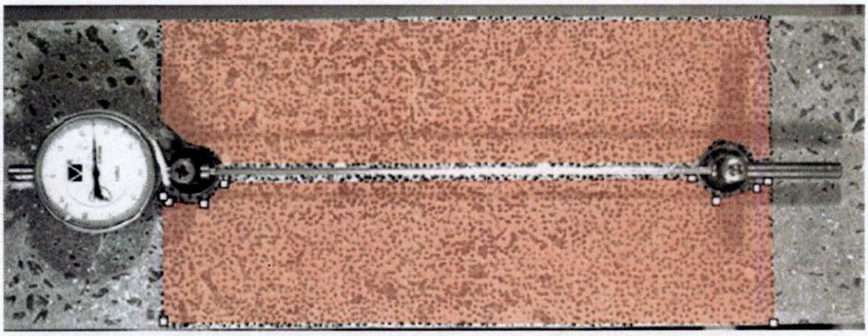

Fig. 4. Identification of the area of interest (AOI) in DIC approach

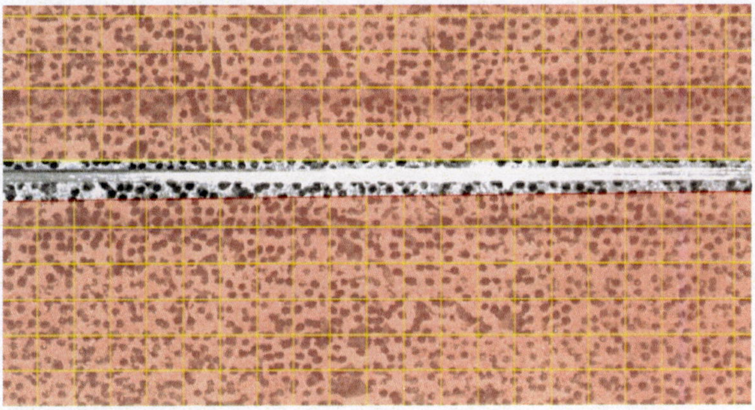

Fig. 5. Dividing the AOI into subsets with unique variation of the gray level

For determination of the strains using the DIC method, virtual extensometers were marked and identified in the corresponding software tab within the middle of the sample with 200 mm base. The process of VIC-2D analysis is shown on Fig. 6.

The digital correlation method makes it possible to record strain values on the entire surface of the sample at all stages of loading, including the destruction stage.

As a result, using the proposed approach, a complete diagram of the destruction of concrete was obtained without the use of specially adapted stands. The "stress-strain" diagrams for concrete prisms with different sizes of dots are shown in Fig. 7.

Th values of DIC measurements for concrete samples with dot sizes of 1, 1.5 and 2 mm were compared with strain values recorded with the use of micro-indicators (gauges). Corresponding "σ-ε" diagrams are shown on the Fig. 7.

Experimental data confirm that the size of dots in the speckle pattern has strong influence on the results' accuracy. The highest concordance was indicated for the sample with 1 mm dots. The strain values recorded with the use of DIC for this sample differed from the gauge data by 3.8%. For the sample with 1.5 mm dots the difference in strain values amounted 12–14%, which is also appropriate for engineering purposes. The experimental investigation has identified the highest deviation for the concrete samples

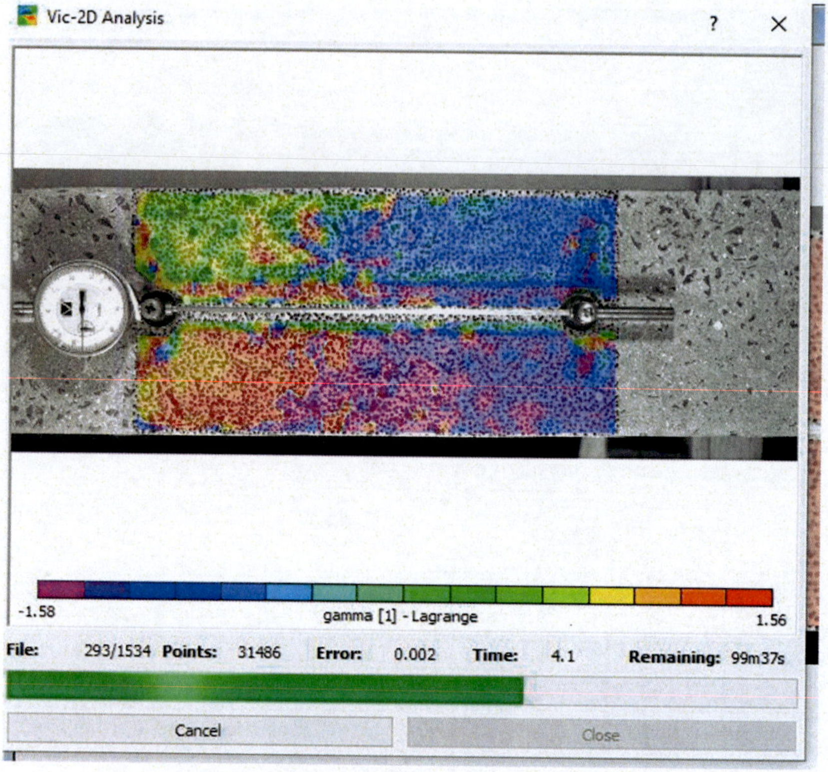

Fig. 6. The process of VIC-2D analysis

with speckle, containing 2 mm-dots, which reached the value of 34%. It is important to note, that the diagram DIC_2 mm (see Fig. 7) covers the process of loading the sample to stages with higher loading than for samples with smaller dot sizes. The probable reason for this is that for this sample the development of cracks on the destruction stage had smaller influence on the DIC measurements. Therefore, although the 2 mm size of dots seems to be unacceptable due to high error, the small number of bigger dots in the speckle pattern could enable to record DIC data at higher loading stages.

5 Scientific Novelty and Practical Significance

The scientific novelty of the conducted research refers to development and deepening of the DIC technique for study of RC and concrete structures. Parameters of speckle pattern and, in particular, the size of dots is highly important for the accuracy of experimental results. Experimental investigation and processing and analysis of obtained data enabled to explore the widely studied issue in a new and innovative way. Results of this study are significant for real-world application of this innovative technology for practical cases of research or real structures.

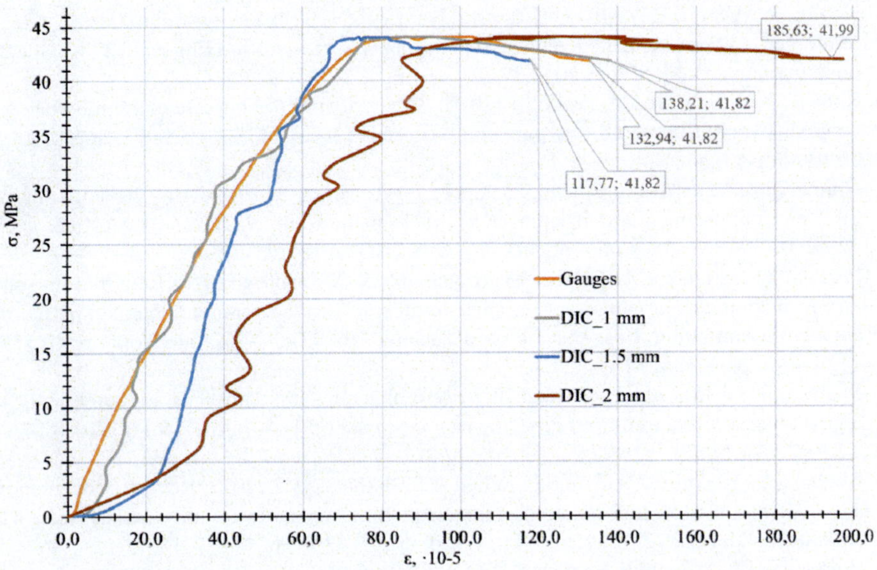

Fig. 7. Complete failure diagrams of concrete samples with different sizes of dots

6 Conclusions

This work presents the results of thorough experimental investigation, aiming to study the parameters of speckle pattern for digital image correlation measurements. The main emphasis was made on identification of the most optimal size of dots, which would ensure the highest accuracy of strain measurements. For this purpose, prism concrete samples with different sizes of dots (1, 1.5, 2 mm) in speckle pattern were tested with the use of DIC and sub-micron indicator. Comparative analysis of the results has shown the highest concordance was obtained for the sample with 1 mm dots. On the other hand, the speckle with 2 mm dots enabled to monitor deformation of the concrete sample until higher levels of loading. This brings us to the conclusion, that the most optimal speckle pattern should contain dots of different sizes, with higher ratio of 1 mm size dots. In addition, compound literature review was made, which confirmed the significance and relevance of the studied issue. It can be summarized, that the Digital Image Correlation method is perspective approach for investigation of stress-strain state of reinforced concrete and concrete structures, subjected to various impacts. It could be concluded, that this innovative technique requires specific attention and additional theoretical and experimental investigation of various factors, which have influence on the accuracy of results. In particular, it could be recommended to study in more detail optimal degree of filling of speckle pattern and effective ways of its creation.

References

1. Blikharskyy Y, Vashkevych R, Kopiika N, Bobalo T, Blikharskyy Z (2021) Calculation residual strength of reinforced concrete beams with damages, which occurred during loading. IOP Conf Ser Mater Sci Eng 1021(1):012012. https://doi.org/10.1088/1757-899X/1021/1/012012

2. Kopiika N, Vegera P, Vashkevych R, Blikharskyy Z (2021) Stress-strain state of damaged reinforced concrete bended elements at operational load level. Prod Eng Arch 27(4):242–247. https://doi.org/10.30657/pea.2021.27.32
3. Koteš P, Vavruš M, Jošt J, Prokop J (2020) Strengthening of concrete column by using the wrapper layer of fibre reinforced concrete. Materials 13(23):5432-1–5432-21. https://doi.org/10.3390/ma13235432
4. Kramarchuk A, Ilnytskyy B, Kopiika N (2023) Ensuring the load-bearing capacity of monolithic reinforced concrete slab damaged by cracks in the compressed zone. Lect Notes in Civil Eng 290:217–229. https://doi.org/10.1007/978-3-031-14141-6_21
5. Poznyak O, Sanytsky M, Zavadsky I, Braichenko S, Melnyk A (2018) Research into structure formation and properties of the fiber-reinforced aerated concrete obtained by the non-autoclaved hardening. East-Eur J Enterp Technol 3(6–93):39–46. https://doi.org/10.15587/1729-4061.2018.133594
6. Manzato S, Di Lorenzo E, Mäckel P (2019) Enhancing standard GVT measurements with digital image correlation. Conf Proc Soc Exp Mech Ser 6:31–41. https://doi.org/10.1007/978-3-319-74476-6_5
7. Blikharskyy Y, Kopiika N, Khmil R, Selejdak J, Blikharskyy Z (2022) Review of development and application of digital image correlation method for study of stress-strain state of RC structures. Appl Sci 12(19):10157. https://doi.org/10.3390/app121910157
8. Wijanarko NP, Arijoeni E, Sentosa BOB (2020) Study on displacement and strain of concrete cubes using digital image correlation (DIC). IOP Conf Series Mater Sci Eng 801(1):012020. https://doi.org/10.1088/1757-899X/801/1/012020
9. Cheng Z, Zhao H, Long G, Yang K, Chen M, Wu Z (2023) The mechanical characteristics of high-strength self-compacting concrete with toughening materials based on digital image correlation technology. Materials 16(4):1695. https://doi.org/10.3390/ma16041695
10. Yuan Y, Li M, Alquraishi AS, Sun H (2021) Experimental study on the novel interface bond behavior between fiber-reinforced concrete and common concrete through 3D-DIC. Adv Mater Sci Eng 2021:1–21. https://doi.org/10.1155/2021/9090348
11. Wang X, Zhou L, Liang Y, Zheng Y, Li L, Di B (2023) Study on shear behaviors and damage assessment of circular concrete short columns reinforced with GFRP bars and spiral stirrups. Polymers 15(3):567. https://doi.org/10.3390/polym15030567
12. Luo T, Pan X, Tang L, Sun Q, Pan J (2022) Research on splitting-tensile properties and failure mechanism of steel-fiber-reinforced concrete based on DIC and AE techniques. Materials 15(20):7150. https://doi.org/10.3390/ma15207150
13. Sharma G, Sharma S, Sharma SK (2021) Non-destructive evaluation of steel and GFRP reinforced beams using AE and DIC techniques. Struct Eng Mech 77(5):637–650. https://doi.org/10.12989/sem.2021.77.5.637
14. Adamczak-Bugno A, Lipiec S, Vavruš M, Koteš P (2022) Non-destructive methods and numerical analysis used for monitoring and analysis of fibre concrete deformations. Materials 15(20):7268. https://doi.org/10.3390/ma15207268
15. Sjölander A, Belloni V, Nascetti A (2022) Dataset to track concrete cracking using DIC with fixed and moving camera. In: Mendeley Data V1. https://doi.org/10.17632/dns97tfdjn.1
16. Cruz H, Aval SF, Dhawan K, Pourhomayoun M, Rodriguez-Nik T, Mazari M (2019) Non-contact surface displacement measurement for concrete samples using image correlation technique. In: Proceedings of the 2019 international conference on image processing, computer vision, and pattern recognition, Las Vegas, NV, USA, pp 151–156
17. Dizaji MS, Harris DK, Kassner B, Hill JC (2021) Full-field non-destructive image-based diagnostics of a structure using 3D digital image correlation and laser scanner techniques. J Civ Struct Health Monit 11:1415–1428. https://doi.org/10.1007/s13349-021-00516-6

18. Schreier H, Orteu JJ, Sutton MA (2009) Image correlation for shape, motion and deformation measurements: basic concepts, theory and applications. Springer Science & Business Media, NY, USA, p 322
19. Ya Ivanytskyi L, Maksimenko OP, Yu Molkov V, Kun PS, Ya Chepil O (2016) Procedure for determination of residual life of concrete bridge beams and development of technical means of deformation monitoring. Technol Diagn Non-Destr Test 3:44–49. https://doi.org/10.15407/tdnk2016.03.07
20. Ivanytskyi YL, Mol'kov YV, Kun' PS, Lenkovs'kyi TM, Wójtowicz M (2015) Determination of the local strains near stress concentrators by the digital image correlation technique. Mater Sci 50:488–495. https://doi.org/10.1007/s11003-015-9746-7
21. Niezrecki C, Baqersad J, Sabato A (2018) Digital image correlation techniques for non-destructive evaluation and structural health monitoring. In: Handbook of advanced non-destructive evaluation, pp 1–46. https://doi.org/10.1007/978-3-319-30050-4_47-1
22. Hu W, Sheng Z, Yan K, Miao H, Fu YA (2021) New pattern quality assessment criterion and defocusing degree determination of laser speckle correlation method. Sensors 21:4728. https://doi.org/10.3390/s21144728

The Influence of the Damage Placement in the Compressed Area of Concrete on the Deformability of Bent Reinforced Concrete Elements

Zinoviy Blikharskyy, Maksym Lobodanov, and Pavlo Vegera(✉)

Department of Building Constructions and Bridges, Lviv Polytechnic National University, Karpinskoho Str. 6, Lviv 79013, Ukraine
Pavlo.I.Vehera@lpnu.ua

Abstract. The study of damaged reinforced concrete elements is difficult at a practical and theoretical level. The work of reinforced concrete elements requires taking into account their complex self-regulating system with a large number of factors. A significant number of these factors affect the operation of the element and come from the composite properties of reinforced concrete. In addition, new factors which directly affects the bearing capacity arise during the appearance of damage. Based on experimental data the article explores the effect of damage in the concrete compressed zone of a reinforced concrete beam, namely, elements with a defect (insufficient cross-section of the working reinforcement) and damage that occurred at a load level equal to 50% of the bearing capacity of the control samples. The choice of the damage load level is based on the fact, that the average operational level of an element is equal to 50–70% of its carrying capacity. Damaged beams are divided into two groups: the first group includes beams with damage in the central cross-section; the second group includes beams with damage that occurs between the central section and a point of applied load. As a result, the article compares and analyze the impact of damage location on the redistribution of stresses in the concrete compressed zone, the nature of the exhausting, and the impact of these factors on the bearing capacity.

Keywords: Bending elements · Reinforced concrete beam · Damages · Defects · Bearing capacity

1 Introduction

Growing trends in cost minimization during reconstruction prompt the development of research determining the residual bearing capacity of reinforced concrete elements. This especially applies to industrial buildings, where reinforced concrete elements are widely used, and their usage changes according to the needs of the market. Thus, to optimize costs when adopting to new tasks, there is a need to adjust existing buildings and structures.

Z. Blikharskyy et al. (Eds.): CEE 2023, LNCE 438, pp. 32–41, 2024.
https://doi.org/10.1007/978-3-031-44955-0_4

Research trends in reconstruction, formed by the needs of the market, are the basis for the development of construction and related industries. Therefore, research determining the residual bearing capacity is relevant for Ukraine and the market of other countries.

The article [1] considers the need to assess the technical condition of building structures and determine the residual bearing capacity of damaged flexural rein-forced concrete elements. The complexity of calculating of damaged elements due to various defects and damage, as well as the composite nature of reinforced concrete, is highlighted.

As an example of economic expediency of combinations of a concrete class, rebar type, and reinforcement ratio in the matter of reliability of pre-stressed reinforced concrete structures according to the shear bearing capacity, is given in the article [2]. Another example is the combination of a classical reinforcing frame and dispersed reinforcement of concrete with steel fiber with the variation of the volumetric distribution of steel fiber [3], as a way of examining such structures by a non-destructive method in non-laboratory conditions, it is possible to consider the use of a non-destructive method of acoustic emission [4]. The basic principles of calculating the reinforced concrete and fiber concrete flexural elements on the bending moment are considered in the paper [5].

The importance of studying the impact of damage on the bearing capacity of reinforced concrete elements in the construction industry is considered in the article [6]. The study focuses on the variability and complexity of studies in the combination of defects and damage in reinforced concrete elements. The article pro-vides an overview of damaged reinforced concrete elements, discusses the influence of damage and additional factors on the elements, and also presents the method of bending tests of reinforced concrete elements with damage to the compressed concrete zone and tensile reinforcement. The study considered the deformability and bearing capacity of the elements under various damage and load conditions and also compared the change of the compressed zone with the load. The conclusions drawn as a result of the research are implemented in practice.

An example of the variability of the vibration load and its effect on the bearing capacity and deformability is given in the article [7]. There, the impact of vibration loads on the bearing constructions of the floor of the five-story factory building was investigated, and further strengthening work was carried out. The study includes measuring the actual level of vibration and assessing structural damage caused by the operation of the equipment.

The article [8] presents the results of an experimental study of reinforced concrete beams damaged in the compressed zone. The main attention is paid to the impact of concrete damage on the bearing capacity of the beams. Four reinforced concrete beams were tested, including one undamaged control beam and three beams with damage in the compression zone at different load levels. The results show that the type of damage and load has a significant effect on the bearing capacity and deformability. Damaged beams under load showed an increase in bearing capacity by 3.8% compared to unloaded damaged beams.

The question of the influence of variation of reinforcement damage in bent elements of rectangular sections is considered in the article [9]. In these articles, an experimental study of reinforced concrete beams with damage to the tensile reinforcement, taking into account the application of the initial load, is considered. The damage included one or five Ø5.6 mm holes, and in one case the damage was extended to a beam failure (up to an

8.4 mm hole). It is also necessary to consider the peculiarities of the manufacture of reinforcement when studying the effect of corrosion, in work [10] such an effect is considered when using thermally strengthened reinforcement. The study of damage to reinforced concrete T-shaped beams was described in [11]. The variability of the following factors was taken into account: the shear span, the ratio of the width of the table overhang to the thickness of the beam edge, the ratio of the thickness of the overhang to the working height of the beam section, reinforcement ratio for shear reinforcement, the level of prestress in the tensile rebar. Another example of the calculation of a bent element is the sections of uncut monolithic reinforced concrete strip beams, which combine pressed or hammered reinforced concrete piles into a single foundation structure. According to the mechanics of reinforced concrete, it is recommended to consider the grid as a continuous reinforced concrete beam [12]. As a method of strengthening damaged beams, one can consider the study in [13], where FRCM reinforced with a composite system is show. The issue of high-quality compatible work of different materials is especially important for reinforcements with composite materials but also pre-stressed bent elements it has a similar critical importance as discussed in the article [14]. Since reinforced concrete is a composite material, one should also pay attention to its features as a material and take into account crack resistance, especially when combined with reinforcement. The issue of crack resistance with the type of load was considered in [15].

2 Aim of the Research

The purpose of the study is to compare the results of testing control samples dam-aged in the center of the pure bending zone and with a displacement from the center of the beam under the action of load. It is achieved by analyzing the effect of the initial load and damage location on the bearing capacity of reinforced concrete beams with a defect in the form of an insufficient area of tensile reinforcement.

3 Methodology of the research

The research was carried out using experimental samples manufactured in factory conditions, which made it possible to idealize the implementation. Reinforcement is made in the form of working tensile reinforcement Ø12 mm, and compressed reinforcement in the zone of maximum transverse force—Ø10 mm. Transverse reinforcement is made with smooth Ø8 mm reinforcement placed in the support zones. When studying the physical and mechanical characteristics of the materials, the following results were obtained: C35/45 grade concrete was used; reinforcement belongs to class A500C. Reinforcement was carried out according to Fig. 1. For comparison, normally reinforced samples with tensile reinforcement Ø14 mm were made, and all other parameters were unchanged. Correspondingly, the reduction of the cross-section of the tensile reinforcement is 26.6%.

Experimental samples are divided into three series, which were marked as follows: BC-beams control; BD.80-0.5-1-beams with damage of 20 × 80 mm at $M = 9.88$ kNm, which is close to 0.5 M_{dest} of the control sample; BD.80-0.5-2-beams with damage of 20 × 80 mm with displacement from the center of the zone of pure bending under the action of load at $M = 9.41$ kNm, which is close to 0.5 M_{dest} of the control sample.

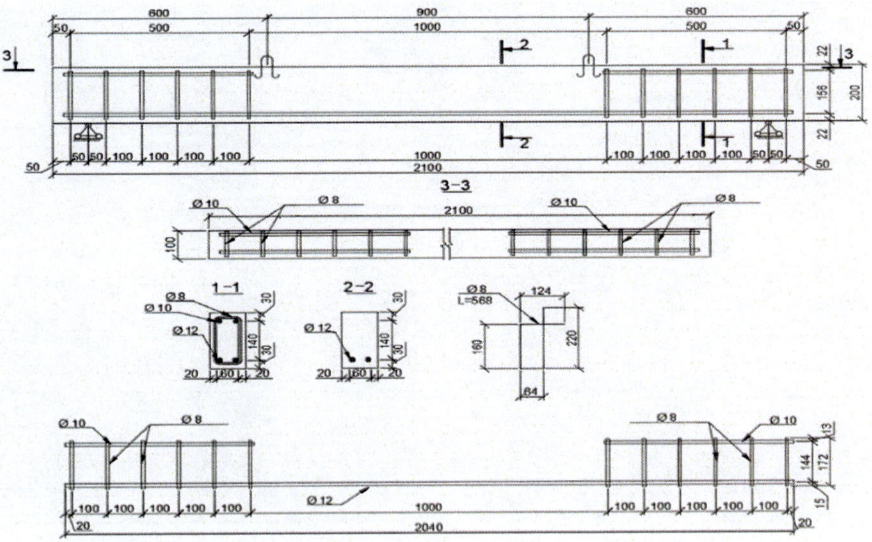

Fig. 1. Reinforcement and dimensions of the experimental sample

To measure concrete strains, dial gauges I1…I8 with a base of 20 cm, located between the applied forces, with a gauging step 40 mm along the height of the beam, were used (see Figs. 2, 3). To study the change in stress during uneven damage to the sample gauges I11…I14 were located in the area of concrete stress concentration. These gauges recorded peak strains and covered the concrete zone damage. Gauges I15 and I16 were used to measure compression strains before damage and maximum strains—after. Indicators I9 and I10 were used to measure reinforcement strains. Subsidences on the stand supports were displayed using P1 and P5. Aistov deflection gauges P2…P4 were used to measure beam deflections (see Fig. 2).

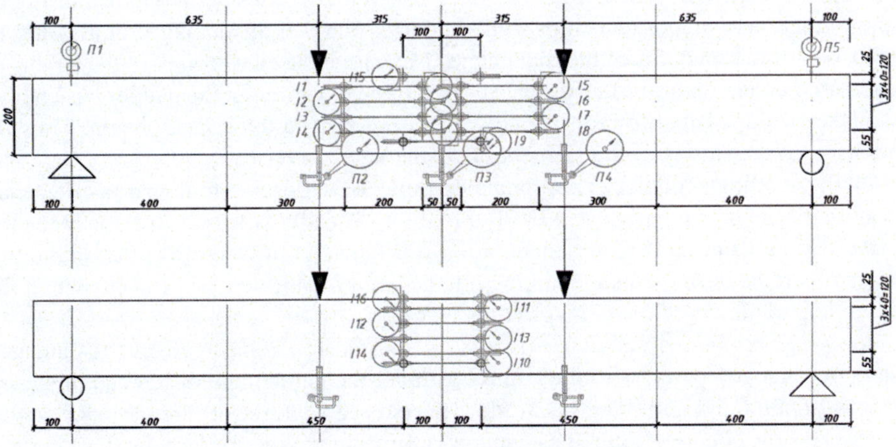

Fig. 2. Scheme of the gauge's arrangement

Fig. 3. Scheme of measuring deformations during and after damaging

The tests are performed in stages, with a load application step of 0.05 M_{dest} load at which the bearing capacity of the control sample was exhausted. This step is per-formed until 0.3 M_{dest} or the appearance of the first cracks. Also, a step of 0.05 M_{dest} at 0.8–0.9 from the calculated destructive. Monitoring of the formation and development of cracks is carried out using an MPB-2 microscope with a resolution of 0.05 mm. For reinforced concrete beams with damage due to load, the damage occurs in three main stages, with a height of damage of 10, 20, and 30 mm (Fig. 3).

4 Analyzing Results of the Experiment

To compare changes in strains due to load for concrete, we consider the indicator I2 due to the formation of cracks and chips from the lower plane of the damage to the points of application of the load. Since the dial gauge I1 cannot be used to compare the strains of compressed concrete due to the absence of part of the concrete in the gauger base plane, this makes it impossible to redistribute what is pre-sent in the control beam. Thus, I1 demonstrates the strains of the element in the damage zone.

Taking readings for the experimental sample BD.80-0.5-1 took place after each execution of the screw with a step of 20 mm, thus, 5 readings were taken for each step of 10, 20, and 30 mm. This implementation was carried out to remove the qualitative change of strains during damage, taking into account the influence of the change in the geometric parameters of the damage (Figs. 4, 6).

The graph in Fig. 4 shows the strains of the tension reinforcement of damaged elements when compared with the control sample, the percentage difference of strains between BD.80-0.5-1 and BD.80-0.5-2 does not exceed 14%, the difference in the strains of compression concrete after the damage is done increases to 24% in the range of 50–70% M_{dest}, at a moment close to 70% M_{dest}, the deviation decreases to 3.1%. At

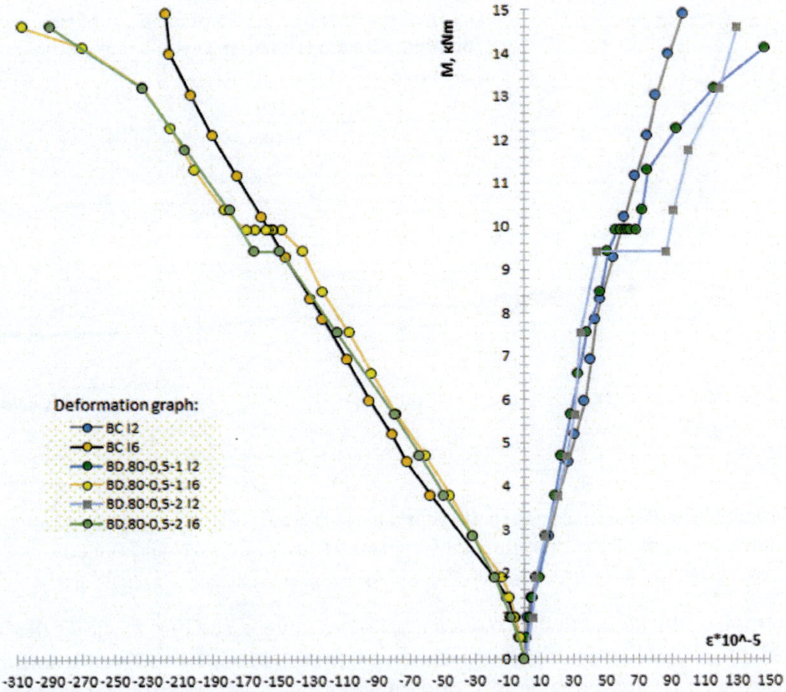

Fig. 4. Graph of averaged reinforcement (left) and compressed concrete (right) strains of the experi-mental samples: BC, BD.80-0.5-1 and BD.80-0.5-2

the bending moment M = 14.59 kNm, the strains of the tension reinforcement for different series of tests $\varepsilon_{s.BC} = -218.21 \times 10^{-5}$, $\varepsilon_{s\ BD.80\text{-}0.5\text{-}2} = -288.96 \times 10^{-5}$ and $\varepsilon_{s\ BD.80\text{-}0.5\text{-}1} = -306.25 \times 10^{-5}$. Strains of concrete compression zone in the central section at moment M = 14.59 kNm is appropriate for series $\varepsilon_{c.BC} = -218.21 \times 10^{-5}$, $\varepsilon_{c.\ BD.80\text{-}0.5\text{-}2} = -288.96 \times 10^{-5}$ i $\varepsilon_{c.\ BD.80\text{-}0.5\text{-}1} = -306.25 \times 10^{-5}$. The moment at which physical destruction occurred in samples BD.80-0.5-2 relative to BC is 15.03% lower, in samples BD.80-0.5-1 by 8.33%. The moment at which the yielding of tension reinforcement occurred in samples BD.80-0.5-2 relative to BC is lower by 11.11%, and in samples BD.80-0.5-1 by 11.25%.

When the damage was performed, BD.80-0.5-1 samples experienced an increase in strains: at $h_d = 10$ mm, ε_s increased by 12.9% and ε_s by 9.5% compared to the readings before damage; at $h_d = 20$ mm, ε_s by 15% and ε_c by 15.8%; at $h_d = 30$ mm, ε_s by 25% and ε_s by 30.5% With the execution of damage $h_d = 0\ldots30$ mm in samples BD.80-0.5-2, the strains of reinforcement and compressed concrete increased by ε_s by 10.7% and ε_c by 115.3%. When comparing the strains of post-skinned elements BD.80-0.5-1 with the control ones at the bending moment M = 13.97 kNm, in compressed concrete they are larger by 45.9%, in reinforcement by 20%.

Measuring the readings of the dial gauges I2…I4 made it possible to determine the neutral axis position in the zone of pure bending with dependence on the bending moment (Fig. 5).

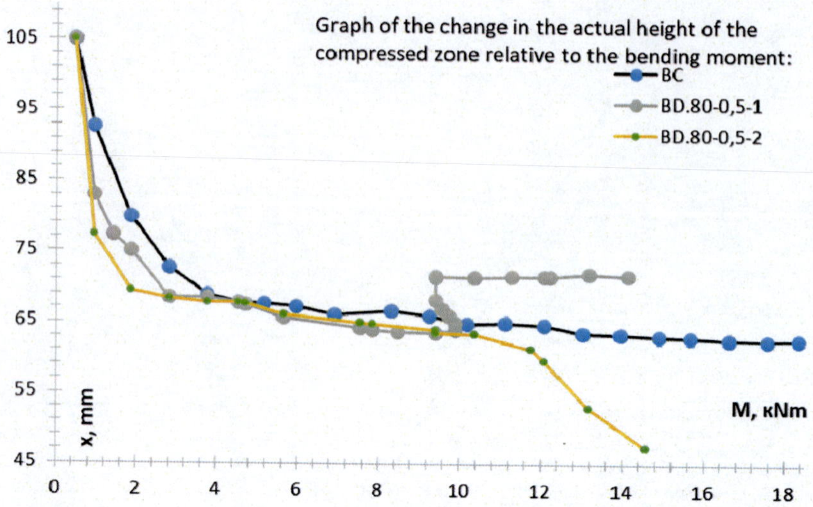

Fig. 5. Graph of the average change in the actual height of the concrete compressed zone relative to the bending moment of the experimental samples: BC, BD.80-0.5-1, and BD.80-0.5-2

The graph of the actual depth of the compressed zone x, see Fig. 5, shows the internal redistribution of the forces of the experimental samples BC, BD.80-0.5-1, and BD.80-0.5-2 when performing damage at a moment of 0.5 M_{dest} of the control sample. Since the damage to the BD.80-0.5-2 was carried out with a shift to the force of application (see Fig. 6), and not in the central cross-section, the main stresses are redistributed with a decrease in the maximum reinforcement stresses in the central cross-section relative to the samples BD.80-0.5-1 under identical loads.

Damage with dimensions of 80 × 30 mm has a dynamic loss of the actual height of the compressed zone at the beginning of the second stage of the stress-strain state and a subsequent gradual decrease position of the neutral axis (see Fig. 6).

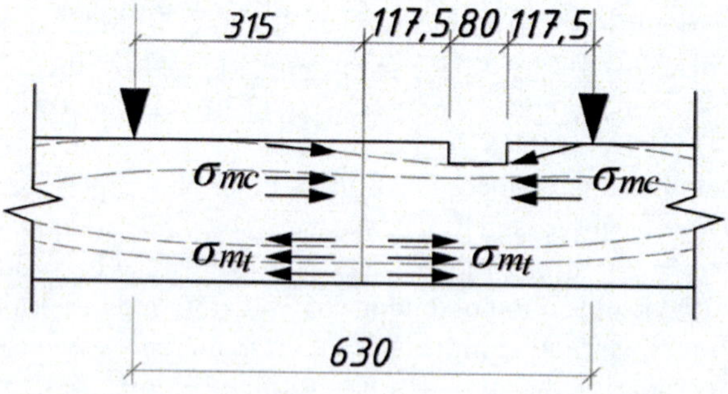

Fig. 6. Schematic representation of the distribution of principal stresses in the zone of pure bending in damaged beams BD.80-0.5-2

The presence of a defect of an insufficient tension reinforcement cross-section leads to greater deformability of the rebar compared to concrete working in compression. This

Fig. 7. Reflection of the distributed cracks and chips in the beam BD.80-0.5-1

type of defect in combination with damage to the concrete working under compression in the central cross-section leads to the early opening of cracks and their significant propagation in the zone of pure bending, which leads to the exclusion of a significant height from the cross-section of concrete from a joint operation and displacement of the zone of maximum compressive stresses concrete, see Figs. 5, 6, 7.

Fig. 8. General view of the formed cracks and chips in the beam BD.80-0.5-2

In case of damage between the central section of the element and the force of application (Fig. 6), there is a shift in the concentration zone of maximum forces. In Figs. 8, 9 shows the effect of changing the distribution of principal stresses in the zone of pure bending in damaged BD.80-0.5-2 beams due to the displacement of the damage from the central axis of the element. A clear demonstration of this is the formation of a through crack from the reinforcement to face damage in samples BD.80-0.5-1 and BD.80-0.5-2 see Figs. 7, 8.

Fig. 9. Splitting of the surrounding concrete of the beam BD.80-0.5-2 from the place of damage

5 Conclusions

Study of the operation of damaged elements with the presence of a defect due to an insufficient cross-section of the tensile reinforcement, when performing damage due to the action of the load, the following are deported:

- the percentage difference in the strains between BD.80-0.5-1 and BD.80-0.5-2 does not exceed 14%, and the difference in the strains of the compression concrete working after performing damage increases to 2% in the range of 50–70% Mdest, at the moment close to 70% of Mdest, the deviation decreases to 3.1%;
- the moment at which physical destruction occurred in samples BD.80-0.5-2 relative to BC is lower by 15.03% and in sample BD.80-0.5-1 by 8.33%. The moment at which yielding of the tension reinforcement took place in sample BD.80-0.5-2 relative to BC is lower by 11.11%, in samples BD.80-0.5-1 by 11.25%;
- when comparing the strains of the damaged BD.80-0.5-1 elements with the control ones at the bending moment M = 13.97 kNm, in compressed concrete, they are larger by 45.9%, in reinforcement by 20%;
- damage with dimensions of 80 × 30 mm has a dynamic loss of the actual height of the compressed zone at the beginning of the second stage of the stress-deformed state and a subsequent gradual decrease in the actual height of the com-pressed zone;
- due to the redistribution of internal stresses in the damage zone, trapezoidal cracks are formed beginning in the lower damage zone, followed by the formation of chips;
- execution of damage with a displacement from the central section of the element to the force of application leads to a change in the distribution of principal stresses and, accordingly, the zones of maximum stress concentrations, which leads to an increase at the moment at which the destruction of the element occurs and changes in its deformability in the central section relative to elements dam-aged in the central section of the element.

References

1. Lobodanov M, Vegera P, Blikharskyy Z (2019) Influence analysis of the main types of defects and damages on bearing capacity in reinforced concrete elements and their research methods. Prod Eng Arch 22(22):24–29
2. Ahaieva O, Vegera P, Karpiuk V, Posternak O (2023) Design reliability of the bearing capacity of the reinforced concrete structures on the shear. In: Blikharskyy Z (ed) Proceedings of EcoComfort 2022. Lecture notes in civil engineering, vol 290. Springer, Heidelberg, pp 1–15
3. Babych YM, Savitskiy VV, Andriichuk OV, Ninichuk MV, Kysliuk DY (2019) Results of experimental research of deformability and crack-resistance of two span continuous reinforced concrete beams with combined reinforcement. In: Panchenko S (ed) TRANSBUD-2019. IOP conference series: materials science and engineering, vol 708, issue no 1. IOP Publishing Ltd., p 012043
4. Adamczak-Bugno A, Lipiec S, Vavruš M, Koteš P (2022) Non-destructive methods and numerical analysis used for monitoring and analysis of fibre concrete deformations. Materials 15(20):7268
5. Kochkarev D, Galinska T, Tkachuk O (2018) Normal sections calculation of bending reinforced concrete and fiber concrete element. Int J Eng Technol 7(3):176–182
6. Blikharskyy Z, Lobodanov M, Vegera P (2022) Investigation of defective reinforced concrete beams with obtained damage of compressed area of concrete. Prod Eng Arch 28(3):225–232
7. Ilnytskyy BM, Kramarchuk AP, Bula SS, Bobalo TV (2019) Study of the vibration influence on load-bearing floor structures in case of machinery operation. In: Panchenko S (ed) TRANSBUD-2019. IOP conference series: materials science and engineering, vol 708, issue no 1. IOP Publishing Ltd., p 012052
8. Lobodanov M, Vegera P, Khmil R, Blikharskyy Z (2021) Influence of damages in the compressed zone on bearing capacity of reinforced concrete beams. In: Blikharskyy Z (ed) Proceedings of EcoComfort 2020. Lecture notes in civil engineering, vol 100. Springer, Heidelberg, pp 260–267
9. Kopiika N, Vegera P, Vashkevych R, Blikharskyy Z (2021) Stress-strain state of damaged reinforced concrete bended elements at operational load level. Prod Eng Arch 27(4):242–247
10. Blikharskyy Y, Selejdak J, Kopiika N (2021) Specifics of corrosion processes in thermally strengthened rebar. Case Stud Constr Mater 15:e00646
11. Karpiuk V, Somina Y, Karpiuk F, Karpiuk I (2021) Peculiar aspects of cracking in pre-stressed reinforced concrete T-beams. Acta Polytech 61(5):633–643
12. Kos Z, Klymenko Y, Karpiuk I, Grynyova I (2022) Bearing capacity near support areas of continuous reinforced concrete beams and high grillages. Appl Sci 12(2):685
13. Kos Z, Blikharskyi Z, Vegera P, Grynyova I (2023) A calculation model for determining the bearing capacity of strengthened reinforced concrete beams on the shear. Appl Sci 13(8):4658
14. Borzovič V, Laco J, Pecník M, Pažma P (2016) The crack development mechanism of pre-stressed girder influenced by different bond between pre-stressed tendons and concrete. Key Eng Mater 691:309–320
15. Golewski GL (2023) The phenomenon of cracking in cement concretes and reinforced concrete structures: the mechanism of cracks formation, causes of their initiation, types and places of occurrence, and methods of detection—a review. Buildings 13(3):765

The Problem of Recycling Construction Waste in Poland

Natalia Brycht(✉)

Faculty of Civil Engineering, Czestochowa University of Technology, J.H. Dąbrowskiego 69, Częstochowa 42-201, Poland
natalia.brycht@pcz.pl

Abstract. In recent years, along with the increase in social environmental awareness, more and more attention is paid to activities related to waste management and environmental protection. Every sector of the economy, including the construction industry, faces the problem of waste management. It is estimated that in 2021 the consumption of natural resources amounted to over 100 billion tones on a global scale, of which about 50% was in the construction sector. Every year, as a result of construction, renovation and demolition works, only in the countries of the European Union, almost 40% of the world's waste is generated, which should be properly managed. The reuse of raw materials obtained from waste processing as additives for the production of new composite materials often contributes to the improvement of the properties of the resulting composites. Recently, the circular economy, assuming the reuse of building materials, is becoming more and more popular. It is estimated that the Polish economy is currently only 10.2% circular. The transition to a circular economy is a long-term and difficult process, not only for organizational and financial reasons, but also because of the legal aspect. The study reviews the literature on the technological possibilities of recycling construction waste and identifies areas of their reuse. It also presents data illustrating the current situation of the Polish and European economy in this regard.

Keywords: Recycling · Construction waste · Environmental protection · Renovation and demolition works · Circular economy

1 Introduction

Recycling is currently one of the key issues in the area of activities undertaken to protect the environment. The level of social environmental awareness is increasing year by year, which translates into the search for and implementation of modern technologies and solutions aimed at reducing the generated waste. Every sector of the economy, including the construction industry, struggles with the problem of their development. Huge amounts of construction waste generated make it necessary to take action to reduce it. It is important to prevent their excessive number already at the planning stage of construction investments. In the study [1], the relationship between the factors contributing to the increase in the amount of waste and those that affect its minimization was determined, which was confirmed by the conducted research. Proper segregation of the generated construction

Z. Blikharskyy et al. (Eds.): CEE 2023, LNCE 438, pp. 42–48, 2024.
https://doi.org/10.1007/978-3-031-44955-0_5

waste is also important. This allows for a more efficient recovery of materials that can be reused [2]. In Norway, research was undertaken on methods for assessing and monitoring the level of construction waste, based on 36 construction projects. It has been shown that an average sorting rate of 89% translates into a low recycling rate (only 32%). In order to better integrate environmental considerations and management of the Group 17 waste generated, an environmental management plan has been developed that can be implemented at every stage of a construction project. It is a set of tools for monitoring waste sources and assessing their management options, in accordance with international standards [3]. The leading issues in terms of proper management of already generated waste are undoubtedly utilization, recycling and other recovery processes. Those from waste that are not harmful to people and the environment should be reprocessed to meet the main assumptions of the sustainable development policy. According to his concept, building objects should be designed, implemented and operated in such a way as not to violate the principles of environmental protection [4, 5].

The purpose of this article was to identify the most frequently generated construction waste and the possibilities of its reuse, as well as to present the current situation of the Polish and European economy in this regard.

2 Recycling of Construction Waste in Poland

According to the Urban Circularity data, the global consumption of natural resources in 2021 amounted to over 100 billion tons, of which about 50% was used by the construction sector, which generates as much as 30% of global waste. This results not only in a constantly deepening shortage of primary raw materials, but also contributes to air pollution and global warming [6]. Every year, as a result of construction, renovation and demolition works carried out in the European Union countries, approximately 40% of the world's waste is generated. According to the Building Performance Institute Europe, around 500 million waste is produced annually in the member states from renovations, modernizations and reconstructions. It is estimated that up to 90% of construction waste can be reused, which would significantly reduce the amount of waste going to landfills and reduce CO_2 emissions. According to the data of the Statistics Poland, in 2021, 820 868.9 Mg of construction and demolition waste was generated, which is municipal waste. Most of them were recorded in Śląskie Voivodship (19.6%), Mazowieckie Voivodship (13.2%) and Wielkopolskie Voivodship (13.1%), and the least in Podkarpackie Voivodship (1.9%) (Fig. 1) [7].

According to the Waste Catalogue, those from construction and renovation projects belong to group 17. Figure 2 shows the amount of construction waste generated and recovered in Poland in 2017–2021. It was noted that the year 2020 was definitely dominant (over 7 352 Gg of waste generated). A significant decrease in the amount of waste recycled during the COVID-19 pandemic period was observed, compared to 2018, when the percentage share of this waste was close to 90% [8]. Despite the continuous increase in environmental awareness, the level of recycling activities in Poland is still low [9]. Among European countries, the leader in this aspect is Switzerland, where only 5% of the generated waste ends up in landfills.

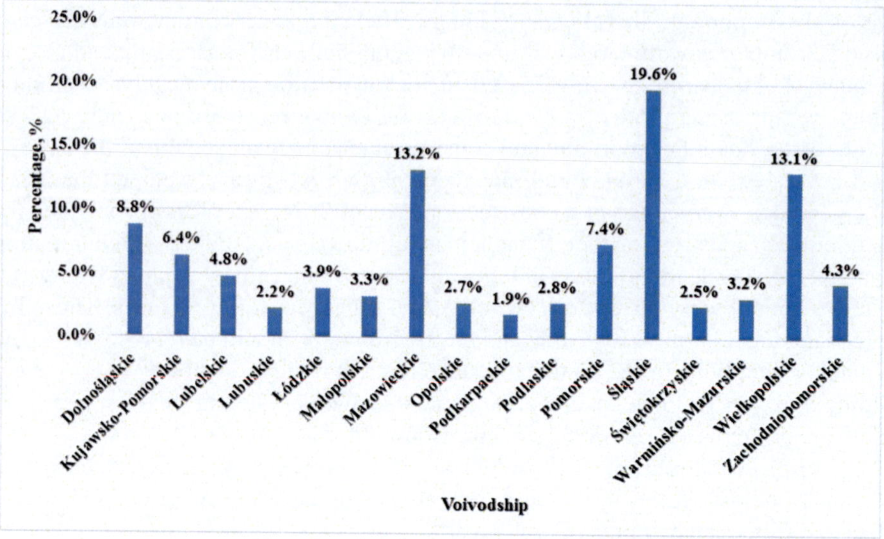

Fig. 1. Collected construction and demolition waste in Poland in 2021 [based on 7]

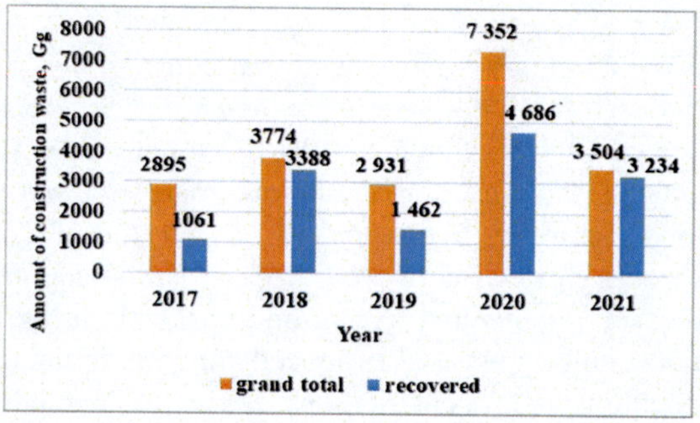

Fig. 2. Construction waste generated in 2017–2021 in Poland [based on 8]

3 Law in Poland

Recently, the circular economy has become increasingly popular. It is an economy model in which products and services are traded in closed loops. Its main assumption is to optimize the process of reusing a given material in such a way as to prevent waste. According to the report "Circular Restart! Polish Circularity Gap Report", the Polish economy is currently 10.2% circular, which means that for every 10 kg of waste, only about 1 kg is recycled. The process of transition to a circular economy is long and quite complicated, as it involves changing business models as well as legal regulations. The transformation of the real estate and construction sectors towards a circular economy is only at the initial stage of implementation, which is to be accelerated by the soon to come into force EU regulations. As much as 70% of construction waste generated in Poland is to be recycled or otherwise recovered, which will significantly reduce the amount of waste generated. According to the estimates of the European Parliament, the circular economy can largely result in an increase in EU GDP of 0.8% by 2030. An important issue in the current legal regulations is also the obligation to segregate construction waste. Currently, waste generated from construction, demolition and renovations is collected in a common container, which is to change from January 1, 2025. Then Art. 101a of the Act on waste [10], which obliges to selective collection and segregation into at least 6 main fractions: glass, metal, wood, plastics, gypsum and mineral waste. The imposition of this obligation is not intended for private persons, but only for companies implementing construction investments. However, the high costs associated with imposing a new obligation may result in the use of cheaper and at the same time environmentally harmful solutions, such as abandoning waste in the forest, which are still often practiced. This results in the formation of the so-called "wild dumps" and is inconsistent with the applicable legal regulations (Fig. 3a). Improper disposal of dismantled asbestos-cement boards is still a frequent phenomenon (Fig. 3b). It is true that a ban on the production and use of asbestos products has been introduced in Poland since 1997, but large amounts of these materials have not been completely neutralized. In 2002, Poland joined the "Program for removing asbestos from the country for 2009–2032", thus committing itself to the complete removal of these products from buildings. The resulting waste, after dismantling by specialized companies, should be deposited in landfills, specially designated for this purpose, which so far is the only legal way to neutralize them. Unfortunately, due to the high costs associated with the dismantling of asbestos-cement boards, and thus also the replacement of structural elements, building owners abandon asbestos waste on their plots or forest areas, violating the provisions on environmental protection [11–14].

a) b)

Fig. 3. Construction waste from demolition abandoned in the forest: **a** brick rubble, **b** asbestos [own study]

4 Management of Construction Waste

One of the most frequently generated construction waste is concrete and brick rubble. The first stage of its processing is separating it from other waste elements, e.g. wood or metal. Then, the cleaned construction rubble is transported to the crusher, where it is crushed to appropriate sizes depending on its intended use. The wide possibilities of reusing construction rubble allow it to be used primarily in road construction. It can be used as a material for making road foundations, bedding for paving stones or stabilizing material for soil reinforcement. It is also a valuable secondary raw material in the production of concrete mix. Rubble resulting from the demolition of concrete structures is also used as a secondary raw material for the production of cement mixes and self-compacting concrete, in the form of a substitute for coarse aggregate [15]. Fly ashes are becoming an increasingly popular substitute for aggregate in the production of concrete. The study [16] presents the results of strength tests of concrete with the addition of fly ash from biomass combustion as a substitute for sand. It has been shown that these concretes do not have a negative impact on the environment, and their strength increases with the amount of added material. A popular solution is also the processing of post-demolition wooden elements into biomass and in order to obtain raw material for the production of new materials, such as chipboards or furniture panels. Along with the rapid increase in the production of plastics at the turn of the last few decades, a simultaneous increase in the amount of waste generated from this plastic has been noticed. Among several recycling methods, the most effective is chemical (raw material) recovery, next to material recovery and energy recovery [17]. Material recovery allows you to obtain the material needed to manufacture a new product, while raw material recovery consists in processing waste into, for example, polymers. Unfortunately, due to the lack of technological possibilities and high costs, some of the accumulated waste is exported outside the European Union for further processing. The most popular plastic used in the construction industry is polyvinyl chloride (PVC), used in the production of window and door joinery and finishing elements. Plastic waste obtained as a result of demolition and renovation, after appropriate treatment, is used in the production of thermal insulation materials, asphalt mixtures, and cement bricks [18]. They are also successfully used in the production of concrete. Introduced in the form of fibers, they increase its tensile strength [19]. Carbon fiber composites (CFC), widely used in industry, and polypropylene fibers are also becoming more and more popular [20, 21]. Concrete elements modified with thermoplastic waste, in the form of a substitute for sand and aggregate, are highly resistant to direct exposure to weather conditions, thanks to which they can be used as elements of street furniture [22]. The study [23] presents the results of bending and compressive strength tests of cement mortars modified with rubber waste. On their basis, it was found that the obtained composites have high strength parameters, and the reuse of waste is undoubtedly in line with the policy of sustainable development. One of the possibilities of using construction waste is the use of shredded polystyrene, glass or mineral wool as an additive for the production of concrete and mortars [24, 25]. Expanded polystyrene waste can be used directly at the place of production, as well as after appropriate processing. The crushed waste is used as an insulating material, for loosening the soil, and as a partial replacement of aggregate in the production of plasters and mortars [26]. The study [27] also paid special attention to the generation of large amounts of metal waste. Ways of their reuse, also in the construction sector, in the production of aggregates were discussed.

5 Conclusions

Recently, we have observed a significant increase in the level of social environmental awareness, which translates into taking actions in line with the idea of sustainable development, also in the construction industry. The increasingly popular slogans regarding the circular economy have made it necessary to take action to transform the real estate and construction sectors towards a circular economy. Poland is currently at the beginning of this process, which, due to the need to make a number of changes, is quite difficult and lengthy. The expected changes in the legal regulations are aimed at reducing the amount of generated waste from group 17 and putting emphasis on recycling activities. The wide possibilities of reusing construction waste fit in very well with the policy of environmental protection and reduction of CO_2 emissions. Construction waste management is constantly being improved, but the biggest problem is still financial and technological limitations, which results in the formation of the so-called "wild dumps", especially in forested areas. Unfortunately, the waste management system in Poland is not functioning properly and the required level of recycling has still not been achieved. There is still a lack of effective tools to monitor the waste market, which makes it difficult to keep reliable statistics and create reliable reports, and thus prevents proper management.

References

1. Nawaz A, Chen J, Su X (2023) Factors in critical management practices for construction projects waste predictors to C&DW minimization and maximization. J King Saud Univ Sci 35(2):102512
2. Fufa SM, Fjellheim K, Venås C, Vevatne JT, Kummen TM, Henke L (2023) Waste free construction site–a buzzword, nice to have or more. Resour Conserv Recycl Adv 18:200149
3. Ismaeel WSE, Kassim N (2023) An environmental management plan for construction waste management. Ain Shams Eng J, pp 102244 (In Press)
4. Ulewicz M (2021) Gospodarka odpadami budowlanymi i rozbiórkowymi w europejskiej strategii zrównoważonego rozwoju—stan i perspektywa (Construction and demolition waste management in the European sustainable development strategy—status and prospects). Przegląd Budowlany 10:49–53
5. Tomov M, Velkoska C (2022) Contribution of the quality costs to sustainable development. Prod Eng Arch 28(2):164–171
6. Bergonzoni M, Melloni R, Botti L (2023) Analysis of sustainable concrete obtained from the by-products of an industrial process and recycled aggregates from construction and demolition waste. Procedia Comput Sci 217:41–51
7. Statistics Poland. Retrieved from https://bdl.stat.gov.pl/bdl/metadane/cechy/4041. Accessed on 18 May 2023
8. Statistics Poland, Ochrona Środowiska (Environmental Protection) 2018–2022
9. Brycht N (2020) Construction waste management in rural areas of the Czestochowa district in the aspect of environmental safety. In: Ulewicz R, Hadzima B (eds) Quality production improvement, vol 2. Sciendo, pp 60–68
10. Act of 14 December 2012 on waste, Dz. U. 2013 poz. 21
11. Ulewicz M, Liszewski W (2020) Influence of public financial support on the process of roof covering replacement and safety of civil structures. Syst Saf Hum Tech Facility Environ 2(1):259–267

12. Ulewicz M, Liszewski W (2023) Problem of utilization of asbestos-cement sheets in Poland. Eng Rural Dev. Jelgava, pp 951–957
13. Brycht N, Ulewicz M (2022) The problem of asbestos-cement sheets disposal in the Kłobuck poviat. Constr Technol Archit 2:65–72
14. Brycht N (2022) Recycling of asbestos-cement waste—an opportunity or a threat? Syst Saf Hum Tech Facility Environ 4(1):10–18
15. Malazdrewicz S, Ostrowski KA, Sadowski Ł (2023) Self-compacting concrete with recycled coarse aggregates from concrete construction and demolition waste—current state-of-the art and perspectives. Constr Build Mater 370:130702
16. Jura J, Ulewicz M (2021) Assessment of the possibility of using fly ash from biomass combustion for concrete. Materials 14(21):6708-1–6708-15
17. Pandey KP, Jha UR, Kushwaha J, Priyadarsini M, Meshram SU, Dhoble AS (2023) Practical ways to recycle plastic: current status and future aspects. J Mater Cycles Waste Manage 25:1249–1266
18. El-Metwally Y, Dewidar K, Ismail M, El-Mahallawi I (2023) Optimization of plastic waste integration in cement bricks. J Eng Appl Sci 70:55
19. Dębska B (2010) Materiały budowlane produkowane z wykorzystaniem odpadów. Cz. 1. Obszary zastosowań (Building materials produced using waste, vol 1, areas of application). Izolacje 5(15):27–33
20. Wang Y, Li A, Zhang S, Guo B, Niu D (2023) A review on new methods of recycling waste carbon fiber and its application in construction and industry. Constr Build Mater 367:130301
21. Helbrych P (2021) Effect of dosing with propylene fibers on the mechanical properties of concretes. Constr Optimized Energy Potential 10(2):39–44
22. Pietrzak A, Ulewicz M (2023) Influence of post-consumer waste thermoplastic elastomers obtained from used car floor mats on concrete properties. Materials 16(6):2231-1–2231-13
23. Dębska B, Krasoń J, Lichołai L (2020) Application of Taguchi method for the design of cement mortars containing waste materials. Constr Optimized Energy Potential 9(1):15–26
24. Szafranko E, Jaromińska M (2020) Recykling odpadów budowlanych na etapie realizacji robot (Recycling of construction waste at the stage of works implementation). Builder 6:32–33
25. Dębska B, Krasoń J, Lichołaj L (2021) The evaluation of the possible utilization of waste glass in sustainable mortars. Constr Optimized Energy Potential 9(2):7–15
26. Iżykowska-Kujawa M (2013) Zagospodarowanie odpadów budowlanych—technologie, z których korzystamy (Management of construction waste—technologies we use). Inż Ekologiczna (Ecol Eng) 33:49–60
27. Lis T, Nowacki K (2022) Pro-ecological possibilities of using metallurgical waste in the production of aggregates. Prod Eng Arch 28(3):252–256

FEM Simulations and Experimental Testing of a Connector for Prefabricated Cylindrical Tank Elements

Lidia Buda-Ożóg, Zdzisław Pisarek, Joanna Zięba(✉), and Damian Nykiel

Department of Building Structures, Faculty of Civil Engineering and Environmental Engineering, Rzeszow University of Technology, Rzeszow, Poland
j.zieba@prz.edu.pl

Abstract. The article presents FEM analysis and experimental testing of a prototype connection of a horizontal, prefabricated cylindrical tank. The connector was designed under a contract between Rzeszow University of Technology and Sienkiewicz MAT-BUD Ltd. The proposed solution for connecting prefabricated components of the tank, does not require wet work on site. The connection is designed for above-ground cylindrical tanks with a diameter of up to 10 m and loaded with a 6-m column of liquid pressure (water or other material of similar or lesser volumetric weight) at 750 mm intervals, but can be successfully used for elements of other dimensions and other spacing. Based on the experimental tests and numerical simulations performed, it can be concluded that the designed horizontal connection of prefabricated elements of a cylindrical tank meets the ultimate limit state conditions for components. Experimental investigations have shown how much of an effect the detailed modelling of all bolt connection components, i.e. nuts, thread lengths, etc., has on the conformity of failure force values obtained from FEM simulations and actual component tests.

Keywords: Fastener · Container · Experimental test · FEM analysis

1 Introduction

The development leap in construction technology and, at the same time, the decreasing number of people willing to do monolithic wet work has made prefabricated constructions increasingly popular in many areas of construction. A real barrier to even more widespread use of prefabricates is the shapes and dimensions of the desired elements, which are often difficult and sometimes even impossible to transport from the prefabrication factory to the site of construction. An alternative in this case is to divide the target construction into smaller prefabricated elements. Such a solution allows efficient transportation and unloading, but requires the design of appropriate connectors that ensure mutual cooperation of the elements to be joined and the transfer of internal forces arising during the various phases of exploitation of the structure. Currently, most systems of connecting prefabricated elements still require wet work on site. Such solutions generate not only execution problems, but also delay the commissioning of the building due to

Z. Blikharskyy et al. (Eds.): CEE 2023, LNCE 438, pp. 49–57, 2024.
https://doi.org/10.1007/978-3-031-44955-0_6

the required 28-day time required for the concrete to reach the target strength. A review of the wall fasteners currently used in volume construction can be found in the work [1], among others.

2 Case study

2.1 Prefabricated Cylindrical Containers

The article presents a system of horizontal connection of the component elements of an aboveground cylindrical tank with a diameter of up to 10 m and a height of 6 m - Fig. 1. The connector was designed under a contract between Rzeszow University of Technology and Sienkiewicz MAT-BUD Ltd. The tank walls, for transportation reasons, divided into four elements horizontally and two elements vertically.

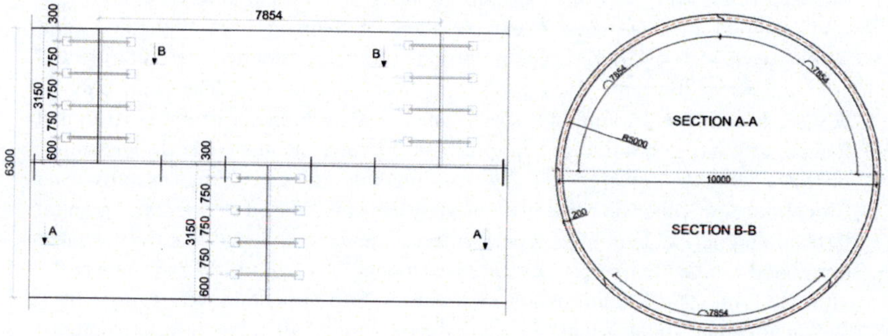

Fig. 1. Division of the tank walls into prefabricated elements: a) view of tank wall in rollout, b) horizontal view of the tank

The tank will be made of C45/55 concrete, reinforced with B500SP steel. According to the information provided by the user of the tank, it was to be a storage element for the pulp obtained from the processed waste. In a typical operating situation, the tank will be filled from 1/3 to 1/2 of its height with liquid with a volume weight of about 8.5 kN/m^3 and elevated temperature. In an exceptional situation, it is possible to fully fill the tank with liquid of ambient temperature. In order to estimate the forces in the analyzed tank, preliminary static calculations, calculations of the reinforcement and checking the serviceability limit state were carried out using the AxisVM program [2]. The numerical model took into account the planned connections of horizontal elements.

2.2 FEM Analysis of a Prototype Connector

Taking into account the possible load cases, a system was designed to connect the individual elements of the tank horizontally. Due to the values of tensile forces and bending moments occurring in the tank shell, the connection system was adopted at a spacing of 750 mm. Steel pipes with a diameter of 48.3 mm and a wall thickness of 3.2 mm were placed in the prefabricated reinforced concrete walls every 750 mm. In the place of grooves in the concrete measuring 200 × 200 × 120 mm, the steel pipes were terminated with a 20 × 120 mm steel plate 200 mm long connected by a weld to the pipe. After assembling the two prefabricated elements, a ϕ 32 mm steel anchor will be guided through the previously prepared pipes sunk into the tank wall. Then, two washers and two nuts will be introduced at the grooves in the concrete to enable the anchor to be bolted to the plate enclosing the pipe. In addition, a rubber gasket is designed at the contact between the two prefabricated components, which will ensure the tightness of the tank after the anchors are tightened. Before proceeding to the experimental verification of the designed connector, a detailed numerical model of the test component was made in the ATENA program according to Fig. 2.

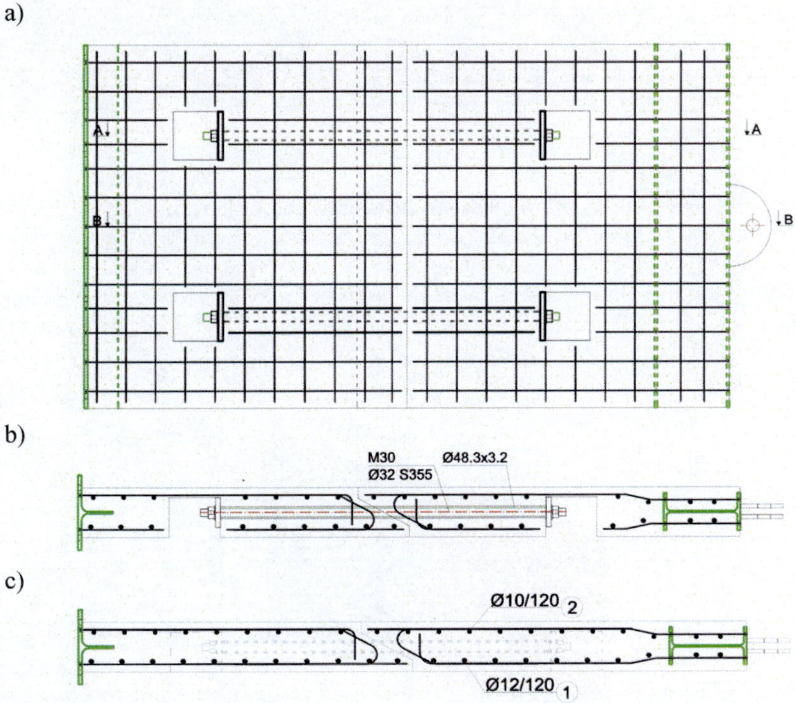

Fig. 2. Prefabricated component for lab tests: a) view of the test component, b) cross-section A-A, c) cross-section B-B

The numerical simulations of research component allowed, among other things, to draw up an optimum plan of experiment with particular emphasis on the method of loading and determining the required points of displacement measurement. A numerical model was constructed using the ATENA Static + GID software package from Cervenka Consulting – Fig. 3. This software has been used to model different material, among other things, reinforced concrete [3]. The description of concrete behavior was defined in the form of an elastic–plastic material model with degradation, i.e., CC3DNonLinear

Cementitious2 [4]. The model was developed for use in the analysis of brittle materials and designed primarily for reinforced concrete structures. The reinforcement was modelled using the "Reinforcement EC2" material model. The reinforcement of the individual members was defined as linear elements, given material properties, and the corresponding diameter. The numerical analysis was preceded by a sensitivity assessment of finite-element mesh dimensions.

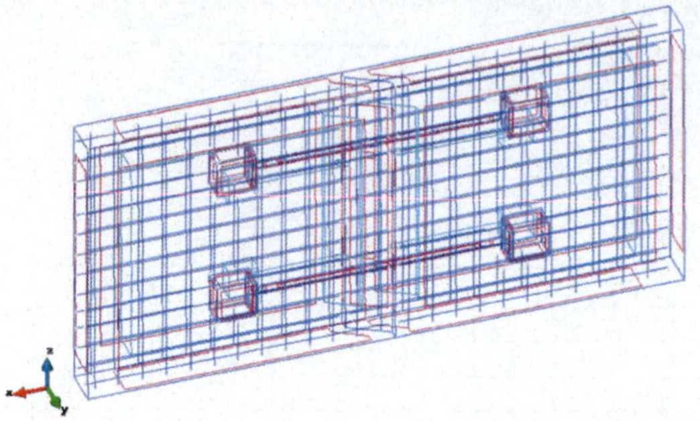

Fig. 3. FEM model of the component for lab tests

In the model, the load was simulated by applying a tensile force at the center of the component's height. The component included two connectors at a spacing of 750 mm. Selected results obtained from the numerical model are shown Figs. 4 and 5.

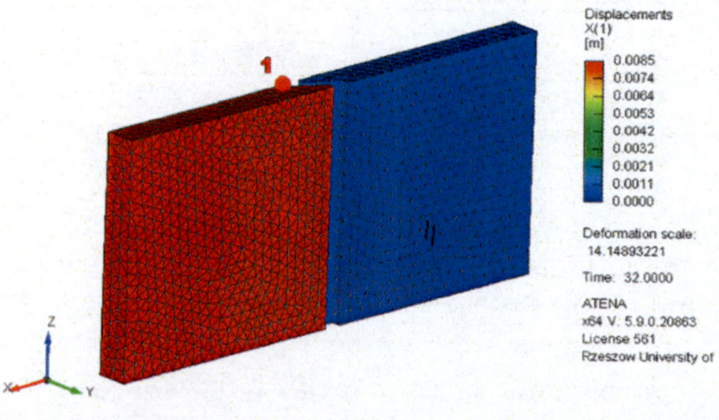

Fig. 4. Displacements of the model: a) displacement map of elements, b) displacement - tensile force relationship of point 1

Based on the results obtained, it was found that the designed connection satisfied the requirements of load-bearing capacity and strain capacity. The conditions of not exceeding the allowable stresses in the concrete and steel connector were met, and the expected displacement of the components at a tensile force of 405 kN was about 8 mm.

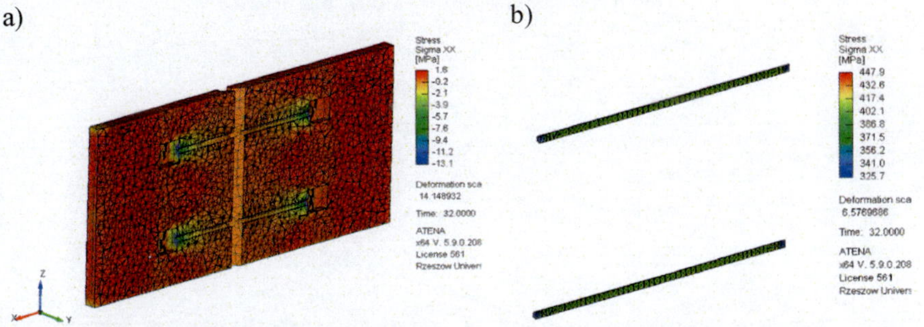

Fig. 5. Maximum stresses: a) stresses in concrete, b) stresses in anchor

2.3 Experimental Test

The next step in verifying the developed connection was experimental testing. For this purpose, three models of the tank wall component shown in Fig. 6 were fabricated and tested. The tests were carried out at the Faculty's Structures Testing Laboratory at Rzeszow University of Technology.

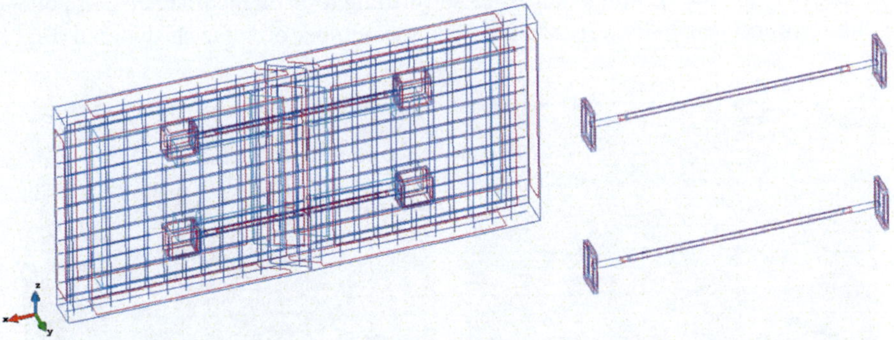

Fig. 6. Prefabricated component for lab tests on the test stand

The tests conducted included a series of static and cyclic loads. In the tests conducted, the tensile force was implemented using an Instron-Schenck actuator in steps of 50 kN until a force of 500 kN was reached. Displacements of individual connected elements were measured by two methods, i.e. using 8 inductive sensors distributed over the height of the element, and using the ARAMIS optical image correlation system. During the conducted tests, for each test model D-1 to D-5, the inductive sensors were placed on the outer side of the component. On the inside, displacement measurements were carried out using the ARAMIS digital image correlation system. This method makes it possible to analyze the displacement of any point on the pre-prepared surface of the test component. Figure 7 shows comparison of the tensile force - displacement relationship of measuring point P-1 for component D-2.

Then 25 cycles were carried out, including unloading to 0 and reloading to 500 kN. After realizing the cyclic loads by increasing the load in increments of 5 kN, the tests were

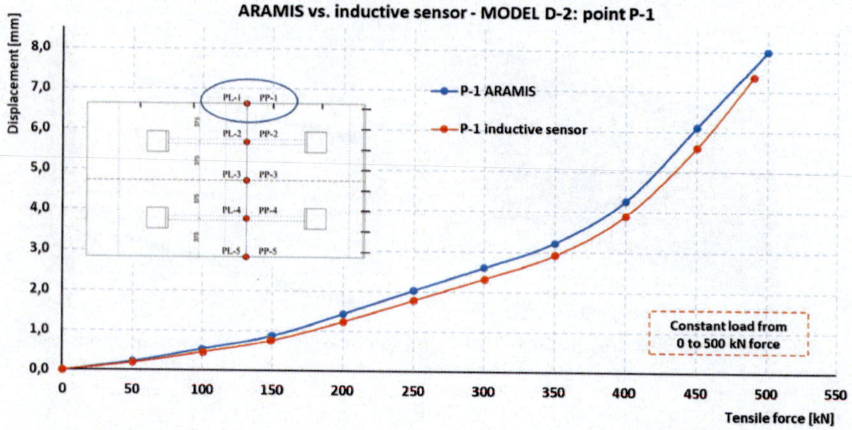

Fig. 7. Comparison of the tensile force - displacement relationship of measuring point P-1 for component D-2 (ARAMIS and inductive sensor)

conducted until the destruction or discontinuation of the test model due to reaching the maximum range of the actuator. Due to the observed failure mechanism of the connection involving only the thread of the bolt, after performing tests on the first two components, the failure connecting bolts were replaced and the number of nuts was doubled (Fig. 8).

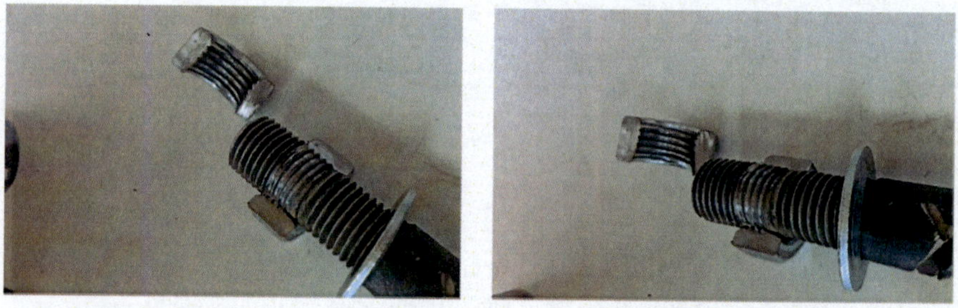

Fig. 8. Destruction of the connecting anchor in the case of a single nut

As a result, tests were carried out on 5 specimens, two with single nuts (designated D-1 and D-2) and three with additional nuts (marked D-3, D-4 and D-5). Cyclic tests were carried out only on the elements with additional nuts D-3, D-4 and D-5. Figure 9 shows comparison of the tensile force - displacement relationship of measuring point P-1 for component D-3.

During the tests, no significant concrete cracks were observed in the tested elements. These are cracks of no more than 0.25 mm in width with a pattern typical of the realized loading and fixing conditions.

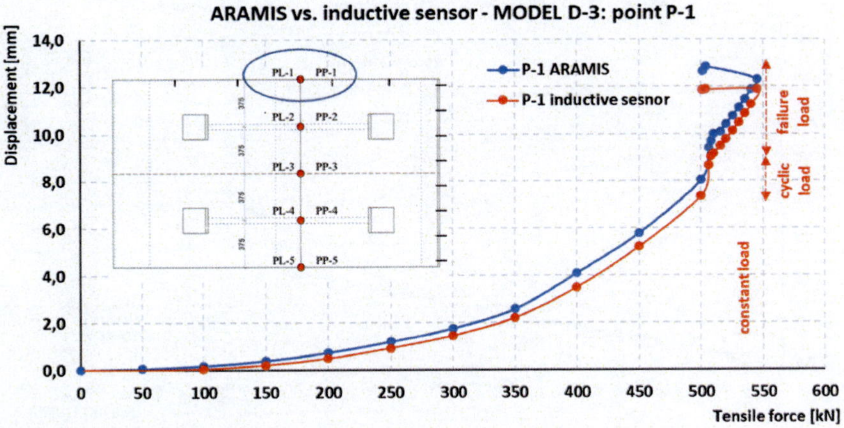

Fig. 9. Relationship of loading and unloading stages vs. displacements of point P-1 in test model D-3 (ARAMIS and inductive sensor)

3 Comparison of Experimental Results and FEM Analysis

Comparing the results obtained by experimental testing and FEM analysis, it can be concluded that satisfactory convergence has been achieved. Figure 10 shows comparison of the tensile force - displacement relationship of measuring point P-1 for component D-2 (results from ARAMIS and inductive sensor) and FEM model. In the FEM model, there was a sharp increase in displacement above force 370 kN, associated with reaching the yield point of the bolted connection. In the experiments, the increase in displacement was milder and therefore a higher connection capacity was obtained, but the displacement (opening) of the connection was the same as in the FEM model and it amounted 8 mm. The difference in the level of maximum force obtained in the FEM model compared to the tests of components D-1 and D-2 (components with single nuts) was about 19%. The value can be considered satisfactory due to the complicated process of numerical modeling of the designed connector.

Figure 11 in orange shows the area of destruction in the FEM model and the destroyed connector in the experimental tests. As can be seen, the same components were destroyed in both cases. Also shown in Fig. 12, the image of the cracks of the test elements was analogous in the FEM model and experimental tests.

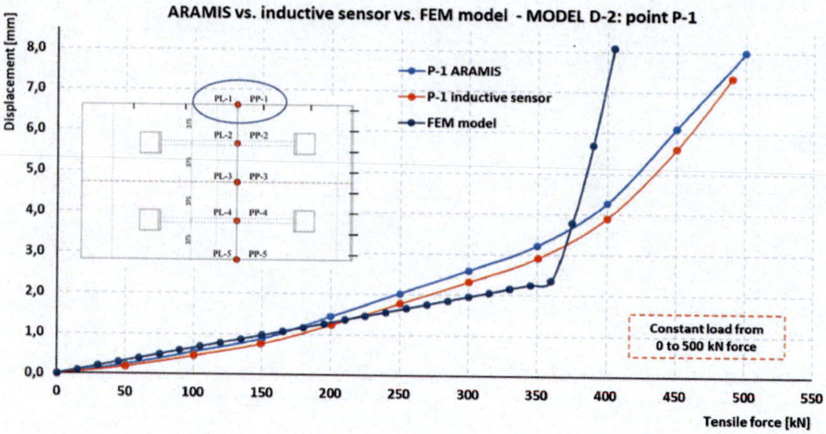

Fig. 10. Comparison of the tensile force - displacement relationship of measuring point P-1 for component D-2 (ARAMIS and inductive sensor and FEM model)

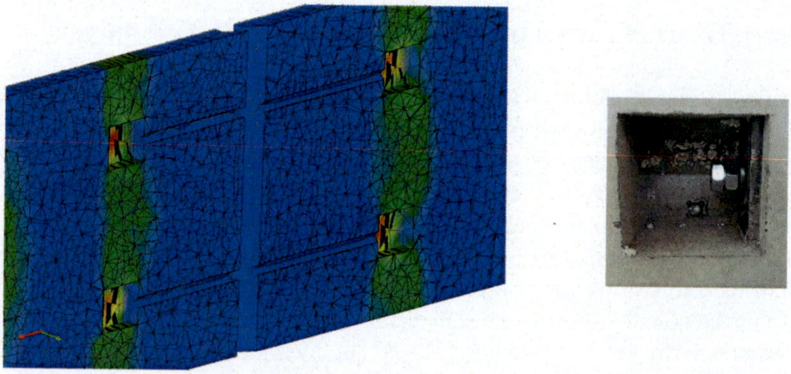

a) b)

Fig. 11. Destruction inside the connection: a) model MES (black areas indicate cracks), b) experimental test

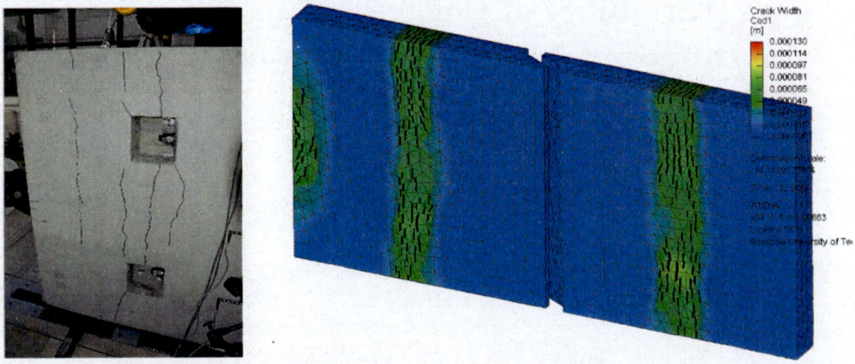

a) b)

Fig. 12. Image of a crack on the outside surface of the tank walls: a) experimental test, b) FEM model (black areas indicate cracks)

4 Conclusion

On the basis of the experimental tests and numerical simulations carried out, it can be concluded that the designed horizontal connection of the prefabricated cylindrical tank meets the load-bearing conditions. In the case of the experimental tests, after doubling the nuts, the joint carried a load of up to 570 kN, i.e. the load capacity of a single joint was 285 kN. The maximum displacement (contact opening) was 15 mm at 570 kN, while the permanent contact opening after 20 load and unload cycles was 5 mm. Carried out cyclic loading for a load of up to 500 kN showed, the possibility of the formation of permanent deformations in the combined contact, values up to 5 mm.

Comparing the FEM results and the experimental studies, the important role of detailed modelling of the bolted connection can be seen. The fastener model adopted in the simulations was simplified, as the nut, thread length, etc. were not modelled in detail. As the tests showed, all the details mentioned had a strong influence on achieving the maximum failure force and they will be subject to further studies and FEM simulations.

References

1. Wit D (2017) Budownictwo kubaturowe ze ścianami nośnymi—możliwości współczesnej prefabrykacji. Cement Wapno Beton 5:414–425 in Polish
2. Nykiel D, Buda-Ożóg L (2023) Wpływ różnicy temperatur wewnętrznej i zewnętrznej powierzchni ściany na wartość naprężeń i obraz zarysowania cylindrycznego zbiornika żelbetowego – porównanie metod obliczeniowych. „Inzynieria i Budownictwo", nr 1–2/2023
3. Zięba J, Skrzypczak I, Buda-Ożóg L (2020) Probabilistic method and FEM analysis in the design and analysis of cracks widths. Eng Struct 209:110022. https://doi.org/10.1016/j.engstruct.2019.110022
4. Cervenka J, Papanikolaou NK (2008) Three dimensional combined fracture–plastic material model for concrete. Int J Plast 24:2192–2220. https://doi.org/10.1016/j.ijplas.2008.01.004

Variability of Darci Friction Factor in Differential Equation for Enforced Flow in Collector-Pipeline

V. V. Cherniuk[1](✉), V. Ye. Fasuliak[1], I. V. Bihun[2], and M. V. Cherniuk[3]

[1] Department of Hydraulic and Water Engineering, Lviv Polytechnic National University, Lviv, Ukraine
v.cherniuk@ukr.net
[2] Lviv City Communal Enterprise "Lvivvodokanal", Lviv, Ukraine
[3] Department of Electronic Engineering, Lviv Polytechnic National University, Lviv, Ukraine

Abstract. Enforced streams with discrete along the path inflow of jets are implemented in collector-pipelines of the following systems: horizontal water-intake; radial water-collectors; purifying structures in water-supply and water-removal; drainage systems in hydromelioration; exhaust ventilation, etc. The fluid inflow into a collector-pipeline through orifices in its wall is variable and increases along the path. The cause of this is the decrease in piezometric head in the collector-pipeline caused by hydraulic resistance of the pipeline and by the increase in kinetic head in it. Reliable and adequate methods of calculation of collector-pipelines are lacking, though such pipelined are widespread. It is known that high accuracy in hydraulic calculation can be achieved basing on methods of differential equation of variable flow rate fluid flow. All authors accept that the hydraulic friction factor and the mean speed along the collector-pipeline are constant. In effect, because jet inflows along the path, they vary. Incomplete taking into account the hydrodynamic characteristics of the variable flow rate fluid flow causes essential errors in designing the collector-pipeline. In this article, relationships for calculation of head losses for overcoming the friction forces along the collector-pipelines are obtained in differential form; zone of laminar flow and three segments of turbulent flow zone, which in sequence emerge along the path, are considered. The variable hydraulic friction factor is expressed in differential form in terms of variable working head in the collector-pipeline.

Keywords: Hydraulic friction factor · Pressure collector-pipeline

1 Introduction

Pressure collector-pipelines (CPs) are used in different branches of engineering; in particular, in water-supply (horizontal water-intakes, radial water-intakes, purification structures), in water-removal (purifying structures), in hydromelioration (drying systems), in ventilation (exhaust ventilation), in energetics (nuclear and thermal power plants), in aviation industry (for jet-engine test chambers), etc. [1].

Though pressure CPs are widely used, no reliable method of their hydraulic calculation has been suggested yet.

Z. Blikharskyy et al. (Eds.): CEE 2023, LNCE 438, pp. 58–69, 2024.
https://doi.org/10.1007/978-3-031-44955-0_7

2 Analysis of Literature Data

Relentless pursuit to explain the physical law of fluid inflow into pressure pipelines is widely observed. Thus Wang et al. [2] numerically investigated inflow of ground water from porous homogeneous media into a CP, and Chen et al. [3] investigated such inflow from heterogeneous porous grounds. Studying the intake of crude oil from porous media Clemo [4] has created a mathematical model of single-phase fluid inflow into a CP. Results of theoretical investigations were compared with data of experimental investigations of inflow of air and water into perforated pipe [4].

Murphy et al. [5] experimentally investigated the operation of a pressure CP layed in sandy ground. They report that the energy losses of the stream for the jet inflow into the CP through small orifices are dominating, and the losses for friction along the CP are insignificant. The results obtained are proposed for application for hydraulic calculation of inflow storm waters into pressure CPs layed in filtrational tranches and under sidewalks covered with water-proof surfaces.

The underground drainage in irrigated territories in dry climate regions is an effective method of increasing the depth of occurring of ground waters and that of reduction of salts in ground. Ghorbani et al. [6] in the course of laboratory investigations have established that 76% of redundant ground water is removed through drainage pipes, and 24% through collectors to which the pipes are connected. Perforated collectors transport water taken from drainage pipes and, at the same time, work in the mode of collector pipeline [6].

Tong et al. [7], by means of numerical modelling, studied fluid flow through parallel tubular channels which connect together the distributing and connecting pipelines. Such streams occur, for example, in solar collectors. The aim of the investigations was to attain the equal fluid flow rates through each channel. The regulation of flow rate was conducted by gates installed in inlets of the collector pipeline.

Moldovanu et al. [8] numerically modeled the decrease in non-uniformity in distribution of speeds of air in cross-section of the CP with a help of artificial roughness. The latter was formed by semi-spherical indents in the inner wall of the collector-pipeline.

As can be seen from the given analysis, the authors of work [2–6] established in different methods the laws of ground water inflow from porous media into a CP. There were investigated individual geometric parameters of pressure collector-pipelines and individual hydraulic characteristics of streams in CP. The contribution of individual resistances upon the total energy of stream in CP or upon the distribution of velocities of the stream in a cross-section of the CP was also studied [8]. The experiments conducted were mainly numerical ones. Mathematical models for calculation of CP are lacking.

The same achievements are in the investigation of work of distributive pipelines [9, 10]. Liu et al. [11] have created a differential equation of momentum of forces which act on a local branch of the jet at an angle of $\beta = 90^{\circ}$ to the stream in the pressure distributive pipelines. The authors of [11] have assumed that hydraulic friction factor λ is invariable (of fixed value) along the distributive pipeline. In the work [11], a series of other simplification are also applied. Based of such assumptions mathematical model of a stream does not present real characteristics of the stream.

However, it is known that that a precise hydraulic calculation of pressure collector-pipelines can be achieved only by means of theory of fluid flow of variable flow rate [12]. In calculation of pressure CPs, the initial equation is the suggested by G. A. Petrov differential equation of fluid flow with variable along the path flow rate [13, p. 18]:

$$d\left(\frac{\alpha_o V^2}{2g}\right) + \frac{dp}{\rho g} + dz + i_f dx + \frac{\alpha_o (V - v\cos\beta)\cdot V}{g}\frac{dQ}{Q} = 0. \quad (1)$$

where β is the angle between the stream velocity vectors $\vec{V}$ inside the CP and $\vec{v}$ of the inflowing jet which enters into the CP through the orifice in the wall; Q is the flow rate of the stream in the CP; $\sin\psi \cdot dx = dz$ is the geometric altitude of position of open area section of the stream above the reference plane; ψ is the angle of inclination of the longitudinal axis of CP to horizon; $dh_x = i_f dx$ are losses in head for friction along the segment dx of the collector-pipeline; i_f is the hydraulic inclination.

A series of approaches to solving Eq. (1) have been suggested. Konstantinov [14] and Smyslov [15] in solving the Eq. (1) assumed that the angle of the inflow of the jets into the main stream $\beta = 90°$, and the CP had been layed horizontally. Having taken $\beta = 90°$, we obtain $v\cos\beta = 0$. Having assumed a constant value of the velocity V of the stream in the CP, the authors of [14, 15] have written Eq. (1) as follows:

$$-dh + \frac{2vdv}{g} + dh_l = 0. \quad (2)$$

It is recommended to apply Eq. (2) for calculation of collector-pipelines in which the distance between inlets is less than ten times the diameter of CP.

Vasylenko and Smyslov [16] have reduced Eq. (2) to dimensionless form (3):

$$-\frac{d^2\overline{Q}}{d\overline{x}^2}\cdot\frac{d\overline{Q}}{d\overline{x}} - A\overline{Q}\frac{d\overline{Q}}{d\overline{x}} + B\overline{Q}^2 = 0, \quad (3)$$

where $A = \alpha_o(2 - m)$; $m = v\cos\beta/V$ is the coefficient of mass joined; $\overline{Q} = Q/Q_{\text{end}}$, $\overline{x} = x/l$, Q_{end} is the flow rate of fluid at the end of CP. The law of joining the flow rate is adapted by the authors of [16] according to Datei [17]: $Q = k \cdot sh(\alpha x)$, where k and α are constants characterizing the CP.

In Eq. (1), Kravchuk [18] recommends to take $m = v\cos\beta/V = 0$. He also assumed that hydraulic friction factor is invariable along the CP, and he proposes to neglect the second and the third terms in Eq. (3).

From the conducted analysis of literature, it can be seen in hydraulic calculation of pressure pipelines that all the authors take a fixed value of the friction factor λ along the whole CP. However, this does not agree with reality. Yegorov [19] warns that incomplete taking into account the design parameters of collector-pipelines and that of hydrodynamic characteristics of a variable flow rate stream causes serious miscalculations in designing a collector-pipeline; that in turn affects the effectiveness of its work.

3 Statement of the Problem

The stream of fluid in a collector-pipeline forms at the expense of the inflow from the jets through the orifices in its walls. The flow rate along the stream in CP increases. Correspondingly, hydraulic losses of head along the CP as well as the energy expenditure

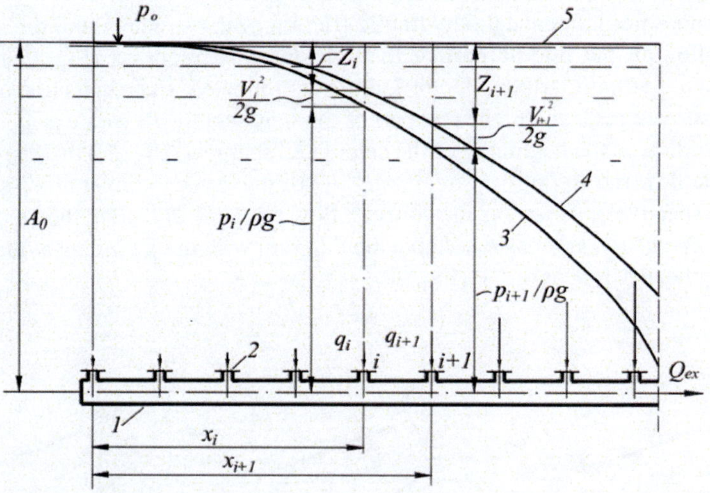

Fig. 1. Schematic diagram of work of pressure collector-pipeline: 1 – CP; 2 – inlet orifices (nozzles); 3 – piezometric curve for stream in CP; 4 – curve of total head; 5 – level of fluid outside CP

for creation of kinetic head $V^2/2g$ of the stream increase. Because of these factors, the piezometric curve 3 and the curve 4 of total head are lowering along the stream in CP (Fig. 1). The difference between the level A_0 of fluid outside the CP and the total head of the stream inside the CP is the working head Z, under the actions of which the jets of fluid flow into CP:

$$Z_{(x)} = A_0 - \left(\frac{p_{(x)}}{\rho g} + \frac{V_{(x)}^2}{2g}\right). \tag{4}$$

Since the working head Z increases along the stream, the inflow of jets into CP increases along the path: $Q_{(x)} = \mu\omega\sqrt{2gZ_{(x)}}$. Therefore, the fluid inflow into CP along the path is non-uniform. In practice, there arises a problem of reduction of the non-uniformity of fluid inflow into CP along the path. To adequately solve this problem, it is necessary to take into account variability of the hydraulic friction factor λ in the differential Eq. (1) of variable flow rate fluid flow. It is also necessary to take into account the fact that the angle β of jets inflow into the fluid stream in CP can be a non-right angle.

Aim of the work: to obtain in differential form a relationship for variable hydraulic friction factor.

4 Derivation of the Differential Equation

Let us consider three variants of jet inflow into a calculated segment of a stream in CP, the length of the segment being dx: (a) at the end of the segment (Fig. 2a); (b) in its middle part (Fig. 2b); (c) at the beginning (Fig. 2c).

The schematic diagram of jet inflow at the end of the segment is of small informability, since the jet inflow at the end of the segment is of small informability, since the jet inflow at the end does not influence the hydraulic friction of the stream along the segment *dx*. The jet inflow in the middle part of the calculated segment dx (Fig. 2b) makes the calculation of hydraulic friction factor λ more complicated, since in its first part the flow rate of the stream is *Q*, and in the rest part of this segment the flow rate is $Q + dQ$. In the case of jet inflow at the beginning of the calculated segment (Fig. 2c) there are taken into account the flow rates *Q* of the stream and *dQ* of the jet in the inlet cross-sections 1-1 and 2-2, respectively, as well as the constant flow rate $Q + dQ$ along the whole segment of the CP. Therefore, as a basis, we take the diagram with the jet inflow at the beginning of the calculated segment of CP (Fig. 2c).

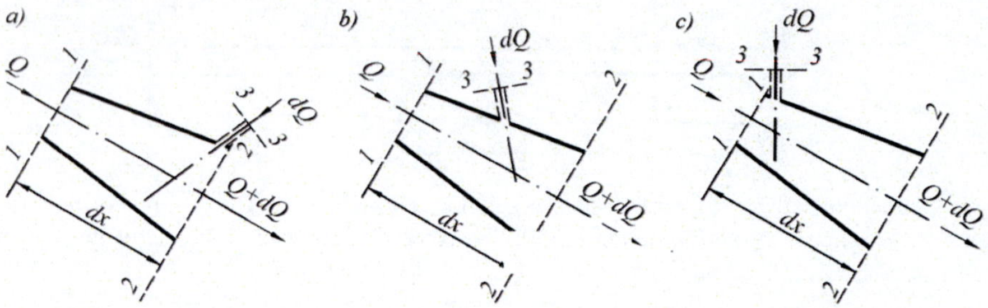

Fig. 2. Schematics diagram of jet inflow into CP in segment of the length *dx*

Basing on the calculational diagram of inclined to horizon collector-pipeline which is washed by an outer stream and dilates with a joined to it jet (Fig. 3) as well as having assumed that the increment of pressure along the axis of CP for infinitely short segment *dx* is positive, we write the hydraulic equation of variation of momentum:

$$\begin{aligned}&\alpha_0\rho(Q+dQ)(V+dV)+\alpha_0\rho dQ\cdot v\cos\varphi-\alpha_0\rho QV\\&=\rho\Omega-(p+dp)(\Omega+d\Omega)+(p+dp/2)d\Omega\\&-\rho g(\Omega+d\Omega/2)\sin\psi\cdot dx-\rho g(\Omega+d\Omega/2)dh_x,\end{aligned} \tag{5}$$

In Eq. (5), there are presented the projections of the following quantities onto the axis *x* of CP (Fig. 3): $\alpha_0\rho(Q+dQ)\cdot(V+dV)$ - momentum of fluid stream in the calculated cross-section 2-2; $\alpha_0\rho dQ\cdot v\cos\beta$ – ditto for the jet inflowing at cross-section 3-3; $\alpha_0\rho QV$ - ditto for fluid stream in the calculated cross-section 1-1; *V* – average speed of stream at the cross-section 1-1; *Q* – its flow rate; *v* – average speed of the jet inflowing at 3–3; β - angle between the vectors $\vec{V}$ and $\vec{v}$ of average speed of main stream in CP and of the jet, respectively; $\rho\Omega$ - momentum of force of hydraulic pressure at cross-section 1-1; $(p+dp)(\Omega+d\Omega)$ - ditto at cross-section 2-2; *p* - pressure at cross-section 1-1; Ω - area of CP cross-section at 1-1; $(p+dp/2)d\Omega$ - longitudinal component of momentum of reaction force *R* of pipe wall that delates in the segment between 1-1 and 2-2; $\rho g(\Omega+d\Omega/2)\sin\psi\cdot dx$ – momentum of the weight *G* of the segment *dx* of the stream, where the minus sign in front of this term agrees the direction of projection of momentum of weight *G* with the direction of axis *x* of the stream under variation of angle ψ within 0°...360°; $\rho g(\Omega+d\Omega/2)dh_x$ – momentum of friction force $T = \tau_o(\chi+(d\chi/2)dx$, where τ_o– average stress of friction at the wall of CP,

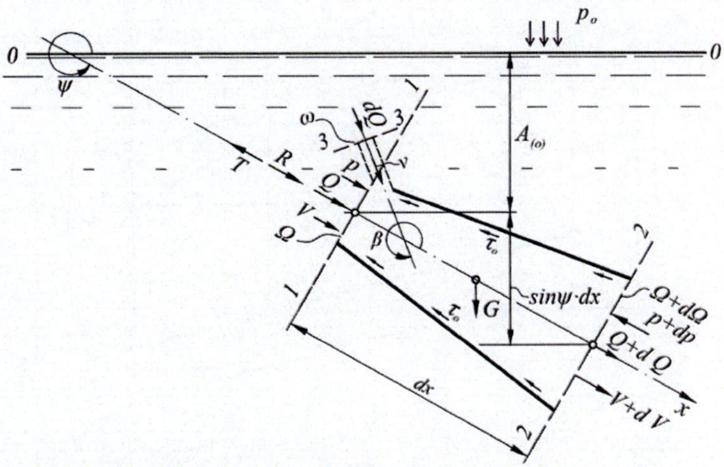

Fig. 3. Schematic diagram of jet inflow into collector-pipeline

$\tau_o = \rho g R(dh_x/dx)$ [20, p. 76]; R hydraulic radius of stream; $\chi = \Omega/R$ - wet perimeter of the stream. Angles β and ψ are referenced counter-clockwise (see. Fig. 3), and their values vary from 0° to 360°.

Regarding the fact that for the cylindric CP $Q = \Omega \cdot V$, after algebraic operations and neglecting the infinitesimals of second order, from Eq. (5) we obtain the differential equation of fluid flow with mass joining along the path:

$$\frac{\alpha_o(2V - v\cos\beta)dV}{g} + d\left(\frac{p}{\rho g}\right) + \sin\psi \cdot dx + dh_x = 0. \quad (6)$$

The differential Eq. (6) relates six variables V, dV, v, dp, dh_x dx. One of the them, namely the length dx, is independent one.

5 Differential of Head Losses for Friction Along CP

5.1 General Statements

In solving the differential Eq. (6) let us express its variables in terms of the total working head $Z_{(x)}$ and the variable independent distance x (Fig. 4), regarding the fact that the stream in CP is formed by discrete feeding by jets which penetrate into CP under the action of total $Z_{(x)}$.

Thus, the flow rate of the jet which inflows into CP through an orifice in it

$$q_{(x)} = \mu\omega\sqrt{2gZ_{(x)}} = bZ_{(x)}^{1/2} \quad (7)$$

The differential of losses of fluid which inflows in the collector-pipeline in its segment with the length dx:

$$dQ_{(x)} = n\mu\omega\sqrt{2gZ_{(x)}}dx = bZ_{(x)}^{1/2}dx, \quad (8)$$

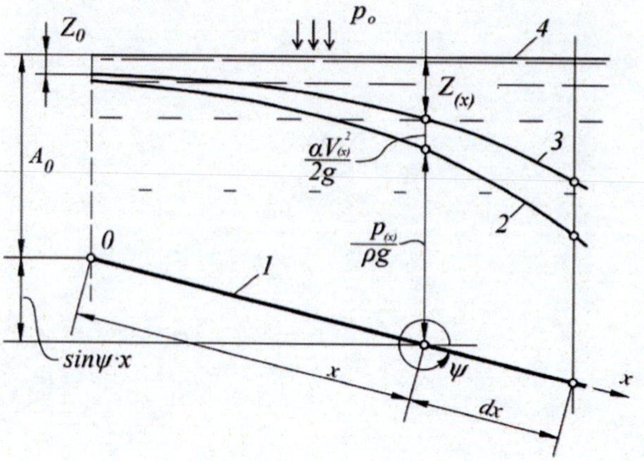

Fig. 4. Heads acting upon collecting-pipeline: 1 – CP; 2 – piezometric curve of stream inside CP; 3 – ditto for total head; 4 – fluid level outside CP; V and v – velocities of stream and inflowing jet, respectively

where n is number of inlet orifices per unit length of the CP, m^{-1}; ω is the area of one inlet orifice; μ is the coefficient of its flow rate; b is the constant multiplier:

$$b = n\mu\omega\sqrt{2g} = const,\ \mathrm{m}^{1.5} / \mathrm{s}. \tag{9}$$

The flow rate of the fluid transported in CP:

$$Q_{(x)} = Q_{tr} + b\int_0^x Z_{(x)}^{1/2}dx, \tag{10}$$

where Q_{tr} is the transit flow rate of the fluid when it enters the calculated segment of CP.
The average speed of stream in CP:

$$V_{(x)} = \frac{Q_{(x)}}{\Omega} = \frac{Q_{tr} + b\int_0^x Z_{(x)}^{1/2}dx}{\Omega}. \tag{11}$$

The differential of the average speed of the stream of CP is determined from (8):

$$dV_{(x)} = d\left(\frac{Q_{(x)}}{\Omega}\right) = \frac{bZ_{(x)}^{1/2}dx}{\Omega}. \tag{12}$$

The differential of head losses in infinitely short segment dx of a cylindric CP, which is situated between neighbouring inlets in CP wall can be sufficiently exactly determined in the same way as in uniform flow of fluid; i. e. by means of Darci-Weisbach formula [20, p. 124]. Taking into account formula (10), we can write:

$$dh_{(x)} = d\left(\lambda_{(x)}\frac{x}{D}\frac{V_{(x)}^2}{2g}\right) = \frac{1}{2g\Omega^2 D}d\left[\lambda_{(x)}\left(Q_{(tr)} + b\int_0^x Z_{(x)}^{1/2}dx\right)^2 x\right], \tag{13}$$

where D is inner diameter of CP.

The Darci factor is the function of the series of variables: $\lambda_{(x)} = f\left(\overline{\Delta};\ Re_{(x)}\right)$, where $\overline{\Delta} = \Delta_{eq}/D$ is the equivalent roughness of the pipe walls; $\mathrm{Re}_{(x)} = f\left(V_{(x)},\ D,\ \nu\right)$ is Reynolds criterion,

$$\mathrm{Re}_{(x)} = \frac{V_{(x)}D}{\nu} = \frac{(Q_{tr} + b\int_0^x Z_{(x)}^{1/2}dx)D}{\Omega\nu}, \tag{14}$$

where ν is the kinematic viscousity.

In fluid flow along CP, the flow rate increases as it is being filled by inflowing jets (Fig. 1). Accordingly, along the stream, there first emerges laminar flow. Then, it is in turn changed by three segments of resistance of turbulent flow. These are the following segments of resistance: of hydraulically smooth pipes; transitive; of hydraulically rough pipes. The Darci factor, depending on Reynolds criterion $\mathrm{Re}_{(x)}$ and on the roughness $\overline{\Delta} = \Delta_{eq}/D$ of the inner walls of CP takes different values of each of these segments of flow (Fig. 5).

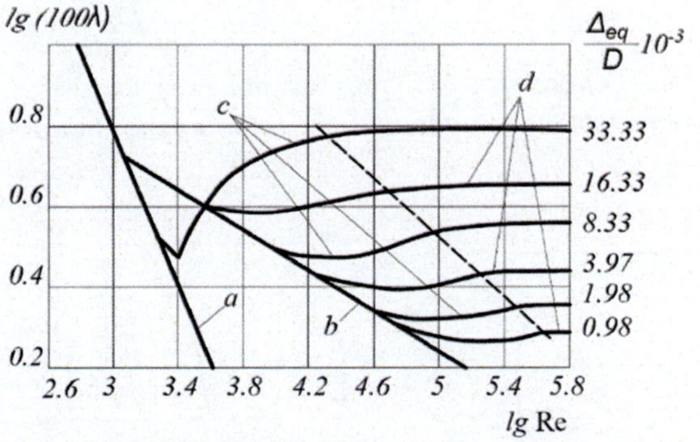

Fig. 5. Dependence $\lambda_{(x)} = f(\overline{\Delta};\ Re_{(x)})$: *a* – zone of laminar flow; *b, c, d* – zone of turbulence flow (segments: *b* – hydraulically smooth pipes; *c* – transitive; *d* – hydraulically rough pipes) [20, p. 94]

5.2 Laminar Mode of Fluid Flow

For laminar flow of fluid (curve 'a' in Fig. 5), that emerge for $\mathrm{Re}_{(x)} \leq 2320$, the Darci factor is calculated according to Poisele formula $\lambda_{(x)} = 64/\mathrm{Re}_{(x)}$; and with taking into account expression (14):

$$\lambda_{(x)} = \frac{64}{\mathrm{Re}_{(x)}} = \frac{64\Omega \cdot \nu}{\left(Q_{(tr)} + b\int_0^x Z_{(x)}^{1/2}dx\right)D}. \tag{15}$$

The substitution of expression (15) into formula (13) gives us the differential $dh_{(x)}$ of head losses for laminar mode of fluid flow CP:

$$dh_{(x)} = \frac{32 \cdot \nu}{g\Omega \cdot D^2}\left(Q_{tr} + b\int_0^x Z_{(x)}^{1/2} dx + bZ_{(x)}^{1/2} x\right) dx. \tag{16}$$

5.3 Segment of Hydraulically Smooth Pipes of Turbulent Flow

For a segment of resistance of hydraulically smooth pipes of turbulent mode of fluid flow (curve b in Fig. 5), that emerges for $\mathrm{Re}_{(x)}(\Delta_{eq_{(x)}} / D_{(x)}) < 10$, the Darci factor is calculated according to Blasius formula $\lambda_{(x)} = 0.3164 / \mathrm{Re}_{(x)}^{0.25}$ and with taking into account expression (14):

$$\lambda_{(x)} = \frac{0,3164}{\mathrm{Re}_{(x)}^{0.25}} = \frac{0.3164(\Omega\nu)^{0.25}}{\left[\left(Q_{tr} + b\int_0^x Z_{(x)}^{1/2} dx\right) D\right]^{0.25}}. \tag{17}$$

After substitution of expression (17) into formula (13), the differential $dh_{(x)}$ of head loss for a segment of resistance of hydraulically smooth pipes of turbulent mode of fluid flow in CP:

$$dh_{(x)} = \frac{0.1582\nu^{0.25}}{g\Omega^{1.75}D^{1.25}} \left[\left(Q_{tr} + b\int_0^x Z_{(x)}^{1/2} dx\right)^{1.75} + 1.75\left(Q_{tr} + b\int_0^x Z_{(x)}^{1/2} dx\right)^{0.75} Z_{(x)}^{1/2} x\right] dx. \tag{18}$$

5.4 Transient Segment of Turbulent Flow

For a transient segment of resistance of turbulent mode of fluid flow (family of c curves in Fig. 5), that occurs for $10 < \mathrm{Re}_{(x)}(\Delta_{eq_{(x)}} / D_{(x)}) < 500$, the Darci factor is calculated according to Altshul formula, that with taking into account expression (14) assumed the form:

$$\lambda_{(x)} = 0.11\left(\frac{\Delta_{eq}}{D} + \frac{68}{\mathrm{Re}_{(x)}}\right)^{0.25} = 0.11\left[\frac{\Delta_{eq}}{D} + \frac{68\Omega\nu}{\left(Q_{tr} + b\int_0^x Z_{(x)}^{1/2} dx\right) D}\right]^{0,25}. \tag{19}$$

By substitution of expression (19) into formula (13) we obtain the differential $dh_{(x)}$ of head losses for the transitive segment of turbulent mode of fluid flow in CP:

$$dh_{(x)} = \frac{0.11}{2g\Omega^2 D}\left[\frac{\Delta_{eq}}{D} + \frac{68\Omega\nu}{\left(Q_{tr} + b\int_0^x Z_{(x)}^{1/2}dx\right)D}\right]^{0,25}$$

$$\left\{\frac{0.25\left(Q_{tr} + b\int_0^x Z_{(x)}^{1/2}dx\right)^2 x}{\left[\frac{\Delta_{eq}}{D} + \frac{68\Omega\nu}{\left(Q_{tr}+b\int_0^x Z_{(x)}^{1/2}dx\right)D}\right]} + \left(Q_{tr} + b\int_0^x Z_{(x)}^{1/2}dx\right)^2 dx\right.$$

$$\left.+2x\cdot\left(Q_{tr} + b\int_0^x Z_{(x)}^{1/2}dx\right)\cdot bZ_{(x)}^{1/2}dx\right\}. \tag{20}$$

5.5 Segment of Hydraulically Rough Pipes of Zone Turbulent Flow

For a segment of resistance of hydraulically smooth pipes of turbulent mode of fluid flow (family of d curves in Fig. 5), that emerges for $\mathrm{Re}_{(x)}(\Delta_{eq_{(x)}} / D_{(x)}) \geq 500$, the Darci factor is calculated according to Shifrinson formula:

$$\lambda_{(x)} = 0.11\left(\frac{\Delta_{eq_{(x)}}}{D_{(x)}}\right)^{0.25}. \tag{21}$$

For invariable values of the roughness Δ_{eq} and those of the diameter D in the calculated segment of CP, the coefficient $\lambda_{(x)}$ is invariable, and its differential is equal to zero:

$$d\lambda_{(x)} = 0. \tag{22}$$

The differential of head losses for the segment of hydraulically rough pipes of turbulent mode of flow is the following:

$$dh_{(x)} = \frac{0.11}{2g\Omega^2 D}\left(\frac{\Delta_{eq}}{D}\right)^{0.25}$$

$$\left[\left(Q_{tr} + b\int_0^x Z_{(x)}^{1/2}dx\right)^2 dx + 2x\cdot\left(Q_{tr} + b\int_0^x Z_{(x)}^{1/2}dx\right)\cdot bZ_{(x)}^{1/2}dx\right]. \tag{23}$$

Notice that the expression (20) can be used instead of the formula (23).

6 Auxiliary Investigations

The rest of variables of the differential Eq. (6), namely V, dV, v, dp, are also expressed in terms of working head $Z_{(x)}$ of fluid flow in CP and the independent distance dx. This is to be published in the next article.

In order to verify the obtained mathematical relations, variation of values of Darci factor along the stream in CP [21] has been experimentally investigated. The effect of Reynolds' criterion $\mathrm{Re}_{(x)}$ [22] and the angle β of jet inflow upon the non-uniformity of fluid inflow into pressure pipelines [22, 23] were studied experimentally in hydraulical laboratory, experiments were conducted for different values of inner diameters D of CP and for diameters d of inlet nozzles. The problem of hydraulic calculation of pressure distributive pipelines [24] is also solved. In particular, the effect of the angle of removal of jets from the distributive pipelines upon the value of the flow rate factor of an outlet cylindric nozzles with lateral inflow jet [25] was experimentally investigated.

7 Conclusions

The differential, dh_x, of head losses for friction along a collector-pipeline is expressed in terms of total working head $Z_{(x)}$ and the independent variable distance dx. The relationships for dh_x have been obtained for laminar mode of fluid flow, that emerges in initial part of CP, and for three segments of the zone of turbulent mode of flow, which are situated one after another along CP, provided the length of the CP is sufficient. These segments are: (a) of hydraulically smooth pipes; (b) transient; (c) of hydraulically rough pipes. The obtained expressions of (16), (18), (20), and (23) are to be substituted into the differential Eq. (6) of enforced fluid flow with joining of masses along the path.

References

1. Ivaniv VV, Bihun IV (2019) Zastosuvannia ta metody hidravlichnoho rozrakhunku napirnykh truboprovodiv-zbyrachiv. Visnyk natsionalnoho universytetu vodnoho hospodarstva ta pryrodokorystuvannia: Zbirnyk naukovykh prats. Tekhnichni nauky 4(88):76–89. https://doi.org/10.31713/vt420197 (in Ukraine)
2. Wang Y, Zhou L, Feng D (2021) An Improved numerical manifold method model for groundwater flow problems. Eng Anal Boundary Elem 125:218–232. https://doi.org/10.1016/j.enganabound.2020.12.017
3. Chen Y, Zheng H, Yin B, Li W (2023) The MLS-based numerical manifold method for Darcy flow in heterogeneous porous media. Eng Anal Bound Elem 148:220–242. https://doi.org/10.1016/j.enganabound.2022.12.030
4. Clemo T (2006) Flow in perforated pipes: a comparison of models and experiments. SPE Prod Oper 21(02):302–311. https://doi.org/10.2118/89036-PA
5. Murphy P, Kaye NB, Khan AA (2014) Hydraulic performance of aggregate beds with perforated pipe underdrains flowing full. J Irrigation Drain Eng 140(8):04014023. https://doi.org/10.1061/(ASCE)IR.1943-4774.0000740
6. Ghorbani K, Wayayok A, Abbaszadeh M, Fikri A (2015) The effect of pipe collectors in reducing the drainage coefficient rate. Water Sci Technol Water Supp 15(4):675–682. https://doi.org/10.2166/ws.2015.014

7. Tong JCK, Sparrow EM, Abraham JP (2007) Attainment of flowrate uniformity in the channels that link a distribution manifold to a collection manifold. ASME J Fluids Eng 129(9):1186–1192. https://doi.org/10.1115/1.2754319
8. Moldovanu D, Csato A, Bagameri N (2019) Flow studies on increasing the efficiency of the inlet manifold. In: IOP conference series: materials science and engineering, vol 568, no 1, p 012050. IOP Publishing. https://doi.org/10.1088/1757-899X/568/1/012050
9. Kale RV, Singh RP, Mahar PS (2008) Optimal design of pressurized irrigation subunit. J Irrig Drain Eng 134(2):137–146. https://doi.org/10.1061/(ASCE)0733-9437(2008)134:2(137)
10. Wang Y, Ju X, Zhu S, Li M (2020) Simulation of local head loss of drip-irrigation tape with integrated in-line emitters as a function of cross section. Span J Agric Res 18(4):e0210–e0210. https://doi.org/10.5424/sjar/2020184-15767
11. Liu H, Zong Q, Lv H, Jin J (2017) Analytical equation for outflow along the flow in a perforated fluid distribution pipe. PLoS One 12(10):e0185842. https://doi.org/10.1371/journal.pone.0185842
12. Navoyan K (1975) Primery gidravlicheskikh raschyotov vodopropusknikh sooruzhenii. Budivelnik, Kiev (in Russian)
13. Petrov GA (1964) Gidravlika peremennoi massy (Dvizhenie zhidkosti s izmeneniem raskhoda vdol puti). Kharkovskii gosudarstvennyi universitet, Kharkov (in Russian)
14. Konstantinov Y, Smyslov VV (1966) K raschetu dirchatykh trub s prisoedineniem raskhoda vdol puti. Gidravlika i gidrotekhnika: Nauchno-tekhnichesiy sbornik 4:118–127 (in Russian)
15. Smyslov VV (1980) K raschetu sbornykh truboprovodov. Gidravlika i gidrotekhnika: Nauchno-tekhnichesiy sbornik 30:60–65 (in Russian)
16. Vasylenko AA, Smyslov VV (1973) Analiz uravneniya dvizheniya zhidkosti v gorizontalnom tsilindricheskom truboprovode s prisoedineniem raskhoda vdol puti. Gidravlika i gidrotekhnika: Nauchno-tekhnichesiy sbornik 17:19–24 (in Russian)
17. Datei Claudio (1961–1962) I condotti emyngenti da un serbatoio. Atti e men Accad patav sci lettere ed arti Parte 2. S. 188–197
18. Kravchuk AM, Chernyshev DO, Kravchuk OA (2021) Hidravlika napirnykh perforo-vanykh truboprovodiv ochysnykh sporud system vodopostachannia ta vodovidvedennia. Kyivskyi natsionalnyi universytet budivnytstva i arkhitektury, Kyiv (in Ukraine)
19. Ehorov AI (1984) Hidravlika napornykh trubchatykh system v vodoprovodnykh ochystnykh sooruzheniiakh. Moskva: Stroiizdat (in Russian)
20. Konstantinov YM (1988) Hidravlika. Kiev: Vyshcha shkola (in Russian)
21. Kravchuk A, Cherniuk V, Kravchuk O, and Airapetian T (2022) Assessing the value of the hydraulic friction factor in pipelines working with a flow connection along the path. East Europ J Enterp Technol 5(7, 119):61–67. https://doi.org/10.15587/1729-4061.2022.265670
22. Ivaniv V, Cherniuk V, Kochkodan V (2021). Non-uniformity of water inflow into pressure collector-pipeline depending on the values of reynolds criterion and of inflow jets angles. In: Proceedings of EcoComfort 2020, lecture notes in civil engineering. Springer Nature Switzerland AG 2021, pp 142–149. https://doi.org/10.1007/978-3-030-57340-9
23. Cherniuk V, Ivaniv V (2017) Influence of values of angle of jet-joining on non-uniformity of water inflow along the path in pressure collector-pipeline. In: 10th international conference on environmental engineering. Vilnius Gediminas Technical University Lithuania. 7 s. eISSN 2029-7092/eISBN 978-609-476-044-0; Article ID: enviro.2017.073
24. Cherniuk V, Hnativ R, Kravchuk O, Orel V, Bihun I, Cherniuk M (2021) The problem of hydraulic calculation of pressure distribution pipelines. East Europ J Enterp Technol 6(7, 114):93–103. https://doi.org/10.15587/1729-4061.2021.246852
25. Cherniuk VV, Ivaniv VV, Bihun IV, Wojtowicz JM (2020) Coefficient of flow rate of inlet cylindrical Nozzles with lateral orthogonal inflow. In: Proceedings of CEE 2019 advances in resource-saving technologies and materials in civil and environmental engineering. Springer, Nature Switzerland, pp 50–57. https://doi.org/10.1007/978-3-030-27011-7_7

Testing of the Ribbed Dome Which is Manufactured by 3D Printing

Bohdan Demchyna[1], Leonid Vozniuk[2], and Mykhailo Surmai[1](✉)

[1] Department of Building Structures and Bridges, Lviv Polytechnic National University, Lviv, Ukraine
mykhailo.i.surmai@lpnu.ua

[2] Department of Architectural Design and Engineering, Lviv Polytechnic National University, Lviv, Ukraine

Abstract. The article describes the testing of the ribbed dome, which was manufactured by the 3D printing method. An analysis of scientific research related to structures made by 3D printing was carried out. To solve the research problem, a prototype of a ribbed dome made of PLA plastic was made using 3D printing technology. In order to conduct experimental research, a device was developed for testing the studied sample of the dome. A gasket was placed on top of the sample, which reproduced the stamp of applying the load along the curvilinear shape of the workpiece. The dome was mounted on a base plate. A hydraulic jack was used to apply the load, and the load value was fixed using a dynamometer located above the die, which reproduced a uniformly distributed load. To determine the calculated values of displacements and stresses, a calculated finite-element model of the ribbed dome was created in the PC LIRA-SAPR. The calculated model of the dome consisted of finite elements and nodes. The conducted experiment made it possible to obtain the nature of the destruction of the ribbed dome model. Based on the results of the experimental study, graphs of horizontal and vertical movements were constructed. The analysis of the obtained results made it possible to establish that the zones of location of the greatest stresses coincide with the experimental area of the dome's destruction. A comparison of the experimental values with the calculated values was performed and a good convergence of the results was obtained.

Keywords: Plastic · 3D printing · Testing · Strength · Displacements · Dome

1 Introduction

Construction using 3D printers is a field in the initial development stage. Construction 3D printers will be improved and widely implemented, as they allow for automating and speeding up the process on the construction site and significantly reducing the cost of construction.

At this stage of the development of 3D printing technologies, the equipment allows you to create various architectural forms and elements of buildings to assemble them

Z. Blikharskyy et al. (Eds.): CEE 2023, LNCE 438, pp. 70–77, 2024.
https://doi.org/10.1007/978-3-031-44955-0_8

at the construction site later or print an entire building directly on the construction site, including a dome.

3D printing on the construction site eliminates time-consuming stages of the design process. With the classic method of construction, the installation of engineering communications requires enormous labour costs on the construction site. However, when 3D printing, some of these features can be incorporated into the 3D printing process. Blank wall printing requires fewer resources, improves insulation, and it makes it possible to use 3D-printed channels to accommodate communications.

2 Analysis of the Latest Research

The appearance of products of individual building structures, which are manufactured by the method of 3D printing, has led to the development of scientific research in this direction. In particular, the works of Ukrainian scientists [1, 2], as well as the works of foreign researchers [3–9] are known.

3D printing is a widespread technology that is planned to be used also in the space sphere during the colonization of other planets or satellites [5–7].

The University of South Carolina (USA) is developing technologies for building materials for the manufacture of complex forms by the method of 3D printing [6].

At this stage of the development of 3D printing technologies, the equipment allows the creation of various architectural forms and elements of spatial structures from metal, concrete [10, 11], and glass [12, 13], as well as parts of buildings for their further installation on the construction site as ready-made modules, but, at the same time, it is possible to print the entire building directly on the construction site [8].

It is worth noting that with the appearance of this construction technology, the field of materials science began to develop rapidly. Special attention is paid to geopolymer concrete and plastics. Manufacturing technology [9] and the study of materials used in the creation of products using 3D printing are also important.

At the moment, high-rise buildings, hotels, and administrative and industrial buildings are already being built in the world with the help of 3D printers, which requires more active development and scientific research in this direction.

3 The Aim of the Study

The aim of the study is to test the experimental model of the ribbed dome made by 3D printing from PLA plastic based on the developed methodology, as well as perform calculations of the experimental sample using the finite element method.

The task of the study is to determine the strength and deformability of the model of the ribbed dome made by 3D printing from PLA plastic and to compare the experimental and theoretical results.

To solve the research problem, a prototype of the dome was made using 3D printing technology. A ribbed dome model was drawn in AutoCAD, based on which a file for 3D printing in STL format was created, and imported into a slice program for preparation for printing (Fig. 1).

As a result, a test sample was obtained, which is shown in Fig. 2.

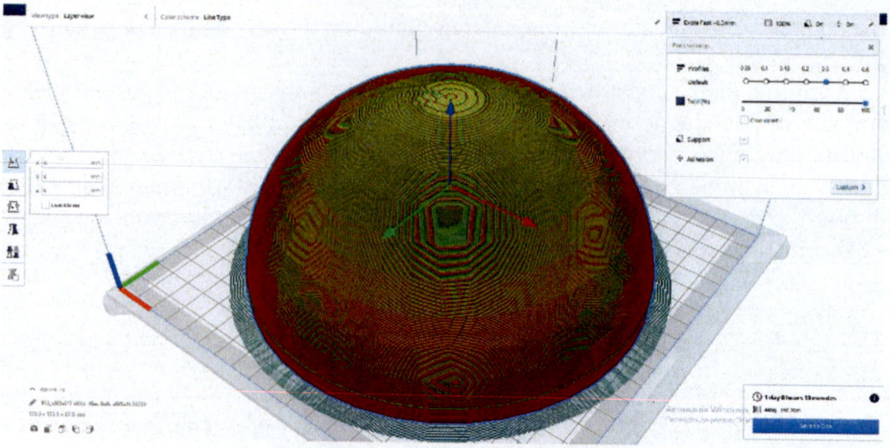

Fig. 1. View of the dome model in the slicing program before 3D printing.

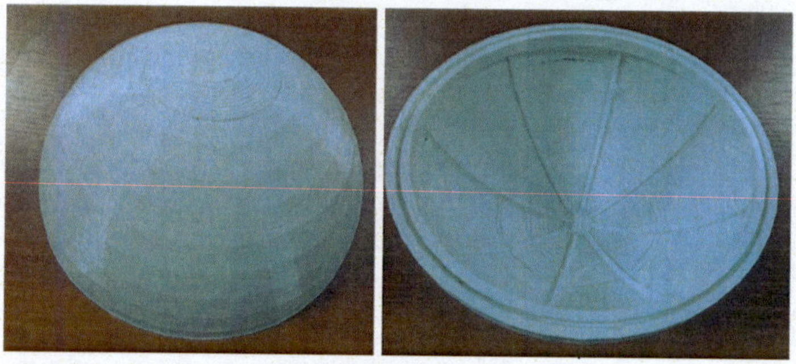

Fig. 2. Experimental sample of a ribbed dome.

The support ring diameter was 200 mm and 10 × 10 mm in cross-section, the thickness of the shell was 3 mm, the dimensions of the ribs were 3 × 6 mm.

4 Methodology of Experimental and Theoretical Research

To conduct experimental studies by the current norms [15, 16], a device was developed for testing the studied sample of the dome. A gasket was placed on top of the tested sample, which reproduced the stamp of applying the load along the curvilinear shape of the workpiece. The dome (1) was mounted on a base plate with a hole for measuring displacements (3) on the inner surface of the sample. A hydraulic jack (5) was used to apply the load, and the load value was fixed using a dynamometer (2) located above the stamp, which reproduced a uniformly distributed load (Fig. 3).

The experimental setup was placed on a PH-125 hydraulic press, which served as the upper and lower support of the setup. The location of the devices is shown in Fig. 4.

Fig. 3. General view of the test facility. 1-experimental model of the dome; 2-dynamometer; 3-beam plate for fastening Aistov protractors; 4-Aistov deflection gauges; 5-jack.

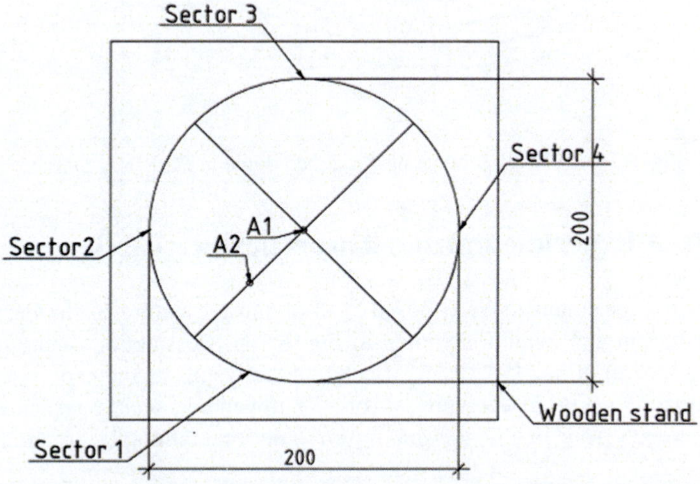

Fig. 4. Scheme of the location of measurement points. A1-point of measurement of vertical movements. A2-point of measurement of horizontal displacement.

A comparator with a division value of 0.001 mm was used to measure meridional deformations. To determine the deflection, an Aistov deflection gauge was used, with a division value of 0.01 mm.

To determine the estimated values of displacements and stresses, an estimated finite-element model of the ribbed dome was created in the PC LIRA-SAPR (Fig. 5).

The basis of the calculation of the PC LIRA-SAPR software complex is the finite element method. The calculated model of the dome consisted of 496 finite elements and 321 nodes. The shell stiffness was modelled using CE-44 with different dimensions, and the fins and support ring were CE-10.

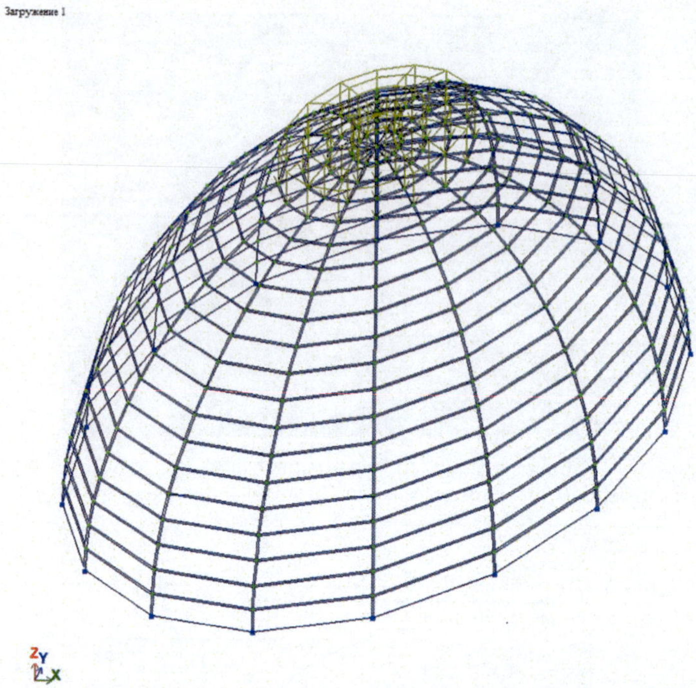

Fig. 5. Calculation scheme of the ribbed dome in PC LIRA-SAPR.

5 Results of Experimental and Theoretical Research

The conducted experiment made it possible to obtain the nature of the destruction of the ribbed dome model, which is shown in Fig. 8. The destruction of the test sample occurred at a load of N = 2456 kN/m^2. Based on the results of the experimental study, graphs of horizontal (Fig. 6) and vertical (Fig. 7) movements were constructed.

As can be seen from Fig. 7, the maximum experimental displacement at point A1 was 2.1 mm.

Based on the created model of the dome, using the PC LIRA-SAPR software, the calculated values of the displacements of the circuit nodes were obtained. The theoretical value of vertical movements in the calculation scheme is shown in Fig. 8.

As can be seen in Fig. 8, the maximum calculated value of displacement at point A1 is 2.42 mm, which exceeds the experimental value by only 13.2%.

As a result of the tests, the destruction of the prototype of the dome was obtained, and based on calculations, the magnitude and character of stresses in the finite elements of the dome model were obtained (Fig. 9).

The analysis of the obtained results made it possible to establish that the zones of the location of the greatest stresses (Fig. 9) coincide with the area of dome destruction obtained in the study. It was also established that according to the calculation in the PC LIRA-SAPR, at a load Nu = 2456 kN/m2, which corresponds to the destructive one, the maximum stresses in the dome model were 13.2 MPa, which is 10.6% more than the

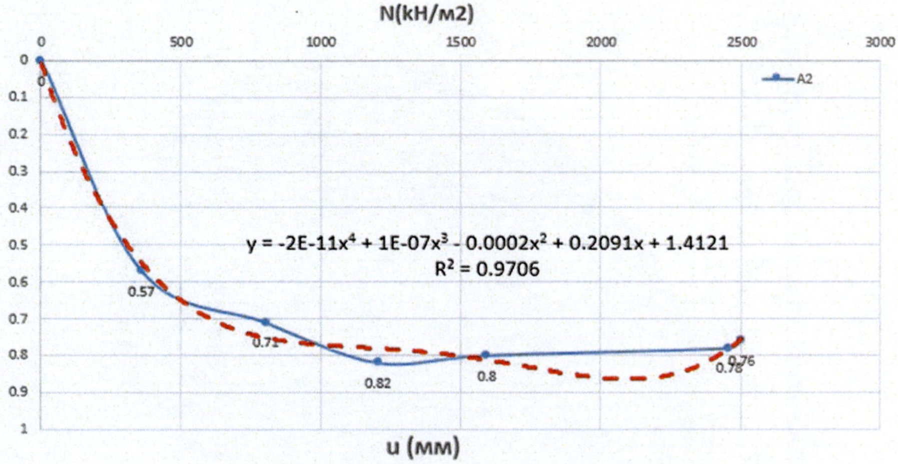

Fig. 6. Diagram of displacements from the load in point A2.

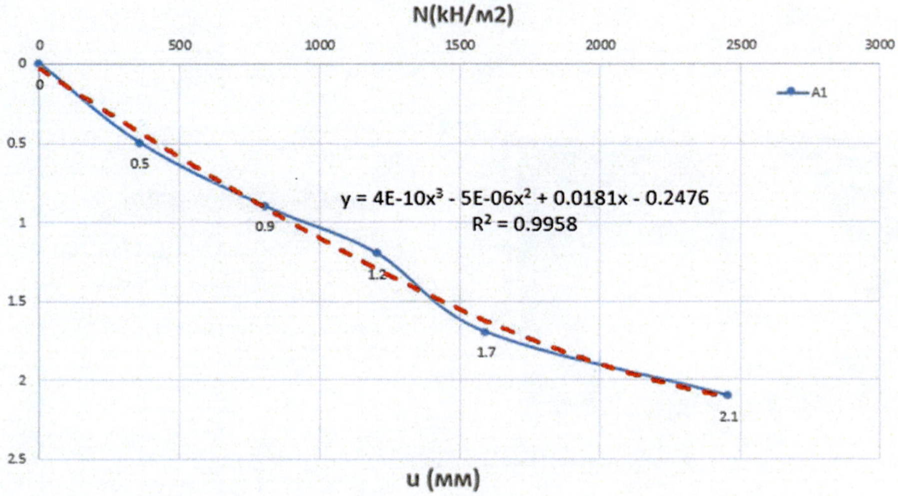

Fig. 7. Diagram of displacements from the load in point A1.

value of the normative resistance of the used PLA plastic, namely the value of 11.8 MPa, which was determined experimentally.

6 Conclusions

The method of testing the ribbed dome proposed in this paper allows for research of experimental samples in laboratory conditions.

As a result of the experimental studies, a ribbed dome made of PLA plastic, which was manufactured by the 3D printing method, was tested and the value of the destructive

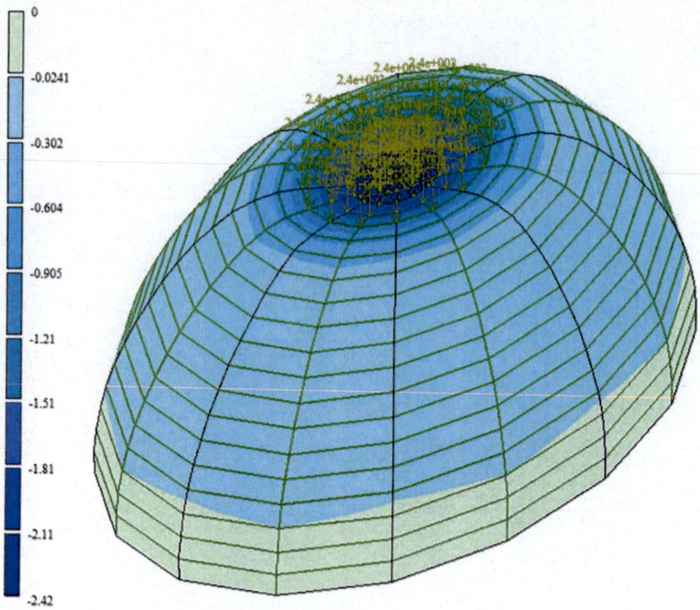

Fig. 8. Estimated displacement values of the dome model.

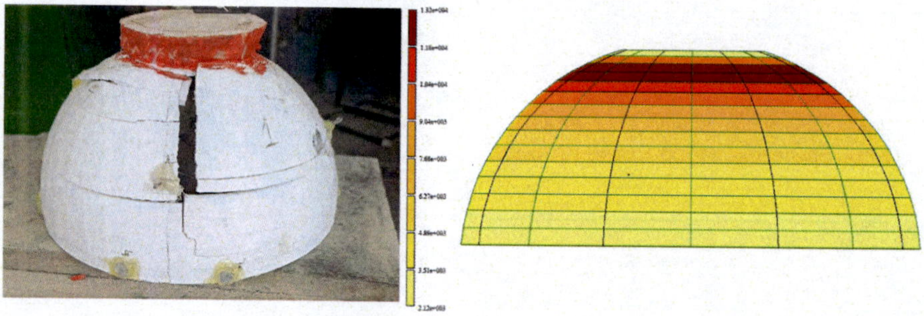

Fig. 9. Experimental sample of a ribbed dome.

load, the value of the maximum displacements in the center of the dome and the ultimate stresses in the zone of destruction were obtained.

The finite element method has shown its sufficient effectiveness for carrying out calculations of ribbed domes, which is confirmed by a good convergence of theoretical and experimental studies, namely, the difference in displacement results - 13.2%, maximum stresses at the point of failure - 10.6%.

The data obtained as a result of experimental research can be used in the creation of models of ribbed domes, which are manufactured by the 3D printing method, as well as in further scientific research.

References

1. Demchyna BG, Surmai MI, Vozniuk LI, Gavrylyak SA (2021) The use of 3D printing in the field of construction (modern science: innovations and prospects: proceedings of II international scientific and practical conference. Stockholm, Sweden 7–9:259–266
2. Gavrylyak SA (2021) New technologies in the field of construction. Using 3d printers Academic Journals and Conferences. JTBP, pp 15–22
3. Williams RL, Albus JS, Bostelman RV (2004) Self-contained automated construction deposition system autom. Constr 13:393–407. https://doi.org/10.1016/j.autcon.2004.01.001
4. Buswell RA, Soar RC, Gibb AGF, Thorpe A (2007) Freeform construction: mega-scale rapid manufacturing for construction autom. Constr 16:224–231. https://doi.org/10.1016/j.autcon.2006.05.002
5. Tomaswick A (2022) Mars and moon dust can be turned into geopolymer cement. Good Enough for Landing Pads and Other Structures, Universe Today Space and astronomy news. http://surl.li/divzd
6. Khoshnevis B, Bodiford MP, Burks KH (2005) Lunar contour crafting: a novel technique for ISRU-based habitat development. In: American institute of aeronautics and astronautics conference, №1, 2005, pp 1–12
7. Cesaretti G, Dini E, De Kestelier X, Colla V, Pambaguian L (2014) Building components for an outpost on the lunar soil by means of a novel 3D printing technology. Acta Astronaut 93:430–450. https://doi.org/10.1016/j.actaastro.2013.07.034
8. Hager I, Golonka A, Putanowicz R (2016) 3D printing of buildings and building components as the future of sustainable construction? Proc Eng 151:292–299. https://doi.org/10.1016/j.proeng.2016.07.357
9. Gosselin C, Duballet R, Roux P, Gaudillière N, Dirrenberger J, Morel P (2016) Large-scale 3D printing of ultra-high performance concrete - a new processing route for architects and builders Mater. Des 100:102–109. https://doi.org/10.1016/j.matdes.2016.03.097
10. Ivanyk I, Vikhot S, Vybranets Y, Ivanyk Y (2018) Theoretical research into spatial work of a steel-reinforced concrete statically indeterminate combined structure. East Europ J Enterp Technol 5(7–95):13–22. https://doi.org/10.15587/1729-4061.2018.143023
11. Vybranets Y, Vikhot S (2023) Spatial calculation of metal truss structure in joint work with reinforced concrete slab. Lecture Notes in Civil Engineering, 290 LNCE, pp 450–458
12. Vihot SI (2015) Strength and deformability metal composite structures based on sustainable design (candidate's dissertation). Available from ProQuest Dissertations and Theses database
13. Demchyna BG, Surmai MI, Tkach RO (2019) The experimental study of glass multilayer columns using digital image correlation. Arch Mater Sci Eng 96(1):32–41
14. Demchyna BG, Surmai MI, Tkach RO, Hula VO, Kozak RP (2020) An analysis of using the method of two-dimensional digital image correlation in glass column research. East Europ J Enterp Technol 4(12–106):52–59
15. DBN V.1.2-2 (2006) System of ensuring reliability and safety of construction objects. Loads and influences. Design standards. Kyiv, Ministry of Construction of Ukraine
16. DSTU B V.1.2-3 (2006) System of ensuring reliability and safety of construction objects. Deflections and movements. Design requirements. Kyiv, Ministry of Construction of Ukraine

Elimination Options of Solar Heat Gains Through Transparent Surfaces–Review

Zuzana Dicka(✉), Dusan Katunsky, and Erika Dolnikova

Institute of Architectural Engineering, Faculty of Civil Engineering, Technical University of Kosice, Vysokoskolska 4, 042 00 Kosice, Slovakia
zuzana.dicka@tuke.sk

Abstract. Buildings protect people from external climatic influences. The envelope structures of the building mitigate the impact of changes in the external environment. However, in recent decades we have been facing climate change, which is affected by many factors and the construction industry is one of them. As a result of long-term climate change, we can expect a more frequent series of warm days. In many cases, existing constructions cannot adapt to sudden climate changes. In summer, buildings overheat due to excessive solar radiation. It causes internal discomfort and temperature stress in the indoor environment. Summer overheating of buildings is influenced by many factors, but the potential source of excessive heat is unwanted solar gains through transparent constructions. This paper presents an overview of the elimination options of solar gains through transparent surfaces of envelope structures. Furthermore, it also deals with assessing the efficiency of the shading systems concerning thermal comfort and sufficient daylight in the workspace.

Keywords: Overheating · Solar heat gains · Shading systems · Glazing

1 Introduction

Building envelope structures form a filter between the building's internal and external environment. Their ability to react to changes in the internal and external environment must be in accordance with the requirements of the users of the internal space. The building envelope significantly affects the architectural expression of the building with its aesthetic effect. Currently, the design of buildings is subject to the influence of modern trends, and the requirements for the indoor climate are increasingly demanding. Facades with a high proportion of transparent surfaces have become an icon of modern times and developing cities. In addition to the aesthetic appearance, the so-called curtain walls primarily ensure maximum daylighting of the interior, which is necessary for efficient work and the health of the building's users. Large-scale transparent surfaces are also popular in terms of the energy efficiency of the building. Using solar gains through highly exposed parts of the perimeter structure reduces energy consumption for heating. However, in some cases, solar gains can disrupt internal thermal comfort and thereby increase space cooling costs [1].

Z. Blikharskyy et al. (Eds.): CEE 2023, LNCE 438, pp. 78–87, 2024.
https://doi.org/10.1007/978-3-031-44955-0_9

As we certainly know, we have been facing global warming for a long time now, and the problem of overheating buildings is now well-known to everyone. This unpleasant phenomenon is occurring more and more frequently as a result of climate change. Even in climate zones where the problem of building overheating was not previously solved. Therefore, it is necessary to pay more attention to this topic. First of all, it is important to focus on mitigating the climate crisis. However, we must also know how to face its consequences [2]. Nowadays, we already know various methods and technologies for mitigating the negative impact of sunlight. The contribution provides an overview of the management of heat gains through transparent surfaces using shading technology and glazing systems. The goal is to understand how to eliminate or reduce solar heat gains through transparent surfaces [3].

2 Solar Gains Through Transparent Surfaces

A solar gain is an effective form of passive heating. Solar radiation is mainly infrared short-wave radiation, which can pass through transparent surfaces and heat the internal structure of the building. Conversely, back-radiated or reflected long-wave radiation is not able to pass back through the glazing. It causes heat accumulation in the interior Re (Fig. 1) [3].

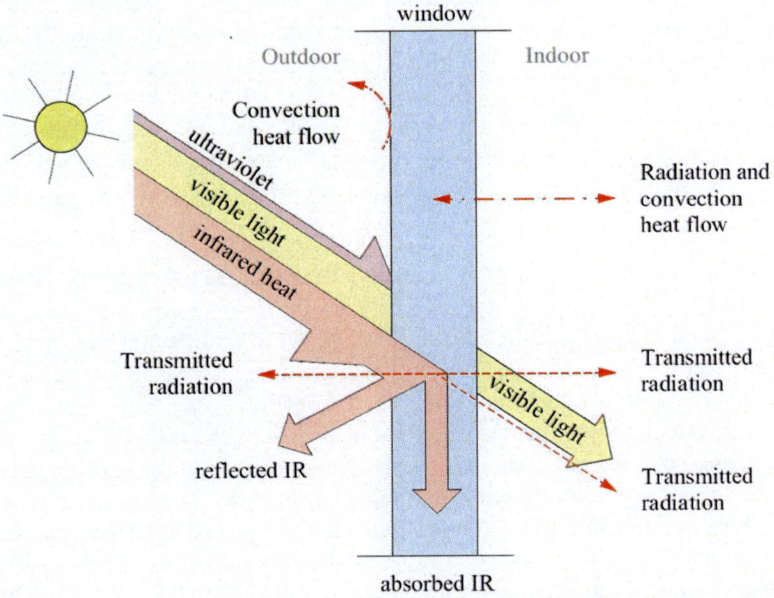

Fig. 1. Reflection, absorption, and transmission of solar radiation through glazing (according to [4]).

Infill structures are the most exposed elements of the building envelope. Most of them are mainly made of glass. They significantly contribute to the creation of an optimal indoor climate and also to the building's energy losses. The primary function is the provision of natural daylight to the interior, the visual connection of the interior and exterior, insolation, and also ventilation of the interior space. In addition, they represent an important role in other functions, so they are the key elements of the building. Unlike non-transparent structures, windows are up to 5 times weaker. They are also the most vulnerable part of the building. The quality of the lighting of the space and the intensity of the solar gains are influenced not only by the size and shape of the transparent surface but especially by its orientation to the cardinal points [5].

Another factor is the shape of the building and its location. It significantly affects the amount of received solar energy. An improperly designed building will be affected by direct sunlight, which will mean overheating of the interior space during warm summer days. Solar energy is considered the largest and most disused source of energy. There are various ways to use the free source of sunlight to save energy in buildings. Solar radiation is used as a passive solution to compensate for the heating needs in the winter. But on the other hand, in the summer season, it can cause overheating of the building due to excessive solar heat gains and thus increased energy consumption for cooling [6].

Solar gain generally refers to the increase in the temperature of an object or structure in the environment. Objects can capture and absorb solar radiation and as a result, their temperature increases. The amount of solar gain depends on the capability of the glazing material to transmit solar radiation into the interior space and on the intensity of solar radiation. The greenhouse effect, which is well-known in the concept of global warming, also works on this principle [7]. Solar heat gains through transparent surfaces are determined as the difference between the transmitted solar energy through the window and its heat loss through conduction. In transparent and translucent buildings, the so-called solar heat gain coefficient has a significant influence on the heating and cooling load. Not too many studies are available in the literature dealing with this coefficient for innovative solutions such as glazing. Measuring methods for solar heat gains, whether through shielded or unshielded transparent surfaces, are still imprecise and often simplified [8].

3 Elimination of Solar Heat Gains Through Transparent Surfaces

Nowadays, when energy is extremely important, we try to design the most energy-efficient structures. The purpose of an energy-efficient structure is to minimize heat loss and, conversely, to ensure maximum heat gain from solar energy in winter [9]. Solar heat gain forms a significant part of the total heat gain of the interior space. However, it is necessary to pay increased attention to its reduction during the cooling season. For this reason, the topic of eliminating solar gains has become a crucial area of scientific research. This issue is important for designing buildings and their energy efficiency. From this point of view, reducing solar radiation is a decisive factor [10]. Nevertheless, it is necessary to think about this problem already in the initial design of the building. We must not forget the correct orientation of the building and transparent surfaces, as well as the disposition of the interior space [11].

In the case of existing buildings, where we can no longer influence the orientation, there are solutions to alleviate this unpleasant phenomenon. There are well-established methods for solving this problem today. The most effective strategy includes modifying transparent surfaces and choosing the right type of glazing or shading systems. There are various shading devices and systems. The researchers also discovered a new type of glazing, the so-called intelligent glazing system that promises controlled regulation of sunlight. We must not forget the shading by the surrounding environment, which also has a significant impact on the internal solar gains in the building [12].

3.1 Shading Systems

Precisely shading systems can significantly limit the overheating of the interior in summer. It can reduce and eliminate unpleasant glare and ensure the visual comfort of the interior space. Another advantage is the provision of privacy for the users of the building. Currently, many types of shading technology are known, and each has advantages and disadvantages. When choosing an adequate kind of shading, it is necessary to consider its advantages and simultaneously avoid its drawbacks. Due to climatic conditions and the diverse functions of buildings, it is impossible to use a generalized manual for shading elements design. For instance, in tropical climate areas is necessary to use year-round shading, but in colder regions, shading is required only during summer. The shading device can regulate and control daylight access to the building [13]. We can categorize shading systems from different points of view. According to the position, control method, material solution, whether they are fixed, movable, or combined elements, etc. Re (Fig. 2) [14].

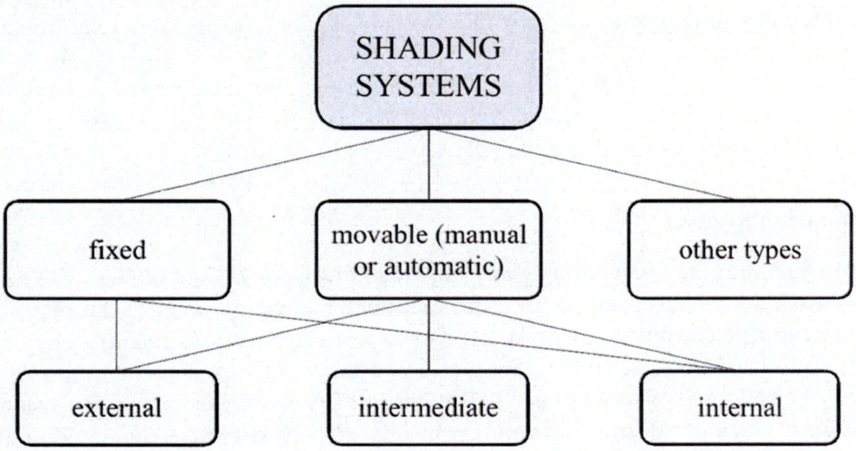

Fig. 2. Solar shading systems for buildings: a possible classification (according to [14]).

The researchers in Estonia in their study analyzed the performance of different types of shading systems on energy consumption in an office building in Tallinn. The results showed the shading system is an effective way to control energy consumption in administrative buildings. Shading efficiency varied with orientation. The evidence points in favor of dynamic systems. They have better results and thus outperform static elements [15]. In a study by Kandar et al., the authors focused on shading the buildings by their own geometry. Specifically, self-shading by the inclined wall of an office building in Malaysia. The self-shading strategy of sloping walls has proven to be significant in reducing direct solar gains. The study found that an inclined wall with an angle of 65° from the horizontal level was the most effective, while a significant decrease in energy consumption for cooling was noted [16]. The authors Joshi M. et al. focused on the integration of fixed exterior shading elements, which represent overhangs above the windows. They observed a significant decrease in solar heat gains. However, they emphasize, the results may vary depending on orientation, location, type of glazing, etc. [17].

According to the authors Babot F. et al., it is necessary to capture direct sunlight before it hits the glazing surface. Pursuant to the study, external shading devices are the most suitable and effective. Energy savings in this case can range from 10 to 40% [18]. In the following study, authors Kraus M. and Senitkova I. compared individual shading elements. In order to calculate the internal thermal comfort, they considered a total of 9 shading variants of the commercial building. The best results were achieved by shading with external semi-transparent window films. The second place was taken by an external blind with automatic control. Internal blinds with manual control had the worst shading efficiency [19]. External shading generally has the best effectiveness in mitigating solar gains before entering the building interior. However, outdoor equipment is exposed to the weather effects and is less durable and maintenance costs are higher. Interior shading, on the other hand, is less effective at protecting against solar radiation but is better in terms of control and availability. It is also not exposed to external conditions, so maintenance is easier and the service life is longer. According to the authors Naboni and Jakic, the most suitable alternative is intermediate shading. It regulates solar radiation and is sheltered against weather influences [20]. The authors Ardabili N. G. et al. also dealt with intermediate shading. Using simulations, they found the system is inefficient for warm climates. Although, in a mild climate, the system could save 15–30% of energy consumption for heating and cooling [21].

3.2 Glazing Systems

Modern buildings are often highly glazed and require a large amount of energy to maintain thermal and visual comfort. In order to reduce the energy costs of building, it is necessary to replace conventional glazing with new energy-efficient glazing systems. In the market, there are many intelligent glazing systems available that can improve the quality of a building's indoor climate [22]. Interesting types of glazing are, for instance, tinted glass, vacuum glazing, low-emissivity glazing, PCM (phase-change material), etc. One noteworthy type of glazing is thermoelectric (TE) glazing, which in addition to reducing energy, can also generate electricity. This glazing was dealt with by the authors Al-Fartoos M. M. R. et al. The study shows that TE glazing reduces waste heat. However, the authors encountered significant disadvantages of the glazing. The most prominent are the toxic materials from which the glazing make and its opacity. The authors want to address these problems in the future and eliminate these shortcomings [23]. Another modern type of glazing is the so-called water-flow glazing. Active transparent facade using this type of glazing was dealt with by researchers Millán B. P. et al. After analyzing the active flow glazing, it is possible to state the saving of cooling energy in the summer and the decline of the heating load in the winter while sufficient transparency is accomplished. It is a new type of glazing that, in addition to saving energy, also optimizes natural lighting. Adding to this also offers architects freedom of design thanks to its multifunctional nature [24].

A modern type of glazing is aerogel glazing, and several researchers are currently working on it. Authors Mohamed A. F. et al. addressed the effectiveness of an aerogel-based glazing system in the hot, dry climate of Egypt. As a result, the aerogel granulate reduced the consumption of cooling energy by 26.3% with adequate transparency. Their study claims that aerogel glazing can make windows more ecological and energy efficient [25]. In further research, Buratti C. et al. compared the effectiveness of aerogel glazing with simple glazing. As a result, the solar gain was reduced by 62–73% compared to simple glazing. Cooling energy savings were in the range of 22–33%. The results differed depending on the location [26]. Another researcher who deals with this glazing is Zhou Y. In his study, he devoted himself to a holistic overview of the aerogel material, its production, use, and optimization. As a result, the new PCM-integrated aerogel glazing systems are a considerably promising candidate for the energy efficiency of buildings Re (Fig. 3) [27].

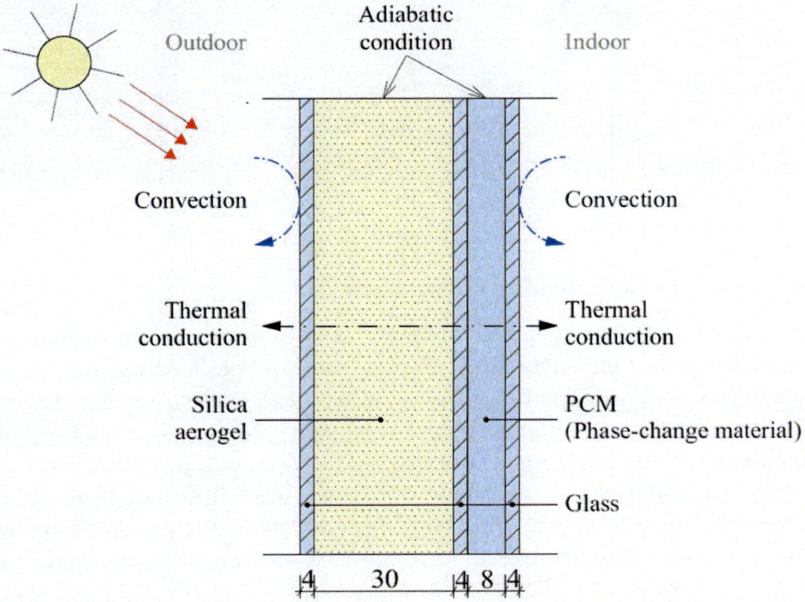

Fig. 3. A PCM and aerogel integrated window glazing system's structural configuration (according to [27]).

Coating systems. Coating solutions form a separate category for glazing systems. Currently, there are several types of coatings, such as low-emissivity foils, anti-reflection, or self-cleaning coatings. The category of colored films shows a high reflectivity of infrared radiation and helps to minimize heat exchange by radiation between the interior and the exterior. However, it will allow the transit of visible light into the interior space. According to the authors Peng Y. et al., the use of colored foil with low emissivity reduced the consumption of cooling energy by 9.87% [28].

This issue was also handled by the authors Armstrong M. M. and Burrows J. Their study was focused on a thermal mirror, i.e. a coating with low emissivity. It is a film glued to glass in a molten state, which is made as a thin transparent coating of metal oxide (tin, silver, zinc). They compared the performance of 2 glazings, namely high solar heat gain (HSG) and low solar heat gain (LSG) glazing Re (Fig. 4). The best total energy savings were reached by HSG, 13–17%, while LSG 8–10%. The energy savings depend primarily on the local climate. The authors recommend HSG glazing in locations with a dominant heating season, and conversely LSG in locations where cooling prevails. Exactly then were energy savings most significant [29].

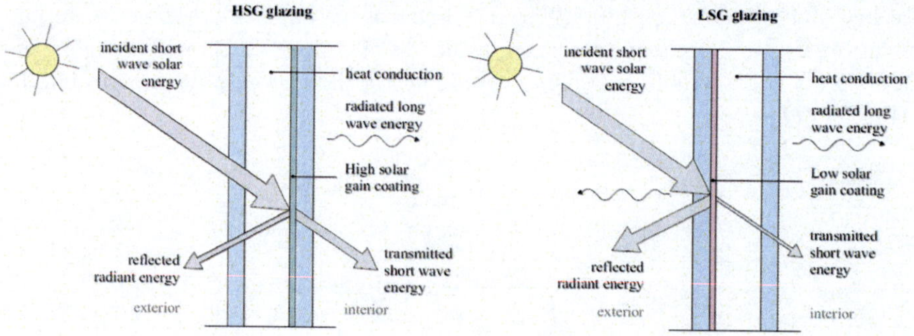

Fig. 4. High solar heat gain (HSG) and low solar heat gain (LSG) glazing (according to [29]).

3.3 Shading by the Surrounding Environment

Last but not least, solar heat gains in buildings are also influenced by shading by surrounding buildings and the environment. For instance, proper landscaping can also be one of the factors affecting thermal gains. However, especially trees and surrounding vegetation effectively shade and thus reduce solar gains. Also, plant evapotranspiration can decrease the ambient air temperature by up to 5 °C. Based on the growth habit, different types of plants can be selected to provide the desired shading [30].

Authors Chagolla M. A. et al. dealt with this issue in their research. Through measurements and simulations, they monitored the impact of shading trees on energy savings for heating and cooling in the climate of Mexico. The total annual energy saving in comparison with an unshielded building was up to 76.6% [31]. Other authors also dealt with the influence of tree shading on the energy saving of buildings. In a study by Hsieh C. et al., the authors compared four situations based on different tree characteristics. They observed the cooling effects of trees on buildings, due to their shading and transpiration. The results showed that it can reduce the total energy consumption of the building by 10.3–15.2%. The topic of reducing the cooling load of buildings is attracting increasing attention. However, there are no general methods, to apply this idea to a specific building or locality [32].

4 Conclusion

The use of solar energy is a fundamental part of energy-efficient strategies. Solar radiation penetrates the building through transparent surfaces, contributing to passive heating. On the other hand, it can cause overheating and glare of the interior space and thus increase costs during the cooling season. This problem is only getting worse with global warming. Therefore, it is necessary to face these consequences. The most important is to design proper sun protection for buildings to ensure visual and thermal comfort. Currently, we know several available options to control this negative phenomenon and thus achieve inner comfort. The correct indoor climate affects not only the energy balance but mainly the health and productivity of building users [33].

This contribution offers an overview of the issue of solar thermal gains and the usual but also the latest elimination options, which are still in the research stage. The paper reports a review of some published scientific papers regarding elimination options, which should be evaluated already in the initial phase of building design. Even with the orientation of the building and its location on the ground, we can influence the resulting energy consumption. Buildings should be designed concerning the given climatic conditions. The resulting optimal strategy for the elimination of solar heat gains, therefore, depends mainly on the location and its climate. There is no general methodology for elimination options, and each building needs an individual solution. Currently, we can use the help of software that generates the optimal geometry of the shading system according to the selected requirements. In any event, we must not forget the importance of this topic in today's world [34].

Acknowledgment. This paper was elaborated with the financial support of the research project VEGA 1/0626/22 of the Scientific Grant Agency of the Ministry of Education, Science, Research and Sports of the Slovak Republic and the Slovak Academy of Sciences.

References

1. Puškár A, Fučila J, Řehák I, Vavrovič B (2002) Obvodové plášte budov. Jaga Group, Bratislava
2. Metz B (2010) Controlling climate change. Cambridge University Press, Cambridge
3. Song Y, Duan Q, Feng Y, Zhang E, Wang J, Niu S (2020) Solar infrared radiation towards building energy efficiency: measurement, data, and modeling. Environ Rev 28:457–465
4. Aminuddin AM, Rao SP, Thing HW (2012) Thermal comfort field studies in two certified energy efficient office buildings in a tropical climate. Int J Sustain Build Technol Urban Dev 3(2):129–136
5. Puškár A, Szomolányiová K, Fučila J (2000) Okná, dvere, zasklené steny. Jaga Group, Bratislava
6. Tamm M, Cid JM, Paramio RC, Baulenas JF, Thalfeldt M, Kurnitski J (2020) Development of a reduced order model of solar heat gains prediction. Energies 13(23):6316
7. Leftheriotis G, Yianoulis P (2022) Glazing and coatings. Compr Renew Energy 3:360–401
8. Hassani AR, Domenighini P, Belloni E, Ihara T, Buratti C (2022) Evaluation of the solar heat gain coefficient of innovative aerogel glazing systems: experimental campaigns and numerical results. J Build Eng 62(4):105354

9. Kon O, Caner I (2020) The effect of shading factor for single and double glazing according to solar heat gain/window heat loss ratio in Turkey for residential buildings. In: IOP Conference series earth and environmental science
10. Taleb S, Yeretzian A, Jabr RA, Hajj H (2020) Optimization of building form to reduce incident solar radiation. J Build Eng 28
11. Kamaruddin M (2021) Investigation of the impact of building orientation on cooling loads in an office building in the tropical climate. J Sci Technol Virt Cult 1(1)
12. Etheridge DW (2010) Ventilation, air quality and airtightness in buildings. Mater Energy Efficiency Therm Comf Build 77–100
13. Hraška J (2020) Tieniaca technika budov. STU, Bratislava
14. Bellia L, Marino C, Minichiello F, Pedace A (2014) An overview on solar shading systems for buildings. In: 6th International conference on sustainability in energy and buildings, Elsevier Ltd., Energy Procedia, pp 309–3017
15. De Luca F, Voll H, Thalfeldt M (2018) Comparison of static and dynamic shading systems for office building energy consumption and cooling load assessment. Manag Environ Qual 29(5):978–998
16. Kandar MZ, Nimlyat PS, Abdullahi MG, Dodo YA (2019) Influence of inclined wall self-shading strategy on office building heat gain and energy performance in hot humid climate of Malaysia. Heliyon 5(7)
17. Joshi M, Buddhi D, Singh R (2022) Effect of overhang shade on the solar heat gain through window in composite climatic in Mid-Western India. J Sci Ind Res 81:1098–1106
18. Babota F, Manea D, Aciu C, Iernutan R, Molnar L (2013) Shading—the way for solar control and reduction of heat gain in buildings. In: C60 International conference "tradition and innovation—60 years of constructions in Transilvania"
19. Kraus M, Šenitková JI, Kučerová L (2022) The impact of solar shading elements on thermal comfort. In: Proceedings of the 26th international meeting of thermophysics 2021, AIP conference on thermophysics 2021, vol 2488, issue 1, Czech Republic
20. Naboni R, Jakica N (2022) Additive manufacturing in skin systems: trends and future perspectives. Rethink Build Skins 425–451
21. Ardabili NG, Feng Y, Wang J (2023) Design and optimization of thermally responsive autonomous dynamic glazed attachment systems for building solar heat gain control. Build Simul
22. Priya AV, Shaik S (2022) Crystal mud transparent slime as a glazing material for net-zero energy buildings: enhanced energy savings, diurnal lighting, and CO_2 mitigation. Energy Sustain Dev 71(2):151–166
23. Al-Fartoos MMR, Roy A, Mallick TK, Tahir AA (2022) A short review on thermoelectric glazing for sustainable built environment. Energies 15(24)
24. Millán BP, Aguirregabiria BL, Vacarezza GO, González JCS, Pérez JML (2015) Active transparent Façades including water-flow glazing. In: VII International congress on architectural envelopes, Spain
25. Mohamed AF, Gomaa MM, Amir AA, Ragab A (2023) Energy, thermal, and economic benefits of aerogel glazing systems for educational buildings in hot arid climates. Sustainability 15(8)
26. Buratti C, Belloni E, Merli F, Mastoori M, Sharifi SN, Pignatta G (2021) Analysis of Nano Silica Aerogel based glazing effect on the solar heat gain and cooling load in a school under different climatic conditions. Environ Sci Proc 12(1)
27. Zhou Y (2021) Artificial neural network-based smart aerogel glazing in low-energy buildings: a state-of-the-art review. Sci 24(12)
28. Peng Y et al (2022) Coloured low-emissivity films for building envelopes for year-round energy savings. Nat Sustain 5:339–347

29. Armstrong MM, Burrows J (2008) Window glazing study at CCHT shows good results. NRCC 14–15
30. Hwang WH, Wiseman PE, Thomas VA (2016) Simulation of shade tree effects on residential energy consumption in four U.S. cities. Cities Environ 9(1)
31. Chagolla MA, Alvarez G, Simá E, Tovar R, Huelsz G (2012) Effect of tree shading on the thermal load of a house in a warm climate zone in Mexico. In: Proceedings of the ASME 2012 international mechanical engineering congress and exposition, Texas, pp 761–768
32. Hsieh CM, Li JJ, Liman Z, Schwegler B (2018) Effects of tree shading and transpiration on building cooling energy use. Energy Build 159(4):382–397
33. Altan H, Ward I, Mohelníková J, Vajkay F (2008) Daylight, solar gains and overheating studies in a glazed office building. Int J Energy Environ 2(2):129–138
34. Aguilar-Santana JL, Velasco-Carrasco HJM, Riffat S (2020) Review on window-glazing technologies and future prospects. Int J Low-Carbon Technol 15(1):112–120

Comparison of Daylight Levels Using Simulation Calculation Tools

Erika Dolníková(✉), Dušan Katunský, and Zuzana Dická

Faculty of Civil Engineering, Institute of Architectural Engineering, Technical University of Kosice, Vysokoškolská 4, 042 00 Košice, Slovakia
erika.dolnikova@tuke.sk

Abstract. The focus of this article is the evaluation of light comfort in a specific attic space. Windows located in the perimeter wall as well as the roof contribute to ensuring this comfort. The objective of this study is to assess and compare the impact of various skylight configurations on the amount of daylight penetration in the designated area, with the aim of optimizing its utilization for human occupancy. Furthermore, indoor air temperature is influenced by the presence of daylight. During winter, thermal insulation properties adequately mitigate the situation as the packaging structures possess sufficient thermal resistance. However, in summer, inadequate heat storage capacity of these structures leads to overheating of the indoor air. This research investigates the relationship between the size of roof windows and the resulting microclimate within the attic space. Four different window size variants are analyzed and compared. The goal is to determine the optimal window size that effectively regulates the microclimate and ensures a comfortable environment in the attic area. The scenario without roof windows presents a viable option, particularly when lighting is solely provided through windows in the perimeter wall, as opposed to the roof. This configuration minimizes overheating but compromises daylight availability. To maximize daylight intake, the skylights should be positioned at the top of the roof. The study presents a graphical comparison of differences in daylight calculations obtained from different simulation tools. The analysis was conducted without considering window shading. The results highlight disparities between the outcomes produced by various simulation tools, accompanied by their justification.

Keywords: Daylighting · Simulations · Illuminance · Daylight factor · Roof windows · Schools

1 Introduction

The sky and the sun are sources of daylight. It is imperative to ensure that buildings are constructed in a manner that allows for ample natural lighting. One should see a large part of the sky throughout the year, such an environment is suitable for him to give 100% power. People are sensitive to whether the interior is well ventilated and well lit in daylight [1, 2]. This is what affects human health, regulates human biorhythms, improves the body's immune system and supports the activity of internal organs. Daylight increases work productivity, reduces fatigue and even reduces energy consumption, as there is no need to switch on artificial lighting [3, 4].

Z. Blikharskyy et al. (Eds.): CEE 2023, LNCE 438, pp. 88–97, 2024.
https://doi.org/10.1007/978-3-031-44955-0_10

Roof windows shall provide at least twice as much light under the same conditions as vertical windows of the same dimensions and three times as much light as vicars of the same dimensions [5]. The roof window also allows for greater variation in light intensity, which increases the visual interest of the room [6, 7]. Furthermore, it is clear that roof windows provide a higher brightness of the walls than the windows in the vicarage and facade windows, which leads to a smoother transition between the high brightness of the window panel and the adjacent wall and thus reduces the risk of glare [8, 9].

Roof windows have the potential to provide significantly more daylight compared to conventional vertical windows or dormer windows. This is due to their direct orientation towards the light source, resulting in minimal diffusion or reflection of light [17, 18]. As a result, roof windows can effectively illuminate areas that would otherwise be dimly lit, offering a greater influx of natural light into the interior spaces. Even in low-pitched roofs, roof windows and skylights can receive direct solar radiation, even when facing north [19–21].

The architectural geometry of a building plays a crucial role in the availability and quality of daylight within the interior spaces. Reliance solely on façade windows for daylighting in deep buildings has its limitations [22, 23]. Despite the inclusion of ample glass in the façade, adequate daylight distribution (DF $>$ 2%) can only be achieved within a few meters from the façade [24–27].

The findings of previous studies have consistently shown that the incorporation of rooflights consistently improves the overall daylighting performance of a space [28–30]. In certain designs, the addition of rooflights has led to a substantial increase in the availability of daylight, ranging from 12% to 45% of the occupied year [31–35]. Moreover, the inclusion of rooflights has significantly reduced the occurrence of periods with insufficient light levels (less than 100 lx) where electric lighting would typically be required [36–40]. Conversely, the presence of rooflights has resulted in an increase in the duration of periods with light levels exceeding 2500 lx [41–44].

2 Materials and Methods

A reference room of a school building was selected for the study. The building location is in Kosice, SK- latitude 49 N, longitude 20.3 E. The room floor dimensions are 5.4 m $\times$ 10.77 m and headroom height 3.03 m. The room daylighting is due to a vertical double-glazed window of dimensions 0.9 m $\times$ 0.86 m and roof windows of dimensions 0.78 m $\times$ 1.18 m. The window sill is 0.9 m. The windows are located in the South orientation. The side windows and roof windows double glazed unit light transmittance is 0.8. The measurement was conducted using two lux meters at nine different control points simultaneously (see Fig. 1). The simulation program considered an exterior horizontal illumination level of 5000 lx. During the specific day chosen for the measurement, the outdoor light intensity varied from 5500 to 7000 lx in the month of February.

The daylight measurements were conducted in accordance with the Slovak standard "Measurements of Daylighting." The equipment utilized for the measurements consisted of two data loggers, specifically the ALMEMO 2690-10A models, along with an illuminance sensor bearing the identification number 15061543 and an accuracy of 5%. The attic space daylighting was simulated for four options, see Fig. 2.

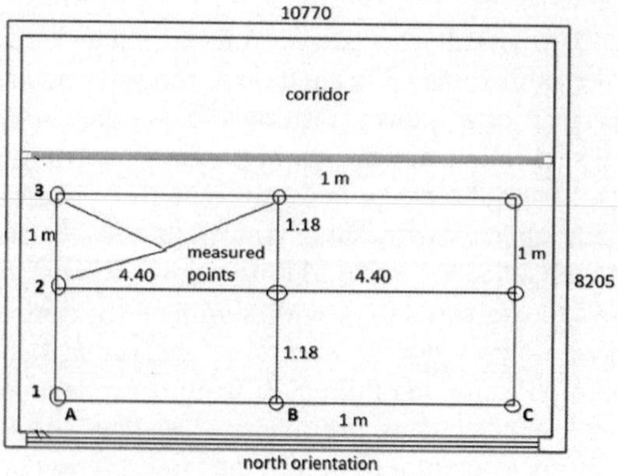

Fig. 1. Floorplan – measured points

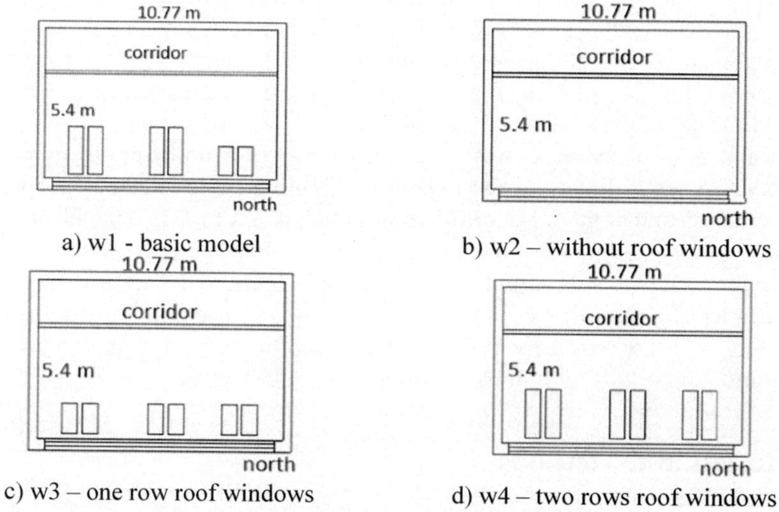

Fig. 2. The reference room with side windows and roof windows selected for daylight simulations

The following coefficients were taken into account in the calculations: (permeability coefficient 0.8, glazing maintenance factor on the outer surface 0.9, glazing maintenance factor on the inner surface 0.85, subsoil reflection factor 0.15 (dark background). The walls and ceiling were painted in a white color with a reflectance coefficient of 0.7, while the floor had a reflectance coefficient of 0.45. The height of the working plane was set at 0.85 m. The coefficient of light loss due to the construction of the window was $\tau = 0.64$. Neighboring objects in the distance yes do not shade the room. The room is usually occupied from 8:00 to 17:00. The room used for medium-precision activities was included in III. Easy - technical class. To meet the lighting requirements of a specific lighting system, the critical point of the functional area on the horizontal plane should adhere to the following criteria: a minimum standard value of daylight factor (DF_{min})

of 1.5%, an average daylight factor ($D_{average}$) of 5%, and a lighting uniformity greater than 0.3, as specified for the given visual task [45, 46].

3 Materials and Methods

Daylighting simulations were performed by Velux Daylight Visualizer and DIALux with the model of tested classroom and 3 variants. The results of measured values and calculated DF from measured values in the points can be seen in Tables 1 and 2 and Fig. 3. The results DF_{min}, DF_{max} and $DF_{average}$ of simulated values can be seen in Tables 3 and 5. The results of DF and illuminance by simulation program can be seen in Fig. 4.

VELUX Daylight Visualizer is an established tool for daylight simulation and visualization in building design and analysis. This validated software allows for accurate prediction of daylight levels and the visual perception of spaces illuminated by natural light, enabling designers to assess and optimize daylighting strategies prior to construction. The software has successfully passed all of the CIE 171:2006 test cases specifically designed for evaluating the accuracy of natural lighting simulations [47, 48].

On the other hand, DIALux is a comprehensive lighting calculation software widely utilized in the European market for assessing the performance of indoor and outdoor lighting systems. It incorporates calculations for electric light, daylight, and energy efficiency. The daylight calculations in DIALux are based on the German standard DIN 5043 and CIE Publication 110. While the geometric input options are limited to certain shapes, the software offers a satisfactory range of sky choices for different weather conditions. For producing visual representations and presentations, DIALux integrates an external radiosity and ray-tracing model called POV-Ray (Persistence of Vision 2010) [49, 50].

The daylight factor (DF) is a measurement expressed as a percentage, representing the ratio of the interior illuminance at a specific point on a designated plane located 0.85 m above the floor, relative to the exterior illuminance. It is typically measured under overcast sky conditions. In the context of classroom lighting, specific standards have been established, specifying a minimum daylight factor of 1.5% and an average daylight factor of 5% [45, 46].

$$DF = E_{indoor}/E_{outdoor} \times 100\,(\%) \tag{1}$$

In the context of daylighting, "E_{indoor}" refers to the illuminance level resulting from daylight at a specific point on a designated indoor plane, measured in lux. On the other hand, "$E_{outdoor}$" represents the concurrent outdoor illuminance on a horizontal plane, also measured in lux.

The simulation outputs show that the attic room is not properly and uniformly daylit in case of both the simulation programs. The attic room daylighting is very low in the distance from this window for all simulation variants. The findings indicate that skylights effectively penetrate light further into the interior space. Good lighting intensity was analysed based on class requirements for Slovak standards. For overcast sky conditions the minimal illuminance of 100 lx is achieved on a very limited attic room area. The rest part of the space is gloomy.

Table 1. Results of DF – base model – w1 – measured and calculated values

Point	A	B	C
1	12.2	12.7	10.6
2	5.4	5.0	3.1
3	2.3	2.8	1.8

Table 2. The outcome of the daylight factor (DF) – base model– w1 – simulated values

Point	1		2		3	
	V.D.V	DIALux	V.D.V	DIALux	V.D.V	DIALux
A	12.3	15.1	4.9	5.5	2.0	2.8
B	12.1	13.5	4.7	4.9	2.5	3.2
C	9.7	8.3	2.8	2.5	1.5	2.0

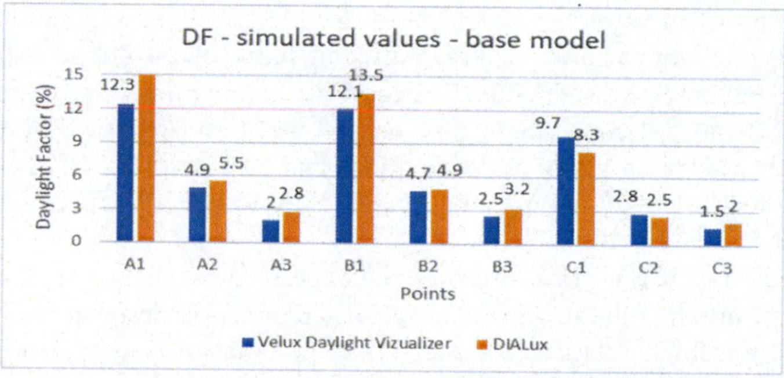

Fig. 3. Results of DF – simulated values

Table 3. Results of calculated DF from measured values

DF	DF_{min} (%)	DF_{max} (%)	$DF_{average}$ (%)	U0 (–)
Measured	0.5	12.3	4.5	0.11
Standardized	1.5	–	5	0.20

The results by simulation program Velux Daylight Visualizer in the basic model (w1), daylight reaches a very high level in areas near the side windows and is reduced along the entire length of the classroom, DF ranging from 0.02% to 23.7%. In variant (w3) it is from 0.16% to 19.6% and in variant (w4) from 0.24% to 22.2%. If we consider variant w2 - without roof windows, it does not meet DF_{min} and $DF_{average}$, although U_0

Table 4. Results of calculated DF by simulation program velux daylight visualizer

DF	DF_{min} (%)	DF_{max} (%)	$DF_{average}$ (%)	U0, (−)
a) w1	0.20	21.73	5.78	0.03
b) w2	0.10	11.53	2.10	0.05
c) w3	0.16	19.55	4.31	0.04
d) w4	0.24	22.15	6.53	0.04

Table 5. Results of calculated DF by simulation program DIALux

DF	DF_{min} (%)	DF_{max} (%)	$DF_{average}$ (%)	U0 (−)
a) w1	2.20	17.92	5.48	0.40
b) w2	0.31	9.05	0.55	0.56
c) w3	1.50	16.63	3.87	0.39
d) w4	2.50	18.14	6.02	0.41

(0.05) is better (see Table 4). The value DF = 1.5% is met only to the middle of the room.

The results by simulation program DIALux in the basic model (w1), daylight reaches a very high level in areas near the side windows and DF ranging from 2.2% to 17.9%. In variant (w3) it is from 1.5% to 16.6% and in variant (w4) from 2.5% to 18.2%. If we consider variant w2 - without roof windows, it does not meet DF_{min} and $DF_{average}$, although U_0 (0.56) is better (see Table 5). The minimal value DF = 1.5% is met in all variants besides variant without roof windows.

4 Conclusions

Schools are facilities in which the visual activity of mostly young people takes place, whose organism is still developing and who is extremely sensitive to incorrect lighting parameters. Visual impairment due to insufficient lighting is extremely relevant during this period. Lighting is one of the basic components of environmental hygiene. It has a direct effect on the eyesight of young people. Poor lighting is a sign of gradual deterioration of vision, while the insidiousness of poor lighting lies in the long-term effect - degeneration of vision often manifests itself after many years. The young child's eye is particularly sensitive to natural light. Hence, it is of paramount importance to consider the occupants' well-being during the space design process. The effect of roof windows on interior daylighting was analysed in this study.

In the article, there is an assessment of daytime illumination using the DF parameter. Due to the fact that the level of illumination decreases with the depth of the space, in the given space away from the window, this parameter was chosen as the evaluation parameter. Of course, it would be appropriate to evaluate the daytime lighting also using

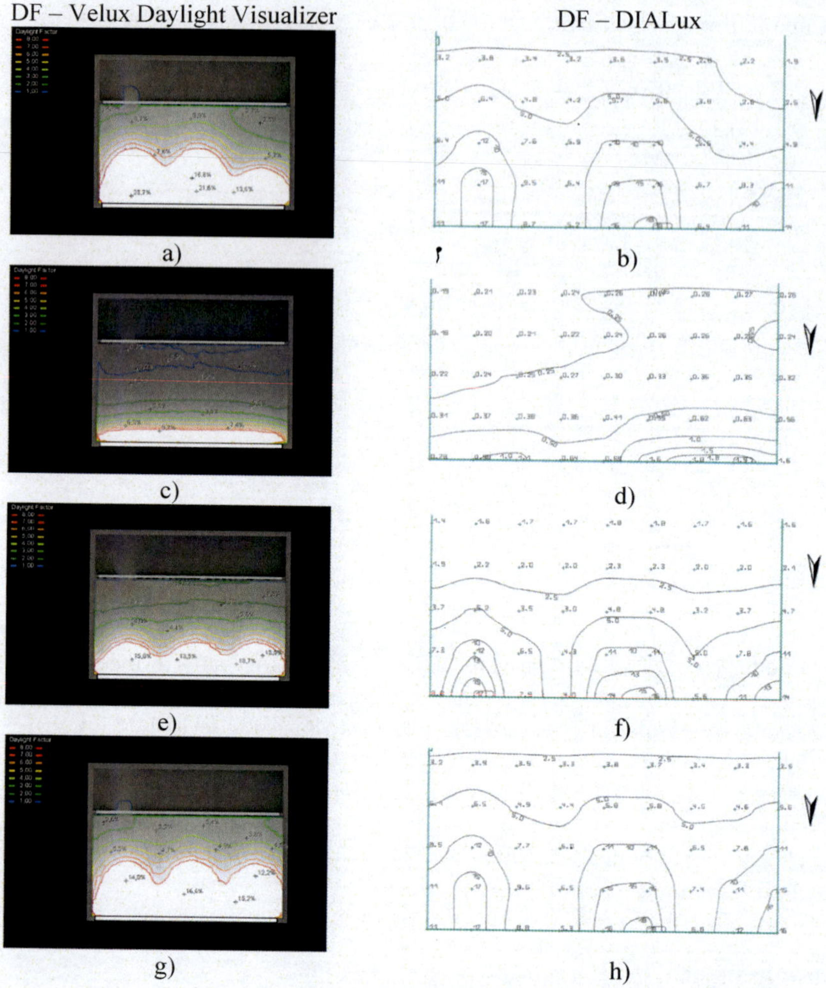

Fig. 4. Results of DF – of simulated values by both simulation programs

the parameter DGP which intolerable glare is higher than 45. DGP is an indication of the percentage of people who would be disturbed by glare. It would also be appropriate to evaluate luminance values (cd/m^2) and potential glare sources.

The paper compares three alternatives of roof windows (basic and two variants) and also case without roof windows. Variant 3 is the best fit, then basic and as optimal is variant 2 for both simulation programs. The results show that the minimum DF value is increased in program DIALux and meets the minimum standard value. Nevertheless, the maximum DF values decreased. Uniformity illuminance is also much higher in DIALux. This is also due to the more detailed boundary conditions of the geometry of the space and windows. When we compare the values at individual points of space, in program DIALux the DF values are also higher. The findings demonstrate the substantial role of roof windows in enhancing the overall effectiveness of daylight as a prominent light source within an attic school environment. And also the possibility of evaluating the light climate in this space with two simulation programs.

5 Acknowlegment

This paper was prepared with the financial backing of the research project VEGA 1/0626/22, granted by the Scientific Grant Agency under the Ministry of Education, Science, Research, and Sport of the Slovak Republic, in collaboration with the Slovak Academy of Sciences.

References

1. Tanner CK (2009) Effect of school design on student outcomes. J Educ Admin 47:381–399. https://doi.org/10.5539/ies.v9n1p175
2. Mohapatra BN, Kumar MR, Mandal SK (2018) Analysis of daylighting using daylight factor and luminance for different room scenarios. Int J Civ Eng Technol 9:949–960. http://www.iaeme.com/IJCIET/index.asp
3. Abdelatia B, Marenne CH, Semidor C (2010) Daylighting strategy for sustainable schools: case study of prototype classroom in Libya. J Sustain Develop 3:60–68. https://doi.org/10.5539/jsd.v3n3p60
4. Li DHW (2010) A review of daylight illuminance determinations and energy implications. Appl Energy 87:2109–2118. https://doi.org/10.1016/j.apenergy.2010.03.004
5. Constanzo V, Evola G, Marletta L (2017) A review of daylighting strategies in schools: state of the art and expected future trends. Buildings 7. https://doi.org/10.3390/buildings7020041
6. Read A (2017) Integration of daylighting into educational (school) building design for energy efficiency, health benefit, and mercury emissions reduction using heliodor for physical modeling. Degree of Master of Architecture, Rochester Institute of Technology
7. http://scholarworks.rit.edu/cgi/viewcontent.cgi?article=10826&context=theses
8. Katunský D, Zozulák M, Kondáš K, Šimiček J (2014) Numerical analysis and measurement results of a window sill. In: Advanced materials research, vol 899, pp 147–150. Trans Tech Publications Ltd. https://doi.org/10.4028/www.scientific.net/AMR.899.147
9. Heschong L, Wright RL, Okura S (2013) Daylighting impacts on Human performance in school. J Illum Eng Soc 31:101–114. https://doi.org/10.1080/00994480.2002.10748396
10. Nicklas MH, Bailey GB (2012) Daylighting in schools: energy costs reduced… student performance improved. Strategic Plan Energy Environ 17. https://doi.org/10.1080/10485236.1997.10530515
11. Yang G, Yonghong Y (2016) Daylighting design in classroom based on yearly-graphic analysis. Sustainability 8:604. https://doi.org/10.3390/su8070604
12. Reinhart CH, Fitz A (2006) Findings from a survey on the current use of daylight simulations in building design. Energy Build 38:824–835. https://doi.org/10.1016/j.enbuild.2006.03.012
13. Edwards L, Torcellini P (2002) Literature review of the effects of natural light on building occupants. Nat Renew Energy Labor Techn Rep 17–26. https://doi.org/10.2172/15000841
14. Nocera F, Faro LA, Constanzo V, Raciti CH (2018) Daylight performance of classrooms in a mediterranean school heritage building. Sustainability 10:3705. https://doi.org/10.3390/su10103705
15. Daylight with roof windows, fat-roof windows and modular skylights. Available online: https: www.velux.com

16. Kuhlenengel M, Waters CE, Konstantzos I (2019) Assessing the impact of outside view on learning: a close look to EN 17037 'view out' practices through the analysis of 220 classrooms. In: Journal of physics: conference series 2019, 1343. IOP Publishing, pp 012159
17. Tureková I, Lukáčová D, Bánesz G (2018) Quality assessment of the university classroom lighting—a case study. TEM Journal 7(4):829–836
18. Mathalamuthu AD, Ibrahim NLN, Ponniah V, Shafiei MWM, Ismail R (2018) Illuminance uniformity using public works department (PWD) standard design for public schools classroom design in Malaysia. J Adv Res Fluid Mech Therm Sci 52(2):205–214
19. Irnawaty I, Rahim MR, Hamzah B, Jamala N (2019) Daylight intensity analysis of secondary school buildings for environmental development. In: IOP conference series: earth and environmental science 2019, vol 382. IOP Publishing, p 012022
20. Ashrafian T, Moazzen N (2019) The impact of glazing ratio and window configuration on occupants' comfort and energy demand: the case study of a school building in Eskisehir, Turkey. Sustain Cities Soc 47:101483. https://doi.org/10.1016/j.scs.2019.101483
21. Qahtan AM (2019) Daylight illuminance in classrooms adjacent to covered and uncovered courtyards under the clear sky of Najran City, Saudi Arabia. Emir J Eng Res 24(4):4. https://scholarworks.uaeu.ac.ae/ejer/vol24/iss4/4
22. Lourenço P, Pinheiro MD, Heitor T (2019) Light use patterns in Portuguese school buildings: user comfort perception, behaviour and impacts on energy consumption. J Clean Prod 228:990–1010. https://doi.org/10.1016/j.jclepro.2019.04.144
23. Kwon CHW, Kang JL (2018) Integrated daylighting design by combining passive method with day sim in a classroom. Energies 11(11):3168. https://doi.org/10.3390/en11113168
24. Cabeza-Lainez J, Almodovar-Melendo JM, Dominguez I (2019) Daylight and architectural simulation of the Egebjerg school (Denmark): sustainable features of a new type of skylight. Sustainability 11(21):5878. https://doi.org/10.3390/su11215878
25. Rubiera ELG, Torija JGS, Frutos CB (2019) Zero cost conditioning techniques to improve the indoor environment of school buildings. Revista de la construcción 18(3):525–535. https://doi.org/10.7764/rdlc.18.3.525
26. Costanzo V, Evola G, Marletta L, Pistone Nascone F (2018) Application of climate based daylight modelling to the refurbishment of a school building in Sicily. Sustainability 10(8):2653. https://doi.org/10.3390/su10082653
27. Galal KS (2019) The impact of classroom orientation on daylight and heat-gain performance in the Lebanese Coastal zone. Alex Eng J 58(3):827–839. https://doi.org/10.1016/j.aej.2019.07.003
28. Fang Y, Cho S (2019) Design optimization of building geometry and fenestration for daylighting and energy performance. Sol Energy 191:7–18. https://doi.org/10.1016/j.solener.2019.08.039
29. Lassandro P, Zonno M (2018) A work-related learning project for energy efficiency evaluation and indoor comfort of school buildings. Ingenierie des Systemes d'Information 23(5):7–27. https://doi.org/10.3166/isi.23.5.7-27
30. Tanić M, Stanković D, Kondić S, Kankhva V (2018) Principles and implementation of daylighting systems in classrooms. In: Energy management of municipal transportation facilities and transport 2018. Springer, Cham, pp 720–731. https://doi.org/10.1007/978-3-030-19868-8_70
31. Zhong L, Yuan J, Fleck B (2019) Indoor environmental quality evaluation of lecture classrooms in an institutional building in a cold climate. Sustainability 11(23):6591. https://doi.org/10.3390/su11236591
32. Bluyssen PM, Zhang D, Kurvers S, Overtoom M, Ortiz-Sanchez M (2018) Self-reported health and comfort of school children in 54 classrooms of 21 Dutch school buildings. Build Environ 138:106–123. https://doi.org/10.1016/j.buildenv.2018.04.032

33. Saraiva TS, Silva EM, Almeida M, Bragança L (2019) Comparative study of comfort indicators for school constructions in sustainability methodologies: schools in the Amazon and the southeast region of Brazil. Sustainability 11(19):5216. https://doi.org/10.3390/su11195216
34. Abdelhakim M, Lim YW, Kandar MZ (2019) Optimum glazing configurations for visual performance in Algerian classrooms under Mediterranean climate. J Daylighting 6(1):11–22. https://doi.org/10.15627/jd.2019.2
35. Ma'bdeh S, Al-Khatatbeh B (2019) Daylighting retrofit methods as a tool for enhancing daylight provision in existing educational spaces—a case study. Build 9(7), 159. https://doi.org/10.3390/buildings9070159
36. Seyedolhosseini A, Masoumi N, Modarressi M, Karimian N (2020) Daylight adaptive smart indoor lighting control method using artificial neural networks. J Build Eng 29:101141. https://doi.org/10.1016/j.jobe.2019.101141
37. Alwetaishi M, Taki A (2019) Investigation into energy performance of a school building in a hot climate: optimum of window-to-wall ratio. Indoor Built Environ 29(1):24–39. https://doi.org/10.1177/1420326X19842313
38. Afacan Y, Ranjbar A (2019) Impact of building massing on energy efficient school buildings. In: Sustainability in energy and buildings 2019. Springer, Singapore, pp 11–22. https://doi.org/10.1007/978-981-32-9868-2_2
39. Tian Z, Lin P, He Y, Jonsson JC (2020) A study of luminous environment with prism daylight redirecting fenestrations in classrooms. Indoor Built Environ. https://doi.org/10.1177/1420326X19895566
40. Kondáš K, Darula S (2014) Daylighting on the working plane in oriented attic rooms under overcast and clear sky. Select Sci Papers J Civ Eng 9(1):33–40. https://doi.org/10.2478/sspjce-2014-0004
41. Katunsky D, Dolnikova E (2019) Assessment of the working environment in terms of visual perception. In: International conference current issues of civil and environmental engineering Lviv-Košice–Rzeszów 2019. Springer, Cham, pp 145–152. https://doi.org/10.1007/978-3-030-27011-7_18
42. Dolnikova E (2019) Assessment of daylight in the selected office through simulation programs: a case study. E-GFOS 2019 10(18):82–92. https://doi.org/10.13167/2019.18.8
43. Dolnikova E, Katunsky D (2019) Visual comfort assessment in the office: a case study. In: SGEM 2019 conference proceedings. 6.2. Nano, bio and green - technologies for a sustainable future: green buildings technologies and materials, green design and sustainable architecture, space technologies and planetary science, Sofia, STEF 92 Technology 2019, pp 597–604. https://doi.org/10.5593/sgem2019/6.2/S27.076
44. Kittler R, Darula S (2013) Determination of time and sun position system. Sol Energy 93:72–79. https://doi.org/10.1016/j.solener.2013.03.021
45. Garcia DLR, Pereira FOR (2021) Daylight glare mitigation by internal shading devices use and effects on building
46. STN 730580 (2000) Daylighting in buildings, Part–1 basic requirements, 1986 Part–2; daylighting of residential buildings. Slovak Republic Office of Standards, Metrology and Testing: Bratislava, Slovakia
47. EN 12464-1 (2012) Light and lighting; lighting of work places-part 1: indoor work places. Slovak Republic Office of Standards, Metrology and Testing: Bratislava, Slovakia
48. https://www.velux.com/what-we-do/digital-tools/daylight-visualizer
49. Hensen JLM, Lamberts R (2019) Building performance simulation for design and operation. Routledge Publisher, London, p 792. ISBN 13-9780-415-47414-6
50. https://www.airfal.com/en/dialux-light-planning/
51. Hnilica R (2012) Porovnanie programov pre výpočty a simuláciu osvetlenia—Dialux a Relux. [online]. In: Světlo. Časopis pro světelnou techniku a osvětlováni, 2, s 25–29. ISSN 2121-0812

The Impact of Vegetative Roof on Heat Energy Demand

Jan Domanicky(✉) and Frantisek Vranay

Faculty of Civil Engineering, Institute of Architectural Engineering, Technical University of Kosice, Vysokoškolská 4, 042 00 Kosice, Slovakia
jan.domanicky@tuke.sk

Abstract. The climate crisis is a topic that affects everyone in some way. Therefore, the goal of this time is to look for ways to mitigate it as much as possible. Recently, the question of energy has been added to this, both from an economic point of view and with the availability of energy. Energy consumption is a common problem, both in the home and in the work environment. As the temperature of the areas increases in the summer and decreases in the winter, thus increasing the problem of urban heat islands, the need for energy for heating and cooling also increases. Often this phenomenon is solved by active measures such as air conditioning units and so on. The goal of my work is one of the ways to use passive elements to reduce the need for energy for heating and cooling. One such element is the vegetation roof. Why passive?

Keywords: Passive cooling · Passive heating · Energy · Renewable energy sources · Vegetative roof · Green roof

1 Introduction

1.1 Building Energy Optimization

The energy efficiency of buildings is a crucial consideration in today's world, as it greatly impacts energy consumption and environmental sustainability. Improving energy efficiency in buildings is essential for reducing our carbon footprint and mitigating climate change.

The issue of global warming remains highly relevant in today's world. Media outlets frequently report on this problem, and individuals are actively aware of its existence. The consequences of climate change have a profound impact on weather patterns, thereby affecting the global population. Urbanization stands out as one of the primary impacts for these changes [1], involving the migration and concentration of people in cities. Long-term observations confirm a continuous rise in urbanization rates, with no indication of a reversal. While fluctuations may occur in some cases, the overall trend is upward. Slovakia serves as an example of this phenomenon. Research data [2] provides insights into the progression of urban lifestyles, presenting the following values in Table 1.

Rising urbanism has far-reaching implications for various aspects of the climate crisis, creating a interconnected chain of climatic effects. Urban development contributes

Z. Blikharskyy et al. (Eds.): CEE 2023, LNCE 438, pp. 98–106, 2024.
https://doi.org/10.1007/978-3-031-44955-0_11

Table 1. The urbanism in world [2]

Place	1990 (%)	2014 (%)	2050 (%) (forecast)
World	43	54	66
Europe	70	73	82
East Europe	68	69	78
Slovakia	56	54	63

to the rise in temperatures within cities, exacerbating the prevalent issue of Urban Heat Islands (UHI). Additionally, it amplifies the local demand for heating and cooling, consequently leading to increased greenhouse gas emissions. Presently, the objective is to counteract this trend by reducing energy consumption. Article is a part of my dissertation research, which primarily aims to identify strategies for minimizing the reliance on additional energy for heating and cooling in our climate. While complete elimination of this energy consumption may not be feasible, it is crucial to explore avenues for significant reduction. Among the potential approaches (such as optimal architectural design, effective shielding, and appropriate building envelopes), the utilization of passive heating and cooling elements emerges as a viable option. In my study, I specifically focus on the implementation of vegetation elements and their impact on interior heating and cooling.

1.2 Vegetation

Today, there is no doubt about the numerous advantages of vegetation elements, particularly vegetation roofs. These elements offer a range of benefits, including reducing rainwater runoff, enhancing urban biodiversity, mitigating the urban heat island effect (UHI), improving air quality, minimizing temperature fluctuations on the roof surface to extend the lifespan of the roof membrane, reducing noise levels, and decreasing energy consumption. The focus of this article is on exploring the impact of vegetation elements on energy consumption [1]. During the summer season, as temperatures rise outdoors and indoors, the demand for cooling increases. In Slovakia, the recommended indoor temperature during the winter is 20 °C, while in the summer, it is 26 °C. However, it is often observed that people tend to desire the opposite—a temperature of 20 °C in summer and 24 °C or higher in winter. Consequently, the demand for both heating and cooling rises. As cooling is frequently achieved through air conditioning units, it leads to an increase in outdoor temperatures at both the local and urban levels, exacerbating the UHI problem. Evapotranspiration is known to convert a substantial amount of solar radiation into latent heat, which does not contribute to temperature rise [3]. This presents yet another justification for implementing passive cooling elements. Why I focus on roofs? In our climate zone, maintaining vegetated exterior facades poses challenges. Additionally, horizontal building structures, such as roofs, endure significant thermal stress during the summer and necessitate measures to prevent heat loss in winter [4]. Given the proven benefits, several jurisdictions have opted to incorporate vegetation elements in their cities and provide support. Toronto, for instance, officially embraced this approach in 2009 [5]. Other cities, including Chicago, have also shown support for this initiative.

2 The Impact of Vegetation Elements on Heat Flow and Thermal Conductivity Coefficient

Multiple factors contribute to the effectiveness of a vegetation roof in meeting a building's energy requirements. These factors include the substrate's thickness and thermal conductivity, volumetric humidity, vegetation height, leaf efficiency, emissivity, leaf coverage, stomatal resistance (which combines leaf water potential and solar flux), as well as the type and thickness of the roof insulation and local climate conditions [1]. Regarding the impact of vegetated roofs on heat flow and the thermal and technical parameters used in energy balances, relevant information was sourced from texts [4, 6]. This particular case study focuses on the analysis conducted during the heating period. In the study [6], three roof types with varying insulation thicknesses were examined, emphasizing the importance of well-insulated roofs for our climate zone and country. This is only study, which can prove my concept to use in our well insulated roofs. Table 2 presents the heat transfer coefficient values for these structures, both with and without a vegetated roof, considering three insulation levels [6].

1. Uninsulated roof:
   a. 25 cm concrete
   b. 10 cm concrete.
2. Slightly insulated roof:
   a. 15 cm of concrete + 5 cm of insulation + 15 cm of concrete
   b. 5 cm concrete + 5 cm insulation + 5 cm concrete.
3. Well-insulated roof:
   a. 15 cm of concrete + 15 cm of insulation + 15 cm of concrete
   b. 10 cm of concrete + 10 cm of insulation + 10 cm of concrete.

Table 2. Heat conductivity coefficient according to research [6]

Type of roof		Heat conductivity coefficient (W/m^2 K)		Saving	
		Without vegetation	With vegetation	W/m^2 K	%
1	A	7.76	1.73	6.03	77.7
	B	18.18	1.99	16.19	89.1
2	A	0.74	0.55	0.19	25.7
	B	0.8	0.59	0.21	26.3
3	A	0.26	0.24	0.02	7.7
	B	0.4	0.34	0.06	15.0

3 Research Focus: Investigating the Direction of Research

The article presents Case Study A, which focuses on an unrealistic object—a warehouse-type hall with dimensions of 100 × 200 × 10 m (width × length × height) - referred to as Part A1. In this case, the roof area accounts for 43.48% of the building's total heat exchange envelope. For the primary assessment, Roof 3a, which closely resembles our conditions, was chosen.When selecting other objects or buildings, it becomes crucial to consider the building's shape and the ratio between the individual areas of the building's envelope and the vegetated area. This article highlights a building with a significant roof-to-wall ratio. To provide a comparison, an apartment building with dimensions of 14.76 × 31.66 × 23.1 m (width × length × height) - referred to as Part A2 - was included, where the roof area represents only 15.13% of the total building area. Figures 1 and 2 display the dimensions of the two objects for comparison.

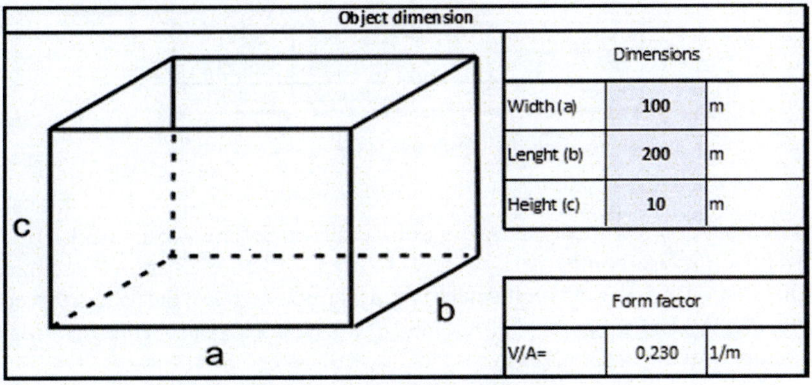

Fig. 1. A figure show dimensions of hall – Part A1.

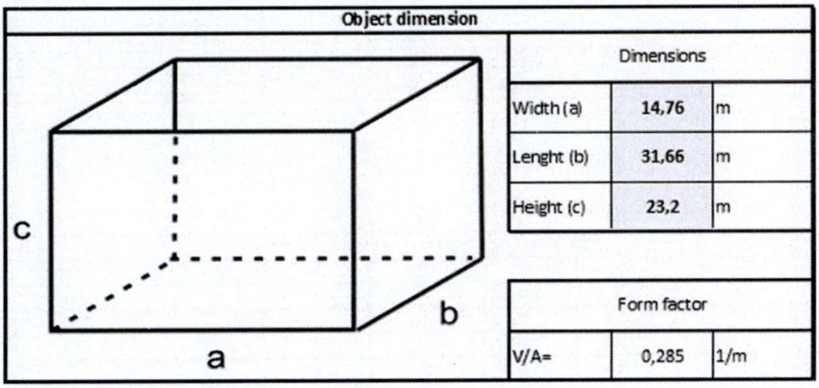

Fig. 2. A figure show dimensions of residential – Part A2.

After the object selection, a comparison was made between the surfaces of the perimeter structures, including the floor, walls, and roof. These relationships are visually represented in Figs. 3 and 4, providing a clear illustration of the proportions and areas involved.

The ratio of the area of the walls to the roof		
Area	m2	%
Walls	6000	13,04%
Roof	20000	43,48%
Object	46000	100,00%

Fig. 3. Building envelopes areas of hall – Part A1.

The ratio of the area of the walls to the roof		
Area	m2	%
Walls	2153,888	69,74%
Roof	467,3016	15,13%
Object	3088,4912	100,00%

Fig. 4. Building envelopes areas of hall – Part A2.

The thermal and technical parameters of the structures are identical for both parts, allowing for a precise comparison of only the roof structures. Part A1 is depicted in Figs. 5 and 6, showcasing the specific details and characteristics of the roof in this context. Similarly, Part A2 is illustrated in Figs. 8 and 9, providing a visual representation of the roof structure in that particular case.

Heat loss of an object without a vegetation element							
	A	U	θi	θe	Δθ	Qi	α
	m2	W/m2.K	°C	°C	K	W	%
SO1	6000	0,144	20	-13	33	28465,80	12%
PDL	20000	0,191	20	3	17	65022,56	27%
S - without vegetation	20000	0,222	20	-13	33	146779,52	61%
						240267,88	

Fig. 5. Heat losses without a vegetated roof – Part A1. (SO1 – walls, PDL – floor and S – roof)

Heat loss of an object with a vegetation element							
	A	U	θi	θe	Δθ	Qi	α
	m2	W/m2.K	°C	°C	K	W	%
SO1	6000	0,144	20	-13	33	28465,80	12%
PDL	20000	0,191	20	3	17	65022,56	28%
S - with vegetation	20000	0,2053	20	-13	33	135477,50	59%
						228965,86	

Fig. 6. Heat losses with a vegetated roof – Part A1. (SO1 – walls, PDL – floor and S – roof)

Part	S - without veg.	S - with veg	Saving	
	W	W	W	%
Construction	146779,52	135477,50	11302,02	7,70%
Object	240267,88	228965,86	11302,02	4,70%

Fig. 7. Object heat balance – Part A1.

Part A1 exhibits a notable reduction in heat loss through the roof structure, as evidenced by the data. This influence is further illustrated in the overall energy balance of the object, as depicted in Fig. 7. The energy balance of Part A is presented in Fig. 10, providing a comprehensive overview of the energy flows and exchanges within the system.

Heat loss of an object without a vegetation element							
	A	U	θi	θe	Δθ	Qi	α
	m2	W/m2.K	°C	°C	K	W	%
SO1	2153,888	0,144	20	-13	33	10218,69	67%
PDL	467,3016	0,191	20	3	17	1519,26	10%
S - without vegetation	467,3016	0,222	20	-13	33	3429,52	23%
						15167,46	

Fig. 8. Heat losses without a vegetated roof – Part A2. (SO1 – walls, PDL – floor and S – roof)

Heat loss of an object with a vegetation element							
	A	U	θi	θe	Δθ	Qi	α
	m2	W/m2.K	°C	°C	K	W	%
SO1	2153,888	0,144	20	-13	33	10218,69	69%
PDL	467,3016	0,191	20	3	17	1519,26	10%
S - with vegetation	467,3016	0,2053	20	-13	33	3165,44	21%
						14903,39	

Fig. 9. Heat losses with a vegetated roof – Part A2. (SO1 – walls, PDL – floor and S – roof)

Part	S - without veg.	S - with veg	Saving	
	W	W	W	%
Construction	3429,52	3165,44	264,07	7,70%
Object	15167,46	14903,39	264,07	1,74%

Fig. 10. Object heat balance – Part A2.

By extrapolating these values to the annual energy balance for heating, we can calculate the corresponding savings in kWh/year or GJ/year. These savings, when combined with current energy prices, allow for the determination of the total annual economic savings. Specifically, for Part A1, these findings are presented in Figs. 11, 12 and 13. Similarly, for Part A2, the corresponding results are displayed in Figs. 14, 15 and 16, providing a comprehensive assessment of the economic benefits associated with the energy savings.

Data of the object:

Town:		Košice	
External calculated temperature:	t_e	-13	°C
Length of heating period:	d	223	days
Average temperature in the heating season:	t_{es}	2,9	°C
Average indoor temperature:	t_{is}	20	°C
Heat loss of the object:	Q_{HL}	240,2679	kW

Calculation of the expected heat demand for heating

d	D	ε	η_r	Q_{HL}	$Q_{ÚK/rok}$	
days	K.days	-	-	kW	kWh/rok	GJ/rok
223	3813,3	0,63	0,98	240,267883	428359,57	1542,09

Fig. 11. Assessment of annual heating energy demand in the absence of a vegetated roof – Part A1.

Data of the object:

Town:		Košice	
External calculated temperature:	t_e	-13	°C
Length of heating period:	d	223	days
Average temperature in the heating season:	t_{es}	2,9	°C
Average indoor temperature:	t_{is}	20	°C
Heat loss of the object:	Q_{HL}	228,9659	kW

Calculation of the expected heat demand for heating

d	D	ε	η_r	Q_{HL}	$Q_{ÚK/rok}$	
days	K.days	-	-	kW	kWh/rok	GJ/rok
223	3813,3	0,63	0,98	228,96586	408209,85	1469,56

Fig. 12. Assessment of annual heating energy demand with the implementation of a vegetated roof – Part A1.

The results of comparison					
	Heat energy demand		Cost comparison		
			Electricity	Gas	Heatpump
	kWh/rok	GJ/rok	€/rok	€/rok	€/rok
S - withou veg.	428359,567	1542,094	68537,53	26772,47	17134,38
S - with veg.	408209,850	1469,555	65313,58	25513,12	16328,39
Saving	20149,72	72,54	3223,95	1259,36	805,99
	4,704%				

Fig. 13. Comparison of heat energy demand – Part A1.

Table No. 3 presents a comprehensive comparison between Part A1 and Part A2, highlighting the percentage savings achieved with a vegetated roof, taking into account the different ratios of roof area. It becomes evident that the utilization of a vegetated roof provides greater justification, particularly in terms of heating, for buildings with larger horizontal construction areas compared to vertical ones. The table provides a clear overview of the energy savings and underscores the advantages of implementing a vegetated roof in such scenarios (Table 3).

Data of the object:

Town:		Košice	
External calculated temperature:	t_e	-13	°C
Length of heating period:	d	223	days
Average temperature in the heating season:	t_{es}	2,9	°C
Average indoor temperature:	t_{is}	20	°C
Heat loss of the object:	Q_{HL}	15,1675	kW

Calculation of the expected heat demand for heating

d	D	ε	η_r	Q_{HL}	$Q_{ÚK/rok}$	
days	K.days	-	-	kW	kWh/rok	GJ/rok
223	3813,3	0,63	0,98	15,1674648	27041,19	97,35

Fig. 14. Assessment of annual heating energy demand in the absence of a vegetated roof – Part A2.

Data of the object:

Town:		Košice	
External calculated temperature:	t_e	-13	°C
Length of heating period:	d	223	days
Average temperature in the heating season:	t_{es}	2,9	°C
Average indoor temperature:	t_{is}	20	°C
Heat loss of the object:	Q_{HL}	14,9034	kW

Calculation of the expected heat demand for heating

d	D	ε	η_r	Q_{HL}	$Q_{ÚK/rok}$	
days	K.days	-	-	kW	kWh/rok	GJ/rok
223	3813,3	0,63	0,98	14,9033921	26570,39	95,65

Fig. 15. Assessment of annual heating energy demand with the implementation of a vegetated roof – Part A2.

The results of comparison					
	Heat energy demand		Cost comparison		
			Electricity	Gas	Heatpump
	kWh/rok	GJ/rok	€/rok	€/rok	€/rok
S - withou veg.	27041,187	97,348	4326,59	1690,07	1081,65
S - with veg.	26570,387	95,653	4251,26	1660,65	1062,82
Saving	470,80	1,69	75,33	29,42	18,83
	1,741%				

Fig. 16. Comparison of heat energy demand – Part A2.

Table 3. Final comparison of heat energy demand

Roof	Area ratio roof/object	Heating energy demand		Saving	
		Without VR	With VR		
		kWh/a	kWh/a	kWh/a	%
Part A1	A	7.76	1.73	6.03	77.7
Part A2	B	18.18	1.99	16.19	89.1

4 Conclusion

While the overall impact of a vegetated roof on a building is influenced by various factors, the findings of this study have confirmed its positive effect on heating. Furthermore, a secondary comparison based on different area ratios has demonstrated its suitability across various scenarios. It is important to note that the focus of this article does not lie in economic principles, as the objective is not solely to compare financial indicators. Instead, the intention is to enhance the understanding of vegetated roofs and highlight their additional advantages, thus providing a more comprehensive perspective on their benefits.

Acknowledgment. This work was supported by Development Agency under the contract (APVV APVV-18-0360 Achieve Active hybrid infrastructure for the sponge city).

References

1. Susca T (2019) Green roofs to reduce building energy use? A review on key structural factors of green roofs and their effects on urban climate
2. https://population.un.org/wup/Publications/Files/WUP2014-Highlights.pdf (available on the Internet: 10 Mar 2023)
3. Takakura T, Kitade S, Goto E (2002) Cooling effect of greenery cover over a building. Energy Build
4. Theodosiou TG (2003) Summer period analysis of the performance of a planted roof as a passive cooling technique. Energy Build
5. https://www.toronto.ca/legdocs/bylaws/2009/law0583.pdf (available on the Internet: 15 Feb 2023)
6. Niachou A, Papakonstantinou K, Santamouris M, Tsangrassoulis A, Mihalakakou G (2001) Analysis of the green roof thermal properties and investigation of its energy performance. Energy Build

Clean and Tasty Water - Analysis of the Effectiveness of Water Filter Jugs

Andżelika Domoń[1](✉), Dorota Papciak[1], and Eleonora Sočo[2]

[1] Department of Water Purification and Protection, Faculty of Civil, Environmental Engineering and Architecture, Rzeszow University of Technology, 35-959 Rzeszow, Poland
adomon@prz.edu.pl

[2] Department of Inorganic and Analytical Chemistry, Faculty of Chemistry, Rzeszow University of Technology, 35-959 Rzeszow, Poland

Abstract. The article evaluates the effectiveness of jug filters in terms of improving the quality of drinking water. Changes in chemical and bacteriological parameters of tap water filtered with jugs from three different manufacturers were analyzed. Organoleptic parameters of filtered water, that is color, turbidity, odor, taste and selected chemical parameters: pH, conductivity, total hardness, chlorine content, total organic carbon, nitrogen, and phosphorus compounds were examined. The assessment of the microbiological quality and sanitary safety of the filtered water was based on changes in the total number of bacteria (samples were inoculated simultaneously on A and R2A agar and incubated at 37 °C and 22 °C) and by bioluminescence analysis (Adenosine triphosphate measurement). The obtained test results were discussed and compared with the declarations of the manufacturers of jug filters. The stability and risk of losing the safety of the filtered water were also assessed.

Keywords: Water quality · Drinking water · Water filter jugs · Filtration · Water pollution

1 Introduction

Public awareness of the impact of diet and water quality on health has increased in recent years [1]. Consumers demand that the water delivered directly to the point of use is clean, tasty, and healthy. However, drinking water distribution systems (DWDS) are complex engineering systems consisting of storage tanks, fittings, valves, and pipes made of various materials that interact with the distributed water. Before water reaches the consumer, it often has to travel long distances through an extensive water supply system. Water quality may deteriorate during the flow to the point of use [2]. The following factors may have an impact on the deterioration of the physicochemical and microbiological quality of water, such as: decomposition of the disinfectant, temperature, hydraulic regime, residence time in the installation, and poor condition of the internal installation [3–6]. It should also be noted that the water supply company is responsible for the quality of water sent to the main water meter, not directly to the consumer's tap [7].

Z. Blikharskyy et al. (Eds.): CEE 2023, LNCE 438, pp. 107–118, 2024.
https://doi.org/10.1007/978-3-031-44955-0_12

A large percentage of the population is not convinced about the quality of tap water supplied by local water treatment plants. As a result, people are increasingly choosing to use home water purification methods [8–11]. To improve the quality of tap water, due to the low price and ease of use, consumers most often decide to buy free-standing filter jugs. Their use does not require a connection to the water supply system, which is a great advantage for many users [12–14]. Manufacturers of filter jugs guarantee the softening of water, as well as the removal of harmful substances (i.e. chlorine, phenols, or heavy metals: mercury, lead, iron) and impurities, i.e. rust, sediment, or sand. Some manufacturers also ensure that pesticides are eliminated from the water, which are common in rural areas. Filter cartridges offered by individual manufacturers differ slightly in the principle of operation and composition. The filling is mainly made of ion exchange resin (responsible for softening the water and reducing the concentration of metal ions) and activated carbon (adsorbing chlorine, pesticides, and organic substances and improving the organoleptic properties of water).

This study was aimed at examining the effectiveness of tap water treatment with the use of three different filter jugs from the most popular manufacturers. The study determined both the changes in the physicochemical and microbiological quality of the filtered water and whether the filtration affects the biological stability of water responsible for secondary bacteriological contamination of water.

2 Method

2.1 Subject of Study

The substrate of the research was tap water, which was taken from the tap in the "K" building of the Rzeszow University of Technology. The tested water was treated with the following processes: preliminary ozonation, coagulation, sedimentation, filtration on anthracite-sand beds, indirect ozonation, active carbon filtration, chemical disinfection, and UV radiation. The point of use point was located 5 km from the water treatment plant.

The tested tap water was filtered using filter jugs from the most popular manufacturers, which for this work were marked as F-I, F-II, and F-III. Physicochemical and microbiological analyzes were performed each time for tap water and after filtering the following volumes of water, i.e. 10, 20, 40, 60, 80, and 100 L. Samples of water taken immediately after rinsing the filter (i.e. after 1 L) were also analyzed. According to the manufacturer, this water is not suitable for consumption and should be rejected. The tests were carried out over a period of one month (i.e. March 2023). After filtration, the analyzed filters were stored at room temperature (approx. 18–20 °C) in a dark place.

2.2 Physicochemical and Microbiological Analysis of Waters

The quality of tap water and water after passing through filter jugs was assessed using the methods presented in Tables 1 and 2. The tests were carried out by the applicable research procedures. The analyzes were carried out in the laboratory of the Department of Water Purification and Protection of the Rzeszow University of Technology.

2.3 Evaluation of Biological Stability of Water

The biological stability of water before and after the filtration process on jug filters was assessed based on the content of biogenic compounds: carbon, nitrogen, and phosphorus.

Table 1. Methods of assessing the microbiological and physicochemical quality of water.

Parameter	Analytical method/standard	Normative value [15]
pH	Pehametr Hach-Lange HQ 40d Multi	6.5–9.5
Turbidity	2100P ISO turbidimeter (Hach, Germany)	1 NTU
Conductivity	Hach-Lange oxygen probe	2500 μS/cm
Total organic carbon (TOC)	TOC analyzer Sievers 5310 C (SUEZ, Boulder, CO, USA)	No invalid changes
Biodegradable organic carbon (BDOC)	TOC = DOC (dissolved organic carbon); BDOC = 9% DOC	–
Nitrite nitrogen	Colorimetric method by Nitrite Test Merck 1.14408 (Germany)	0.15 mg $N\text{-}NO_2^-$/L
Ammonium nitrogen	Spectrophotometric method Hach-Lange DR 500 spectrophotometer (Germany)	0.39 mg $N\text{-}NH_4^+$/L
Nitrate nitrogen		11.23 mg $N\text{-}NO_3^-$/L
Phosphates		–
Total chlorine		0.3 mg Cl_2/L
Absorbance UV		–
Color		Acceptable by consumers; No invalid changes
Carbonate hardness	Titration method with EDTA	–
Total and Calcium hardness		1.2–10 mval/L
The heterotrophic plate count (HPC) method at 22 °C and 37 °C (A agar)	HPC method using A Agar (P-0231) manufactured by the Department of Enzymes and Peptons (Poland) (HPC at 22 °C (incubation 3 days) and 37 °C (incubation 2 days)	*37 °C: 20 CFU/ml *22 °C : 100 CFU/ml in water entering the water supply network; 200 CFU/ml at the consumer's tap)
HPC method at 22 °C and 37 °C (R2A agar)	HPC method at 22 °C and 37 °C using R2A Agar (CM0906) manufactured by Oxoid Thermo Scientific (UK) (incubation 7 days)	–

(*continued*)

Table 1. (*continued*)

Parameter	Analytical method/standard	Normative value [15]
Escherichia coli	Membrane filtration procedure using Chromocult® Coliform Agar (MERCK, Poland)	0 CFU/100 ml
ATP analysis	LuminUltra Photonmaster Luminometer (BacTiter-Glo Microbial Cell Enviability Assay, Promega)	–

The following threshold values determining the biological stability of water have been adopted [16]:

- BDOC $\leq$ 0.25 mg C/L,
- $\sum N_{inorg.} \leq$ 0.2 mg N/L,
- $PO_4^{3-} \leq$ 0.03 mg PO_4^{3-}/L.

3 Results and Discussion

3.1 Physicochemical Quality of Water

The results of the tests of physical and chemical parameters obtained for tap water and water purified with the use of filter jugs F-I, F-II, and F-III are presented in Table 2. The obtained test results allowed to conclude that the tap water met the requirements of the Regulation of the Minister of Health on the quality of water intended for human consumption except for the values obtained for turbidity. The recommended value of this parameter, according to the guidelines of the regulation, should be acceptable to consumers and should not exceed 1.0 NTU. In the tested tap water, the average turbidity was 1.38 NTU, while after filtration, higher values were obtained, i.e. 1.68 NTU (F-I), 1.58 NTU (F-II) and 1.38 NTU (F-III) (Table 2).

The main purpose of using filter jugs is to eliminate impurities present in water, improve the organoleptic properties of water and soften water. High water hardness can cause technical problems such as scaling in pipes and kitchen appliances, which in turn increases the cost of maintaining water systems and can lead to breakdowns. Hard water can also affect the quality of prepared food and drinks, so in such cases, the use of water softeners can be beneficial. For health reasons, the general hardness of tap water should be in the range of 1.2–10 mval/L (Table 2). The tested tap water in the analyzed period reached an average hardness value of 4.40 mval/L (medium hard water).

In turn, after filtering the water through filter jugs, a decrease in the value of this parameter was observed (the average values were 3.70, 2.45, and 2.02 mval/L for F-I, F-II, and F-III, respectively). The study also showed a decrease in calcium hardness by 18% (F-I), 46% (F-II), and 55% (F-III). In the case of the F-II and F-III filters, the highest water softening efficiency was found in the initial analyzes (after filtering 1 L, the efficiency was 91 and 84%). Water hardness is related to the presence of mineral salts, mainly calcium and magnesium. As the water hardness decreases, the electrical conductivity of the water also decreases. The average value of the conductivity of tap water during the tests was 474 μS/cm (Table 2). After the filtration process, a decrease in the content of ions dissolved in water was observed, the average conductivity values were 332, 372, and 285 μS/cm for F-I, F-II, and F-III, respectively). The highest efficiency, which amounted to 40%, was obtained for the F-III filter.

When analyzing the change in the pH value of tap water within 4 weeks, fluctuations in its value in the range of 7.38 to 7.85 were observed (Table 2). The filters lower the pH value of the water as a result of the ion exchange process. The exchange of magnesium and calcium ions on the cation exchangers can result in the formation of acids or carbon dioxide, which contributes to the acidification of the water. It is noticeable that the F-II and F-III filters lower the pH of water much more than the F-I filter. The filling of the filter jug insert is a mixture of ion exchange resin and activated carbon. Cartridge content varies by manufacturer and jug model and is a trade secret.

Chlorine values in tap water ranged from 0.04 to 0.22 mg Cl_2/L (Table 2). According to the guidelines contained in the Regulation of the Minister of Health, the concentration of free chlorine at the consumer should not exceed 0.3 mg Cl_2/L. The obtained test results allow to conclude that the removal of chlorine compounds occurred only in the case of filtration on F-II (reduction by 46%). Filtering the water through F-I and F-III did not reduce the chlorine concentration.

Total organic carbon is one of the parameters of drinking water quality, which refers to the content of all organic compounds dissolved in water. A high TOC content may indicate contamination of the water with organic substances such as pesticides, chemicals, or organic breakdown products. According to the regulation, the acceptable level of TOC is defined by the term "no invalid changes", this parameter does not have to be marked in the case of water production of less than 10 000 m^3 per day. TOC analysis showed that tap water reached values in the range of 1.18–1.76 mg C/L (Table 2). The efficiency of TOC removal was variable, i.e. after filtering 1 L, leaching of organic compounds was found, and thus an increase in this parameter (manufacturers do not recommend using the first 2 L for drinking purposes). The analysis performed after 10 L obtained the best results for all 3 analyzed filters (efficiency ranged from 74 to even 99%). In subsequent analyses, the effectiveness was successively reduced, i.e. after 20 L the effectiveness was 43–60%, and after 80–100 L only 0–15%.

Another indicator of organic substances is UV absorbance. After filtering 1L, an increase in this parameter was also observed in the filtered water. Taking into account the monthly period of filters operation, it was found that they contributed to the reduction of the UV absorbance value by an average of 10% (F-I), 37% (F-II), and 0.3% (F-III) (Table 2). The F-III filter exhausted its sorption properties the fastest - after 40 L, the same value was obtained as for tap water, while each subsequent measurement indicated secondary water contamination with organic compounds. The research presented in the paper [17] also confirms that the efficiency of pitcher filters significantly decreases during use. After 3–4 weeks of use, they stop fulfilling their functions in terms of retaining selected components contained in tap water. In the work [13], an additional chromatographic analysis of water was carried out after one and three months of filter operation, which showed several organic micropollutants that got into the water as a

Table 2. Summary of physicochemical results of the analyzed water.

Water volume, L	pH	Conductivity, µS/cm	Turbidity, NTU	TOC, mg C/L	Absorbance, UV^{254}	Ammonium nitrog., mg $N\text{-}NH_4^+$/L	Nitrite nitrogen, mg $N\text{-}NO_2^-$/L	Nitrate nitrogen, mg $N\text{-}NO_3^-$/L	Phosphates, mg PO_4^{3-}/L	Chlorine, mg Cl/L	Color, mg Pt/L	Total hardness, mval/L	Carbonate hardness mval/L	Calcium hardness mval/L
Tap water														
	7.85	486	2.47	1.39	1.58	0.00	0.00	0.90	0.12	0.13	–	5.24	4.30	4.60
	7.76	630	0.80	1.76	2.96	0.04	0.00	1.10	0.12	0.04	–	5.40	5.20	4.04
	7.75	449	1.36	1.42	2.00	0.01	0.00	2.10	0.10	0.11	–	4.20	3.50	3.80
	7.69	321	0.70	1.38	2.14	0.00	0.00	1.60	0.24	0.05	–	3.92	2.20	2.60
	7.80	501	1.63	1.63	2.32	0.02	0.00	1.20	0.75	0.16	3	4.88	4.30	4.20
	7.78	465	1.48	1.63	2.38	0.00	0.00	1.70	0.05	0.22	3	3.60	4.00	5.00
	7.38	468	1.14	1.18	2.72	0.03	0.00	1.60	0.61	0.08	8	3.84	4.30	6.00
Jug filter - F-I														
1	7.87	500	0.90	1.61	1.80	0.03	0.00	0.80	2.46	0.09	–	4.56	4.20	3.92
10	7.80	538	1.70	0.42	2.62	0.05	0.00	2.30	2.62	0.02	–	4.72	4.60	3.68
20	7.50	277	1.46	0.65	1.80	0.09	0.00	2.00	1.35	0.24	4	3.28	3.60	3.00
40	7.61	337	0.93	1.03	1.96	0.08	0.00	1.80	1.12	0.10	–	2.80	1.90	2.24
60	7.70	175	3.63	1.51	2.80	0.10	0.00	1.70	0.57	0.10	6	3.84	3.90	2.88
80	7.70	99	1.78	1.51	1.76	0.02	0.00	1.20	0.02	0.18	4	3.40	4.10	4.40
100	7.31	401	1.38	1.00	1.70	0.01	0.00	1.40	0.07	0.14	–	3.32	4.00	4.56

(*continued*)

Table 2. *(continued)*

Water volume, L	pH	Conductivity, μS/cm	Turbidity, NTU	TOC, mg C/L	Absorbance, UV^{254}	Ammonium nitrog., mg $N\text{-}NH_4^+$/L	Nitrite nitrogen, mg $N\text{-}NO_2^-$/L	Nitrate nitrogen, mg $N\text{-}NO_3^-$/L	Phosphates, mg PO_4^{3-}/L	Chlorine, mg Cl/L	Color, mg Pt/L	Total hardness, mval/L	Carbonate hardness mval/L	Calcium hardness mval/L
Jug filter - F-II														
1	7.77	447	3.7	1.65	1.82	0.30	0.00	0.40	1.98	0.08	11	0.48	1.50	0.30
10	7.30	407	1.1	0.05	1.00	0.00	0.00	1.00	2.27	0.03	–	2.60	2.80	1.60
20	7.10	197	1.43	0.57	0.00	0.00	0.00	1.40	1.57	0.01	–	2.50	2.10	2.00
40	7.20	389	0.77	1.02	1.50	0.00	0.00	1.90	0.57	0.05	–	2.40	1.20	1.76
60	7.20	381	1.06	1.21	1.62	0.00	0.00	1.00	0.09	0.07	–	3.40	3.20	2.80
80	7.50	342	1.54	1.42	1.90	0.00	0.00	1.20	0.03	0.03	–	2.96	3.40	3.88
100	7.30	443	1.24	1.02	2.44	0.04	0.00	1.40	0.15	0.18	–	2.84	4.80	4.12
Jug filter - F-III														
1	7.92	426	1.42	4.46	2.18	1.05	0.00	1.10	0.45	0.10	–	0.84	3.30	0.48
10	7.20	412	1.60	0.45	2.14	0.78	0.00	1.20	0.16	0.13	–	1.24	3.00	0.76
20	7.20	320	0.93	0.80	0.42	0.20	0.00	0.70	0.64	0.11	5	1.60	1.70	0.92
40	7.54	150	0.72	0.41	2.14	0.06	0.00	1.80	1.14	0.09	–	1.92	1.10	1.08
60	7.40	182	2.02	1.63	2.72	0.00	0.00	1.40	0.17	0.10	4	2.92	2.30	2.2
80	7.20	85	1.62	1.71	2.36	0.05	0.00	1.40	0.06	0.12	2	2.52	3.60	3.8
100	7.25	422	1.38	1.12	3.58	0.10	0.00	1.50	0.83	0.15	–	3.10	3.50	4.24

result of direct contact with the user of the filter jug. The presence of trace amounts of substances included in detergents and personal care products (ethylene brassylate, thymol, tributyl citrate), derivatives of pharmaceutical compounds (diclofenac methyl ester) and industrial additives (diisooctyl phthalate, dibutyl phthalate) was found.

3.2 Microbiological Quality of Water

The results of the microbiological analysis of the tap water and after filtering it through jug filters F-I, F-II, and I-III are presented in Table 3. The test results obtained show that the analyzed tap water was of very good bacteriological quality. The highest number of microorganisms incubated at 37 °C was 2 CFU/ml and 30 CFU/ml on A and R2A agar. In turn, the number of microorganisms incubated at 22 °C reached 6 CFU/ml and 30 CFU/ml on A and R2A agar. The guidelines of the Regulation on the quality of water intended for consumption indicate that the total number of microorganisms at 22 °C should not exceed: 100 CFU/ml in water introduced into the water supply network and 200 CFU/ml in the consumer's tap (in Poland, the reference agar is nutrient A agar). However, the number of bacteria incubated at 37 °C should not exceed 20 CFU/ml [15].

To reduce the risk of bacteria multiplication on the filter inserts, the jugs were placed in a dark place without access to light immediately after taking water for analysis, and microbiological analyzes were performed immediately after filtering the water. After filtration, the bacteriological quality of water was worse: the number of bacteria incubated at 37 °C oscillated between 0–42 CFU/ml (agar A) and 0–68 CFU/ml (agar R), while the number of bacteria incubated at 22 °C reached from 0–70 CFU/ml (A agar) and 0–380 CFU/ml (R agar) (Table 3). The worst microbiological quality was found in the F-II filter (the number of bacteria in the filtered water increased with the life of the cartridge). The activated carbon that filled the jug began to promote the accumulation and proliferation of bacteria. While bacteria incubated at 22 °C are a natural part of the microflora found in tap water, the presence of bacteria that prefer higher temperatures can be dangerous. Some strains of bacteria growing at 37 °C can cause various diseases, especially in people with a weakened immune system [18]. In one sample of water after filtration, the normative value of 20 CFU/ml for bacteria incubated at 37 °C was exceeded (Table 3). The conducted tests showed that no presence of *coliform bacteria* was found in the analyzed waters (the tested waters had no contact with feces) (Table 3).

During the control of the microbiological quality of the analyzed waters, two methods of detecting microorganisms were used: the classic culture plate method and the bioluminescent measurement of the content of adenosine triphosphate (ATP). During the study of the phenomenon of bioluminescence, the device measures light levels. The higher the ATP level, the more light is produced in the bioluminescence reaction, indicating the presence of microbial or organic contamination. The measurement result is given in RLU (Relative Light Unit) [19]. The analysis of the bioluminescence measurement showed that the ATP concentration for tap water was in the range of 7,266–24,851 RLU, while for filtered water it was 6,443–95,245 RLU (Table 3). The average values obtained for the entire research period indicate a slight increase in the level of ATP in waters in contact with the filter cartridge. The highest level of ATP in the cells was recorded for the F-II jug filter (which is consistent with the results of the culture methods). The research presented in the paper [20] shows that tap water had lower values of 827–15,489 RLU (average 9,609 RLU), in the analyzed example, the average value was 15,486 RLU. Due to the short waiting time for the result (5 min per sample) and the easy determination procedure, the bioluminescent ATP measurement method has become an attractive tool for monitoring the microbiological quality of water. In addition, this method also allows the detection of viable but non-culturable bacterial cells (VBNC) [19].

Manufacturers list numerous advantages of filter jugs, however, their effectiveness will also depend on how the filter cartridge is used and the quality of the water being purified. Manufacturers rarely provide information on the removal of bacteria and viruses from water. It is worth remembering, however, that the bacteriological composition of water will depend on the type and number of bacteria in the treated water and the age of the filter cartridges. There is a risk that as a result of the use of jugs, secondary microbiological contamination of the consumed water may occur. Bacteria retained on the filter have ideal conditions for development and multiplication, which may result in the entry of microorganisms into the water. The source of bacteria can be both the tested water and the person who uses the filters (contamination can occur when installing a new cartridge and when cleaning the jug). Storing jugs on kitchen counters, at room temperature and with access to light, also increases the risk of microbial growth.

So can you trust the manufacturers that drinking water directly from the filter jugs without boiling it is safe? In the analyzed case, even though the tap water was of good microbiological quality, the number of bacteria in the filtered water increased. In the research conducted by Kruszelnicka, it was observed that the number of bacteria incubated at 37 °C rapidly increased after three days of using the filters (180–450 CFU/ml) [21]. Gizińska et al. also noted a gradual increase in bacteria with the time of use of the jugs [17].

3.3 The Influence of the Use of Jug Filters on the Stability of Tap Water

Obtaining biologically stable water during treatment processes is not a simple task, because microorganisms can multiply even at low concentrations of nutrient substrates. The literature on the subject states that the threshold values of nutrients limiting the secondary growth of microorganisms in distribution systems should be for: BDOC < 0.25 mg C/L, nitrogen compounds < 0.2 mg N/L, and phosphorus compounds < 0.03 mg PO_4^{3-}/L [16]. Analyzing the obtained test results, it was found that tap water is not biologically stable, and the use of filter jugs additionally increases the content of nitrogen and phosphorus compounds in the water. For nitrogen, there was an increase of 0.18 and 0.15 mg N_{inorg}/L for F-I and F-II, while for phosphorus there was an increase of 0.89, 0.67, and 0.21 mg PO_4^{3-}/L for F-I, F-II, and F -III (Table 4).

This phenomenon can be explained by the impregnation of activated carbons, which consists in adding chemicals to them that improve their sorption properties. Typically, chemical compounds such as compounds of sulfur, phosphorus, nitrogen, heavy metals, and others are used for this purpose. Impregnation of activated carbons can increase their ability to remove specific contaminants from water, such as organic compounds, heavy metals, pesticides, and chlorine. However, the choice of the type of impregnation should depend on the type of impurities that are present in the water and the intended use of the activated carbon. In this case, an increase in the concentration of nitrogen and phosphorus may increase the possibility of proliferation of microorganisms (Table 4).

Table 3. Summary of microbiological results of the analyzed water.

Water volume, L	Bacteria incubated at 37 °C		Bacteria incubated at 22 °C		*E. coli* CFU/100ml	ATP measurement RLU
	A agar CFU/ml	R agar CFU/ml	A agar CFU/ml	R agar CFU/ml		
Tap water						
	0	1	0	10	0	7,266
	0	14	0	1	0	24,851
	0	4	0	22	0	16,589
	2	4	6	5	0	10,576
	0	30	0	30	0	17,681
	2	3	0	26	0	23,250
	2	1	0	2	0	17,558
Jug filter - F-I						
1	0	0	1	3	0	9,934
10	0	1	0	0	0	19,927
20	4	20	6	17	0	12,489
40	1	7	1	86	0	14,374
60	5	68	16	250	0	14,847
80	1	18	15	70	0	36,114
100	42	60	70	65	0	17,558
Jug filter - F-II						
1	0	7	0	58	0	7,963
10	0	10	0	25	0	11,073
20	0	11	7	105	0	11,522
40	16	30	20	98	0	8,757
60	8	42	14	190	0	8,842
80	1	13	2	270	0	32,213
100	6	30	17	350	0	95,245
Jug filter - F-III						
1	1	1	0	2	0	6,443
10	1	2	0	1	0	15,942
20	0	3	0	8	0	12,349
40	1	9	0	23	0	8,587
60	2	23	1	90	0	16,880

(continued)

Table 3. (*continued*)

Water volume, L	Bacteria incubated at 37 °C		Bacteria incubated at 22 °C		*E. coli* CFU/100ml	ATP measurement RLU
	A agar CFU/ml	R agar CFU/ml	A agar CFU/ml	R agar CFU/ml		
80	13	50	0	20	0	34,189
100	1	4	0	10	0	20,272

Table 4. Average values of nutrients in the analyzed waters.

Nutrient	Tap water	F-I	F-II	F-III	Limit value [16]
N	1.47	1.65	1.23	1.62	0.2
P	0.28	1.17	0.95	0.49	0.03
BDOC	0.13	0.09	0.08	0.14	0.25

The highest content of biogenic substances was found in the F-II filter, for which the worst bacteriological quality was recorded.

4 Conclusion

1. During the tests, the water before and after filtration did not exceed the permissible standards of physicochemical indicators contained in the Regulation of the Minister of Health on the quality of water intended for human consumption. The exception was water turbidity, which values were 0.7–3.63 NTU.
2. The effectiveness of water softening and removal of organic impurities decreases during the use of filters. Removal of chlorine compounds was found only in the case of one analyzed jug filter.
3. The analysis of microbiological parameters showed that the use of filter jugs contributes to the deterioration of the bacteriological quality of water. In one water sample, the number of microorganisms incubated at 37 °C was exceeded. The type of filter cartridge had an impact on the intensity of microorganism development. The highest number of bacteria was recorded for the F-II filter.
4. The use of filter jugs also contributes to the deterioration of water biostability as a result of leaching nitrogen and phosphorus compounds from the filter cartridges.

References

1. Lin L, Yang H, Xu X (2022) Effects of water pollution on human health and disease heterogeneity: a review. Front Environ Sci 10:880246
2. Tchórzewska-Cieślak B, Pietrucha-Urbanik K, Papciak D (2019) An approach to estimating water quality changes in water distribution systems using fault tree analysis. Resources 8:162
3. Zhang J, Li W, Chen J, Wang F, Qi W, Li Y (2019) Impact of disinfectant on bacterial antibiotic resistance transfer between biofilm and tap water in a simulated distribution network. Environ Pollut 246:131–140
4. Liu G et al (2017) Potential impacts of changing supply-water quality on drinking water distribution: a review. Water Res 116:135–148
5. Chen L et al (2020) Assessing the transition effects in a drinking water distribution system caused by changing supply water quality: an indirect approach by characterizing suspended solids. Water Res 168:115159
6. Jachimowski A (2017) Factors affecting water quality in a water supply network. J Ecol Eng 18:110–117
7. Act of 7th June 2001 on the Collective Water Supply and Collective Wastewater Disposal [Dz. U. 2001 nr 72 poz. 747]
8. Wang BB (2021) Research on drinking water purification technologies for household use by reducing total dissolved solids (TDS). PLoS ONE 16:e0257865
9. Bosscher V, Lytle DA, Schock MR, Porter A, Del Toral M (2019) POU water filters effectively reduce lead in drinking water: a demonstration field study in Flint, Michigan. J Environ Sci Health, A 54:484–493
10. Carrière A, Brouillon M, Sauvé S, Bouchard MF, Barbeau B (2011) Performance of point-of-use devices to remove manganese from drinking water. J Environ Scie Health, A 46:601–607
11. Delpla I, Legay C, Proulx F, Rodriguez MJ (2020) Perception of tap water quality: assessment of the factors modifying the links between satisfaction and water consumption behavior. Sci Total Environ 722:137786
12. Doré E et al (2021) Effectiveness of point-of-use and pitcher filters at removing lead phosphate nanoparticles from drinking water. Water Res 201:117285
13. Puszczało E, Kudlek E, Marszałek A (2020) Assessment of the possibility of secondary water pollution during its purification in filtering jugs. Desalination and Water Treatment, 290–296
14. Jezierska K, Podraza W, Domek H, Szwed J (2014) A method for evaluating the effectiveness of water purification in household jug filters. Ecol Eng 37:62–71 (In Polish)
15. Regulation of the Minister of Health of 7 December 2017 on the quality of water intended for human consumption [Dz. U. 2017 poz. 2294]
16. Wolska M (2015) Removal of nutrients in water purification technology for human consumption. Publishing House of the Wroclaw University of Technology, Wroclaw (In Polish)
17. Gizińska M, Pytka A, Skwarzyńska A, Micek A, Jóźwiakowski K, Marzec M, Sosnowska B (2014) Comparison of efficiency and life of jug filters for water. Water Technol, 25–29
18. Cabral JPS (2010) Water microbiology. Bacterial pathogens and water. Int J Environ Res Public Health 7:3657–3703
19. Rygała, A.: Usefulness of ATP measurement in microbiological research. Food Ind 69(10):20–23 (2015) (In Polish)
20. Zamorska J (2019) Evaluation of the microbiological quality of water using culture methods, flow cytometry and luminometry. Publishing House of the Rzeszow University of Technology, Rzeszow (In Polish)
21. Kruszelnicka I, Michałkiewicz M, Ginter-Kramarczy KD (2017) Is it worth using jug filters? (Part 2). Sci Technol, 30–35 (In Polish)

Accumulation of Thermal Energy in the Thermal Tank Storage with the Help of the Sun

Michal Gorás(✉) and Frantisek Vranay

Faculty of Civil Engineering, Institute of Architectural Engineering, Technical University of Kosice, Vysokoškolská 4, 042 00 Kosice, Slovakia
michal.goras@tuke.sk

Abstract. The propose of study is a convert the capture of solar energy into thermal energy as one of the renewable energies. Surface-selective hot water collectors with an area of 225 m^2 are used to collect solar energy. This energy captured by the collectors is transferred by technologically modified water, which transfers the heat in the technical room with the help of an exchanger. On the other side of the exchanger is water in which energy is stored. Water with a volume of 160 m^3 is stored in three different sizes of thermal tank storage. The smallest thermal tank storage with a volume of 24 m^3. This thermal tank storage was measuring during 2 years which charging and discharging cycles took place. In the first year, there were large heat losses through evaporation from the water surface. And in the second year of research, the water surface was covered, which helped to eliminate heat losses and extend the accumulation period.

Keywords: Thermal tank storage · Renewable energy · Solar energy · Long-term accumulation · Thermal energy

1 Introduction

In Slovakia and similarly situated countries, the demand for heat in the heating season is 10 times higher than in the non-heating season. Paradoxically, at the same time when solar radiation reaches its maximum, the need for heat for heating is the lowest. From May to August, solar radiation in our country is most intense [1]. In this summer season, there is no need to use another fossil source, but you can heat and heat hot drinking water directly. However, during the non-heating period, there is a minimal demand for heat, which is also a problem, Fig. 1. This heat captured during the summer needs to be accumulated in order to reduce energy consumption in the heating season [2].

2 Thermal Energy Tank Storage

2.1 Thermal Energy Tank Storage Measured

Surface-selective hot water collectors with an area of 225 m^2 are used to collect solar energy. This energy captured by the collectors is transferred by technologically modified water, which transfers the heat in the technical room with the help of an exchanger. On

Z. Blikharskyy et al. (Eds.): CEE 2023, LNCE 438, pp. 119–126, 2024.
https://doi.org/10.1007/978-3-031-44955-0_13

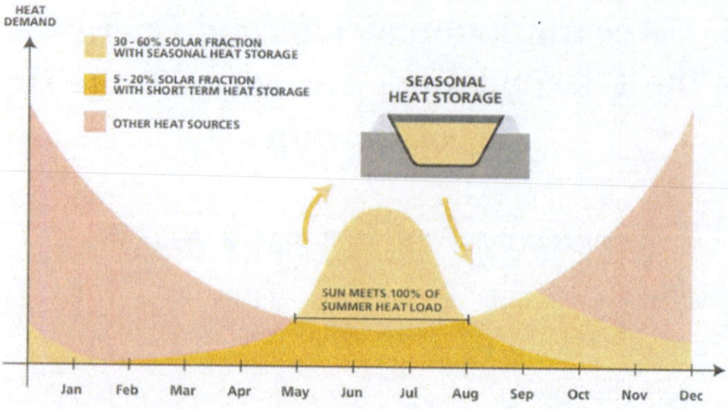

Fig. 1. Energy production and energy demand in year [2].

the other side of the exchanger is water in which energy is stored. Water with a volume of 160 m^3 is stored in three different sizes of thermal tank storage, Fig. 2. Different sizes of thermal tanks allow them to be used for different purposes and at different times. The structure of the thermal tank storage is made of reinforced concrete with thermal insulation and they are stored under the floor of the hall. Collectors are also installed on the roof of the hall.

Fig. 2. Three sizes on thermal energy storage

2.2 Thermal Energy Tanks Storage in Different Types

We can divide the thermal tank storage according to the media in which the energy is accumulated into sensible heat and latent heat. Performance, efficiency and other parameters can be divided according to media, method of accumulation and use. But we basically divide them into thermal and chemical accumulation, Fig. 3 [3].

We will not discuss chemical energy that comes from fossil fuels because it is not a renewable resource. Thermal storage tanks are used today. These thermal tanks storage can be divided into sensible heat and latent heat storage tanks. Latent heat is generated when the state changes. E.g. when switching from ice to water, just as much energy is needed as heating the water to 80 °C. Sensible heat represents the behavior of the

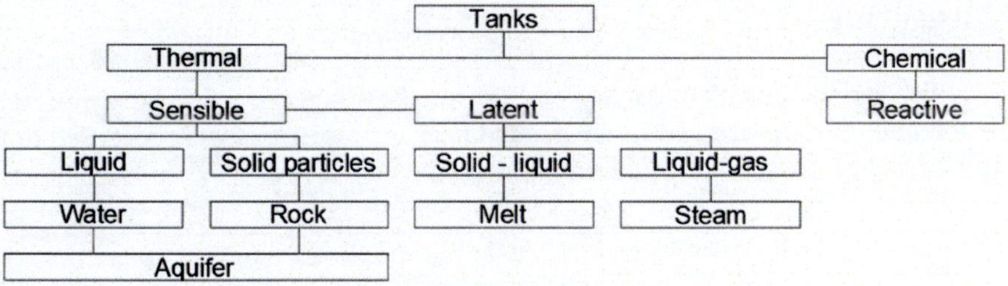

Fig. 3. Types of tanks [3].

temperature during the supply and consumption of energy. Sensible heat is derived according to the Eq. (1) [4] (Table 1)

$$Q_{sp} = m.c.(\varnothing_2 - \varnothing_1). \tag{1}$$

Table 1. Specific and volumetric heat capacity of materials for heat accumulation at 20 °C [4]

Medium	Temperature range (°C)	Specific heat capacity (Wh/kg K)	Volume heat capacity (Wh/m³ K)	Density (kg/m³)
Water	0–100	1.16	1160	998
Air	−273 -> 1000	0.28	0.31	1.1
Oil	0–400	0.44–0.5	350–450	800–900
Gravel sand	0–800	0.2	360–390	1800–2000
Granite	0–800	0.21	570	2750
Concrete	0–500	0.24	460–560	1900–2300
Brick	0–1000	0.23	330–440	1400–1900
Iron	0–800	0.13	1000	7860
Gravel backfill (37% water)	0–100	0.37	810	2200
Mix of salts ($53KNO_3$ + $4NaNO_2$ + $7NaNO_3$)	150–450	0.36	480–550	2561–2243

The heat in the Q_{Sp} storage tank depends on the weight of the medium (m), the heat capacity (c) and the temperature difference. If we have a 200 L tank and needed to heat it from 10 °C to 60 °C, the energy needed would be:

$$Q_{sp} = 200 \cdot 1.16(60 - 10) = 11600 \text{ Wh} = 11.6 \text{ kWh}. \tag{2}$$

According to these values, water availability and specific heat were selected as the medium [5].

3 Measuring

Were collected data took place for 2 years. During these 2 years, the water temperatures in the thermal tank storage and in the ground near the reservoirs were recorded to find out if there is heat loss to the environment, Fig. 4.

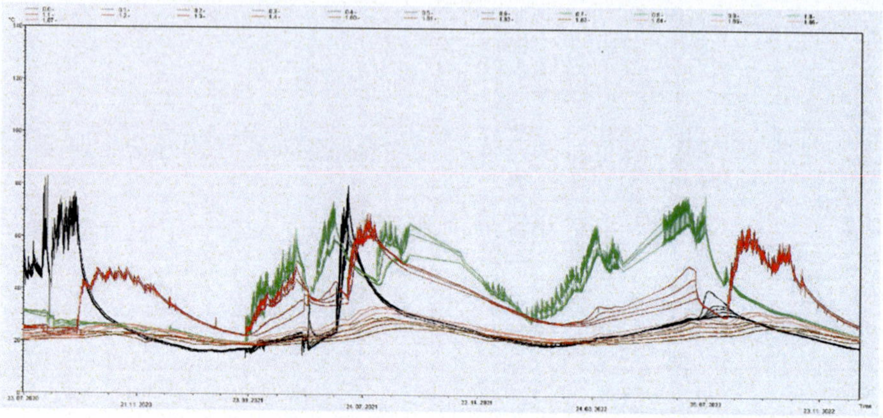

Fig. 4. Graph of 2-year temperature measuremets in tanks

A total of 27 temperature sensors are located in all 3 thermal tank storage and the ground. These sensors recorded temperatures for two years, which gave us the graphic representation in Fig. 4. In Fig. 4, the color-coded temperature curves according to location correspond to the colors according to Fig. 5.

4 Comparison AN1 Tank Between Seasons

So far, the thermal tank storage AN1 with a volume of 24 m^3, which is the smallest of them, has been investigated. This tank showed heat losses caused by evaporation. This graphic representation is the course of measurement over two years, Fig. 6.

4.1 Accumulation Tank Without Cover

Evaporation through the water surface caused large heat losses in the first year, which are graphically shown in Fig. 7.

The real measurement data shows us how quickly the energy was discharged from the reservoir. The discharge process is quadratic without energy consumption, Fig. 7.

The graphic percentage representation is from the maximum medium temperature of 67 °C to the initial temperature of 18 °C, which corresponds to the ambient temperature. The discharge took place in 85 days. The effectiveness of such discharge is very low, as it would only be enough for a short time, Fig. 8.

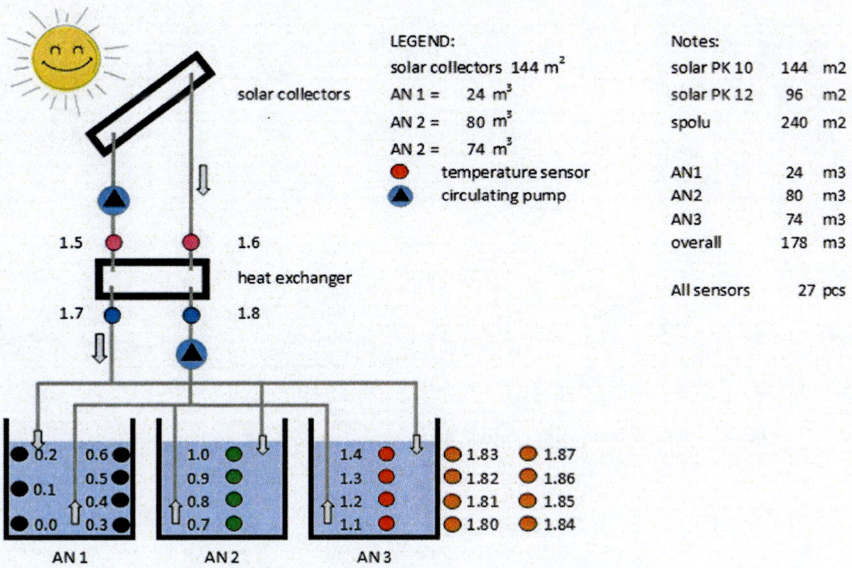

Fig. 5. Placement of temperature sensors

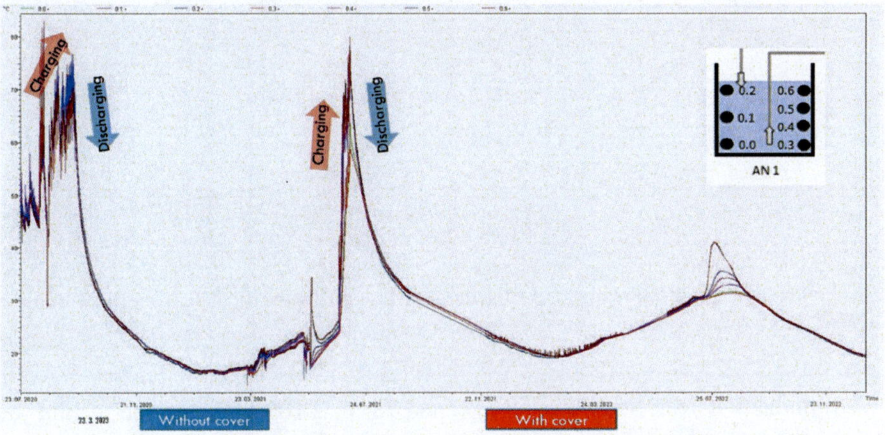

Fig. 6. Graphs of 2-year temperature measurements in AN1-tank

4.2 Accumulation Tank with Cover

In the second year, we improved the AN1 research thermal tank storage by covering the water surface. The covering consisted of thermal insulation made of polystyrene with a thickness of 20 mm. It can be proven from the collected temperature values that the discharge is not so fast. This discharge passes at a certain temperature from quadratic to linear discharge, Fig. 9.

In the second year, the tank was charged up to a temperature of 73 °C, but 100% of the temperature value for comparison between the first and second year was set to

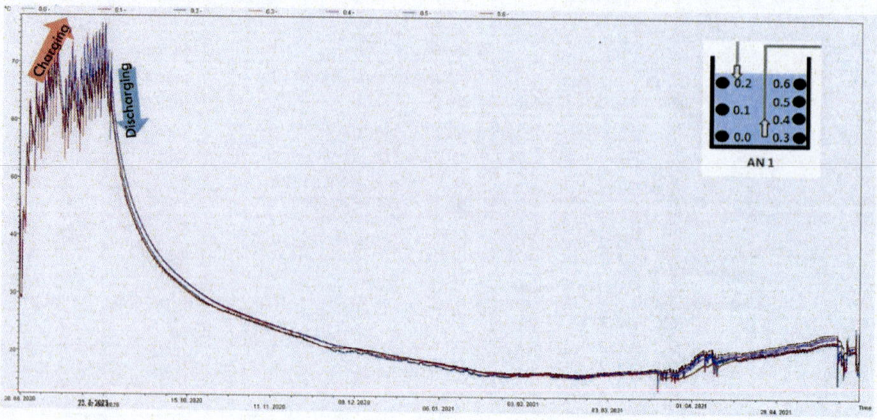

Fig. 7. First year of temperature measurements in tank without the cover

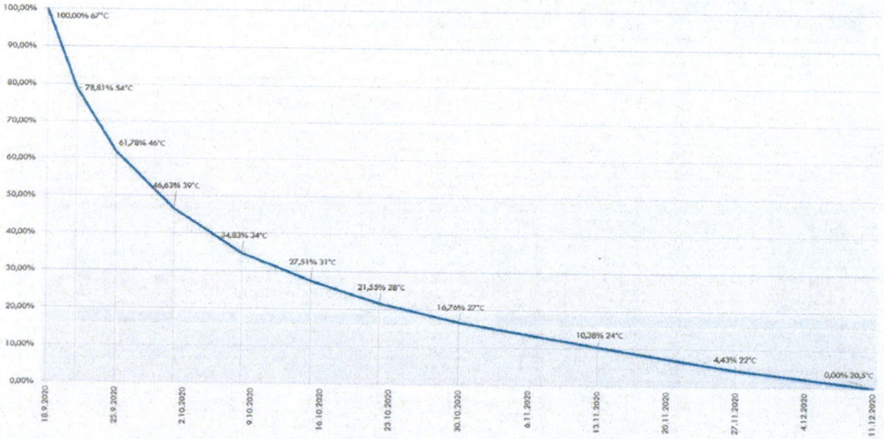

Fig. 8. Percentage graph showing tank discharging energy in the first year, vertical column - 0%–100%. Horizontal column - time data in days.

67 °C from the first year. So 100% is 67 °C and 73 °C is 108,11%. For the second year, therefore, the covered water surface shows a more effective solution for long-term energy accumulation. Comparing the first and second years, in the third week the temperature before covering was 34 °C, and after covering it was 46 °C. In the second year, the tank was discharged to the original temperature of 18 °C, which was 183 days, Fig. 10.

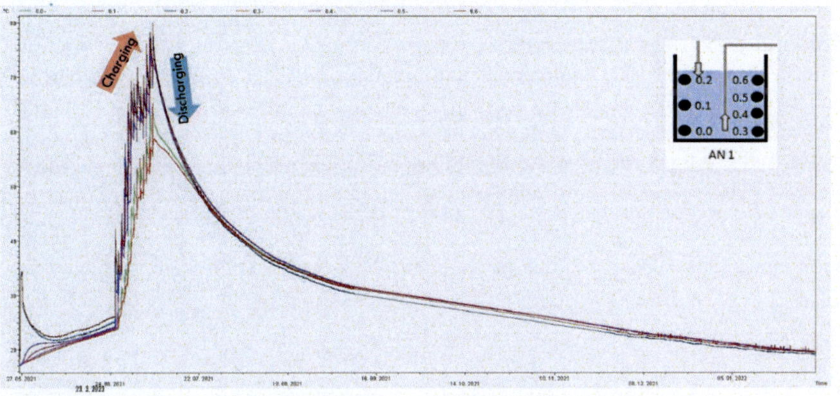

Fig. 9. Second year of temperature measurements in tank with the cover

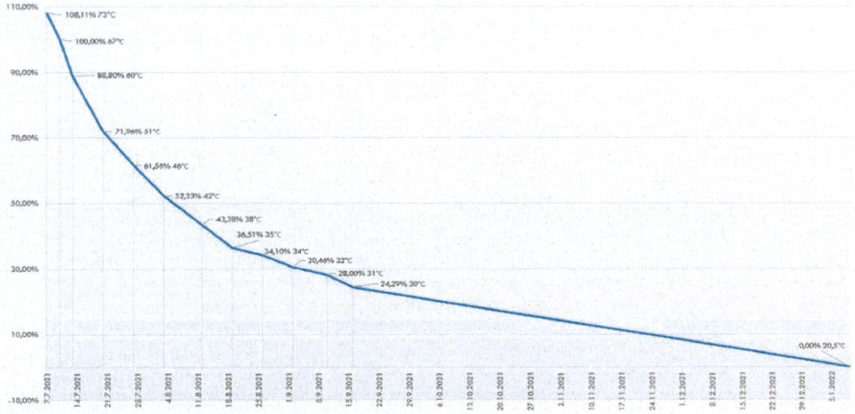

Fig. 10. Percentage graph showing tank discharging energy in the second year, vertical column - 0%–100%. Horizontal column - time data in days.

5 Conclusion

This research pointed out how to prevent heat losses, eliminate them and thereby increase the efficiency of the accumulation. In the future, we will focus on examining physical processes such as heat loss, stratification, etc.

Acknowledgements. The article was created with the support of the project Hungary-Slovakia-Romania-Ukraine ENI CBC Programme 2014–2020 HUSKROUA/1702 New Energy Solutions in Carpathian area.

References

1. Riahi L, Martinez C (2019) District energy in cities. UNEP. http://www.districtenergyinitiative.org

2. Trier D, Bava F, Kok Skov C, Stendorf Sorensen S (2018) Solar district heating trends and possibilities. Helsinge, Denmark
3. Ladener H, Späte F (2001) Solaranlagen (Staufen bei Breisgau). ISBN 80-247-0362-9
4. Understanding renewable energy systems. Carl Hanser Verlag Gmbh Co KG (2005). ISBN 1-84407-136-7
5. Technology roadmap energy-efficient buildings: heating and cooling equipment. International Energy Agency – IEA 9 rue de la Fédération, 75015 Paris, France, Printed in France by Corlet (2011)

Energy-Efficient Ventilation Systems for Underground Fortification Structures

Bogdan Gulai(✉), Stepan Shapoval, Olena Savchenko, and Hanna Klymenko

Lviv Polytechnic National University, Lviv 79013, Ukraine
bogdan.i.gulai@lpnu.ua

Abstract. An important task is the development of effective, energy-saving ventilation systems, taking into account the modern achievements of science and technology to create inconspicuous ventilation of underground engineering and construction structures, both military and civilian. In the construction of underground fortification structures, the introduction of ventilation systems with heat exchange and removal of exhaust air for a certain distance will improve comfortable conditions and safety. The purpose of the research was to compare the efficiency of the heat exchanger with and without flow swirlers, as well as with a different number of their installation. Based on the study of the proposed model of the ventilation system, a linear relationship between the efficiency of the heat exchanger and the number of installed flow turbulators per 10 m of the air duct. According to the results of the study, the installation of flow swirlers ensured an increase in efficiency by 6% compared to the same design without additional elements. Therefore, based on the research results, it can be concluded that the use of swirl generators in the "pipe-in-pipe" ventilation system is an effective method for increasing efficiency both for underground fortification structures and for civil facilities using energy-saving technologies.

Keywords: Underground shelters · Underground fortification structures · Soil heat exchanger · Soil temperature · Ventilation system · Flow swirlers · Swirl generators

1 Introduction

Energy-saving systems and technologies are becoming increasingly important in today's world not only from an environmental and economic point of view, but also from the point of view of solving geopolitical and security problems. In the context of war or the potential for future military conflicts, the importance of energy-saving systems can hardly be overstated. In the construction of underground fortifications, it is important to implement ventilation systems, including heat exchange and effective removal of exhaust air at a safe distance from these shelters, to ensure their inconspiuousness for the enemy. Such systems have the potential to significantly increase the comfort and safety of both military personnel and civilians seeking shelter in these underground structures. In addition, during wartime, the development of energy-efficient ventilation systems that operate quietly

Z. Blikharskyy et al. (Eds.): CEE 2023, LNCE 438, pp. 127–133, 2024.
https://doi.org/10.1007/978-3-031-44955-0_14

and unobtrusively is extremely important. It is imperative to focus on the improvement of existing ventilation systems, and to strive to develop innovative solutions that are highly efficient, energy-saving and technologically advanced, and especially accessible to the public. This includes taking into account the latest advances in science and technology to ensure that these systems meet the specific requirements of underground engineering and construction structures. By prioritizing the development and implementation of energy-saving ventilation systems, we can increase the resilience and self-sufficiency of critical infrastructure in the face of potential disruptions or attacks, with little visibility. In summary, the importance of energy saving systems in the context of underground fortifications, both for military and civilian purposes, cannot be overstated. By investing in energy-efficient technologies, we can ensure the safety and well-being of the people who seek shelter in these structures, and promote resilience in the face of new security challenges.

2 Objectives the Formulation of the Problem

The purpose of the article is to develop an energy-efficient ventilation system for underground fortifications to increase the efficiency of heat energy recovery by installing vortex generators that cause air turbulence. Determination of the value of the change in the efficiency of the heat exchanger, taking into account the introduced innovations.

3 The Analysis of Recent Research and Publications

In the conditions of war, when the enemy can use long-range and high-precision weapons both for surveillance and for attack, as well as unmanned aerial vehicles with thermal imaging devices capable of determining the position of objects, both military and civilian, the effective masking of these structures becomes of critical importance, and in particular engineering networks. To maintain the accepted sanitary and hygienic conditions in underground fortifications, it is necessary to build a system of mechanical ventilation using heat exchangers (heat energy recuperators) [1, 2], for optimal use of buildings of various purposes, various energy-saving technologies are used [3–6]. Today, ventilation systems with heat recovery are considered one of the primary factors of energy efficiency of buildings and structures [7–9]. For preliminary heating of the outside air, it is advisable to use both low-potential base energy and exhaust air energy [10, 11].

Underground fortifications can be effective against long-range weapons such as artillery and unmanned aerial vehicles, provided they are invisible, both visually and to thermal imagers, which can detect even the slightest temperature difference. The main requirements for construction are strength of the building (use of strong building materials and deeper construction), invisibility of the building (use of camouflage, etc.), safe and comfortable use (providing life support systems such as ventilation and heating). Therefore, as an important safety measure, it is important to divert the exhaust ventilation air, which is from the surrounding air and can unmask the shelter, to a certain distance from it [12], and also to reduce the temperature difference between the exhaust air and the environment by using heat exchangers. This is especially relevant in the cold or warm season, when the temperature difference between the internal and external air is significant, and therefore visible to the enemy's thermal imagers. The requirements for the effective use of ventilation systems in underground fortifications in wartime require large-scale parametric studies, the development and implementation of engineering calculation methods, a comprehensive study allows to increase the efficiency of the functioning of this system. Also, an important factor in the selection and design of such systems is their price and general availability, with an understanding of their application effect.

4 The Main Material

It is proposed to use the design of the supply-exhaust ventilation system, which is made of cheap publicly available materials and will perform such functions as air replacement in the building, heat exchange between supply and exhaust air, removal of exhaust air to a certain distance from the building, application of design solutions to increase the efficiency of the system.

In this proposed energy-efficient ventilation system, a soil heat exchanger of the counter-flow type is used. To improve the performance of the heat exchanger, additional elements of air flow swirl are used. Also, the goal of using publicly available and cheap materials is the basis for the creation of the above-mentioned scheme.

To solve the above-mentioned problems, a design (Fig. 1) with an installed counter-current heat exchanger of the "pipe-in-pipe" type and additional elements for turbulating the flow is proposed.

The proposed ventilation system consists of a combination of supply and exhaust air ducts (vent channels) based on the countercurrent heat exchanger "pipe in a pipe" principle for exchanging heat between the used (heated) and fresh (cold) air. Additional flow turbulence elements are installed to improve performance (Fig. 2).

The study involved calculations for 20% more supply air than exhaust air, without taking into account the heat from the stove. The proposed ventilation system was designed to use materials that are widely available and low-priced to ensure its economic feasibility.

The objective of the study was to compare the efficiency of the heat exchanger with and without flow swirlers, and with varying numbers of them installed. Computer modeling, specifically the SolidWorks software with FloEFD and KeyShot modules for visualization, was used during the research.

Table 1 shows the input parameters used during the computer modeling.

As a result of calculations using the FloEFD module, graph-analytical dependencies and values of supply air temperature were obtained. The obtained temperature values and efficiency data are presented in Table 2 (Fig. 3).

Based on the study of the proposed model of the ventilation system, a linear relationship between the efficiency of the heat exchanger and the number of installed flow turbulators per 10 m of the air duct was obtained (Fig. 4).

It should be noted that only the efficiency coefficient of the heat exchanger was taken into account during the research. However, when installing air flow swirlers in the ventilation system, this causes additional resistance to air movement, which requires the installation of more powerful fans, which will increase the energy consumption of the

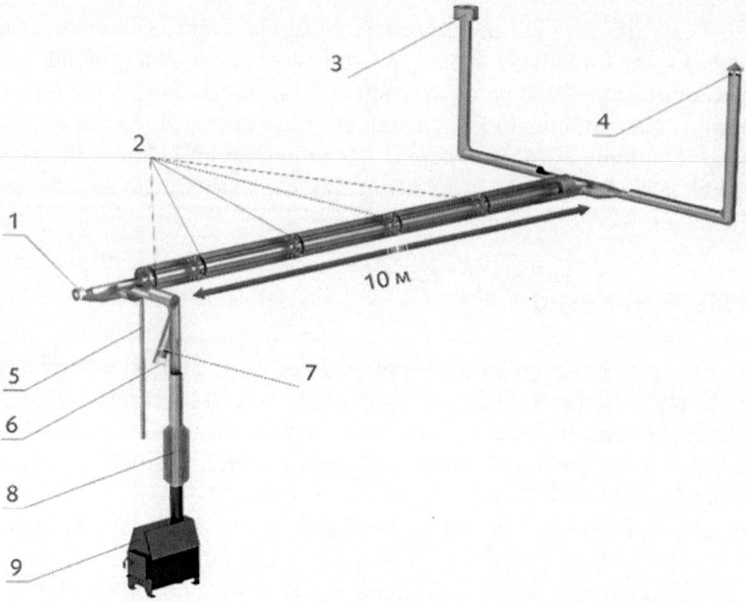

Fig. 1. Scheme of ventilation of the underground storage: 1-supply air duct (with installed confusor and supply fan); 2-flow swirlers; 3-deflector of the exhaust duct (exhaust ventilation / smoke removal); 4-umbrella of the supply duct; 5-condensate drain pipe; 6-internal exhaust duct with fan and/or non-return valve; 7-return valve and/or exhaust fan; 8-heat exchanger of the stove; 9-stove for heating with a flue channel.

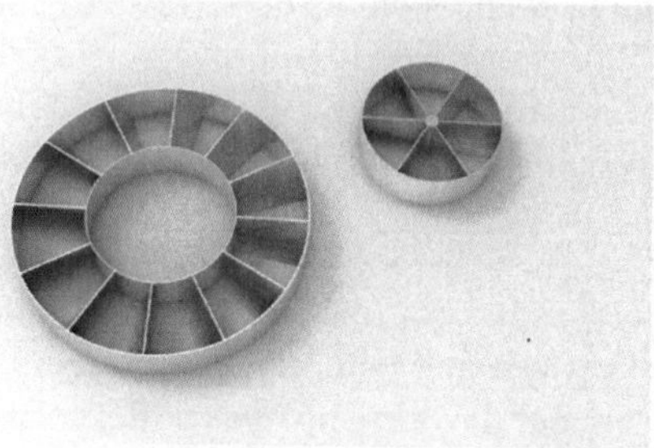

Fig. 2. Image of flow swirlers for the outer and inner pipe

ventilation system. The overall efficiency of the ventilation system, such as the ratio of the power of the fans in the system to its performance, should be taken into account in further studies.

Table 1. Input parameters of the study.

Input parameters	Parameter values
The temperature of exhaust gases from the furnace, °C	250
Soil temperature, °C	−5
Temperature of the outside air (*t°street*), °C	−10
Indoor air temperature (*t°room*), °C	+20
Productivity of the exhaust ventilation system, m^3/h	160
Productivity of the inflow ventilation system, m^3/h	200
The number of people who will be in the dugout, people	7

Table 2. The results obtained.

The number of central offices	0	2	4	6	8	10
Supply air temperature, (*t°rec*), °C	1.21	1.61	2.16	2.63	2.9	3.26
Internal (exhaust air) temperature	25.02	24.63	24.21	23.7	23.29	22.85
Efficiency*, %	46.31	47.45	49.02	50.37	51.14	52.17

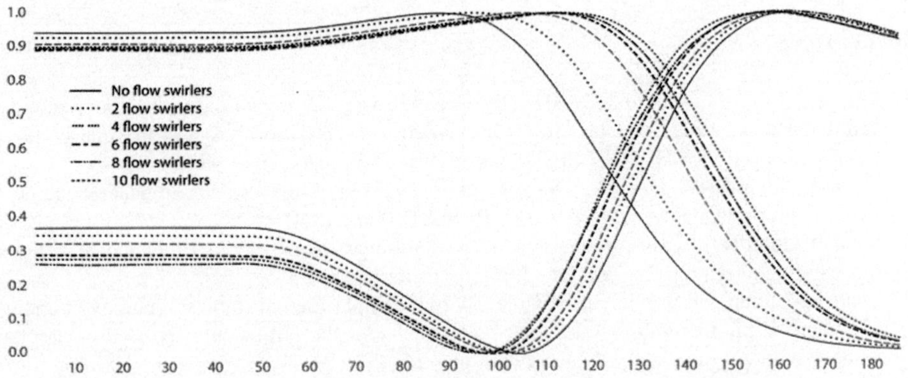

Fig. 3. Graphs of the ratio of temperatures to the number of iterations.

5 Conclusions

According to the conducted research, the use of flow swirlers in the pipe-in-pipe ventilation system led to a noticeable increase in efficiency by 6% compared to a design without these elements. The dependence of the change in the efficiency coefficient on the installation of a different number of swirlers of repeated flows is also deduced. This conclusion convincingly confirms the effectiveness of swirlers in increasing the efficiency of both underground fortification structures and civil objects using energy-saving technologies.

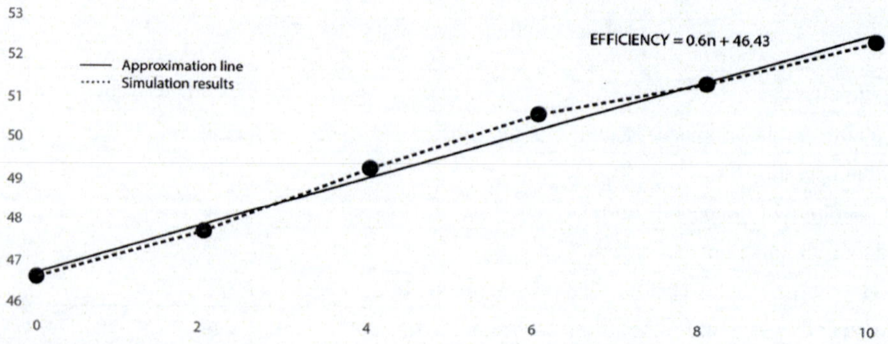

Fig. 4. Linear dependence of the change in heat exchanger efficiency on the number of swirlers on a 10-m section

According to the research results, it was established that the use of swirlers in the ventilation system is a highly effective method of optimizing energy efficiency. This applies to a variety of applications, including underground fortifications and civil structures, which contribute to the overall goal of promoting energy conservation practices.

Thus, the study confirms the significant benefits of introducing flow swirlers in a pipe-in-pipe ventilation system, highlighting their crucial role in increasing efficiency and supporting the implementation of energy-saving technologies, as well as diverting exhaust air some distance from the underground storage to ensure their invisibility.

References

1. Mukhtar A, Ng KC, Yusoff MZ (2018) Design optimization for ventilation shafts of naturally-ventilated underground shelters for improvement of ventilation rate and thermal comfort. Renew Energy 115:183–198. https://doi.org/10.1016/j.renene.2017.08.051
2. Gulay B, Sukholova I, Dzeryn O, Shepitchak V (2021) Investigations of compact recuperators acoustic properties. https://doi.org/10.1007/978-3-030-57340-9_16
3. Pona O, Gulai B (2015) A comparison of the combined solar collector efficiency at different modes of its operation. Eastern-Eur J Enterp Technol 2(8):53–57
4. Venhryn I, Shapoval S, Zhelykh V, Gulai B (2023) Experimental studies of energy efficiency of a thermal photovoltaic hybrid solar collector under the influence of wind flow. Lecture notes in civil engineering, vol 290 LNCE, pp 424–431
5. Mandrika AS, Antonenko SS, Husak OG et al (2021) Energy-efficient technologies (Mandrika AS, ed). Sumy State University, Sumy, 330 p
6. Bobrovnikova KYu (2020) Methods of ensuring energy efficiency and energy saving in the smart home system. Comput Syst Inf Technol 1:53–58
7. El Fouih Y, Stabat P, Rivière P, Hoang P, Archambault V (2012) Adequacy of air-to-air heat recovery ventilation system applied in low energy buildings. Energy Build 54:29–39. https://doi.org/10.1016/j.enbuild.2012.08.008
8. Bai H, Liu P, Justo AM, Mathisen H (2022) A review of heat recovery technologies and their frost control for residential building ventilation in cold climate regions. Renew Sustain Energy Rev 162:112417. https://doi.org/10.1016/j.rser.2022.112417
9. Shapoval S, Zhelykh V, Venhryn I, Kozak K, Krygul R (2019) Theoretical and experimental analysis of solar enclosure as part of energy-efficient house. Eastern-Eur J Enterp Technol 2(8–98):38–45

10. Trząski A, Zawada B (2011) The influence of environmental and geometrical factors on air-ground tube heat exchanger energy efficiency. Build Environ 46(7):1436–1444. https://doi.org/10.1016/j.buildenv.2011.01.010
11. Michalak P (2022) Hourly simulation of an earth-to-air heat exchanger in a low-energy residential building. Energies 15(5):1898. https://doi.org/10.3390/en15051898
12. Rogach IM, Palko AI, Virag MV, Feger OV (2021) Methodological recommendations for independent preparation for practical classes in the discipline "Training of reserve officers in the field of "Health care" specialty "Medicine" - "Military hygiene" Uzhhorod

Analysis of Natural and Mechanical Ventilation in the Classroom During the Pause of Teaching

Peter Kapalo[1], Hanna Klymenko[2](✉), Vasyl Zhelykh[2], and Bogdan Gulai[2]

[1] Institute of Architectural Engineering, Technical University of Kosice, Kosice, Slovakia
[2] Institute of Civil Engineering and Building Systems, Lviv Polytechnic National University, Lviv, Ukraine
Hanna.M.Klymenko@lpnu.ua

Abstract. The topic of classroom ventilation in schools has been very topical in recent years. Several scientific and professional studies points to insufficient ventilation in classrooms, which results in lower concentration of pupils on lessons and earlier fatigue. In our study were analyzed the three methods of classroom ventilation during the pause between teaching hours. Air quality investigated was using the measured course of carbon dioxide (CO_2) concentration. CO_2 concentration measurements provided were by in three stages. In the first stage, ventilation provided was only by infiltration through leaks in the building structure and by opening the door of the classroom to the corridor during the pause between school hours. In the second stage, ventilation provided was by opening the windows of the classroom doors during the pause between school hours. In the third stage, ventilation was provided by a ventilation recuperation unit installed in the closed window through the between school hours. Based on the measured evaluation values of CO_2 concentration and calculated values of CO_2 production it is possible to state that in our case was from the point of view of air exchange very effective natural ventilation. From an energy point of view however, is effective mechanical ventilation using heat recovery.

Keywords: Mechanical ventilation · Natural ventilation · Classroom · Measurement · Carbon dioxide (CO_2) · Indoor air quality

1 Introduction

Currently, the company focused is for reducing energy consumption in buildings. In this paper, we fiddle with buildings for education. One of the possibilities is also the reduction of energy consumed for heating or cooling ventilation air in buildings, and too for driving fans. Reducing energy consumption should not be at the expense of creating a healthy indoor environment in classrooms. The numerical simulations show energy savings with a natural ventilation system of between 11% and 26% [1] Therefore, the effort is to prioritize natural ventilation over mechanical ventilation or the use of hybrid ventilation.

According to study Gao et al. [2] the classroom with automatically operable windows and an exhaust fan that ensured adequate ventilation at all times performed better in the

Z. Blikharskyy et al. (Eds.): CEE 2023, LNCE 438, pp. 134–142, 2024.
https://doi.org/10.1007/978-3-031-44955-0_15

case of the perceptions and symptoms reported by pupils. According to study Ding et al. [3] the personalized ventilation including personalized air supply systems and personalized air exhaust systems, can efficiently decrease the exposure risk of exhaled contaminants of the source person within a close contact, and improve local indoor air quality for the co-occupants.

Ventilation is highly sharing on creating a healthy internal environment. In classroom guaranteed should be by ventilation removal of degraded air from the room and replacing it with fresh ventilation air. In the classroom, mainly the people in the room, building materials and room equipment, degrade the air. The most known well harmful substances degrading the indoor environment are dust particles, odor, CO_2, temperature, relative humidity, noise and various harmful gases resulting from evaporation from classroom equipment and building materials.

The study Pellegatti et al. [4] showed a negative effect of fan noise on students, but the natural sounds from open windows had a positive effect on students' learning and comfort. According to study Krawczyk et al. [5], in buildings with high heat gains from people, like classrooms, during estimation of the CO_2 concentration it strongly recommended is to take into account changes of indoor air temperature. Based on experimental research Teleszewski et al. [6], a model of relative humidity developed was in classrooms equipped with natural ventilation. Based on the comparison of the moisture balance in the classroom with experimental data, it can be concluded that the average humidity generated by one person in the classroom is approximately 45 g/(h person). Regarding the production of dust particles, neither mechanical ventilation nor natural ventilation systems sufficiently lower PM2.5 in classrooms [7].

The CO_2 concentration level is one of the produced harmful, that are can representatively point to the level of air quality in the room. With these considerations, it is necessary to remind, that the CO_2 concentration level is not the only condition to determine the required air exchange. Results of study Kapalo et al. [8] showed the insufficient ventilation intensity in classrooms as well as obvious rise of CO_2 concentration during the exams. Regarding CO_2 production by respondents, it can be see, that it visibly increased with increasing physical activity. The study Canha et al. [9] focused on examination of ventilation rates in classrooms with two different types of ventilation systems: natural and mechanical. Registered was a mean mechanical ventilation rate of 13.3 L/s per per-son, which is higher than the recommended ventilation standards and displayed short-term CO_2 levels above 1,200 ppm was. During the natural ventilation, the classrooms showed low ventilation rates with mean values of 2.4 L/s per person, which is markedly lower than the minimum recommended value. The analysis Burridge et al. [10] confirmed elevated levels of CO_2 during the autumn and winter months. According to study Krawczyk et al. [11] the percentage of students satisfied with indoor microclimate decreased during classes more significantly, where CO_2 concentration increased above the recommended maximum level 1,000 ppm. The case study of Kapalo et al. [12] confirms that the rate of CO_2 formation is linearly dependent on the increase in the level of physical activity. In the study Fantozzi et al. [13] created was a method to predict the infection probability from CO_2 concentration, based on a cost effective CO_2 monitoring system, what will help to enhance the indoor air quality in the room. The study Rosbach et al. [14] and the studies of Twardella et al. [15], and Wargocki and

Wyon [16] show that field experiments are not comparable with laboratory experiments and that it can be challenging to execute the study as designed. Altogether, the study shown has, that classrooms CO_2 levels can be significantly reduced, by installing a CO_2 controlled mechanical ventilation system.

2 Aim

The aim of the work is to determine the suitability of the mode of ventilation of the classroom during the pause between lessons.

3 Method

In order to determine the appropriate mode of classroom ventilation, were performed the three experimental measurements in a selected room at the Institute of Civil Engineering and Building Systems, Lviv Polytechnic National University. During the measurement, the following recorded were: air temperature, relative humidity, air pressure and CO_2 concentration. In this article, presented are only the results of CO_2 concentration measurement.

3.1 Characteristics of the Room

In the room, in which the experimental measurement took place is located in a seven-story building on the fourth floor. The dimensions of the classroom are length: 5.90 m, width: 6.30 m and height: 3.30 m. The volume of the room is 123 m^3. The windows of the room are oriented to the northwest. A ventilation device with heat recovery installed was in one closed window (Fig. 1).

Fig. 1. Classroom with ventilation unit and measuring device.

During the lesson, the doors and windows closed were. At that time, the ventilation of the room ensured was by leaks in the building structure.

3.2 Used Ventilation Equipment and Measuring Devices

For classroom ventilation was used the ventilation device with heat recovery PRANA 150. For measuring indoor air parameters was used a temperature and humidity sensor S3541, a carbon dioxide concentration sensor C-AQ-0001R and a pressure gauge for measuring atmospheric pressure (Fig. 2).

Fig. 2. Ventilation device with heat recovery PRANA 150.

The ventilation and recuperation unit PRANA-150 designed is for one room with a floor area of up to 30 m^2. Our classroom has a floor area of 37 m^2. It ensures the inflow and outflow of air at the same time. Air exchange volume up to 105 m^3/h. It controlled is using a remote control.

Thermo-hygrometer S3541which designed is for recording temperature and humidity used was simultaneously with carbon dioxide concentration sensor C-AQ-0001R, with whom he was during measurement interconnected (Fig. 3).

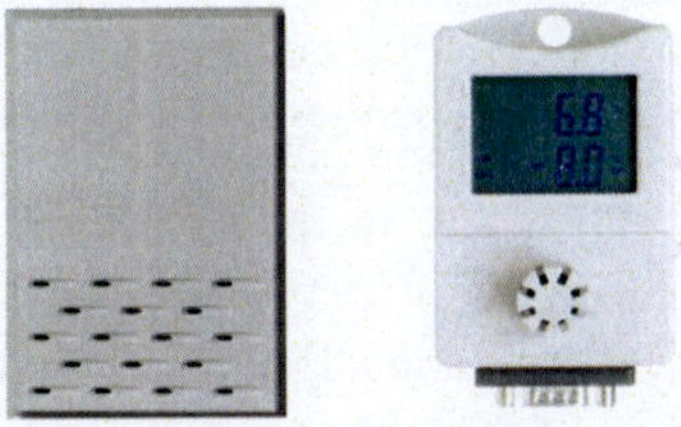

Fig. 3. Thermo-hygrometer S3541and CO_2 concentration sensor C-AQ-0001R.

Technical parameters of the device C-AQ-0001R and S3541 are CO_2 concentration measurement range (from 0 to 5,000 ppm), accuracy of CO_2 concentration measurement (±75 ppm), the effect of temperature on measurement (±0.5% ppm for 1 °C), temperature measurement range (from −30 to +70 °C), temperature measurement accuracy (±0.4 °C). Humidity measurement range (from −30 °C to +70 °C) and humidity measurement accuracy (±2.5% relative humidity for temperature 25 °C).

3.3 The Experimental Measurements Process

Experimental measurements performed were on three different days in the month of March. Each measurement had a different mode of ventilation of the classroom during the pause between lessons. On indoor air quality in the classroom, it is particularly influential the air temperature, relative humidity, odor, fumes from furniture and building construction and gaseous parts produced by humans, of which also is CO_2. On the worsening indoor air quality in the classroom the biggest has influence is CO_2 concentration. Therefore, it evaluated is in this work the CO_2 concentration.

Air quality investigated was using the measured course of CO_2 concentration. CO_2 concentration measurements provided were by in three stages.

In the first stage, ventilation provided was only by infiltration through leaks in the building structure and by opening the door of the classroom to the corridor during the pause between school hours. In this work, this mode called is for the sake of simplicity "Without ventilation". In the second stage, ventilation provided was by opening the windows of the classroom doors during the pause between school hours. In this work, this mode called is for the sake of simplicity "With natural ventilation". In the third stage, ventilation was provided by a ventilation recuperation unit installed in the closed window through the between school hours. In this work, this mode called is for the sake of simplicity "With mechanical ventilation".

4 Results and Discussions

Resulting values of CO_2 concentration measured during individual experimental measurements documented are in Fig. 4. In blue color is marked the course of CO_2 concentration during the first stage of measurement – without ventilation. From the course of CO_2 concentration can be observed that on that day took place in the room four teaching hours. During the pause of teaching, it is only small decrease in CO_2 concentration, because at that time the room ventilated was only through the open door to the corridor.

In green color is marked the course of CO_2 concentration during the second stage of measurement – with natural ventilation. From the course of CO_2 concentration can be observed that on that day took place in the room three teaching hours. During the pause is big decrease in CO_2 concentration, because at that time the room ventilated was with open windows to the exterior and open doors to the corridor.

Orange color is marked the course of CO_2 concentration during the third stage of measurement – with mechanical ventilation. From the course of CO_2 concentration can be observed that on that day took place in the room also three teaching hours. The he is small decrease in CO_2 concentration during pause, because at that time the room was ventilated only the ventilation unit to the exterior.

Given that the most important boundary condition in the production of CO_2 is the weight of a person and their number in the room, so for another analysis are processed measurements with approximately the same weight of persons. The results of measurements and calculations presented are in Table 1.

In the next step was calculated mass flow of CO_2 per person and for the whole room according to Eq. (1) [17].

$$q_m = 0.120406.m_p, \tag{1}$$

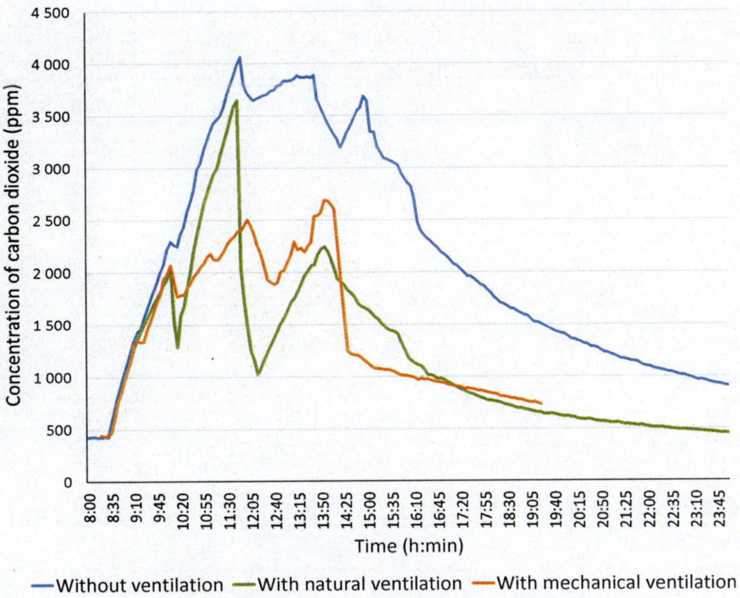

Fig. 4. The measured CO_2 concentration.

where: q_m is mass flow of CO_2 (mg/s) and m_p is the average weight of the person in the room (kg).

Necessary volume flow of ventilation air per person and for the whole room calculated was according to Eq. (2) [17].

$$q_V = 0.401353.m_p, \tag{2}$$

where: q_V is necessary volume flow of ventilation air (m^3/h) and m_p is the average weight of the person in the room (kg).

The results of calculations presented are in Table 2.

In Fig. 5 is documented the decrease CO_2 concentration calculated per person for compared ventilation modes.

From Fig. 5 it observed can be that the smallest decrease concentration of CO_2 was in the room during a pause when ventilation not provided was. The biggest drop CO_2 concentration was in the room during the pause when natural ventilation ensured was. The mechanical ventilation used is insufficient for this case. When we want to use mechanical ventilation type PRANA 150, is necessary in the case of this classroom at least 11 pieces.

5 Scientific Novelty and Practical Significance

In the course of research on ventilation of classroom was an experimental measurement and calculation determined mass flow rate of CO_2 and the required volumetric flow air rate. Measured and evaluated data used may be in another research, which is dealing with indoor air quality.

Table 1. Measured and calculated data of CO_2 concentration.

	Without ventilation	With natural ventilation	With mechanical ventilation
Start CO_2 concentration (ppm)	430	1,175	432
Finish CO_2 concentration (ppm)	2,296	2,038	2,068
Increase CO_2 concentration (ppm)	1,866	863	1,636
Time of increase CO_2 concentration (min)	95	60	100
Increase CO_2 concentration (ppm/min)	19.64	14.38	16.36
Decrease CO_2 concentration (ppm/pause)	47	754	282
Time of decrease CO_2 concentration (min)	10	10	20
Decrease CO_2 concentration (ppm/min)	4.7	75.4	14.1
Decrease CO_2 concentration (ppm/(min.pers))	0.52	10.77	1.41

6 Conclusions

Based on the measured values of CO_2 concentration and calculated values of CO_2 production it is possible to state that in our case was from the point of view of air exchange very effective natural ventilation. From an energy point of view however, is effective mechanical ventilation using heat recovery. The mechanical ventilation used is insufficient for this case. When we want to use mechanical ventilation type PRANA 150, is necessary in the case of this classroom at least 11 pieces. The results presented in this article only be used can in the case of the investigated classroom.

Table 2. Calculated data of mass flow of CO_2 and necessary volume air flow.

	Without ventilation	With natural ventilation	With mechanical ventilation
Number of persons (–)	9	7	10
Average weight of persons (kg)	77.89	77.57	78.50
Mass flow of CO_2 per person (mg/s)	9.38	9.34	9.45
Mass flow of CO_2 per classroom (g/stay)	481.11	235.37	567.11
Volume air flow per person (m^3/h)	31.26	31.13	31.51
Volume air flow per classroom (m^3/h)	281.35	217.93	315.06

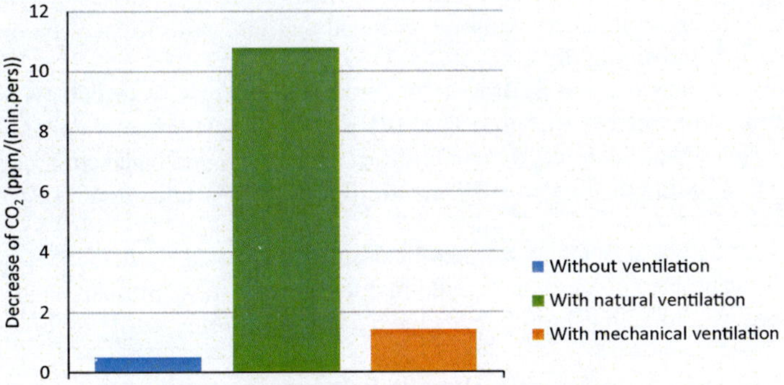

Fig. 5. The decrease CO_2 concentration.

Acknowledgments. This work supported was by the project run by the Slovak Research and Development Agency VEGA 1/0492/23 Transformation of existing buildings into sustainable buildings - the ecological potential of flat roofs.

References

1. Gil-Baez M, Barrios-Padura A, Molina-Huelva M, Chacartegui R (2017) Natural ventilation systems in 21st-century for near zero energy school buildings. Energy 137:1186–1200. ISSN 0360-5442. https://doi.org/10.1016/j.energy.2017.05.188
2. Gao J, Wargocki P, Wang Y (2014) Ventilation system type, classroom environmental quality and pupils' perceptions and symptoms. Build Environ 75:46–57. ISSN 0360-1323. https://doi.org/10.1016/j.buildenv.2014.01.015

3. Ding E, Zhang D, Bluyssen PM (2022) Ventilation regimes of school classrooms against airborne transmission of infectious respiratory droplets: a review. Build Environ 207, part A:108484, ISSN 0360-1323. https://doi.org/10.1016/j.buildenv.2021.108484
4. Pellegatti M, Torresin S, Visentin Ch, Babich F, Prodi N (2023) Indoor soundscape, speech perception, and cognition in classrooms: a systematic review on the effects of ventilation-related sounds on students. Build Environ 236:110194. ISSN 0360-1323. https://doi.org/10.1016/j.buildenv.2023.110194
5. Krawczyk DA, Żukowski M (2020) Experimental verification of the CO_2 and temperature model. Int J Vent 19(2):127–140. https://doi.org/10.1080/14733315.2019.1592333
6. Teleszewski TJ, Gładyszewska-Fiedoruk K (2020). Characteristics of humidity in classrooms with stack ventilation and development of calculation models of humidity based on the experiment. J Build Eng 31:101381. ISSN 2352-7102. https://doi.org/10.1016/j.jobe.2020.101381
7. Cai Ch, Sun Z, Weschler LB, Li T, Xu W, Zhang Y (2021) Indoor air quality in schools in Beijing: field tests, problems and recommendations. Build Environ 205:108179. ISSN 0360-1323. https://doi.org/10.1016/j.buildenv.2021.108179
8. Kapalo P et al (2019) Investigation of CO_2 production depending on physical activity of students. Int J Environ Health Res 29(1):31–34. https://doi.org/10.1080/09603123.2018.1506570
9. Canha N, Almeida SM, Freitas MC, Täubel M, Hänninen O (2013) Winter ventilation rates at primary schools: comparison between Portugal and Finland. J Toxicol Environ Health A 76(6):400–408. https://doi.org/10.1080/15287394.2013.765372
10. Burridge HC, Bontitsopoulos S, Brown Ch, Carter H, Roberts K, Vouriot C, Weston D, Mon-Williams M, Williams N, Noakes C (2023) Variations in classroom ventilation during the COVID-19 pandemic: insights from monitoring 36 naturally ventilated classrooms in the UK during 2021. J Build Eng 63, Part B:105459. ISSN 2352-7102. https://doi.org/10.1016/j.jobe.2022.105459
11. Krawczyk DA, Rodero A, Gładyszewska-Fiedoruk K, Gajewski A (2016) CO_2 concentration in naturally ventilated classrooms located in different climates—measurements and simulations. Energy Build 129:491–498. ISSN 0378-7788. https://doi.org/10.1016/j.enbuild.2016.08.003
12. Kapalo P, Vojtasko L, Vasilisin D, Domniţa F, Bacoţiu C, Kandrac R, Batorova M (2021) Investigation of the influence of the level of physical activity on the air exchange requirements for a gym. Build Environ 204:108123. ISSN 0360-1323. https://doi.org/10.1016/j.buildenv.2021.108123
13. Fantozzi F, Lamberti G, Leccese F, Salvadori G (2022) Monitoring CO_2 concentration to control the infection probability due to airborne transmission in naturally ventilated university classrooms. Archit Sci Rev 65(4):306–318. https://doi.org/10.1080/00038628.2022.2080637
14. Rosbach JT, Vonk M, Duijm F, Ginkel JT, Gehring U, Brunekreef B (2013) A ventilation intervention study in classrooms to improve indoor air quality: the FRESH study. Environ Health 12:110. https://doi.org/10.1186/1476-069X-12-110
15. Twardella D et al (2012) Effect of classroom air quality on students' concentration: results of a cluster-randomized crossover experimental study. Indoor Air 2012(22):378–387. https://doi.org/10.1111/j.1600-0668.2012.00774.x
16. Wargocki P, Wyon DP (2013) Providing better thermal and air quality conditions in school classrooms would be cost effective. Build Environ 2013(59):581–589. https://doi.org/10.1016/j.buildenv.2012.10.007
17. Kapalo P, Domnita F, Bacotiu C, Podolak M (2018) The influence of occupants' body mass on carbon dioxide mass flow rate inside a university classroom – case study. Int J Environ Health Res 28(4):432–447. https://doi.org/10.1080/09603123.2018.1483010

Model of Accumulation of Rainwater from Surface Runoff in a Tank for Use in Building

Daniela Kaposztasova(✉) and Danica Košičanova

Department of Building Engineering, Institute of Architectural Engineering, Faculty of Civil Engineering, Technical University of Kosice, Kosice, Slovakia
daniela.kaposztasova@tuke.sk

Abstract. There are various rainwater accumulation systems suitable for any building. Rainwater harvesting systems can be integrated into the building's design, using rooftops and downspouts to collect rainwater. When applying the simplified method of design of the storage tank, only the required water intake for the selected storage period is considered. The average amount of precipitation that falls on the catchment area and is subsequently diverted and retained in the storage tank is not considered. The main goal of the article is to demonstrate the need for individual processing of rainwater balance using multi-year reductions of consumption in an apartment building and then to estimate the tank size. The use of a detailed method to design the volume of the storage tank reflects its model virtual filling and consumption of water. The results show that the simplified method is advantageous to use for smaller objects, where the captured rainwater can be used for several activities. For larger objects, it is more advantageous to use a detailed design method with its features.

Keywords: Rainwater · Water tank · Simplified method · Consumption · Apartment building

1 Introduction

Rainwater has physical, chemical, and microbiological characteristics that can vary depending on different factors. Raw rainwater is soft and free of pollution, but the collected rainwater already contains pollution present in the atmosphere and washed away from roofs and other surfaces [1]. Rainwater is slightly alkaline with a mean pH higher than 5.8, and chloride and bicarbonate are the most abundant ions. The chemical characteristics of seasonal rainwater samples indicate the dominance of Ca^{2+}-Mg^{2+}-HCO^{3-} and mixed Ca^{2+}-Mg^{2+}-Cl^{-} facies during different periods [2]. Rainwater could be used as an alternative source of water in dry seasons, but their presence of reversible microorganisms and pollution should be considered. It can contain various contaminants, including bacteria, viruses, parasites, dust, smoke particles, and other chemicals it can also contain per- and polyfluoroalkyl substances (PFAS), which are synthetic chemicals

Z. Blikharskyy et al. (Eds.): CEE 2023, LNCE 438, pp. 143–153, 2024.
https://doi.org/10.1007/978-3-031-44955-0_16

used in non-stick pans, fire-fighting foam, and water-repellent clothes [3–5]. Other harmful substances such as asbestos, lead, and copper can also be introduced into rainwater from roofing materials, gutters, piping, and storage materials. Dirt and germs can be washed into collected rainwater from the roof, especially when rain follows several days of dry weather [4]. Rainwater harvesting (RWH) has many challenges. These include ensuring water quality and safety, design considerations such as optimum tank size and efficiency of the system, water quality related issues and available technologies for disinfection, water-related legislations, inadequate financial support, and lack of national coordination.

RWH is an effective way to conserve water in urban areas. To make the most of this method, apartment dwellers should assess the potential for rainwater harvesting, collaborate with building management, design an efficient system, and ensure regular maintenance. By doing so, they can positively impact the environment [6].

RWH is a sustainable practice that can help address water scarcity issues in urban areas. Here are some facts about rainwater harvesting potential in apartment buildings:

- RWH potential of a particular catchment area is the amount of rainwater that can be effectively harvested from the said catchment area.
- RWH systems can be installed in apartment buildings to collect and store rainwater for various uses such as toilet flushing, laundry washing, and irrigation.
- a study attempted to find out the potential of rainwater harvesting in large premises with high water demand as well as in few residential buildings [7].
- RWH can be done from residential rooftops in Jordan, and an integrated approach is needed to investigate potential water savings and optimal tank sizing [8].
- renters and non-renters can also engage with, harvest, cycle, and enhance local resources to live sustainably [9].
- assessing the RWH potential in urban areas is important to address water scarcity issues for residential buildings and factories of the future [10].

1.1 Materials and Methods

All apartment buildings built in the last century and some newly built apartment buildings in Slovakia use high-quality drinking water for toilet flushing. A lot has been written and analyzed about this issue, but the results have been weak and have not had a significant impact on the solution of water distribution in apartment buildings. During the year we have dry periods with a lack of rainwater, but we also have periods when this water is plentiful and is not used at all. Instead, it is sent away without being used. Many apartment buildings could set up rain gardens (Fig. 1) in their surroundings to support sustainability. However, there are also housing estates with higher density and without suitable greenery where a rain garden cannot be placed.

The question of collecting rainwater in a tank may seem like a high investment, so we decided to compare costs and usage in apartment houses. Based on previous practice and inspections of apartment buildings in Slovakia, many apartment buildings are suitable for using rainwater.

Fig. 1. Example of newly built rain garden at TUKE

1.2 Suitability of Rainwater Use

The main goal of the article is to demonstrate the need for individual processing of rainwater balance using multi-year reductions of consumption in an apartment building and then to estimate the tank size. Selected parameters and properties of an apartment building should be monitored if we want to use the water from the surface drain to flush the toilet. It is necessary to consider:

- dimensions of the apartment building - apartment buildings with a large roof area are more suitable,
- number of inhabitants - apartment buildings with a smaller number of inhabitants are more suitable
- location of storm drains - apartment buildings with an internal storm drain are more suitable
- space in the basement - near the rain drain, the possibility of placing a container for collecting water
- electrical connection - the possibility of connecting a discharge pump
- sewage waste - the possibility of connecting the overflow to the sewer or to the terrain. In the case of discharge of the safety overflow into the public sewer, there is a problem with determining the amount of drained water with the subsequent billing of the sewage.

In our research, we compared two apartment buildings. Apartment Building No. 1 (AB 1) (Fig. 2) is a five-story building with two entrances. Four floors are currently inhabited, and the roof area measures 450 m^2. There are five apartments on each floor, consisting of two three-room apartments, two two-room apartments, and one one-room apartment. In total, there are 48 residents living in the building. The building has two internal storm drains.

Apartment Building No. 2 (AB 2) (Fig. 3) is a nine-story tower house with eight floors currently inhabited. The roof area measures 318 m^2. There are five apartments on each floor, consisting of four two-room apartments and one studio apartment. In total, there are 72 residents living in the building. The building has two internal storm drains.

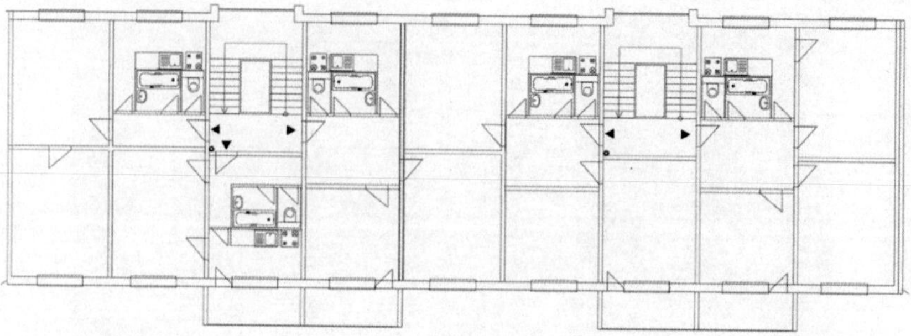

Fig. 2. Typical floor in AB1

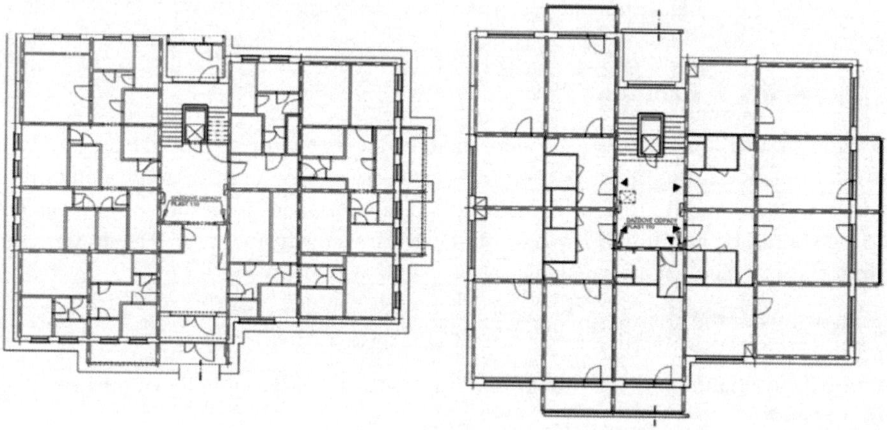

Fig. 3. Typical floor in AB 2 3a and ground floor

In the case of four-story buildings, we often encounter sloping roofs and external waste, which increases the cost of implementation. An underground tank is needed, preferably resistant to the pressure below the ground, i.e., made of waterproof and impermeable concrete. The use of a plastic tank is also suitable, especially regarding the weight of the device and assembly in one piece.

Common parameters of AB 1 and AB 2

Space in the basement:

- both buildings have the option of placing a tank with filtration in the basement/ground floor near the waste.

Electrical connection:

- both buildings have the option of connecting a discharge pump.

Sewage waste:

- both buildings have the option of connecting the overflow to the terrain.

The advantage of drainage to the terrain is clarity in billing for rainwater drainage without discharge to the public sewer. In the case of discharging the safety overflow into the public sewer, there is a problem with determining the amount of drained water with subsequent settlement of the sewage system.

1.3 Simplified Method Approach

A simplified method is properly used for sizing rainwater harvesting tanks using long day-resolution rainfall time series. This method uses daily rainfall and water demand data to calculate the necessary storage capacity of the tank. When designing these systems, the most important calculation is to correctly size the water tank to provide sufficient storage capacity. Six different calculation methods have been analyzed by Santos and Pinto-Taveira [11], including both simplified procedures and detailed approaches with varying criteria. An equation has also been developed to determine the efficiency of a rainwater tank, considering factors such as demand, roof area, tank size, and pluvial regime.

To summarize, the use of simplified methods for sizing rainwater storage tanks have several advantages. They are easier and quicker to use compared to detailed methods that require daily simulations over a long period. Additionally, they use long day-resolution rainfall time series, which is readily available and easy to obtain. However, there are also some disadvantages to using simplified methods. They may not take into account all the factors that affect the size of the tank, such as maximum rainwater use and 80% efficiency criteria. They may not be suitable for all types of catchment areas and may not provide the most optimal tank size for a specific residential housing unit. Furthermore, they may not be accurate enough for larger or more complex systems [11–14].

1.4 Balance of Consumption and Rainwater – Preliminary Calculations

The calculations represent an approximate recalculation of the amount of rainwater collected from the roof during the year, its use for flushing during the year, the approximate price level of implementing the collection tank and savings on water per year. Since we reuse water, sewage will be paid for this water. To obtain the sewage value, it is necessary to install an invoicing water meter at the outlet of the tank. All calculations are balance calculations since it is impossible to determine exactly how much water falls on the roof area or how many times a person flushes the toilet per day. Nevertheless, there are professional or scientific articles available that deal with these values, and therefore the balances will be based on this data.

- for one person, it is assumed that the toilet is flushed 6–8 times a day.
- the flushing tank is 6–8 L, older tanks up to 10 L.

Table 1 presents the values of precipitation totals in Eastern Slovakia and throughout Slovakia. The data from the SHMÚ website (13.1.2023) were used (updated once a month, usually at the end of the first decade of the following month) [15].

The used values are usable for the whole of Slovakia, i.e., the average value of precipitation. The next column in the table is the number of days in each month, followed by three columns with data for the first and second apartment buildings. In principle, except for four cases, there is a lack of rainwater in both cases. The smaller apartment building (AB 1) has an excess of water four times over the entire period of 3.5 years, which is a negligible amount. Therefore, connecting a safety overflow to the terrain next to the apartment building will not harm it. Rainfall can vary in duration and intensity in different locations. We can use an average value to help with this, but this means a strong dispersion for areas with high and low precipitation. This figure will be purely indicative and intended for balance on average for Slovakia. At the same time, we will know the indicative value of the installation.

Design of rainwater tanks

$$\text{Tank size (in m}^3\text{)} = \text{annual rainwater consumption (in m}^3\text{)} \times \text{supply time (in days)}/365 \text{ (days per year)}. \quad (1)$$

Result for AB 1:

supply for 2 weeks: tank size (14 days) = (1415/3.5) × 14/365 d = 15.30 = 15 m^3.
supply for 3 weeks: tank size (21 days) = (1415/3.5) × 21/365 d = 23.01 = 20 m^3.

Result for AB 2:

supply for 2 weeks: tank size (14 days) = (2123/3.5) × 14/365 d = 23.2 = 20 m^3.
supply for 3 weeks: tank size (21 days) = (2123/3.5) × 21/365 d = 34.8 = 30 m^3.

Based on these calculations, two reservoirs can be selected with half the volume, each for one waste, or one common tank according to spatial possibilities. This reservoir will not be refilled according to the balance. The reservoirs are filled by automatic leveling, which is connected directly for cold water supply.

2 Results and Discussion

2.1 Analysis of Water Prices for 2023 in Slovakia

In Table 2, a summary of water and sewage prices without VAT for 2023 is presented. Some water companies have not yet updated their prices, so the prices listed are for 2022. The information was obtained from individual websites and announcements made by water companies. A brief analysis has been conducted on the average price in Slovakia, considering the simple return without being affected by price increases.

2.2 Short analysis for AB 1 and AB 2

Result for AB 1:

- Storage tank 15 m^3 – 1,530 euros
- Automatic water filling + fittings - 800 euro
- Pump + fittings - 600 euro
- Pipes, construction modifications - 250 euro

Table 1. Balances of drinking water needs and rainfall for 42 months (according 15)

Precipitation data				Roof AB 1			Roof AB 2		
Month	R [mm]	R [mm]	Number	Roof	Need	+ Reserve	Roof	Need	+ Reserve
	vs	SR	Days	(m^3/month)		– Shortage	(m^3/month)		– Shortage
XII.2022	83	77	31	21	34	**−13**	15	51	**−37**
XI.2022	30	26	30	7	33	**−26**	5	50	**−45**
X.2022	29	26	31	7	34	**−27**	5	51	**−46**
IX.2022	132	118	30	32	33	**−1**	23	50	**−27**
VIII.2022	43	62	31	17	34	**−17**	12	51	**−40**
VII.2022	85	73	31	20	34	−15	14	51	−37
VI.2022	32	48	30	13	33	−20	9	50	−41
V.2022	19	34	31	9	34	−25	6	51	−45
IV.2022	50	48	30	13	33	−20	9	50	−41
III.2022	38	27	31	7	34	−27	5	51	−46
II.2022	31	44	29	12	32	−20	8	48	−40
I.2022	35	36	31	10	34	−25	7	51	−44
XII.2021	51	54	31	15	34	−20	10	51	−41
XI.2021	55	53	30	14	33	−19	10	50	−40
X.2021	7	13	31	4	34	−31	2	51	−49
IX.2021	57	54	30	15	33	−19	10	50	−39
VIII.2021	152	140	31	38	34	4	27	51	−25
VII.2021	107	101	31	27	34	−7	19	51	−32
VI.2021	40	33	30	9	33	−24	6	50	−43
V.2021	104	121	31	33	34	−2	23	51	−28
IV.2021	67	55	30	15	33	−18	10	50	−39
III.2021	22	19	31	5	34	−29	4	51	−48
II.2021	66	53	29	14	32	−18	10	48	−38
I.2021	76	65	31	18	34	−17	12	51	−39
XII.2020	60	60	31	16	34	−18	11	51	−40
XI.2020	23	23	30	6	33	−27	4	50	−45
X.2020	134	161	31	43	34	9	31	51	−21
IX.2020	76	89	30	24	33	−9	17	50	−33
VIII.2020	113	99	31	27	34	−7	19	51	−32

(continued)

Table 1. (*continued*)

Precipitation data				Roof AB 1			Roof AB 2		
VII.2020	120	94	31	25	34	−9	18	51	−33
VI.2020	160	140	30	38	33	5	27	50	−23
V.2020	78	68	31	18	34	−16	13	51	−38
IV.2020	15	12	30	3	33	−30	2	50	−47
III.2020	46	48	31	13	34	−21	9	51	−42
II.2020	55	72	29	19	32	−13	14	48	−34
I.2020	21	20	31	5	34	−29	4	51	−48
XII.2019	72	69	31	19	34	−16	13	51	−38
XI.2019	143	135	30	36	33	3	26	50	−24
X.2019	42	36	31	10	34	−25	7	51	−44
IX.2019	61	68	30	18	33	−15	13	50	−37
VIII.2019	106	91	31	25	34	−10	17	51	−34
VII.2019	80	71	31	19	34	−15	14	51	−38
				739	1415	**−676.608**	522	2123	**−1601**

Table 2. Overview of water and sewage prices without VAT for the year and 2023 in Slovakia.

Water company	Water price	Sewage price
Bratislava Water Company	1.2025	1.125
Liptovská water company	1.043	1.579
Orava Water Company	0.986	1.6316
Podtatranska Water Company (2022)	1.096	1.233
Považská Water Company (2022)	1.829	1.4745
North Slovak waterworks and sewers	1.0094	1.0947
Central Slovakian Water Operating Company	1.3351	1.3143
Ružomberok Water Company (2022)	0.8024	1.0415
Trenčín waterworks and sewers	1.1369	1.2875
Trnava water company	0.8409	1.0752
Turčianska water company	0.7978	0.9907
Water company Ružomberok	0.8024	1.0415
East Slovak Water Company	1.6314	1.3626
West Slovak Water Company	1.27	1.4783
Average without VAT	1.1274	1.2664

- Total price = approx. 3,180 euros.

Price for water from the roof - the amount of rainwater that falls will drain into the sewer.

The water value for the given volume is saved. $1.124 \times 210 \times 1.2 = 283$ euros.

Assuming stable prices from 2023, the simple payback is between 11 and 12 years.

Result for AB 2:

- Storage tank 2 pcs 10 m^3 – 1,450 euro
- Automatic water filling + fittings - 1,600 euros
- Pump + fittings - 1000 euro
- Pipes, construction modifications - 450 euro
- Total price = approx. 4,500 euros.

Price for water from the roof - the amount of rainwater that falls will drain into the sewer. The water value for the given volume is saved. $1.124 \times 150 \times 1.2 = 202$ euros.

Assuming stable prices from 2023, the simple payback is from 22 to 23 years.

2.3 Advantages and Disadvantages

Reusing rainwater has both advantages and disadvantages. On the positive side, it feels good to flush with rainwater instead of drinking water, and it can result in faster payback if water prices rise. It is also an environmentally friendly solution that helps protect water sources. However, there are also some disadvantages to consider. For example, the smaller and taller an apartment building is, the more advantageous it is to implement rainwater reuse. Additionally, the calculations for implementing such a system must take into account the need for water for flushing toilets and the installation of large reservoirs that may not be filled during the year. These reservoirs must be automatically replenished from the potable water distribution system. The disadvantage of the monthly balance is the strong simplification of the calculation, the need for water for flushing does not include the number of people who go to work and school and the number permanently staying in the apartment.

Depending on these inputs, it is possible to assume that the resulting collection tank would be at the level of 50–60% of the calculated value.

3 Conclusion

Rainwater quality in apartment buildings must be ensured through several methods. Builders must provide an in-built rainwater harvesting system that collects water from rooftops and surface run-off and directs it into underground tanks. These systems must be properly planned, designed, and maintained to ensure that the collected water is free from contaminants. Rainwater can be used for non-drinking purposes during times of scarcity, reducing the need for potable water. The collected rainwater can also be used for direct usage or to replenish natural aquifers. Correctly sizing the rainwater tank is important for ensuring the efficiency and effectiveness of the system. Flushing toilets with rainwater in apartment buildings can be a good idea for several reasons. It can help conserve water by reducing the demand for potable water and can also reduce water bills. Rainwater can be collected and stored in underground tanks or cisterns and used for non-drinking purposes such as flushing toilets. A first-flush system can be installed to prevent debris and contaminants from washing off the roof and directly into the rainwater tanks. Proper planning, design, and maintenance of rainwater harvesting systems are important to ensure that the collected water is free from contaminants.

The simulation of the filling and use of an apartment building would certainly refine the values in favor of the use of rainwater, because the reservoirs would certainly be smaller and the return faster. To determine the size of the storage tank and the balance of rainwater, they must therefore be processed individually with the help of multi-year consumption deductions in the apartment building.

Acknowledgements. VEGA 1/0492/23 Transformation of existing buildings into sustainable buildings – the ecological potential of flat roofs.

References

1. Al-Batsh N, Al-Khatib IA, Ghannam S, Anayah FM, Jodeh S, Hanbali G, Khalaf B, van der Valk M (2019) Assessment of rainwater harvesting systems in poor rural communities: a case study from Yatta Area, Palestine. Water
2. Valappil NKM, Viswanathan PM, Hamza V (2020) Chemical characteristics of rainwater in the tropical rainforest region in northwestern Borneo. Environ Sci Pollut Res 27:36994–37010. https://doi.org/10.1007/s11356-020-09542-1
3. Cousins IT, Johansson JH, Salter ME, Sha B, Scheringer M (2022) Outside the safe operating space of a new planetary boundary for per- and polyfluoroalkyl substances (PFAS). Environ Sci Technol 56(16):11172–11179. https://doi.org/10.1021/acs.est.2c02765
4. CDC. Centers for Disease Control and Prevention. https://www.cdc.gov/healthywater/drinking/private/rainwater-collection.html. Last accessed on 15 May 2023
5. Euronews.green: Rainwater everywhere on Earth unsafe to drink due to 'forever chemicals', study finds https://www.euronews.com/green/2022/08/04/rainwater-everywhere-on-earth-unsafe-to-drink-due-to-forever-chemicals-study-finds. Last accessed on 15 May 2023
6. RMS (2019) Rainwater harvesting at an apartment complex in Blacksburg, Virginia. https://rainwatermanagement.com/blogs/news/rainwater-harvesting-at-apartment-complex-in-blacksburg-virginia. Last accessed on 12 May 2023
7. Dakua M, Akhter F, Biswas PP, Siddique MdLR, Shihab RM (2014) Potential of rainwater harvesting in buildings to reduce over extraction of goundwater in urban areas of Bangladesh. European Scientific Institute, ESI. https://doi.org/10.19044/esj.2013.v9n10p
8. Al-Qawasmi O (2021) Feasibility of rainwater harvesting from residential rooftops in Jordan. Appl Water Sci 11(2):1–6. https://doi.org/10.1007/s13201-021-01365-w
9. Lancaster B (2019) Rainwater harvesting for drylands and beyond, vol 2, 2nd edn
10. Maqsoom A, Aslam B, Ismail S, Thaheem MJ, Ullah F, Zahoor H, Musarat MA, Vatin NI (2021) Assessing rainwater harvesting potential in urban areas: a building information modelling (BIM) approach. Sustainability
11. Santos C, Taveira-Pinto F (2013) Analysis of different criteria to size rainwater storage tanks using detailed methods. Resourc Conserv Recycl 71:1–6. ISSN 0921-3449. https://doi.org/10.1016/j.resconrec.2012.11.004

12. Galarza-Molina S, Torres A (2017) Método De Dimensionamiento Para Tanques De captación De Agua Lluvia Utilizando Series De resolución Diaria De Lluvia Y De Demanda De Agua». Luna Azul, n.º 45, julio de 2017, pp 107–22. https://doi.org/10.17151/luaz.2017.45.7
13. Fezai S, Belhadj J (2017) Sizing optimization of a stand-alone photovoltaic system using genetic algorithm. In: 2017 18th international conference on sciences and techniques of automatic control and computer engineering (STA), pp 499–504
14. Sameer S, Alawna S (2021) Optimal sizing of rooftop rainwater harvesting tanks for sustainable domestic water use in the West Bank, Palestine. Water 13(4):573. https://doi.org/10.3390/w13040573
15. SHMU (2023). https://www.shmu.sk/sk//?page=838. Last accessed on 16 May 2023

Study of the Strength of Concretes Based on M500 Cements with the Addition of Berament A2 at an Early Age

Ihor Karkhut[1] (✉) and Roman Havryshkiv[2]

[1] L'viv Polytechnic National University, Bandery St, 12, L'viv 70013, Ukraine
karkhoot1@gmail.com

[2] Institute of Construction and Engineering Systems, L'viv Polytechnic National University, L'viv, Ukraine

Abstract. The article presents the results of experimental studies of strength development at the age of up to 28 days with concretes based on M500 cements produced by three Ukrainian plants (PJSC VOLYN-CEMENT, PODILSKIY CEMENT JSC, Ivano-Frankivsk Cement), modified with the Slovak-made BERAMENT A2 additive. The data on the materials used are given. The strength of concrete was determined by testing cubes with an edge of 100 mm for uniaxial compression at the age of 1 day, 7 and 28 days. Tables and graphs of concrete strength gain with different doses of additives (0.6; 1.2; 1.8; 2.4%) by weight of cement are given. A comparison was made of the rates of curing concretes depending on the type of cement and the amount of additive. The upper and lower limits, as well as the optimal dose of the additive for each of the three studied cements, were established from the conditions for ensuring the required stiffness of the mixture. Recommendations have been developed for the use of this additive for prefabricated monolithic and prefabricated reinforced concrete structures on M500 cements from these manufacturers. The main criterion for the use in the production of prestressed prefabricated reinforced concrete structures is the achievement of at least 50% of the strength of the concrete class at the age of 1 day.

Keywords: Shrinkage of the cone · Steaming · Tempering strength · Additions

1 Introduction

For the operation of concrete and reinforced concrete structures, the indicators of concrete density and its uniformity are important, which should be maximum, especially for protective structures. This affects both the protective properties of the structure during commissioning, and their change throughout the entire service life, as well as their behavior in design and emergency cases. The change in protective properties over time is associated with the removal of free water from concrete and a decrease in its density and specific gravity. Under emergency conditions, including temperature ones [1], in pores with a minimum amount of water, the pore pressure and the risk of sudden destruction

Z. Blikharskyy et al. (Eds.): CEE 2023, LNCE 438, pp. 154–163, 2024.
https://doi.org/10.1007/978-3-031-44955-0_17

of the surface layers of concrete are significantly reduced. Many works are devoted to the study of the use of additives to concretes based on Portland cements, which make up 95% of all protective structures [2–8]. One of the main characteristics of plasticizers is the degree of reduction in water consumption and the nature of interaction with cement. It is noted that the use of additives requires high-quality technological support, which is impossible without knowledge of the characteristics of the cements used, their quality, and therefore requires preliminary studies of the interaction of additives with each of the cements. Studies have shown that the relative fluidity index can depend on factors such as the chemical composition, mineralogical characteristics and specific surface area of the cement. It was pointed out [4] that it is important to carry out preliminary tests, especially if a cement other than type I (no additives) was used for the manufacture of cast concrete. The conducted studies did not reveal a clear relationship between the initial value of the cone draft and the rate of its loss. The rate of loss of mobility depended on the type of cement; this could be influenced by many factors, such as the content of C3A, alkali, gypsum, and C4AF.

The authors [5, 6] concluded that high initial strength can be achieved by using an increased amount of cement, however, due to more intense heat release during its hydration, cracks and excessive shrinkage, and, subsequently, excessive creep can occur in concrete. The research results showed that the introduction of plasticizers and a decrease in water consumption reduced the diffusion permeability by 4–6 times, transferring concrete into the category of denser ones. It has been established that in order to obtain high-quality concrete with additives, it is necessary to solve two main tasks:

- the choice of the right additives or their combinations, which reduce the stickiness of concrete when smoothing surfaces in cases where large doses of plasticizers are used;
- preparation of the optimal concrete formulation, taking into account the composition of the cement and the required quantities of fine components (cement, additives in concrete and fine parts of sand - below 0.125 mm).

In Ukraine, the cement industry is well developed, has high-quality raw materials and production traditions. But they differ significantly in chemical composition from European cements. Since the amount of additives required for optimal use depends on these characteristics of cements, the recommendations [9] developed by scientists and European manufacturers of plasticizers based on experiments with European cements are approximate for Ukrainian cements and require clarification.

2 Experimental Studies

In NU "Lviv Polytechnic" studies were carried out, the purpose of which was to study the interaction of M500 cements with the complex additive BERAMENT A2 (Slovakia). Studies of concrete mixtures and standard samples of concrete cubes with an edge of 100 mm were carried out in order to establish indicators of strength development (design class), technological properties of concrete with a complex (superplasticizer with antifreeze properties) plasticizing additive BERAMENT A2.

During the research, the following research tasks were set:

1. To carry out the selection of the optimal composition of concrete using high-quality local inert components (sand and two fractions of crushed stone).
2. Experimentally investigate the effect of the amount of additive on the degree of reduction in W/C when using various domestic cements M500 and establish the maximum possible reduction in W/C while maintaining the convenience of laying the concrete mix.
3. Determine the optimal doses of the BERAMENT A2 additive with various domestic cements M500 under steaming and under normal conditions.
4. Determine the M500 cements that most effectively interact with the BERAMENT A2 additive and develop recommendations for the use of the BERAMENT A2 additive with domestic M500 cements.

Portland cements from the following enterprises were used as a binder for the preparation of the studied concrete mixtures: JSC Ivano-Frankivsk Cement (designation IFC); PJSC VOLYN-CEMENT (designation VC); PODILSKIY CEMENT JSC (designation KPC) with up to 20% additives. For the preparation of the studied concrete mixtures, quartz sand from the Davydivsky quarry (Lvov region) and crushed granite from the Klesivsky quarry (Rivne region) were used. The selection of the optimal composition included the determination of the nominal composition, the calculation and correction of the working composition, the calculation of the working dosages of the concrete mixture, taking into account the volumes of batches for the manufacture of concrete samples. The normalized indicators of concrete mix and concrete were established as follows:

1) standardized class of concrete for compressive strength C25/30;
2) concrete mix components:

   - binders - Portland cements M500 produced by cement plants in Ukraine;
   - large aggregate - granite crushed stone of two fractions 2.5–10 mm and 10–20 mm;
   - fine aggregate – quartz sand;
   - chemical additive - complex superplasticizer BERAMENT A2 with a variable dose, (0.6; 1.2, 1.8 and 2.4% by weight of cement);

3) very mobile normalized consistency of the concrete mix with shrinkage of a standard cone CS = 13.5–14.5 cm;

When optimizing the composition of concrete, the following consumption of components was adopted based on 1 m^3 of the mixture: cement – 470 kg, sand – 550 kg, crushed stone of a fraction of 2.5–10 mm – 550 kg, crushed stone of a fraction of 10–20 mm – 750 kg. The water consumption was taken in the amount at which the accepted value of the standard cone of the concrete mixture was reached.

Portland cements were used as a binder for the preparation of the studied concrete mixtures, the indicators of which are given in Table 1. In the process of preparing concrete mixtures, the main rheological properties were studied: uniformity, water-holding capacity, stratification and ease of placement and the effect of the amount of consumption of a chemical additive on the above properties was also studied. Table 2 shows the indicators of the draft of the cone for different doses of the additive. The values presented in the tables can be compared as they are obtained with approximately the same cone shrinkage values. When preparing a concrete mixture on all cements with an additive consumption of 2.4% by weight of cement, a homogeneous and non-separating concrete mixture was obtained when the shrinkage of the cone was up to 10 cm.

Table 1. Indicators of cement M500 for the manufacture of concrete cubes.

Manufacturer's designation	Type of cement	% (by mass)					
		Main components					Additional components
		Clinker	Blast furnace granulated slag	Pozzolana	Fly ash	Limestone	
VC-500	CEMII/A-C-500	80–94	Total 6–20				
IFC-500	CEMI-500R-N	–	–	–	–	–	0–5
KPC-500	CEMII/A-C-500	80–94	Total 6–20	–	–	–	–

Table 2. Strength of M500 cements during steaming.

Name of indicator	Designation of cement		
	VC-500	IFC-500	KPC-500
Guaranteed class	500	500	500
Additives, %	Flask, granulated slag up to 20	–	Granulated slag 9.7; flask 8.5
Steaming activity, N/mm^2	29.9	–	–
Steaming efficiency group	II	II	II
Activity at age (days), N/mm2	20.7(2)	–	29.0 (3)
Normal density of cement paste, %	26.2	–	26.25
Presence of signs of false setting	No	No	No
Activity at the age of 28 days, N/mm^2	50.5	47.4	51.5

Table 3 shows the values of W/C and shrinkage of the cone at different doses of the additive. With an increase in water consumption in order to obtain the shrinkage of the cone CS = 13… 15 cm, the following phenomena were observed:

- when mixing the mixture in a concrete mixer, it had the form of a homogeneous and non-separable mixture;
- when the mixer was stopped, the concrete mixture in it easily spread (quickly self-leveling) and at the same time the process of air capture was observed;
- when compacting the concrete mixture by bayonet method in a standard cone, as well as in a metal mold (CS = 13…15 cm), it behaved as rigid;

- in all concrete mixtures, the phenomenon of false setting and water separation (mixtures of water and additives) was observed in the form of a yellow-brown film on the surface of concrete mixtures;
- after vibrocompaction of concrete mixtures in a metal mold, water separation occurred on the surface (mixtures of water with an additive);
- after heat treatment in a steaming chamber, a peeling film 1.0–1.5 mm thick formed on the upper surfaces of all concrete samples.

Thus, it can be concluded that with an additive consumption of 2.4% by weight of cement, it is impossible to obtain a homogeneous non-separable concrete mixture from CS = 13… 15 cm. Such a mixture is unsuitable for use.

Table 3. Strength of M500 cements during steaming.

Type of cement	Storage conditions	Index	Additive dose, % cement mass/(l/m^3)				
			0/0	0.6/2.35	1.2/4.7	1.8/7.05	2.4/9.4
VC	Normal conditions	W/C	0.49	0.46	0.37	0.38	0.37
		CS, cm	15.0	13.6	14.4	13.8	13.8
		Water, l	230	215	175	180	175
	Heat treatment	W/C	0.49	0.46	0.38	0.36	0.37
		CS, cm	15.0	13.6	14.6	14.0	13.8
		Water, l	230	215	180	170	175
KPC	Normal conditions	W/C	0.51	0.47	0.44	0.40	0.36
		CS, cm	14.4	14.2	13.8	14.5	13.8
		Water, l	240	220	205	190	170
	Heat treatment	W/C	0.52	0.47	0.44	0.39	0.35
		CS, cm	14.7	14.4	13.6	14.4	13.6
		Water, l	245	220	205	185	165
IFC	Normal conditions	W/C	0.51	0.47	0.40	0.37	0.35
		CS, cm	14.0	13.8	14.4	14.0	14.0
		Water, l	240	220	190	175	165
	Heat treatment	W/C	0.51	0.47	0.40	0.37	0.35
		CS, cm	14.1	13.6	14.1	14.0	14.2
		Water, l	240	220	190	175	165

Comparison of M500 cements is shown in the graphs (Figs. 1, 2, 3, 4 and 5). As can be seen from Fig. 1, without the use of heat treatment and in the absence of additives on the first day, it is impossible to carry out the operations of stripping products from ordinary reinforced concrete.

Only when using KPC and IFC cements and an additive consumption of at least 1.8% (Figs. 1 and 2), products from such concretes can be stripped in a day. When using heat treatment, the strength of the samples on all types of M500 cements, even without the use of additives, exceeds 70% of the class. When using an additive, this indicator is in the same range of 70–85%, which indicates the inefficiency of the additive in the heat treatment of products on M500 cements. During heat treatment at a water-cement ratio of 0.35–0.44, KPC-500 cement exceeded the indicators obtained from the logarithmic curve both under normal hardening conditions and during heat treatment. The same can be said about IFC-500 cement. Without the use of heat treatment, the strength value of the samples on this cement is much lower than the values of the logarithmic curve.

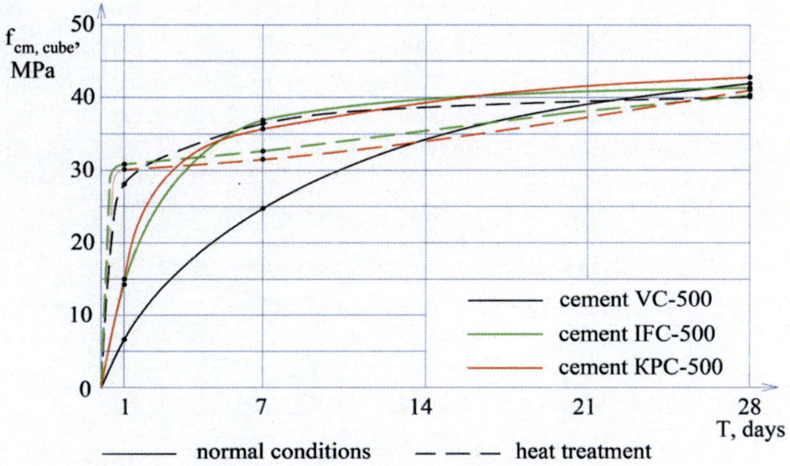

Fig. 1. Curing of concrete without BERAMENT A2 additive.

According to this indicator, M500 cements in order of increasing quality can be placed in this order – IFC-KPC-VC, although in general, despite the linear nature of the strength gain at the age of 7–28 days, concrete on VC-500 cement showed the highest strength indicators in at the age of 28 days compared with IFC-500 and KPC-500 cements (by 4–9%, Fig. 2). Heat treatment did not reveal the benefits of any of the three cements studied. Their control values differed by 3–4% throughout the study period (Figs. 2, 3, 4 and 5). The results of studies of VC-500 cement are presented in Table 4. For the first day, the cubic strength was 6.63–18.97 MPa (normal conditions) and 28.02–44.12 MPa (heat treatment). Under normal storage conditions, the strength tripled, which indicates a significant effect of the additive. For seven days of age, a decrease in the spread of strength values by one and a half times for samples stored under normal conditions (24.69–43.78 MPa) is typical.

At the age of 28 days, a very slight increase in the values (no more than 8.0 MPa) of the strength of the samples after steaming was obtained. For normal storage conditions, this indicator was in the range of 7–30 MPa.The greatest growth was observed at an additive consumption of 1.2%. With an increase in the amount of the additive, the maximum rate of strength development shifted towards the first day. Tests of samples on IFC-500 (Table 5) cement at the age of 1 day showed that the strength reached 14.36–38.06 MPa under normal storage conditions and 30.50–50.36 MPa during heat treatment.

Table 4. Results of studies of concrete on cement VC-500.

Age, days	Storage conditions	Dose of BERAMENT A2 (% by weight of cement)				
		0	0.6	1.2	1.8	2.4
1	n/c	6.63	8.35	10.99	11.47	18.97
	h/t	28.02	36.92	45.45	51.57	44.12
7	n/c	24.69	23.05	27.69	29.9	43.78
	h/t	36.38	44.73	51.93	56.95	54.71
28	n/c	41.91	41.85	57.54	48.14	50.97
	h/t	40.06	47.2	59.84	61.57	53.31

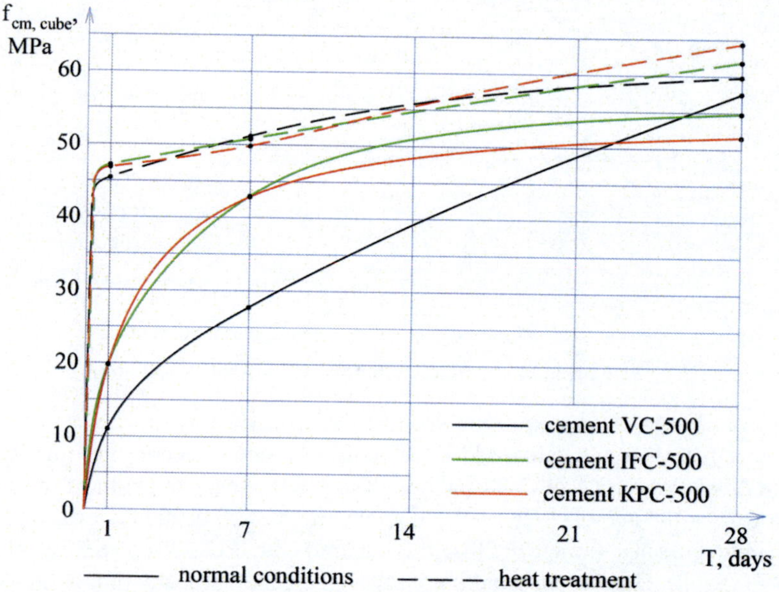

Fig. 2. Curing of concrete with 0.6% additive BERAMENT A2.

At the age of 7 days, the strength index reached values of 36.87–66.15 MPa and 32.62–55.93 MPa, respectively, for samples stored under normal conditions and for samples subjected to heat treatment. At the age of 28 days, the strength of concrete cubes was 41.34–71.92 MPa (normal conditions) and 40.31–69.0 MPa (heat treatment). The results of studies of KPC-500 cement are presented in Table 6.

On the first day after concreting, a very rapid increase in strength was observed both for samples subjected to heat treatment and for samples stored under normal conditions and amounted to 30.08–50.36 MPa and 15.03–39.04 MPa, respectively.

For the second period of time from the first to the seventh day, the scatter decreased, especially for the samples stored under normal conditions – 20.31–24.16 MPa. For

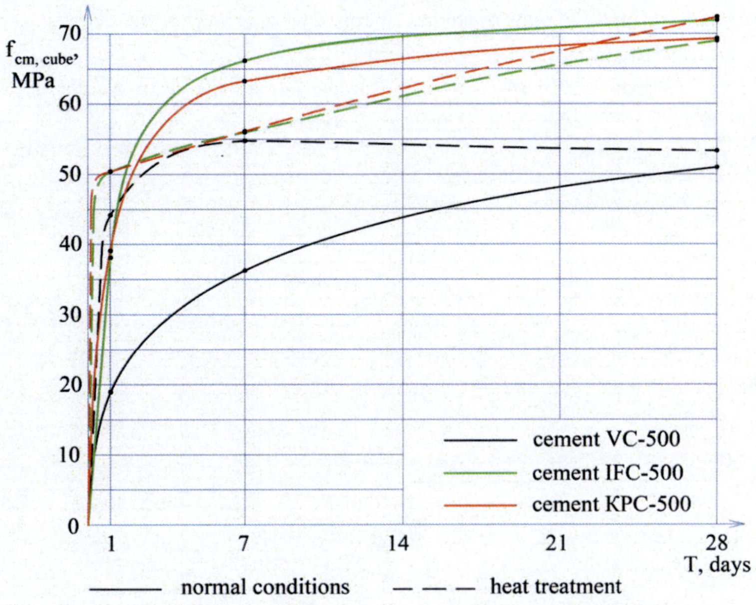

Fig. 3. Curing of concrete with 1.2% additive BERAMENT A2.

Table 5. Results of studies of concrete on cement FC-500.

Age, days	Storage conditions	Dose of BERAMENT A2 (% by weight of cement)				
		0	0.6	1.2	1.8	2.4
1	n/c	14.36	16.17	19.85	33.02	38.06
	h/t	30.81	30.5	47.21	46.73	50.36
7	n/c	36.87	37.15	42.93	46.32	66.15
	h/t	32.62	36.7	50.81	53.5	55.93
28	n/c	41.34	41.51	54.82	58.9	71.92
	h/t	40.31	46.48	61.86	68.03	69.0

samples that underwent heat treatment, it was 1.42–5.7 MPa, and the absolute values of the strength of steamed samples were lower than those stored under normal conditions by 5–8 MPa. Samples on this cement gained strength rather under normal conditions. Such results were also recorded for IFC-500 cement.

3 Conclusions

The studied M-500 cements most effectively interact with the BERAMENT A2 additive in the range of 0.6–1.2% (Tables 4, 5 and 6). As a result of the research, the following differences were revealed. A linear dependence in the entire studied range of 0–2.4% is

Table 6. Results of studies of concrete on cement KPC-500.

Age, days	Storage conditions	Dose of BERAMENT A2 (% by weight of cement)				
		0	0.6	1.2	1.8	2.4
1	n/c	15.03	16.42	20.25	33.1	39.04
	h/t	30.08	30.68	46.87	47.0	50.36
7	n/c	35.68	36.7	42.81	46.89	63.2
	h/t	31.5	34.92	49.85	54.23	56.07
28	n/c	42.76	46.9	51.58	68.3	69.32
	h/t	40.98	50.36	64.42	72.27	72.38

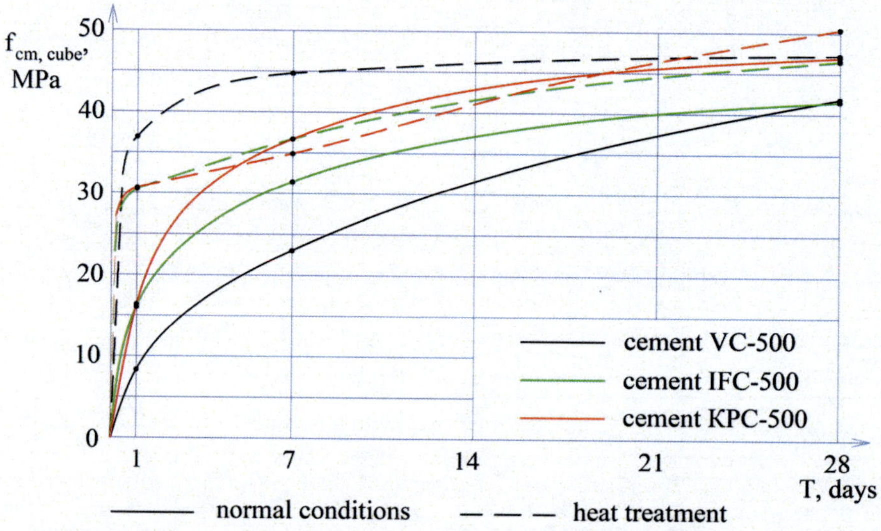

Fig. 4. Curing of concrete with 1.8% additive BERAMENT A2.

typical for cement KPC-500. VC-500 and IFC-500 cements sharply reduce W/C with an increase in additive consumption from 0.6 to 1.2% (from 0.46–0.47 to 0.38–0.4). With a further increase in the consumption of the additive, W/C remains almost constant. The minimum value of W/C (which is very important when taking into account strength indicators) can be achieved using both KPC-500 cement and IFC-500 cement. The M500 cements were compared with respect to the rate of strength development (Figs. 1, 2, 3 and 4) and the strength level at which the strength of samples on VC cement was taken as 100%. Strength fluctuations on the 28th day without the use of the additive did not exceed 2% and 12% when using the additive in the amount of 0.6–1.2%. At high additive contents, the spread was more than 40%. Heat treatment did not significantly change the described trend. Only at high concentrations of the additive, the spread of strength somewhat decreased and did not exceed 35%. In general, while ensuring the convenience

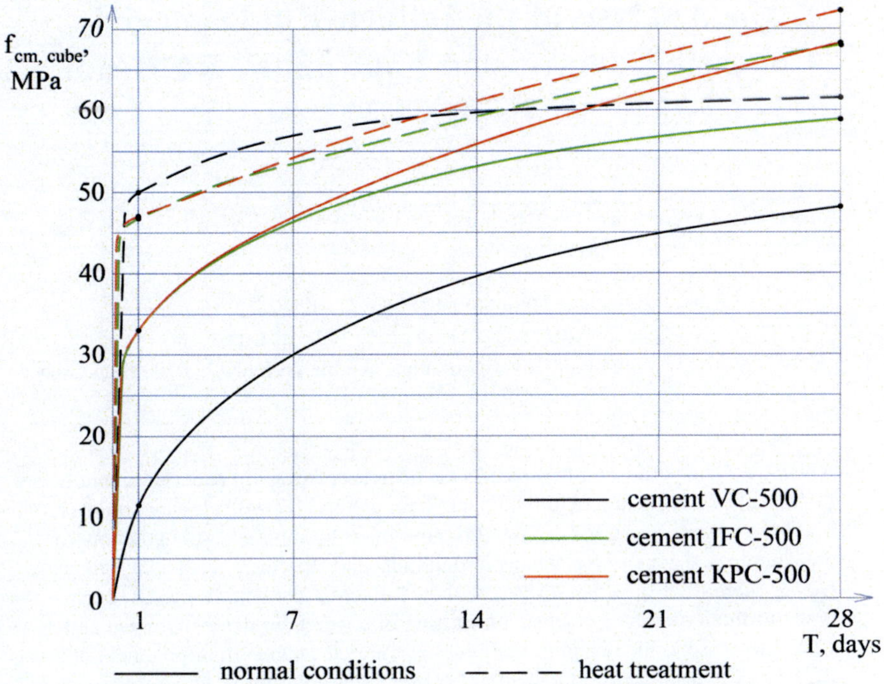

Fig. 5. Curing of concrete with 2.4% additive BERAMENT A2.

of laying concrete mixtures for all cements using the BERAMENT A2 additive, the W/C was in the range of 0.3–0.34.

References

1. Karkhut I (2021) Determining the stressed-strained state of concrete in the zone of exposure to local laser radiation. Eastern-Eur J Enterp Technol 111(3/7):24–36. https://doi.org/10.15587/1729-4061.2021.232671
2. Chandra S, Bjornstorm J (2002) Influence of cement and superplasticizer type and dosage on the fluidity of Portland cement mortars. Part 1. Cement Concr Res 32:1605–1611
3. Chandra S, Bjornstorm J (2002) Influence of cement and superplasticizer type and dosage on the slump loss of Portland cement mortars. Part 2. Cement Concr Res 32:1613–1619
4. Grabiec A (2005) Some physicomechanical characteristics of self-compacting concretes in dependence on kind of cement and curing conditions. In: Fracture mechanics and physics of construction materials and structures, issue 6. Lviv, pp 311–320
5. Grabiec A, Piasta Z (2000) Some studies on compatibility of cement-superplasticizers with use of multycriteria statistical optimisation. In: Proceedings of the sixth international conference on superplasticizers and other chemical admixtures in concrete, Nice, pp 379–393
6. Grzeszczyk S (2002) Beton samozageszczalny – proektowanie, wlasciwosci, kierunki rozwoju. Inzynieria I Budownictwo 9:465–468
7. Kaszynska M (2004) Beton samozageszczalny – rozwoj technologii I wyniki badan. Dni Betonu. Conferencija, Wisla, s. 96–110
8. Okamura H, Ouchi M (2003/04) Self-compacting concrete. J Adv Concr Technol 1(1):5–15
9. Specification and guidelines for self-compacting concrete. EFNARC, 2002/02

The Analysis of the Influence of Damaged Concrete Compression Zone on the RC Beam Using FEM

Andriy Klym[1], Yaroslav Blikharskyy[1](✉), Oleksandr Panchenko[2], and Yuriy Sobko[1]

[1] Lviv Polytechnic National University, Lviv 79013, Ukraine
yaroslav.z.blikharskyy@lpnu.ua

[2] Kyiv National University of Construction and Architecture, Kyiv 03037, Ukraine

Abstract. The purpose of the study is the analysis of stress in concrete and reinforcement of a RC beam by the finite element method (FEM) software complex. The current state and trends in the direction of restoration and strengthening of the compressed zone of RC beams in buildings and structures were analyzed, as well as scientific works that investigated the influence and behavior of RC beams in the formation of deep cracks and splitting of concrete in the compressed zone with a decrease in the height of compressed concrete beams. Reinforcement of a RC beam can be done by a large number of methods and approaches, but for each beam it is necessary to apply the most appropriate and practical one, depending on the reinforcement. For this, it is necessary to analyze the physical nature of a RC beam in a large range when its compressed zone is destroyed. After modeling in Femap and receiving results from NX Nastran solvers, it was visually demonstrated how the stress in concrete and reinforcement increases sharply approaching the limit value when the compressed zone of concrete is reduced by 50 mm. This research is an urgent problem and a significant direction of scientific development, which has significant advantages in the construction industry and requires constant additional research and scientific development for improvement.

Keywords: Damage · Compressed zone of concrete · RC beam · Restoration · High-strength solution · Femap

1 Introduction

The topic of research on the load-bearing capacity of RC beams with damage to the compressed zone of concrete is very relevant for the construction industry [1–4]. This issue concerns the safety of construction structures and is of great importance to engineers and designers. Damage to the compression zone of concrete can occur for a variety of reasons, such as chemical exposure, physical damage, or incorrectly designed parts [5–7]. This can lead to a decrease in the load-bearing capacity of the beam and an increase in the risk of the structure collapsing [8, 9].

Z. Blikharskyy et al. (Eds.): CEE 2023, LNCE 438, pp. 164–177, 2024.
https://doi.org/10.1007/978-3-031-44955-0_18

Research in this area is aimed at understanding the mechanisms of damage and determining its impact on the bearing capacity of a RC beam. This allows engineers to develop effective methods of diagnosing damage and calculating the load-bearing capacity of beams taking into account the compressed zone of concrete. Damage to the compression zone of concrete can lead to a reduction in the strength and stiffness of the structure, which can pose serious threats to human safety and the environment. However, replacing or reconstructing damaged structures is quite a complex and expensive process that can take a lot of time and cost a lot of money [10–12]. Therefore, studying the problem of restoring and strengthening the compressed area of concrete of RC structures is very important, as it can allow to effectively and economically restore damaged structures and preserve their safety and strength. But for this, it is necessary to improve existing methods of analysis and introduce various innovative options for obtaining results and analyzing RC beams.

The results of such studies can be used to establish safety standards and develop recommendations for the design and repair of various RC structures and beams. They also make it possible to improve existing methods of analysis and design, ensuring optimal load-bearing capacity of beams and increasing the overall safety of construction structures.

In summary, it can be concluded that studies aimed at analyzing the load-bearing capacity of RC beams with damage to the compressed zone of concrete are of great practical importance for improving the safety and efficiency of building structures, therefore, in this study, we will consider the analysis of a RC beam with a damaged compressed zone using the Femap software complex based on NX Nastran solvers.

2 Literature Review

The analysis of a RC beam that has damage in the compressed zone of concrete is a current topic in the field of structural engineering, so there is a lot of research in this direction. When the compressed zone of concrete is damaged and the height of compression is reduced, the bearing capacity of the beam drops significantly, and the deflections and stresses in the reinforcement increase to the yield point instantly.

Having considered articles [13–15] in which the influence of cracks in the compressed zone of concrete on the load-bearing capacity of RC beams and their influence on the mechanical characteristics of structures are investigated. These articles are substantiated by numerical and experimental research methods. The main focus is on the analysis of the effect of cracks on changes in the strength of beams, in particular, various factors such as the depth and length of cracks, reinforcement and material properties are taken into account. The results of the research on the formation of cracks and the destruction of the compressed zone of concrete contribute to a better understanding of this effect on the performance of RC beams and their load-bearing capacity.

The use of numerical analysis of the effects of damage in the compression zone of concrete in RC beams provides important information for understanding the behavior of RC beams with damaged concrete in the compression zone [16]. Numerical methods are used to analyze the impact of concrete damage on the strength of beams and study the mechanical characteristics of such structures.

Sometimes there is a problem of reinforcing RC beams on the section, so in this case, a strengthening option using material with near-surface fixed polymer tapes (NSM FRP) is proposed [17]. This article describes an experimental study conducted on four beam specimens that were strengthened with NSM FRP strips. As a result of the study, the data showed that the use of NSM FRP tapes can significantly improve the strength and shear stiffness of RC beams, and the method of optimal parameters to ensure the maximum effect was obtained.

The paper [18] describes an experimental study using a developed fiber-reinforced composite system to improve the strength of RC beams. The authors investigated different options for strengthening the beams using the AFRP system, which consisted of thin fiber-reinforced polymer sheets and a sealer-primer. The effectiveness of the reinforcement was evaluated through deflection and failure tests, and the results showed that the use of the AFRP system improves the strength and stiffness of the concrete beams, which is important in terms of their preservation and future use.

To evaluate the structure of a RC beam and obtain important information about the change in the strength of RC beams in the presence of damage in the compressed zone is considered in the articles [19, 20]. The articles describe experimental research, including the development of special methods and techniques, to create damage in the compression zone of RC beams and determine their effect on the load-bearing capacity experimentally. The research is also aimed at studying the interaction between the damaged concrete in the compressed zone with the RC beam as a whole and determining their influence on the mechanical properties of the structures.

It is also possible to consider the combined damage of reinforcement and compressed concrete at the same time and what effect this has on the load-bearing capacity of the RC beam. Such research is described in the article [21, 22], where the process of corrosion of reinforcement was simulated and damages were artificially applied in the compressed zone of concrete. The results of the experiment make it clear how the corrosion of the reinforcement in compatibility with the damaged compressed zone of concrete affects the load-bearing capacity of the RC beam, which is of practical importance for the design and assessment of the technical condition of existing structures.

Damage to the compressed zone of concrete in a RC beam can occur under the action of external forces and the aggressive environment of their operation. Concrete has high resistance to compression, but when critical stresses are exceeded, it breaks down in the form of cracks and crumbling. This can be the result of a variety of factors, such as excessive load on the beam, defects in the concrete, insufficient reinforcement, and other design and fabrication issues. In such cases, various methods can be used to restore and strengthen RC beams, such as the use of reinforcing materials such as composite materials, FRP bars, repair mortars, and various combinations of means of strengthening and restoring the compressed zone of the RC beam.

Analyzing a series of scientific publications in the direction without indirect reinforcement, there are many different methods and approaches in the implementation of reinforcement of the compressed zone of a RC beam. When analyzing a real structure on the object, it is possible to introduce the most suitable and expedient method and technique of reinforcement. The paper [23] investigated an innovative hybrid system that uses glass fiber reinforced polymer (FRP) and steel to flexurally strengthen RC beams. As a result of the experiment, the effectiveness of such a system of reinforcing RC beams for bending was confirmed, which leads to an improvement in their productivity and stability under the influence of loads. An evaluation of the effectiveness of the system in strengthening RC beams for bending under the action of external loads using the proposed installation method was also carried out.

Reinforcement of the compressed zone of RC beams with the help of composite materials based on polymer fibers has been the subject of research in a number of articles [24, 25]. In these articles, the influence of various parameters, such as the angular orientation of the fibers, the number of layers, and the surface coating of the beam, on the effectiveness of strengthening the compression zone of RC beams was studied. Experimental studies have shown that the use of composite materials can successfully improve the strength and stability of RC beams in the compressed zone.

In the articles [26, 27] research was carried out on the strengthening of the compression zone of RC beams using reinforcement in the form of strips of ultra-long and high-modulus fibers reinforced with polymer (UHM CFRP) and carbon strips of infinite length (NSM CFRP), which were fixed in the compressed zones of the beams. The research results showed that both when using UHM CFRP strip and NSM CFRP carbon tapes, there is an increase in the bearing capacity of the compressed zone of concrete, as well as an increase in the strength and stiffness of RC beams. Also, the method of restoring the compressed zone of RC beams with the use of tension-reinforced composite strips with reinforced glass fabrics (TRC) turned out to be very effective, as indicated by the result of an experimental study in the article [14].

A well-known innovative method of strengthening the compressed zone of a RC beam is the use of high-strength mortar, as described in the study [28]. This article examines the effect of different types of high-strength solutions, in particular cement and polymer, for strengthening the compressed zone of RC beams. The results show that high-strength solutions can successfully increase the strength and service life of RC beams, and also have numerous advantages for the further operation of structures.

Also effective is the use of high-strength solution Sika MonoTop-412-Eco, this solution shows high resistance to wear and has properties that make it ideal for use in conditions with a high level of loads and water-resistant operating conditions. Sika MonoTop-412-Eco can be used on both vertical and horizontal surfaces, as well as for both indoor and outdoor applications [29]. To confirm the excellent adhesion of this mortar, a study was carried out that included tensile tests on concrete samples pre-treated with the repair mortar [30]. The results of these studies confirmed the effectiveness of the use of Sika MonoTop-412-Eco repair solution to improve the bond between the concrete base and the repair solution.

The same solution, Sika MonoTop-412-Eco, was used in a complex system to strengthen the compression zone of RC beams, together with the use of CFRP-reinforced strips. Articles [31] investigated the effect of these two materials on the behavior of RC beams of different geometries and reinforcement during experimental bending tests.

A book by Kumar offers a detailed overview of the analysis and design of RC structures using FEMAP and NASTRAN software [32]. The author considers methods of modeling, analysis and design of RC elements and provides a description of the process of creating models, determining loads, performing analysis and assessing the strength and stability of RC structures. The author also considers in detail the use of software products FEMAP and NASTRAN for calculations and optimization of structures.

Among the publications, an extensive overview of the modeling and analysis of RC beam using Femap and NASTRAN software products is shown [33]. The article describes the method of constructing a geometric model of a RC beam in Femap, including the determination of dimensions, material properties and loading conditions, and also considers the process of analyzing a RC beam in NASTRAN, including setting boundary conditions, entering material properties and performing numerical analysis. The paper also offers examples of modeling and analysis of RC beams to demonstrate the effectiveness of using Femap and NASTRAN. The results of the analysis indicate the possibility of using these software products to determine the behavior of RC beams under applied loads.

3 Aims

The purpose of the study is the analysis of literary sources and the process of modeling a RC beam in the Femap software complex by the method of finite element modeling (FEM). Based on modeling, calculate and thoroughly analyze the RC beam, how the presence of damage to the compressed zone of concrete affects its bearing capacity and deformability.

4 Research Methodology

For research and modeling, a RC beam from the university's testing laboratory was used, which will be experimentally tested in further studies, and will be compared with the results obtained from FEM modeling. The RC beam has a geometric size of 2100 × 200 × 100 mm. The longitudinal reinforcement of the beam consists of rolled steel Ø20A500S in the stretched zone and 2Ø6A240S in the compressed zone. The transverse reinforcement is made in the form of U-shaped clamps Ø6A240S with a step of 75 mm (see Fig. 1).

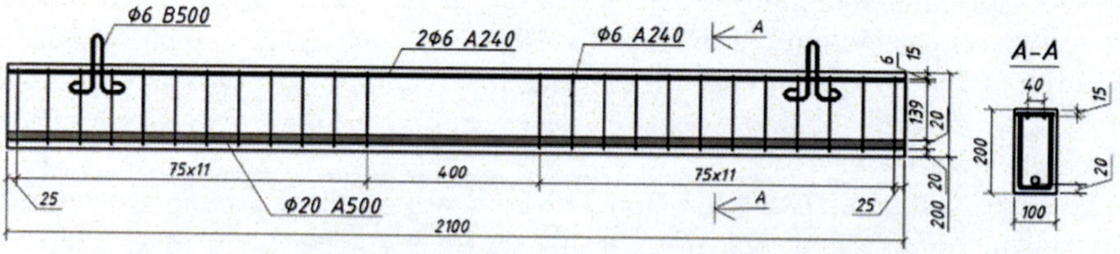

Fig. 1. The general view of reinforcement in the sample.

The RC beam was so designed to be installed on the stand, that its destruction is expected to occur in the zone of pure bending by the normal sections (see Fig. 2). The constructive reinforcement was joined in the reinforcing cage with the use of contact welding in industrial conditions. The concrete was of C30/35 strength class.

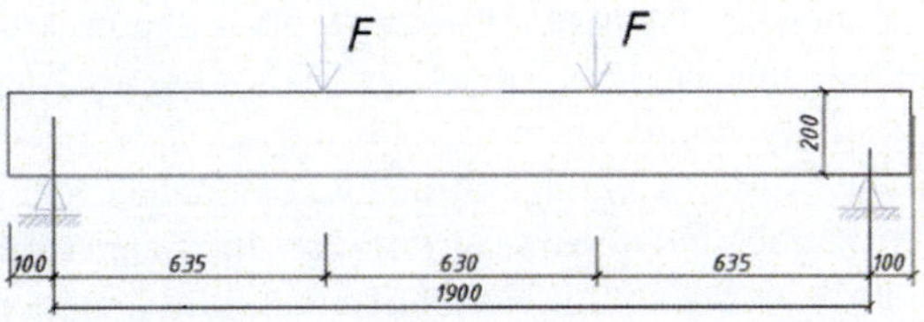

Fig. 2. The general view of the sample.

For the analysis of the damaged RC beam from a technical and economic perspective, it was proposed to cut out a compressed zone of concrete with a depth of 50 mm and a length of 900 mm (symmetrically centered at 450 mm) (see Fig. 3). The next step involved cleaning the beam from dust and moistening the dry concrete for better adhesion with Sika MonoTop-412-Eco and monolayering the compressed zone with a high-strength solution.

Fig. 3. Cutting and restoration of the compressed zone of concrete in a RC beam using a high-strength solution Sika MonoTop-412-Eco.

To create the model and perform calculations for the RC beam, the widely used and modern software package Femap was employed. Femap is an engineering analysis software that facilitates the construction of finite element models for intricate engineering problems (referred to as "pre-processing") and the visualization of solution outcomes (known as "post-processing"). Femap acts as an intermediary between the user and the solvers, which are the core components responsible for conducting calculations in engineering analysis tasks. While Femap supports various solvers, the most seamless integration is achieved with NX Nastran solvers, which were utilized to perform the static calculations for this particular RC beam.

The Femap system, built upon NX Nastran solvers, provides the capability to analyze the deformability and strength of structures, address building mechanics concerns, and obtain solutions for complex, non-stationary, nonlinear processes described by systems of nonlinear differential equations. Additionally, numerous engineering analysis tasks were completed, including linear static analysis, assessment of natural frequencies and modes of free oscillations, evaluation of structural stability, and key aspects of nonlinear analysis. The fundamental nonlinear analysis accounts for geometric nonlinearity (deformations, displacements, load tracking) and physical nonlinearity (considering nonlinear materials such as plasticity, creep, and viscoelasticity).

The Femap software offers several significant advantages, such as the ability to calculate finite element models that accurately simulate the behavior of complex technical systems. It seamlessly integrates with NX Nastran, and its independence from specific solvers allows for the utilization of other solvers when dealing with intricate engineering analysis tasks. Furthermore, Femap provides customization tools (e.g., program files, VB) that automate frequently repeated calculation tasks.

The beginning of modeling begins with the creation of 3 layers (the third layer will be a conditionally damaged compressed zone of concrete, which is unable to bear the load in the first case and in the second case will be restored with a high-strength Sika MonoTop-412-Eco solution). Creation of 3 layers that make up a RC beam, namely concrete, reinforcement and high-strength solution Sika MonoTop-412-Eco. In the concrete layer, using Solid Primitives, a rectangular block of shape was modeled according to the dimensions of the drawing (see Figs. 1 and 3), later the volume of the compressed zone of conditional damage of the RC beam was cut out (for solution or simulation of destruction, namely the absence of concrete) Creation of the metal frame, the reinforcement layers were started by drawing the points of the future clamps, after which they were connected into one clamp and copied along the entire length with a step of 75 mm, and at the corresponding points of the clamps, longitudinal lower and constructive upper reinforcement were modeled according to the drawing (see Fig. 1).

The next stage is setting the physical characteristics of concrete, high-strength mortar and steel (A500S-longitudinal and A240S-constructive, clamps) such as (see Fig. 4): Young's modulus, Poisson's ratio, tension, density, shear, compression. Next, three-dimensional elements were created, which will appear during the calculation of the model based on the materials of concrete, high-strength mortar and steel reinforcement (A500S and A240S). In the define property mode, we select the element type solid for concrete, and for the reinforcement we assign two values to the beam with shape circular bar R10 steel A500S and shape circular bar R3 steel A240S.

The next stage was the creation of a finite element mesh, which was designed in such a way that the edges and nodes of the concrete finite element mesh coincide with the edges and nodes of the finite element mesh of the reinforcement and the edges of the high-strength solution, if it is present in the models. The formation of the grid of elements included the creation of a plane on the front side of the beam, which was divided into a grid with appropriate dimensions. This mesh was then stretched in a given direction using the "extrude element" function, which allows you to incrementally create a finite element mesh across the entire beam. In this model, concrete and high-strength mortar were mainly divided into cubes of 15×10 and 20×10 mm, and longitudinal and structural reinforcement of 15 mm (horizontal rods) and 5 mm for vertical clamps. This ensured that the edges of the cubes coincided in the final mesh for all materials. After that, all the hanging nodes and temporary elements that were involved in the process were removed. Additionally, all nodes of the model were merged with the condition tolerance = 1 mm in the "merge coincident" function, since the distance between the nodes of the finite elements was greater than 1 mm. This made it possible to combine

Fig. 4. Physical characteristics: a)—concrete; b)—high strength solution; c)—Steel A500S; d)—steel A240S

the nodes of concrete and reinforcement at the points of their connection to ensure the interaction of the finite elements of the RC beam.

Fixation of RC beam, located at a distance of 100 mm from its edge at the lower nodes of the beam along its entire width. on one side the hinged support was with the prohibition of movement along the axes TX, TY, TZ, whereas the rotation is possible. On the other side was created the hinged movable support with the prohibition of movement along the axes TX, TY while the rotation and movement along TZ axes are possible.

Loading with the magnitude F (kN) was applied on the RC beam in 1/3 and 2/3 of the span (see Fig. 2), so that the pure bending was ensured. For more even load distribution on the steel plate and upper concrete face it was applied to modes along the width of the beam and distributed among 7 rows of nodes along the length of the beam of 120 mm width. In each of the nodes was created the loading was applied along the Z axis with the value of F' (N), equal to the destruction load, distributed among 77 nodes and multiplied by the acceleration of gravity. As the result of modelling with the use of software Femap the model of reinforced RC beam was obtained (see Fig. 5), with steel (see Fig. 6), without a compressed zone of concrete (see Fig. 7), RC beam beam with a restored compressed zone with a high-strength solution Sika MonoTop-412-Eco (see Fig. 8). In the case of hypothetical failure of the compressed zone of concrete, the load was applied in the region where a portion of the beam was removed at a depth of 50 mm from the upper surface of the RC beam.

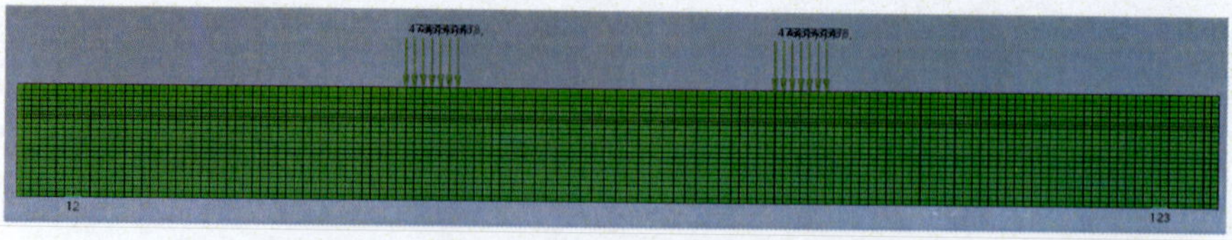

Fig. 5. Modeled general view of a RC beam.

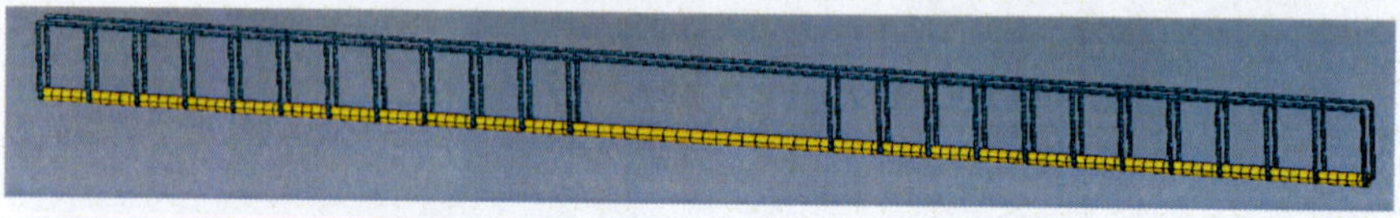

Fig. 6. Modeled spatial metal frame of a RC beam.

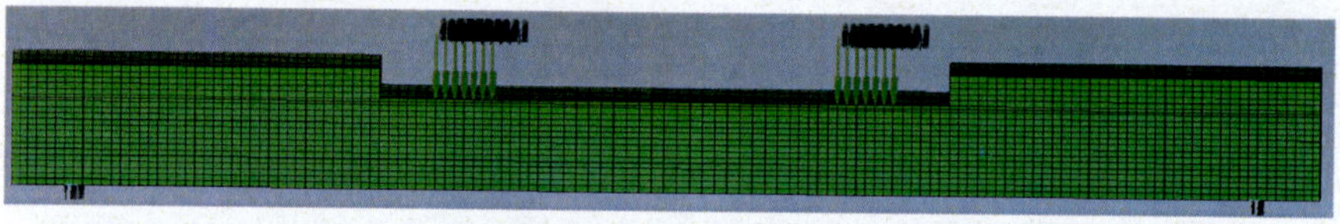

Fig. 7. Modeled RC beam without a compressed zone of concrete.

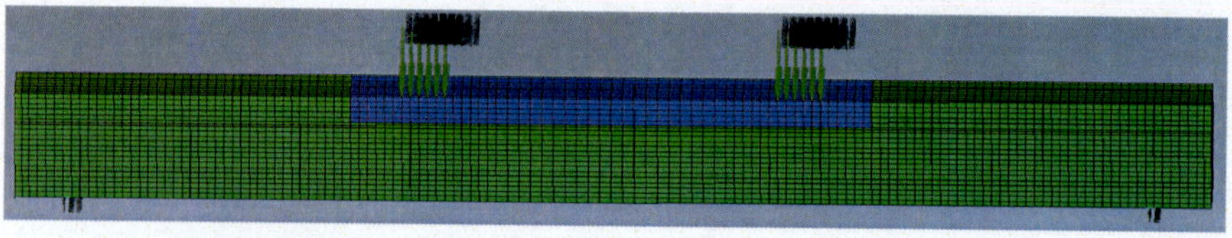

Fig. 8. Modeled RC beam with a restored compressed zone.

5 Results and Discussion

The experimental testing of the intact RC beam, according to the drawing (see Fig. 1), showed failure at a critical applied force on a single ring (see Fig. 2) of F = 37.5 kN (M = 23.74 kNm).

The experimental testing of the damaged RC beam in the compressed zone with geometric dimensions of 900x100x50 mm and restored with the high-strength solution Sika MonoTop-412-Eco (see Figs. 1 and 3), (was used to restore the previous beam that was brought to failure with a force of F = 37.5 kN on a single ring). The test result showed failure of the RC beam when a force was applied to a single ring (see Figs. 2 and 8) of F = 30.03 kN (M = 19 kNm). As a result of inputting all the initial data and loads, and based on the experimental tests in Femap, the following stress results were obtained in the concrete and reinforcement.

Undamaged RC beam. Under the load F = 37.5 kN, the stress in the reinforcement is σ_s = 127.75 MPa, the stress in the most compressed zone of the concrete is σ_c = 33.74 MPa, and the deflection of the RC beam is f = 3.26 mm (see Fig. 9).

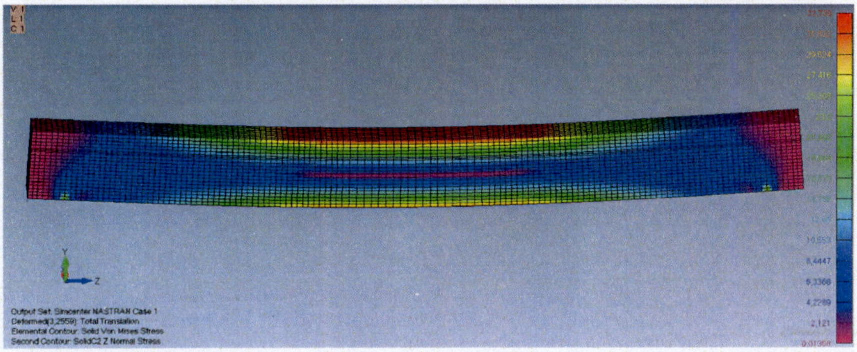

Fig. 9. Stress isofield in concrete and RC beams under load F = 37.5 kN.

A RC beam reinforced with a high-strength solution. Under the load F = 30.03 kN, the stress in the reinforcement is $\sigma_s = 95.3$ MPa, the stress in the most compressed zone of the concrete is $\sigma_c = 22.77$ MPa, and the deflection of the RC beam is f = 2.59 mm (see Fig. 10).

Despite the fact that the beam was brought to failure, it experienced significant residual deformations in both the concrete and the reinforcement. However, after reinforcement and successful application of a high-strength solution, it was possible to restore the load-carrying capacity up to 80% of F = 30.03 kN. To achieve the stress in the most compressed zone of the high-strength solution is $\sigma_c = 33.74$ MPa, the load needs to be applied F = 52.75 kN. As the load approaches such a level, the stresses in the compressed zone is growing slowly, while in the tensile zone, is growing slowly. Therefore, the stress in the compressed zone is $\sigma_c = 33.74$ MPa, and in the tensile zone, it is $\sigma_c = 39.9$ MPa under a load of F = 52.75 kN (see Fig. 11).

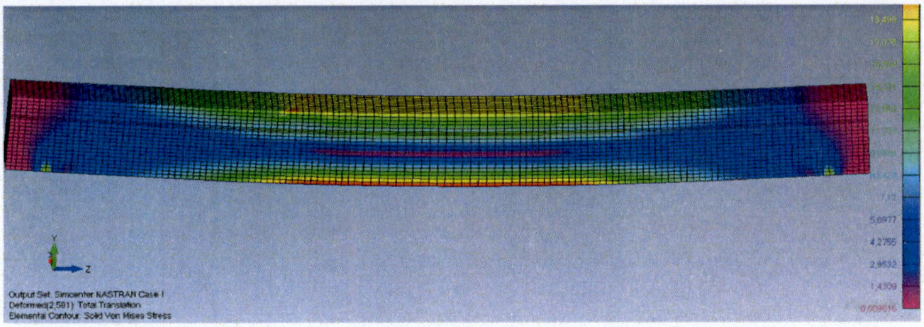

Fig. 10. Stress isofield in concrete and RC beams under load F = 37.5 kN.

The damaged compressed zone of the RC beam. In case the compressed zone of concrete becomes non-functional due to damage and is excluded from the analysis, the following calculation scheme is applied (see Fig. 7). To obtain the stress $\sigma_c = 33.74$ MPa in the compressed zone of concrete at a depth of 50 mm, a load needs to be applied F = 21.75 kN, the stress in the reinforcement is $\sigma_s = 114.5$ MPa, and the deflection of the

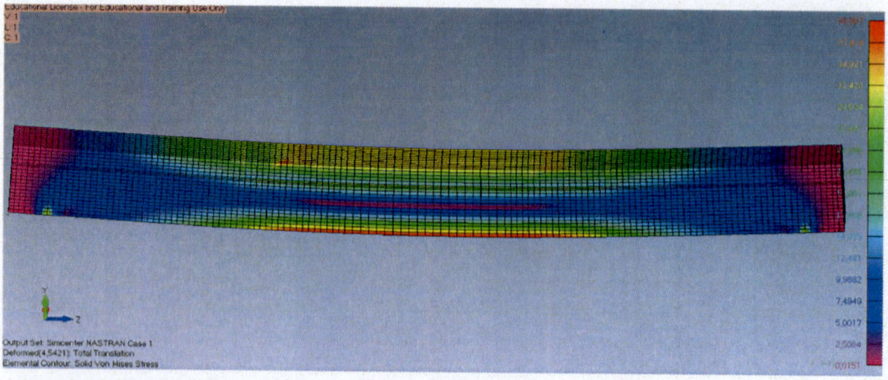

Fig. 11. Stress isofield in concrete and RC beams under load F = 21.75 kN

RC beam is f = 3.88 mm. (see Fig. 12). There is also stress concentration occurring in the regions of compressed height variation, which is equal to $\sigma_c{}'=\sigma_c\bullet 12\% = 37.2$ MPa.

If we consider the stresses in the tensile zone of the concrete that are smaller than the characteristic value $f_{ctk} = 3.6$ MPa the concrete was of C30/35 strength class (see Fig. 12). In this case, it is evident that the concrete essentially loses its functionality, and the beam can experience failure at much lower loads (Fig. 13).

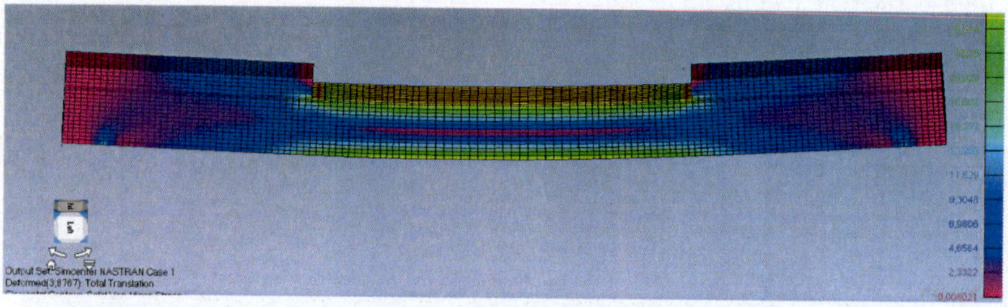

Fig. 12. Stress isofield in concrete and RC beams under load F = 21.75 kN

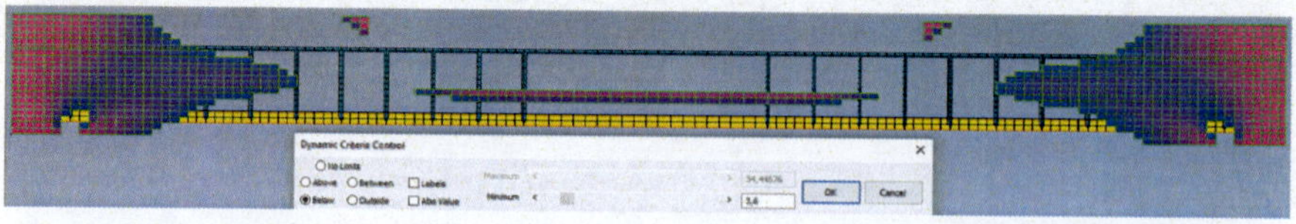

Fig. 13. Stress isofield in concrete and RC beams under load F = 21.75 kN

6 Conclusion

The study involved a detailed investigation of stress in concrete and reinforcement depending on the damage to the compressed concrete zone in the reinforced concrete beam using FEM in Femap. The experimental data (which were attributed) was obtained with the use digital image correlation technique and with the use of sub-micron contactless gauges. Results of FEM also confirmed the experimental data.

The obtained results demonstrate the significant influence of the compressed zone on the overall reinforced concrete beam, as each centimeter of its damage reduces the effective working height of the concrete and diminishes its load-bearing capacity. As the results have shown, the restored compressed zone with a high-strength solution was able to restore the load-bearing capacity by 80%, considering the presence of residual deformations. If the beam is restored before reaching the point of complete failure, caused by external or aggressive factors, it could potentially regain a load-bearing capacity exceeding 100%. Based on the favorable outcome with the high-strength solution Sika MonoTop-412-Eco, it can be inferred that it withstands stress closely approaching a load of $F = 52.75$ kN according to the modeling; however, experimental confirmation is still required. The absence of the compressed zone, which can occur due to various factors, significantly affects its load-bearing capacity, reducing it by 40% instantly. Therefore, it is necessary to monitor the process and speed of failure to facilitate timely restoration and strengthening of the structure. The article also includes the thorough literature review, on restoration of the compressed zone of concrete RC beams.

It can be concluded that this study is not complete and requires further refinement in this direction for a better and more effective approach in the engineering practice of assessing the load-bearing capacity and reliability of reinforced concrete beams in the concrete compression zone. Additionally, since the FEM results demonstrated good correspondence with experimental data, the study can be considered reliable and practical for application.

References

1. Czajkowska A, Raczkiewicz W, Bacharz M, Bacharz K (2020) Influence of maturing conditions of steel-fiber reinforced concrete on its selected parameters. Construct Optim Energy Potential (CoOEP) 9(1):47–54. https://doi.org/10.17512/bozpe.2020.1.05
2. Golewski GL (2023) The phenomenon of cracking in cement concretes and reinforced concrete structures: the mechanism of cracks formation, causes of their initiation, types and places of occurrence, and methods of detection—a review. Buildings 13(3):765. https://doi.org/10.3390/buildings13030765
3. Koteš P, Strieška M, Bahleda F, Bujňáková P (2020) Prediction of RC bridge member resistance decreasing in time under various conditions in Slovakia. Materials 13(5):1125. https://doi.org/10.3390/ma13051125
4. Vegera P, Vashkevych R, Blikharskyy Y, Khmil R (2021) Development methodology of determinating residual carrying capacity of reinforced concrete beams with damages tensile reinforcement which occurred during loading. Eastern-Euro J Enterprise Technol 4(7–112):6–17. https://doi.org/10.15587/1729-4061.2021.237954
5. Blikharskyy Z, Shnal T, Khmil R (2017) The influence of the damaged reinforcing bars on the stress-strain state of the reinforced concrete beams. Prod Eng Arch 14(14):23–26. https://doi.org/10.30657/pea.2017.14.06
6. Lipiński T (2017) Roughness of 1.0721 steel after corrosion tests in 20% NaCl. Prod Eng Arch 15(15):27–30. https://doi.org/10.30657/pea.2017.15.07
7. Lipiński T (2021) Investigation of corrosion rate of X55CrMo14 stainless steel at 65% nitrate acid at 348 K. Prod Eng Arch 27(2):108–111. https://doi.org/10.30657/pea.2021.27.13

8. Blikharskyy Y, Selejdak J, Bobalo T, Khmil R, Volynets M (2021) Influence of the percentage of reinforcement by unstressed rebar on the deformability of pre-stressed RC beams. Prod Eng Arch 27(3):212–216. https://doi.org/10.30657/pea.2021.27.28
9. Poznyak O, Sanytsky M, Zavadsky I, Braichenko S, Melnyk A (2018) Research into structure formation and properties of the fiber-reinforced aerated concrete obtained by the non-autoclaved hardening. Eastern-Euro J Enterprise Technol 3(6–93):39–46. https://doi.org/10.15587/1729-4061.2018.133594
10. Blikharskyy Z, Bobalo T, Kramarchuk A, Ilnytskyy B, Vashkevych R (2021) Bearing capacity of stone beam reinforced by GFRP. Lecture Notes Civil Eng 100:42–52. https://doi.org/10.1007/978-3-030-57340-9_6
11. Koteš P, Farbák M, Kotula P, Brodňan M, Čavojcová A (2013) Using CFRP lamellas for strengthening of dynamically loaded beams. Procedia Eng 65:302–310. https://doi.org/10.1016/j.proeng.2013.09.047
12. Koteš P, Vavruš M, Jošt J, Prokop J (2020) Strengthening of concrete column by using the wrapper layer of fibre reinforced concrete. Materials 13(23):1–21, 5432. https://doi.org/10.3390/ma13235432
13. Li S, Zhang N, Zhao I (2018) Influence of cracks in the compressed zone on the strength of reinforced concrete beams. J Struct Eng 144(6):04018105. https://doi.org/10.1061/(ASCE)ST.1943-541X.0002085
14. Lu Y, Li S, Zhang Y (2017) Experimental study on the repair of compression zone of RC beams with pre-stressed TRC strips. Constr Build Mater 149:123–134. https://doi.org/10.1016/j.conbuildmat.2017.05.165
15. Yang H, Ma L, Li G, Zhao S (2020) Influence of cracks in the compressed zone on the load-carrying capacity of reinforced concrete beams. Eng Struct 203:109817. https://doi.org/10.1016/j.engstruct.2019.109817
16. Guo C, Sun L, Tan Z, Liu G (2019) Numerical analysis of the influence of concrete damage in the compressed zone on the strength of reinforced concrete beams. Adv Struct Eng 22(6):1419–1431. https://doi.org/10.1177/1369433218818156
17. Al-Ani MSMS, Al-Qaisi IH, Abdullah AN (2020) Strengthening of reinforced concrete beams in shear with near-surface mounted FRP strips. Materials 13(16):3549. https://doi.org/10.3390/ma13163549
18. Maranan RS, Kim DH, Kim JW (2009) Behavior of concrete beams strengthened with an advanced fiber-reinforced polymer (AFRP) composite system. Compos Struct 88(4):634–643. https://doi.org/10.1016/j.compstruct.2008.07.014
19. Kim S, Kwon O, Yun I, Lee I (2017) Experimental investigation on the influence of concrete damage in the compressed zone on the load-carrying capacity of reinforced concrete beams. Mag Concr Res 69(11):574–588. https://doi.org/10.1680/jmacr.15.00367
20. Smith A, Davis K, Johnson M (2017) Experimental investigation on the influence of concrete damage in the compressed zone on the load-carrying capacity of reinforced concrete beams. Constr Build Mater 136:629–637. https://doi.org/10.1016/j.conbuildmat.2017.01.055
21. Huang R, Chen L, Li S (2016) Influence of reinforcement corrosion on the strength of reinforced concrete beams with damaged concrete in the compressed zone. Struct Concr 17(5):733–743. https://doi.org/10.1002/suco.201600051
22. Blikharskyy Z, Sobol K, Markiv T, Selejdak J (2021) Properties of concretes incorporating recycling waste and corrosion susceptibility of reinforcing steel bars. Materials 14:2638. https://doi.org/10.3390/ma14102638
23. Al-Ani ASMS, Al-Taie AA, Abdulrahman R (2018) Flexural strengthening of reinforced concrete beams using an innovative hybrid FRP/steel system. Eng Struct 175:194–207. https://doi.org/10.1016/j.engstruct.2018.08.033

24. Abdelrahman M, Salam A (2017) Experimental investigation on strengthening of reinforced concrete beams under compression using CFRP sheets. Ain Shams Eng J 8(3):443–455. https://doi.org/10.1016/j.asej.2016.03.002
25. De Lorenzis L, Nanni A (2004) Strengthening of reinforced concrete beams in compression with fiber-reinforced polymer composites. ACI Struct J 101(2):174–183. https://doi.org/10.14359/13179
26. Torres L, Araújo M (2017) Experimental study on the flexural behaviour of RC beams strengthened with NSM CFRP strips in compression zone. Constr Build Mater 140:80–93. https://doi.org/10.1016/j.conbuildmat.2017.02.073
27. Zhu Y, Liu Z, Guo Z (2019) Experimental study on the flexural behavior of RC beams strengthened by UHM CFRP strips in compression zone. Constr Build Mater 197:455–467. https://doi.org/10.1016/j.conbuildmat.2018.11.108
28. Hoshnavis N, Golsorkhi MA (2012) Strengthening of reinforced concrete beams using high-strength concrete overlay. Int J Concrete Struct Mater 6(3):155–163. https://doi.org/10.1007/s40069-012-0028-6
29. Sika Monotop®-412 Eco, https://www.sika.com/dms/getdocument.get/a93a5743-0396-41b4-9e1f-4f741fac2242/glo-waste-water-treatment-plants.pdf, Last accessed 05 May 2023
30. Islam KMK, Hasan MR, Islam MN (2016) Experimental investigation of the bond strength of Sika Monotop 412 Eco repair mortar to concrete substrate. Int J Eng Technol 8(6):2828–2833. https://doi.org/10.21817/ijet/2016/v8i6/160806265
31. Rahman MM, Rahman MH, Razzak MA (2017) Effect of Sika Monotop 412 Eco on the mechanical properties of reinforced concrete beams strengthened with CFRP sheets. J Mater Civ Eng 29(10):04017148. https://doi.org/10.1061/(ASCE)MT.1943-5533.0002042
32. Kumar S (2020) Finite element modeling of reinforced concrete structures using Femap and NASTRAN. Int J Struct Anal Design 7(4):104–110. https://doi.org/10.5121/ijsad.2020.7409
33. Zhang H, Li G (2019) Modeling and analysis of reinforced concrete beams in Femap and NASTRAN. J Struct Eng 145(6):04019039. https://doi.org/10.1061/(ASCE)ST.1943-541X.0002359

Check for updates

Principles of Modeling the Work of Damaged Columns Made of Compressed Reinforced Concrete

Yevhenii Klymenko[1], Iryna Grynyova[2](✉), Olena Maksiuta[1], and Zeljko Kos[3]

[1] Department of Reinforced Concrete Structures and Transport Facilities, Odessa State Academy of Civil Engineering and Architecture, Didrihson Str. 4, Odesa 65023, Ukraine

[2] Department of Architecture Structure, Odessa State Academy of Civil Engineering and Architecture, Didrihson Str. 4, Odesa 65023, Ukraine
irene.grinyova@gmail.com

[3] Department of Civil Engineering, University North, University Centre of Varaždin, 104. Brigade 3, 42000 Varazdin, Croatia

Abstract. The article presents the results of computer modeling of damaged test samples. Experimental elements in the form of H-shaped reinforced concrete columns subjected to eccentric compression were utilized. Damage was induced through three variables: depth, inclination angle, and length of damage. A three-level three-factor Box-Behnken plan has been created, resulting in statistically substantiated results within the variation of variable factors, using only 15 experimental columns. Samples of compressed elements were manufactured and tested as per the experimental plan. The stress-strain state of reinforced concrete columns is described, and the impact of damage on bearing capacity is established. The modeling of experimental samples was carried out within the software complex "LiraSAPR." Recommendations are provided concerning the procedure for model implementation and calculation. The obtained modeling results were analyzed, and their convergence with experimental data was established, showcasing a high level of agreement. Significant convergence between the results of computer modeling using the software complex and experimental data was observed. Recommendations for creating models of H-shaped reinforced concrete columns subjected to eccentric compression are offered.

Keywords: H-section · Reinforced concrete columns · Computer modelling · Carrying capacity

1 Introduction

In the construction industry, reinforced concrete holds a leading position as the most commonly used material. The durability of reinforced concrete structures can exceed 100 years under favorable conditions, surpassing the intended service life during design. As stated in [1], in industrial buildings, physical deterioration of structures may occur

Z. Blikharskyy et al. (Eds.): CEE 2023, LNCE 438, pp. 178–186, 2024.
https://doi.org/10.1007/978-3-031-44955-0_19

100–120 years after construction. However, under certain circumstances (improper operation, design errors, changes in environmental conditions, increased loads during reconstruction or equipment modernization, mechanical and various damages, including war-related damages), construction elements can be driven to a state of emergency, requiring repairs even before the anticipated time frame. During the manufacturing process, like any other material, reinforced concrete structures can have defects (such as shrinkage cracks, chips, etc.), and during operation (or in case of military action), they can sustain damage. Examples of such damages are shown in Fig. 1.

In [3], Toryanik and his colleagues, as well as Vakhnenko [4], provide examples of the calculation of double-T section reinforced concrete columns subjected to complex deformations, specifically inclined eccentric compression. Flat eccentric compression is a special case of inclined eccentric compression where the load is applied in one of the principal planes of inertia. Inclined eccentric compression occurs due to the eccentric application of the load in the principal planes of inertia (principal axes of the cross-section). Additionally, it is necessary to ensure a long service life, and the maintenance of the structures should be as straightforward as possible.

All the aforementioned factors necessitate more frequent repairs than prescribed by the regulations [5, 6].

The problem lies in accurately determining the technical condition of damaged structural elements of buildings and assessing their residual load-carrying capacity.

The existing methodology for calculating reinforced concrete structures for the first group of limit states is well-established and reliable. It is grounded in experimental and theoretical research carried out on centrally and eccentrically compressed elements, including cases of flat eccentric compression. However, there are no recommendations for calculating the load-bearing capacity of eccentrically compressed reinforced concrete elements in current standards. Currently, the behavior of concrete and reinforced concrete structures with more generalized profiles, such as I-sections, is not yet thoroughly explored, particularly in cases where eccentric compression arises due to damage that is asymmetric in relation to the principal axes of the cross-section. The current normative documents lack specific recommendations regarding the calculation of damaged I-section elements. Consequently, a number of issues arise that require resolution. One such issue involves the improvement and development of new calculation methods for assessing the residual load-bearing capacity of damaged reinforced concrete elements. This entails evaluating their strength in order to determine the most suitable strengthening scheme. This approach is rational as the costs associated with reinforcement are not always justified. However, in many cases, the reconstruction of load-bearing elements proves to be economically advantageous and less labor-intensive, as it can be carried out without the need for specialized equipment and without interrupting production processes.

Therefore, it is crucial to make informed and timely decisions regarding the reconstruction, modernization, and strengthening methods for damaged elements. Given these circumstances, it can be deduced that the subject matter of this study holds significant relevance in both scientific and practical contexts in the present era.

In the article [7], an illustration of the fluctuation in vibration load and its influence on bearing capacity and deformability was presented. The investigation focused on

Fig. 1. Damage to reinforced concrete columns

the impact of vibration loads on the structural components of the five-story factory building floor, followed by subsequent reinforcement measures. The study involved the measurement of actual vibration levels and the evaluation of structural damage resulting from equipment operation.

In the article [8], the outcomes of an experimental investigation on reinforced concrete beams with damage in the compressed zone were presented. The primary focus is on analyzing the influence of concrete damage on the bearing capacity of the beams. Four reinforced concrete beams were subjected to testing, consisting of one undamaged control beam and three beams with damage in the compression zone at varying load levels. The results indicate that both the type of damage and the applied load significantly impact the bearing capacity and deformability. Specifically, the damaged beams under load exhibited a 3.8% increase in bearing capacity compared to unloaded damaged beams.

The article [9] addresses the question of how variations in reinforcement damage affect bent elements with rectangular sections. The experimental study focuses on reinforced concrete beams with damage to the tensile reinforcement, taking into consideration the application of initial load. The damage includes one or five Ø5.6 mm holes, and in one case, the damage extends to beam failure with an 8.4 mm hole. Additionally, it is important to consider the peculiarities of reinforcement manufacturing when studying the effects of corrosion. Work [10] explores such effects when using thermally strengthened reinforcement.

In [11], the study examines damage to reinforced concrete T-shaped beams, considering various factors like shear span, overhang width-to-beam edge thickness ratio, overhang thickness-to-beam section working height ratio, shear reinforcement's reinforcement ratio, and pre-stress level in the tensile rebar.

Another example pertains to the calculation of bent elements in the form of uncut monolithic reinforced concrete strip beams, which combine pressed or hammered reinforced concrete piles into a single foundation structure. In line with the mechanics of reinforced concrete, it is recommended to consider the grid as a continuous reinforced concrete beam [12].

For the strengthening of damaged beams, one can explore the study in [13], which demonstrates the use of FRCM (fiber-reinforced cementitious matrix) reinforced with a composite system. The issue of achieving high-quality compatibility among different materials is particularly important for reinforcements with composite materials. Similar critical importance is discussed in [14] concerning prestressed bent elements.

Considering reinforced concrete as a composite material, attention should also be given to its properties and crack resistance, especially when combined with reinforcement. The issue of crack resistance under different loading conditions is examined in [15].

2 Aim of the Research

The objective of this research is to develop a model for determining the residual load-carrying capacity of damaged double-T-shaped reinforced concrete columns. The model will be based on the fundamental principles of current standards and will extend their applicability to the behavior of obliquely compressed reinforced concrete elements, where the appearance of a complex stress state is attributed to the damage occurring in a portion of the concrete cross-section along the height of the element.

3 Methodology of the Research

Based on the conducted [10] experimental tests and analysis of the obtained results, the parameters of the stress-strain state of double-T-shaped reinforced concrete columns with damage were analyzed, and the main prerequisites for calculating the residual load-carrying capacity were formulated.

The design calculation of reinforced concrete structures for load-carrying capacity, according to the current standards [8], should consider the physical nonlinearity of concrete behavior using the real deformation-stress diagram ε–σ. The calculation method proposed in this study is intended for use as a verification calculation for damaged double-T-shaped elements. It allows for the simplified assumption of a rectangular stress distribution across the compressed zone. Such an assumption is permissible based on the cross-section shape and is consistent with Clause 3.1.7.2 [6].

The position of the neutral axis in inclined eccentric compression is determined by the compressed zone height (x) and the inclination angle of the neutral axis (γ). Flat eccentric compression, where $\gamma = 0$, represents a specific case in which the problem solution is simplified as there is no need to search for one of the unknown variables, and the neutral axis is parallel to the cross-section line. Analyzing the results obtained from testing damaged double-T-shaped specimens, it can be concluded that the neutral axis may exhibit some rotation. This rotation can either increase or decrease the angle γ, depending on the proportions of the cross-section, the inclination angle of the force plane, and the placement of reinforcement within the cross-section. This phenomenon can be explained by the fact that the internal force plane passing through the points of application of the equivalent forces in tensioned reinforcement, the equivalent forces in compressed concrete and reinforcement, and the point of application of the external force in inclined eccentric compression all lie on the same line. The prerequisite for this alignment is ensured by the appropriate position of the neutral axis. During the laboratory testing of damaged double-T-shaped specimens and the visualization of calculations using the PC "Lira SAPR" software, it was observed that the neutral axis takes the form of a straight line.

The calculation of elements subjected to inclined eccentric compression is generally recommended to be performed under the condition of force plane parallelism [6]. This means that the point of application of the external force, the equivalent forces in compressed concrete and reinforcement, and the equivalent forces in tensioned reinforcement should lie on the same line.

In addition, it is necessary to consider the exposure of reinforcement during the calculation.

Based on the conducted experimental tests [10] and the analysis of the obtained results, it was possible to analyze the parameters of the stress-strain state of double-T-shaped reinforced concrete columns with damage and formulate the fundamental prerequisites for calculating the residual load-carrying capacity.

The calculation of elements subjected to inclined eccentric compression is generally recommended to be performed under the condition of force plane parallelism [8]. This means that the point of application of the external force, the equivalent forces in compressed concrete and reinforcement, and the equivalent forces in tensioned reinforcement should lie on the same line.

In addition, it is necessary to consider the exposure of reinforcement during the calculation.

The methodology takes into account the recommendations [8] regarding the limitation of flange width: the value of b'eff, which is used in the calculation, is assumed under the condition that the width of the overhang of the flange on each side of the rib should not exceed 1/6 of the span of the element, and in the case of cantilevered flanges, it should not exceed:

1. $h'_f \geq 0.1$h—we accept $6h'_f$;
2. $0.05h \leq h'_f < 0.1h$—we accept $3h'_f$;
3. $h'_f < 0.05h$—overhangs are not taken into account.

Two groups are considered: with direct (Fig. 2a) and angled (Fig. 2b) section damages. Such distribution is determined by the peculiarities of the calculation described below.

Calculation prerequisites:

1. We assume the hypothesis of planar sections.
2. Stress in the compressed zone is uniformly distributed with intensity [specified value].
3. Stress in the reinforcement is assumed based on the height of the compressed zone of concrete. The forces in the tensile zone are carried by the reinforcement and do not exceed the design tensile resistance ft.
4. The work of the tensile concrete is not taken into account.
5. We assume the condition of parallelism of the force planes: the plane of external and internal forces coincides or is parallel depending on the design case.
6. We consider the exposure of the reinforcement bars.

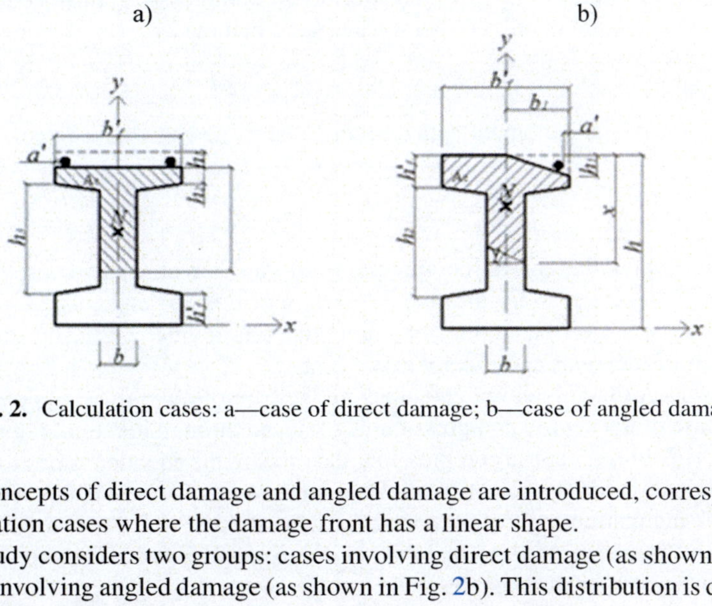

Fig. 2. Calculation cases: a—case of direct damage; b—case of angled damage.

The concepts of direct damage and angled damage are introduced, corresponding to the calculation cases where the damage front has a linear shape.

The study considers two groups: cases involving direct damage (as shown in Fig. 2a) and cases involving angled damage (as shown in Fig. 2b). This distribution is determined by the specific characteristics of the calculation described further.

Columns with an angled damage front experience oblique eccentric compression not due to the eccentricity of the applied load in two directions, but because of the angled damage front. Under conditions of flat eccentric compression, the column operates when the damage front is angled (as depicted in Fig. 2a) or when it is parallel to one of the principal axes of the inertia of the normal section (as shown in Fig. 2b). In the case of oblique eccentric compression, it is necessary to solve a system of equations consisting of: the equilibrium equation of force projections along the longitudinal axis of the element, the moment equations with respect to the x-axis, the moment equations with respect to the y-axis, the static moment equation with respect to the compressed concrete zone along the x-axis, and the static moment equation with respect to the compressed concrete zone along the y-axis. The system of equations for the case of oblique eccentric compression is presented below.

1. The equation of equilibrium of force projections along the longitudinal axis of the element.

$$N - f_{cd} \cdot A_c + \sum \sigma_{1-4} \cdot A_{1-4} = 0 \tag{1}$$

2. The equation of moments about the x-axis.

$$N \cdot e - f_{cd} \cdot A_c \cdot x_c - \sigma_{1-2} \cdot A_{1-2} \cdot h_0 + \sigma_{3-4} \cdot A_{3-4} \cdot a = 0 \tag{2}$$

2. The equation of moments about the y-axis.

$$\begin{aligned} N \cdot b \cdot \frac{1}{2} - \sigma_1 \cdot A_1 \cdot a - \sigma_2 \cdot A_2 \cdot (b - a) + \sigma_3 \cdot A_3 \cdot a + \\ + \sigma_4 \cdot A_4 \cdot (b - a) - f_{cd} \cdot A_c \cdot y_c = 0 \end{aligned} \tag{3}$$

4. The equation of static moment with respect to the compressed zone of concrete along the x-axis (assuming parallel or coincident force planes).

$$S_x = A_1 \cdot y_1 - A_2 \cdot y_2. \tag{4}$$

5. The equation of static moment with respect to the compressed zone of concrete along the y-axis.

$$S_y = A_1 \cdot x_1 - A_2 \cdot x_2 \tag{5}$$

In the case of a flat centric compression, the number of equations reduces from (1…5) to (1, 2). It is necessary to solve the system of Eqs. (1, 2), which consists of the equation of force projection equilibrium along the longitudinal axis of the element and the equations of moments about the x-axis.

The solution of the system of equations, in the case of oblique centric compression and in the case of flat centric compression, is achieved through iteration. For the system of equations in oblique centric compression, the roots of the equations need to be found: R—residual load-carrying capacity of the column, hc—height of the compressed zone of concrete, γ—inclination angle of the neutral axis. For the case of flat centric compression, the roots to be found are: R—residual load-carrying capacity of the column, hc—height of the compressed zone of concrete.

The normal stresses in the longitudinal reinforcement in Eqs. (1–3) are determined using an empirical formula (6).

$$\sigma_{si} = \frac{\sigma_{sc,u}}{1 - \frac{\omega}{1.1}} \cdot \left(\frac{\omega}{\xi_i} - 1 \right), \tag{6}$$

where: $\sigma_{sc,u}$—the ultimate value, which is assumed to be 400 MPa, ω—the coefficient of the completeness of the compressed zone diagram of concrete, which is calculated using formula (7), ξ_i—the relative height of the compressed zone of concrete, which is determined by expression (8).

$$\omega = \alpha - 0.008 \cdot f_{cd} \tag{7}$$

where α is the coefficient that is taken according to the guidelines [8], f_{cd} is the compressive strength of concrete, and the value of the compressive strength of concrete is substituted into the expression in MPa.

$$\xi_i = \frac{x}{h_0} \tag{8}$$

where x is the height of the compressed zone of concrete, h_0 is the distance from the center of mass of the reinforcing bar to the most compressed face of the cross-section.

The reduction in the diameter of the working reinforcement and its consideration in the methodology for determining the residual load-carrying capacity of damaged I-section columns is adopted according to the recommendations obtained from Professor Blikharsky and his colleagues and students [5, 6].

As a result of solving the system of equations, we obtain pairs of roots that need to be analyzed, and there may be several of them. If the roots do not satisfy these requirements, it is necessary to recalculate the problem by changing the form of the compressed concrete zone.

The result of the solution for column C15 (+1 + 1 + 1) is as follows:

$$\begin{aligned} N &= 475\,\text{kN} \\ x &= 0.15\,\text{m} \\ y &= -0.48\,\text{rad} \end{aligned} \tag{9}$$

Analyzing the obtained result, we conclude that this pair of roots satisfies the solution to our problem because the values and are positive, and the value N and x is less than or equal to $x \leq h_2 + h_3$, indicating that it lies within the web. The "−" sign of the inclination angle of the neutral line indicates the direction of measurement $\gamma = 0.48\text{rad} \approx 30°$. Therefore, we conclude that the chosen form of the compressed zone is correct.

The ultimate load obtained from experimental studies is $N_{\exp} = 530$ kN, and the ratio is $\frac{N_{calc}}{N_{\exp}} = 0.90$.

4 Conclusions

The developed model provides a determination of the residual load-carrying capacity of damaged double-T section reinforced concrete columns. It is based on the fundamental principles of the existing codes and extends their application to the behavior of inclined compressed reinforced concrete elements, considering the complex stress state resulting from damage occurring along a portion of the concrete cross-section height.

References

1. Klymenko L, Vegera YP, Blikharskyy Z (2019) Influence analysis of the main types of defects and damages on bearing capacity in reinforced concrete elements and their research methods. Prod Eng Arch 22(22):24–29

2. Ahaieva O, Vegera P, Karpiuk V, Posternak O (2023) Design reliability of the bearing capacity of the reinforced concrete structures on the shear. In: Blikharskyy Z (ed) Proceedings of EcoComfort 2022, Lecture Notes in Civil Engineering, vol 290. Springer, Heidelberg, pp 1–15
3. Babych YM, Savitskiy VV, Andriichuk OV, Ninichuk MV, Kysliuk DY (2019) Results of experimental research of deformability and crack-resistance of two span continuous reinforced concrete beams with combined reinforcement. In: Panchenko S (ed) TRANSBUD-2019, IOP conference series: materials science and engineering, vol 708(1). IOP Publishing Ltd, p 012043
4. Torkhov MS, Vakhnenko PF, Faleev LV et al (1974) Calculation of reinforced concrete structures under complex deformations. Moscow, Stroyizdat, p 297
5. Vakhnenko PF (1992) Modern methods of calculating reinforced concrete structures under complex types of deformations. Kyiv, Budivelnik, 112, p., ill
6. Blikharskyy Y, Pavliv A (2021) Formation of a complete stress-strain curve of concrete using digital image correlation. Eastern-Euro J Enterprise Technol 3(7):111
7. Krainskyi P, Blikharskyy Y, Khmil R, Blikharskyy Z (2018) Experimental study of the strengthening effect of reinforced concrete columns jacketed under service load level. MATEC Web Conf 183:02008. https://doi.org/10.1051/matecconf/201818302008
8. State Building Norms of Ukraine (DBN) B.2.6–98:2009 (2011) Structures of buildings and structures. Concrete and reinforced concrete structures. Basic provisions. Ministry of Regional Development, Construction and Housing and Communal Services of Ukraine
9. Klymenko YV, Mustafa HM (2014) Damaged compressed concrete structures: work, calculation. Odessa, Odessa National University named after I. I. Mechnikov, p 169
10. Klymenko Y, Kos Z, Grynyova I, Maksiuta O (2021) Operation of damaged H-shaped columns. In: Blikharskyy Z (ed) EcoComfort 2020, LNCE 100. Springer Nature Switzerland AG, pp 192–201. https://doi.org/10.10007/978-3-030-57340-9_24

Green Facades and Their Benefits in a Sponge City

Mária Kocúrková(✉), Pavol Knut, and Zuzana Vranayová

Institute of Architectural Engineering, Faculty of Civil Engineering, Technical University of Kosice, Kosice, Slovakia
maria.kocurkova@tuke.sk

Abstract. The problem of today's cities is drought and overheating. More and more scientists are addressing this problem. Since cooling cities is energy intensive, alternative ways of cooling cities and retaining water are being pursued. One of them is the creation of green spaces in cities. We are putting greenery inside blocks and creating parks. However, these natural ways of greening cities are not enough. The first alternative green spaces were created as green roofs, and later green facades were also explored. Compared to a green roof, the facade of a building usually has a larger surface area. However, we also encounter a number of disadvantages and structural but also operational obstacles. When designing a green wall, a thorough analysis of where, when and how the façade will be built is needed. As it is a living structure, all impacts need to be reviewed. Once the green facade has been designed and constructed, other unexplored aspects come into play. Such as what impact the façade will have on the surrounding environment, how and if the proposed green façade will thrive. These and many other questions are currently the subject of research. However, one thing is certain. In every place in the world, designing a green façade is a multidisciplinary task and the measured results can be different even with the same design.

Keywords: Green structures · Green facades · Sponge cities · Humidity · Temperature · Weather · Climatic conditions

1 Introduction

In densely built-up city centres, the temperature is higher than in suburban or less built-up to unbuilt environments. The density of built-up areas creates an urban heat island, resulting in a reduction in the quality of life in cities [1]. Population density causes the building of larger areas of impervious terrain and less natural vegetation. These aspects cause less evapotranspiration, overheating of surfaces and overheating of cities [2]. Building structures and surface materials absorb solar radiation and reflect heat energy into the environment. Buildings also trap infrared radiation between buildings due to the limited view factor of the sky [3]. The Intergovernmental Panel on Climate Change (Intergovernmental Panel on Climate Change (IPCC), 2013) projects that the global average surface air temperature will increase by 0.3–4.8 °C by 2100, depending

Z. Blikharskyy et al. (Eds.): CEE 2023, LNCE 438, pp. 187–197, 2024.
https://doi.org/10.1007/978-3-031-44955-0_20

on the specific emissions scenario and climate model [11]. Related to this is the energy consumption of colds, with frequent heat waves [2]. Green facades is a passive way to improve the energy performance of buildings. By using green facades, we shade the envelope, increase the insulation of the envelope and increase the evapotranspiration rate. These aspects help to reduce the need for cooling in summer and also reduce heating in winter [7]. One of the ways that help cooling is by increasing the greenness index in cities by adding parks and gardens, but also by creating green spaces on buildings. If we look back in history, we have found climbing greenery on buildings for a long time, but we have modern solutions that allow us to create vertical vegetation gardens [4]. Green areas on buildings can be directly on facades and roofs, for example. On façades they occur as vegetated facades or green walls. In addition, greenery on buildings occurs on terraces, balconies or loggias. When combining the benefits of greenery and substrate, we get an even higher evaporation rate. Facades have larger surfaces than roofs and terraces. The thermal comfort of the urban environment is achieved by lowering the temperature, air flow and changing the humidity [5]. Nowadays, greening has become a common practice in commercial and residential areas. In large cities, real estate projects focus on promoting the greening of roofs and walls. In different climates, feasibility analyses of greening facades are carried out [5]. Stormwater from impermeable areas puts a strain on storm drainage systems. Cities use GSI systems such as rain gardens, green walls and green roofs to reduce this runoff. Evapotranspiration (ET) (i.e., evaporation from plants and the soil surface plus transpiration from plants) is an important element of the hydrologic cycle [6]. Thus, understanding the correlations between man-made urban structures and the urban thermal environment is essential to mitigating the urban heat island and developing sustainable and healthy urban environments [8]. The creation of green walls is a complex process and their proper functioning depends on many components. One of the most important is the type of construction and abstract used and the irrigation of vertical gardens [9]. Growing at height also brings with it complications in routine maintenance such as weed clearing or adding substrates and fertilizing [10]. The aim of this review article is to introduce the issue of green facades and to bring together knowledge from different studies. Green facades have many benefits such as promoting biodiversity, mitigating heat islands, reducing urban noise and storm water management, and improving air quality [11]. Despite all the benefits of green facades, there are still few implementations due to imperfect systems. Currently, but also in the past, quite a few studies have been carried out in different climates. Scientists are trying to investigate all the elements affecting the functionality of a green façade. However, as a living organism, the green façade needs an individual approach for each design.

2 Methods

This article is the beginning of a study that aims to create an ideal green wall for temperate conditions, namely Slovakia. At the beginning of this study, a general overview of the issues and benefits of green walls is important. We have drawn from studies around the world in different climates and even from different study focuses. We searched these studies on databases such as Science Direct, Google Scholar, SpringerLink and many others. We searched these databases for the phrase "vertical green structures" because this was the title under which most of the publications were found. We used so-called mind maps and schemas to clearly display all the data. A mind map is an outline in which the major categories emanate from a central image and the smaller categories are shown as branches of larger branches [24]. The first step in this research is to create a mind map, focusing on the General Overview of Green Walls and their categorization. The second step is to categorize green walls and green facades according to their design. The next step is to categorize the benefits of green facades and the processes that take place on them. In the last step, we want to create a scheme that will show the ideal type of green facade for Slovak conditions according to the facts studied so far.

3 Review

In the architectural design of built-up areas, the greenness index is important. In urban planning, most cities are already focusing on the regulation of built-up areas and the greenness index. Green structures such as green roofs and green facades or walls in Fig. 1 are now more a matter of the architectural appearance of the building. However, raising awareness of green structures contributes to the conscious use of greenery to improve the quality of the indoor environment of buildings through the efficient use of energy. Green facades have a significant impact on the environment, but also bring benefits in the public space and in the economic aspect. Greenery helps to improve the face of the city and has a psychological impact on the inhabitants. The relations between the green and the city can be seen in Fig. 2.

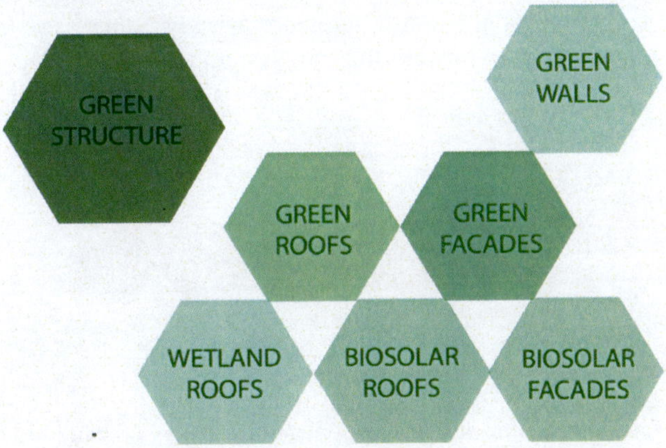

Fig. 1. Diagram of the distribution of green structures

3.1 History

The first of the examples of vertical gardens we consider the gardens of Baby-lon from 500 BC. Later, in the Mediterranean area, plants such as vines were used to provide greenery to cool overheated facades. In central Europe, we encounter green facades in

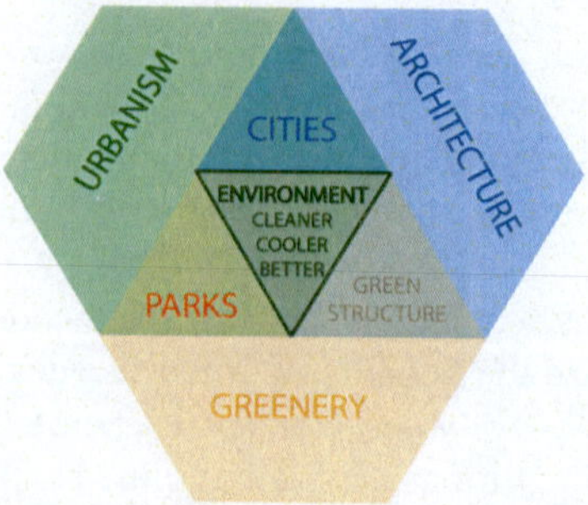

Fig. 2. Scheme of connecting urbanism, architecture and greenery

the 17th and 18th centuries, when they played a mainly aesthetic function [3]. Vines were mainly used, as were fruit trees, fruit espaliers, which could also be appreciated for their fruit, and climbing roses were also popular. Vertical greenery was seen as climbers, which climbed directly on the façade of the building, or greenery placed on terraces and balconies. Since the first social housing projects were designed, many engineers and scientists have become interested in this topic. Plants such as English and Boston ivy were used to green the brick façades. The timeline of history can be seen on the cos axis in Fig. 3. The first of the researches that focused on this topic came from Germany [13]. The first of these was put into practice at the beginning of the 20th century. At that time, approximately 246,000 m2 of green facades were created in Berlin. In Germany today, 10% of buildings can be considered green [14]. In developed countries such as the USA, Australia, Canada and many others, new rules for renovation using green elements are being introduced. Currently, green structures are already being used in the design of green elements on buildings to prevent the root system from disturbing the façade.

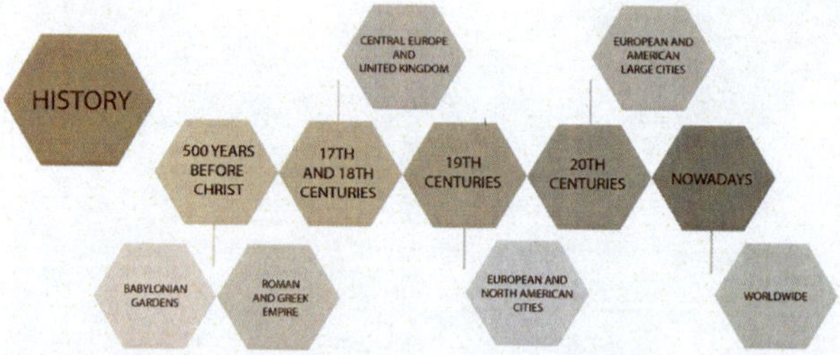

Fig. 3. Timeline of the development of greenery on buildings

Vertical green façades have evolved since climbing plants used the façade as a support and nowadays we know several design solutions to create green walls [20]. The double skin green façade creates a so called green curtain that is not placed directly on the building. This can be climbing plants that grow from the ground upwards along the structure and hanging planters from which plants grow vertically downwards. In these cases, modular trellis systems, wire and net structures are used to support the growing greenery. They are in most cases made of lightweight load-bearing metal elements anchored into the supporting skeleton of the façade. The green facades available on the market today fall into two categories. The first are plants planted in soil and the second are plants rooted in an artificial substrate. The advantage of plants directly in the soil is that there is no need to deal with artificial irrigation [15].

Living wall system (LWS). It is a direct structure placed on the façade into which greenery is planted. Their advantage is that they can be installed on almost any building and are not limited in size. A greater variety of greenery can be used in the selection of planting, which can help return insects and biodiversity to cities. A significant disadvantage is the maintenance of greenery on such schemes. Artificial substrates are used in combination with artificial irrigation and the use of artificial fertilizers.

Prefabricated green walls are created from prefabricated panels or boxes that are placed on a support frame. The use of vegetation is varied and herbs, ground cover plants, ferns and various species of flowers are recommended. The panels are designed to be permeable so that in the event of rainfall, water can drain away to prevent overflowing plants. In periods of insufficient rainfall, such a wall needs to be irrigated, for example with drip irrigation. In addition to drinking water, non-potable rainwater or grey water can be used for irrigation.

In situ living walls are green systems that are attached to buildings as finished facades. This system consists of pockets that are placed on a PVC board. This entire façade is fixed to a steel frame, which is a load-bearing element fixed against the building. In this system, it is important to choose the right kind of plants that have limited space in the pockets for the root system.

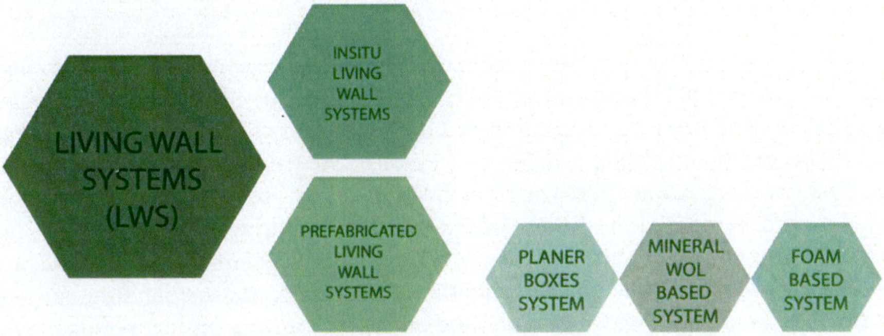

Fig. 4. Scheme of distribution of green walls

Planter boxes system is a system whose dimensions are based on the width of the pots. Basically, we can cover an entire wall with modules. The planters are designed to restrict the choice of plants as little as possible, they have additional depth. The dimensions, color design and composition of the modules depend on the manufacturer. They can be used for interior as well as exterior green walls. The modules are suspended on steel profiles. Mineral wol based system is made of aminoplastic resin foam. It has a white growth medium with pH neutralization. As in the Planter boxes system it can be applied indoors but also outdoors. The plants are irrigated with a drip irrigation system to which fertilizer is added. The supporting structures are aluminum on which the steel baskets are placed. Foam based system, whose growing medium is usually mineral or stone wool. It is suitable for thermally stable climates where there is not a large temperature difference between day and night. As in previous systems, the supporting element is an aluminum frame. This system acts as a plant stand into which the greenery is inserted [15]. The differentiation of the green wall systems can be seen in Fig. 4.

4 Results

4.1 Benefits of Vegetated Facades- Thermal Output

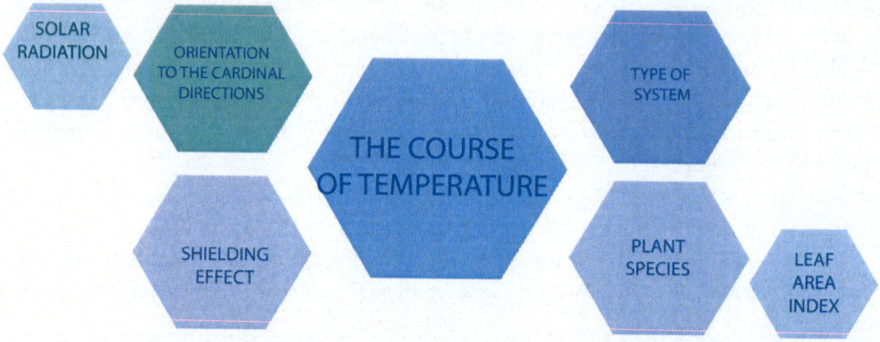

Fig. 5. Diagram of the effects on the temperature course

Figure 5 shows us that in different climatic conditions, researches have been carried out that have proven that the green facade is important in insulating the building and protecting the building from overheating in summer and from cooling and wind in winter. In each of the examined references, the temperature rose from the morning and reached its maximum around noon. On research in Australia, they proved that solar radiation is dependent on the orientation of the world's sides [16]. During the research in China, they focused on comparing the course of temperatures depending on the sides of the world and also took into account the shielding effect from the surrounding buildings. After the measurement, it was found that the orientation to the cardinal points plays an important role in the design of green facades, because each facade reaches its maximum and minimum temperatures at a different time and the shading of the surrounding buildings affects the temperature course [17]. Another study that focused on the course of temperatures comes from Austria. When evaluating the results of this study, it was found that higher temperatures were measured on the facade that was not shaded by greenery than on facades with green systems. In this study, two facades were created, the first from aluminum troughs and the second using a cassette system. When comparing the course of temperatures on these two systems, it was found that even if the facades were oriented to one side of the world, the temperatures in the aluminum troughs were more constant than in the case of the cassette system. This knowledge is especially important for facades in climate zones with large deviations between night and day temperatures. Finding out which system is more suitable for our climatic conditions can help to achieve lower vegetation mortality in the design of a green facade [18]. A study of a green wall in Tallinn focused on the effect of leaf area on the cooling effect [19].

4.2 Benefits of Vegetated Facades- Reducing Air Pollution, Reducing Noise and Improving Quality of Life

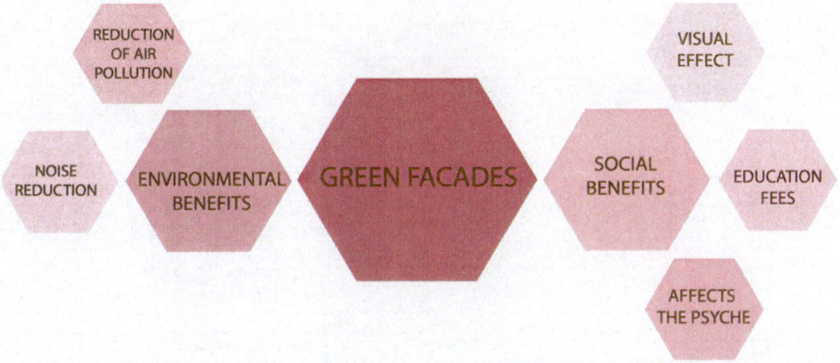

Fig. 6. Diagram of environmental and social benefits of green facades

The environmental and social benefits of green facades can be seen in Fig. 6. Studies around the world demonstrate the impact of green elements on air cle-anliness. This means that just like trees or other greenery, green facades can act as a filter. Plants take carbon dioxide out of the air and replace it with oxygen [21]. Plant roots are able to remove toxic substances and organic compounds from the soil. The rate of air purification is envious of the type of plant used [22]. In a study in Chile, it was found that the most effective plants were *S. album*, *S. reflexum*, *S. palmeri* and *L. spectabillis*. According to this study, the ability of plants to clean the air should be used in decontamination [23].

The main noise attenuating component in green facades is the substrate and differences in noise insulation were found when comparing different growth me-dia. Other important influences were the thickness of the growing medium and the water content of the structure, which had a bad influence on the measurement results. However, it logically follows that the denser the greenery was, the more noise was absorbed by the green façade. Studies such as Perez et al. [25] and Romanova et al. [26] demonstrate the positive effect of green facades on the isolation of noise from traffic or various disturbing elements in cities.

In architecture, greenery is a rewarding element for the completion of the environment, whether in residential buildings or in public or office spaces. Gre-enery is natural to man and visually brings a perfection that man, with his abili-ties, is not able to create. This is why greenery is also used in educational insti-tutions, as it encourages creativity and helps in teaching about the environment. In public spaces, these walls bring many social benefits such as stress reduction, regenerative effects, alleviate fear levels and have a positive impact on reducing crime [22].

4.3 Benefits of Vegetated Facades- Hydrology and Evapotranspiration Rate

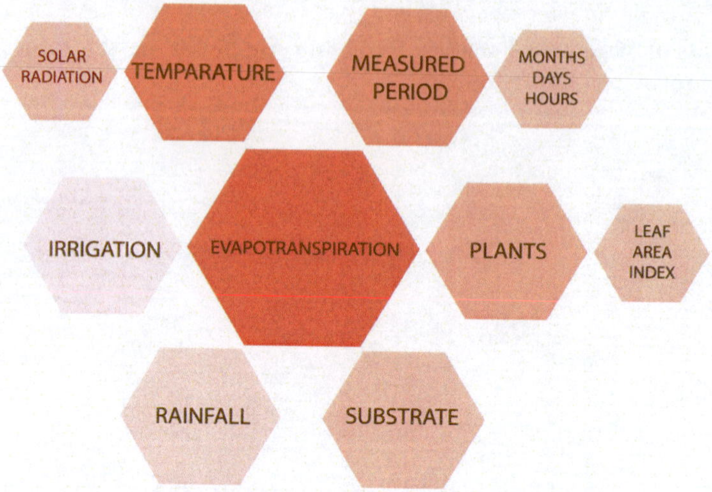

Fig. 7. Diagram hydrology and rate of evapotranspiration

As illustrated in Fig. 7 green walls play an important role in urban water cycling and improving the local microclimate. Walls absorb rainwater in the substrate and in the plants, thus reducing investment in rainwater harvesting. The second advantage of rainwater harvesting is its use in irrigating green walls and thus helps in urban water management. Excess water from the walls can be collected in storage tanks or captured using a drainage system and reused for green wall irrigation using pumps. In addition to rainwater, the green walls can also be watered with grey water, as for example in research in Australia [16]. The biggest advantage of green walls combined with water is the evaporation of water into the environment and its cooling. According to research by Wan et al. annual ET rates in cities are low and contribute the least to ET rates compared to non-urban environments at the scale of the whole country [26]. According to a study by Bakhshoodeh et al., evapotranspiration of green facades that have been irrigated depends to varying degrees on the strength of solar radiation. Also in this research it was shown that the evapotranspiration rate depends on the orientation to the cardinal directions. When comparing north and west facing facades, it was found that the maximum ET was measured on the west facade and the north facade was more stable compared to the west facade [16]. For the measurements in Vienna, the evaporation rate was measured over a period of two months. The results showed differences in maximum values and in days with zero evaporation rate between the two months [18].

5 Summary

Based on the findings, we must take these facts into account when designing green facades in Slovakia. The first is the course of temperature in summer and winter. The main problem is the winter months, which drop below zero at night and thaw during the day when the weather is sunny. Due to the fact that the tem-peratures in winter are unstable, some construction solutions of green facades such as foam based system or mineral wool based system are unsuitable, because the plants would be exposed to high temperature stress and their frequent death would occur. Studies show that the temperature with the planer boxes system is more stable, and thus the vegetation can survive better even in the winter months. The second is the required cooling in dry and hot summers. The main goal is to retain as much rainwater as possible in the substrate in order to reduce the runoff of rainwater into the canals. According to the information on green facades studied so far, the most ideal way to create a so-called sponge city is by using the planter boxes system. This system can capture the most rainwater and is less demanding on irrigation and maintenance. The size of the boxes allows the planting of different types of plants with a larger root system and this also en-sures a larger leaf area. The rate of evapotranspiration, which ensures the cooling of the city, is closely related to water. All of these impacts are discussed in Fig. 8 as factors influencing the design of green facades in Slovakia.

This study, in addition to all the mentioned benefits, has one main task and that is the restoration of a natural, natural environment in places that have be-come unnaturally full areas. Habitat, insects, birds and especially greenery are disappearing from them. Vegetation walls can offer the growth of natural plants, the return of animals and improve the connection between the landscape and nature. Green walls are a key element in solving the global crisis in every aspect. Whether it is an economic, energy or ecological crisis, green facades appear to be one of the key elements in solving them.

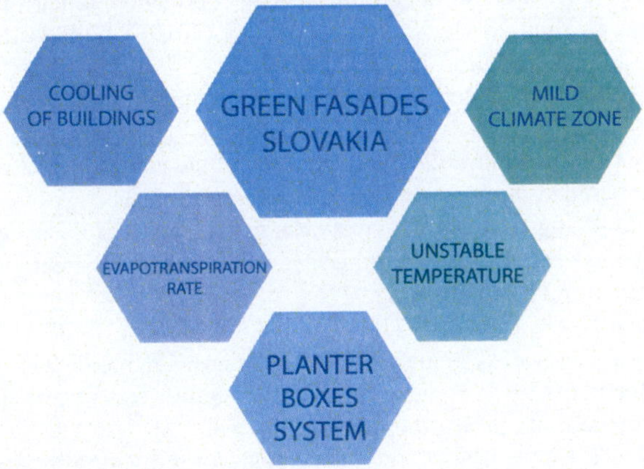

Fig. 8. Schematic selection of the best facade for Slovakian conditions

6 Conclusion

The aim of this article was to point out the advantages and importance of green facades in the urban environment and to evaluate the best possible variant for Slovak conditions. The mentioned studies prove that green facades in the urban environment have their place and purpose. The main idea of this article is to raise awareness of the issue and find the ideal green facade for Slovak conditions. Since almost no studies have been conducted in Slovakia, there is an obvious need for research. It is clear from the studied information that the ideal green facade is from the planter boxes system. This system was also chosen for the experimental green facade in Košice, which will be placed on one of the three experimental cells. Another goal of this study is to find out how our chosen facade reacts to the conditions and whether our assumption was correct. The results of the entire study will subsequently influence the design of green facades in Slovak cities and dispel doubts about their proper functioning.

Acknowledgments. This work was supported by the project run by the Slovak Research and Development Agency VEGA 1/0492/23 Transformation of existing buildings into sustainable buildings—the ecological potential of flat roofs.

References

1. Wong NH, Tan CL, Kolokotsa DD et al (2021) Greenery as a mitigation and adaptation strategy to urban heat. Nat Rev Earth Environ 2:166–181
2. Zheng X, Hu W, Luo S, Zhu Z, Bai Y, Wang W, Pan L (2022) Effects of vertical greenery systems on the spatiotemporal thermal environment in street canyons with different aspect ratios: a scaled experiment study. Sci Total Environ 859
3. Ahmet BB, Cuce E (2017) Green roofs and facades: a comprehensive review 82
4. Chàfer M, Pérez G, Coma J, Cabeza LF (2021) A comparative life cycle assessment between green walls and green facades in the Mediterranean continental climate 249
5. Peng LLH, Jiang Z, Yang X, He Y, Xu T, Chen SS (2019) Cooling effects of block-scale facade greening and their relationship with urban form
6. Ebrahimian A, Wadzuk B, Traver R (2019) Evapotranspiration in green stormwater infrastructure systems 688(9)
7. Convertino F, Vox G, Schettini E (2021) Evaluation of the cooling effect provided by a green façade as nature-based system for buildings. Appl Energy 203
8. Chen G, Wang D, Wang Q, Li Y, Wang X, Hang J, Gao P, Ou C, Wang K (2020) Scaled outdoor experimental studies of urban thermal environment in street canyon models with various aspect ratios and thermal storage 726
9. Dede OH, Mercan N, Ozer H, Dede G, Pekarchuk O, Mercan B (2021) Thermal insulation characteristics of green wall systems using different growing media 240
10. Kazemi F, Mohorko R (2017) Review on the roles and effects of growing media on plant performance in green roofs in world climates 23
11. Lai D, Liu W, Gan T, Liu K, Chen Q (2019) A review of mitigating strategies to improve the thermal environment and thermal comfort in urban outdoor spaces 661
12. Vox G, Blanco I, Convertino F, Schettini E (2022) Heat transfer reduction in building envelope with green façade system: a year-round balance in Mediterranean climate conditions 274:112439, ISSN 0378–7788
13. Köhler M (2008) Green facades—a view back and some visions. Urban Ecosyst 11, Part 4
14. Vijayaraghavan K (2016) Green roofs: a critical review on the role of components, benefits, limitations and trends 57
15. Mir MA (2011) Green facades and Building structures, 39–49
16. Bakhshoodeh R, Ocampo C, Oldham C (2022) Evapotranspiration rates and evapotranspirative cooling of green façades under different irrigation scenarios 270

17. Salonen T, Hollands J, Sesto E, Korjenic A (2022) Thermal effects of vertical greening in summer: an investigation on evapotranspiration and shading of façade greening in Vienna. Buildings 12:1705
18. Convertino F, Schettini E, Blanco I, Bibbiani C, Vox G (2022) Effect of leaf area index on green facade thermal performance in buildings. Sustainability
19. Pérez G, Coma J, Martorell I, Cabeza LF (2014) Vertical greenery systems (VGS) for energy saving in buildings: a review. Renew Sustain Energy Rev 39
20. Timur ÖB, Karaca E (2013) Vertical gardens. Adv Landscape Arch
21. Radić M, Brković Dodig M, Auer T (2019) Green facades and living walls—a review establishing the classification of construction types and mapping the benefits. Sustainability
22. Viecco M, Vera S, Jorquera H, Bustamante W, Gironás J, Dobbs C, Leiva E (2018) Potential of particle matter dry deposition on green roofs and living walls vegetation for mitigating urban atmospheric pollution in semiarid climates. Sustainability
23. Budd JW (2004) Mind maps as classroom exercises, 35–46
24. Pérez G, Coma J, Barreneche C, Gracia AD, Urrestarazu M, Burés S, Cabeza LF (2016) Acoustic insulation capacity of vertical greenery systems for buildings 110
25. Romanova A, Horoshenkov KV, Hurrell A (2019) An application of a parametric transducer to measure acoustic absorption of a living green wall 145
26. Wan H, McLaughlin D, Shao Y, Eerden BV, Ranganathan S, Deng X (2021) Remotely-sensed evapotranspiration for informed urban forest management 210

Check for updates

Peculiarities of the Load-Bearing Structures Restoration of the All Saints Church, a 17th-Century Architectural Monument in Lviv Region

Andrii Kramarchuk(✉) and Borys Ilnytskyy

Institute of Civil Engineering and Building System, Lviv Polytechnic National University, Lviv 79013, Ukraine
kramarchyk76@gmail.com

Abstract. The article presents the peculiarities of the preservation of the All Saints Church, an architectural monument of the XVII century in the village of Hodovytsia, Lviv region (Ukraine). The object of the instrumental research is the All Saints Church, which was built in 1751–1758 according to the project of a famous architect Bernard Meretyn in the Baroque style. This building is a monument of architecture of national importance. The interior of the church is made according to the best examples of Italian baroque churches. The inner walls of the nave and transept are divided vertically by the double pilasters with Ionic capitals, with deep arched niches between them. In the plan, the building is cruciform, single nave, with a shortened transverse transept. At the time of the survey, the church building was in a very dilapidated state, the roof was missed, the vaults were destroyed, and there were significant cracks in the walls. Speaking about the load-bearing structures of the church, the foundations and walls are the most preserved (except for the sacristy and precinct), despite the large number of cracks in the last ones. According to the instrumental survey of the foundations, walls and preserved vaults, the causes of all cracks in the brick walls were analyzed in detail and the factors of significant deformatiions of the building were determined. Unsatisfactory operation of the building was continued after the fire that broke out in 1963, which resulted in the complete destruction of all supporting structures of the roof. Urgent anti-emergency works with the necessary measures to eliminate the consequences of significant deformations of the church building and to prevent its further destruction with the fall of the load-bearing structures have been proposed.

Keywords: Church · Architectural monument · Foundations · Brickwork · Deformations · Vaults · Spacer · Heaving · Arch · Cracks · Emergency technical condition

1 Introduction

The All Saints Church (an architectural monument of the 17th century) is located in the northern part of the village of Hodovytsia, Lviv region (Ukraine), not far from the main road, on the bank of a pond. Oriented to the northwest—southeast (altar—from

Z. Blikharskyy et al. (Eds.): CEE 2023, LNCE 438, pp. 198–213, 2024.
https://doi.org/10.1007/978-3-031-44955-0_21

the northwest side). The church was built in 1751–1758 according to the project of the famous architect Bernard Meretyn in the Baroque style. Prominent artists and sculptors were also the authors of interior decoration works. The polychrome wall paintings were made by the artist Oleksandr Rolinskyi. Wooden sculptures on the altar were carved by the brilliant sculptor Johann Pinzel, in collaboration with Macei Poleiovskyi. The founder of the church was Lviv archdeacon Shchepan Mikulski. The construction of the church and interior works continued until the 1770s. The solemn opening and consecration of the church in honor of All Saints took place in 1774.

Fig. 1. A fragment of the facade of axes "4"–"6"

Fig. 2. A general appearance of the facades of axes "1"–"6" and "6"–"1"

The interior of the church is made according to the best examples of Italian baroque churches. The inner walls of the nave and transept are divided vertically by double pilasters with Ionic capitals, between them there are deep arched niches (axes "3"–"4"). Along axis "2" in the presbytery, on the level of the 2nd floor, there is a tall narrow niche with a hemispherical ceiling (conh). It contained the Pinzel's Crucifixion—at the time of the survey it was a wooden cross that had been preserved. In the right part of the nave there was a carved wooden pulpit with sculptures of angels in the upper part.

Fig. 3. A general appearance of the facades of axes "1"–"6" and "6"–"1"

In Ukraine, particularly in Lviv, there is a large number of buildings and structures, including architectural monuments, which need urgent strengthening. The authors of the article have already presented scientific works related to the strengthening and reconstruction of architectural monuments in Lviv. In particular, the publication [2] highlights the instrumental research of the foundations of the architectural monument of national significance of the seventeenth century, a stone building Holubovychivska—a back tenement house No 32 on the Virmenska street in Lviv. The scientific work [19] provides the arrangement of foundations with their conservation, as well as the exhibition of open fragments of the architectural monument. In the specified section, it is proposed to strengthen the foundations of the hotel complex at the expense of drill press piles. Also, full-scale tests of drill press piles for axial load are carried out on the construction site.

In our case, the load-bearing structures of the All Saints Church suffered significant destruction as a result of a great fire in 1963, which led to a partial collapse of the structures of the roof and vaults of the main nave, the presbytery, and the nave (see Figs. 1, 2 and 3). Further destruction of the church, which still continues up to this day, was caused by atmospheric influences and a large number of cycles of alternating moistening-freezing-thawing (see Figs. 8, 9, 10 and 11). The presented scientific article is devoted to a detailed instrumental study [13, 16] of the load-bearing structural elements of the architectural monument, which were preserved at the time of the inspection of the church (foundations, walls, brick cross vault, which serves as an attic floor for the sacristy). According to the results of the surveys, the authors proposed emergency work with the necessary measures to eliminate the consequences of significant deformations of the church building [6, 16, 18].

Most often, during reconstruction, as well as unsatisfactory operation of buildings and structures, including architectural monuments, the question of the strengthening of the load-bearing structures arises, which work according to different load schemes, and therefore for different types of efforts [1]. The need to strengthen buildings and structures with different structural schemes can also be associated with a change in the seismic resistance of the construction site [7, 19], an increase in the load as a result of a change in the structural scheme of the load-bearing structures, the aggressive effect of the temperature and humidity parameters of the environment on concrete and reinforcement, violation of the work performance technology [1, 3–5], and the influence of dynamic load [15]. A key role in reinforcements during reconstruction and unsatisfactory operation is occupied by bent elements, which is reflected thorough research with the possibilities of ensuring the compatible operation of existing and additional reinforcement elements [3–5].

2 Materials and Methods

The church building under investigation is a cruciform in plan with a single nave and a shortened transverse transept. The dimensions in plan along the outer faces of the walls of axes "1"–"6" "A"–"E" are 33.956 m*17.495 m (see Fig. 4). The maximum height is 29.9 m. The main nave of axes "3"–"4" "A"–"E" along axis "4" changes into the narthex, and along axis "3" changes into the presbytery (altar). To the narthex along the "5" axis, an almost square chapel with dimensions along the inner faces of the walls of 3.977*3.765 m is attached, to the presbytery along the "2" axis is attached a rectangular sacristy with the inner dimensions of 3.949*4.438 m (see Fig. 5). The dimensions of the presbytery (altar) and narthex within the axes "2"–"3" "4"–"5" and "B"–"E" (the inner faces of the walls) are practically the same and are 10.60*5.573 and 10.57 *6.658 m, respectively. (see Fig. 4).

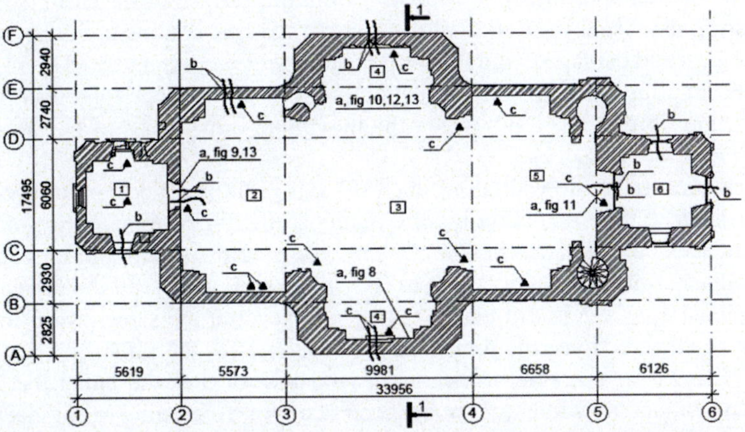

Fig. 4. The plan of the ground floor of the church on 0.000 with the designation of numbers of the photofixations, main defects and damages, explication of premises. 1—sacristy (axes "1"–"2" "C"–"D"); 2—presbytery, altar (axes "2"-"3" "B"–"E"); 3—the main nave (axes "3"–"4" "A"–"F"); 4—transepts (axes "3"–"4" "A"–"B", "3"–"4" "E"–"F"); 5—narthex (axes "4"–"5" "B"–"E"); 6—the antechamber (axes "5"–"5" "C"–"D"); a—photofixations of defects and damages (Figs. 9, 10, 11, 12, 13, 14 and 15); b—cracks in brick walls; c—destruction of brickwork.

The dimensions of the main nave with the side arms of the transept within the axes "3"–"4" "A"–"E" (inner faces of the walls) are 19.945*9.981 m (see Fig. 4). The outer contour of the church construction plan has plastic contours due to cut and rounded corners. The main facade is fixed with double pilasters. The decorative stone vases complete the Baroque pediment. The courtyard is finished with a triangular pediment. All internal volumes are covered with brick vaults: the main nave, narthex and presbytery are covered with cross vaults on wide transverse arches resting on the walls of axes "3"–"B", "4"–"B", "3"–"D" ", "4"–"D". Above the intersection of the nave and the transept there is a vaulted vault, the side arms of the transept have barrel vaults. The sacristy and the antechamber are covered with cross vaults resting on the load-bearing walls of axes "1", "2", "5", "6", "C", "D". Above all load-bearing structures of the attic floor of the church, wooden roof structures with a roof in the form of metal sheets are provided.

The roofs were gabled, three-gabled over the altar, sacristy and antechamber. We want to note that at the time of the survey, only the cross brick vaults of the antechamber and sacristy and one wide transverse arch resting on the walls of the axes "B"–"3" and "D"–"3" were preserved. Moreover, this arch needs immediate reinforcement to prevent it from collapsing to the bottom. All the load-bearing wooden structures of the roof and the roof itself were also destroyed, except for the sacristy and the antechamber (see Figs. 1, 2 and 3).

3 Research Results

According to engineering and geological studies and archival data, the foundations are strip, rubble, the depth of laying is 1800 mm from the surface of the earth (see Figs. 6 and 7). During the visual inspection of the body of the foundations in the open pits, no strength defects and damage to the foundation structures were found, and there were no strength deformations: tilting, shifting, bending. Pit № 1 was made for a wall (pillar) at the intersection of axes "2"–"B", where the maximum dimensions of the wall (pillar) are 1.8*2.47 m.

The deviation from the wall along the "B" axis is 0.6 m. Taking into account the same protrusion of the foundation in all directions, the dimensions of the sole for the pillar of the axes "2"–"B" will be 2.4*3.07 m. For pit No 2 shows the foundations of the wall (pillar) at the intersection of axes "3"–"F" (see Fig. 7) with the dimensions of the wall (pillar) 1.27*2.5 m. Similarly, taking into account the same protrusion of the foundation in all directions, the dimensions of the sole for the pillar of axes "3"–"F" will be 1.72*2.95 m. The sole of the strip foundation of the wall along the "F" axis has no extension and its width does not exceed 600 mm. According to engineering and geological studies, the basis of all strip foundations is the soil layer №2 of semi-hard, heavily ironed, glazed, yellow-brown and yellow-gray loam with the following physical and mechanical characteristics: specific gravity $\gamma = 18.9$ km/m^3, deformation modulus E = 19 MPa, angle of internal friction $\varphi = 22^0$, specific adhesion C = 22 MPa. The maximum thickness of the bearing layer of the soil under the soles of the foundations is 3.4 m. Only within axes "1" "C"–"D" the foundations of the sacristy ("axes "1"–"2" "C"–"D" rest on soil layer №3 (medium-grained, medium-density sand, wet and saturated with water). The layer of loose sand with thickness of 0.15–0.25 m (serves as sand preparation) with the modulus of deformation of at least 30 MPa lies directly under the soles of the foundations. The average modulus of deformation under the soles of the foundations is 1.3 times higher than the modulus outside the foundation, which corresponds to a possible shrinkage of 30–80 mm.

A visual inspection of the walls along the perimeter of the building, both from the outside and from the inside, showed that there are a large number of vertical cracks in

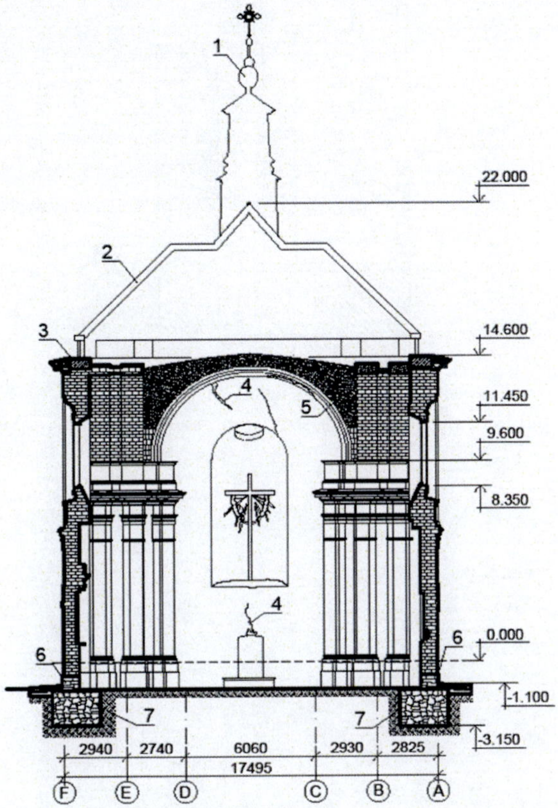

Fig. 5. Cut 1–1. 1—arrangement of wooden structures of the ridge turret based on research materials and archival materials; 2—arrangement of wooden roof structures based on research materials and archival materials; 3—installation of a monolithic steel-concrete belt along the crown of the walls under the roof and floor structures of the building; 4—arrangement of local metal tightening of vertical cracks in the walls, followed by injection with a strengthening solution; 5—installation of foam concrete vault shells suspended from metal floor beams; 6—socles of walls made of cut limestone; 7—rubble stone foundations

the walls, especially in its central part (see Figs. 8, 9, 10, 11, 12 and 13). The opening width of the cracks is insignificant in the socle and increases towards the top, most of the cracks are concentrated in the window arches, the maximum opening width of which can exceed 20 mm. We want to note that the building is without a roof, as a result of the negative impact of atmospheric influences the roof and supporting structures of the floor over the main part of the axes "2"–"5" "A"–"F" have completely collapsed, all the upper layers of the walls are covered by erosion.

The calculation of the bearing capacity of the foundations showed a sufficiently large value of the maximum compressive load (this was influenced by the high calculated resistance of the soil R = 4.2 kg/cm^2 and the significant dimensions of the soles of the foundations under the pillars). Taking into account the maximum height of these pillars of 16.97 m, their mass will be 136 t for pit №1 and 97 t for pit No 2, respectively. Taking

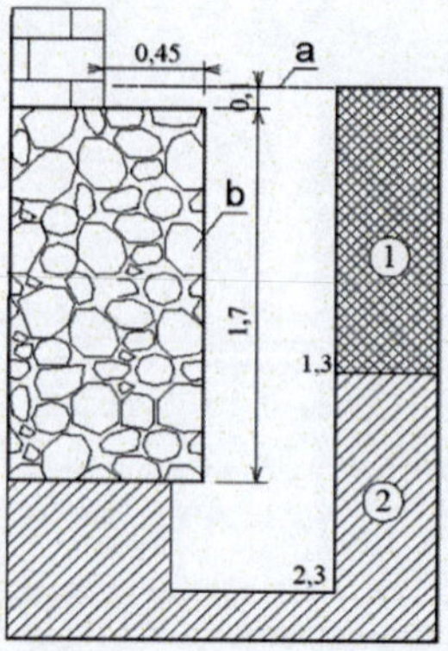

Fig. 6. The results of excavating of the foundations of pit No 2. A—the surface of the earth; b—the existing foundation of the church; 1—soil layer No 1; 2—soil layer No 2.

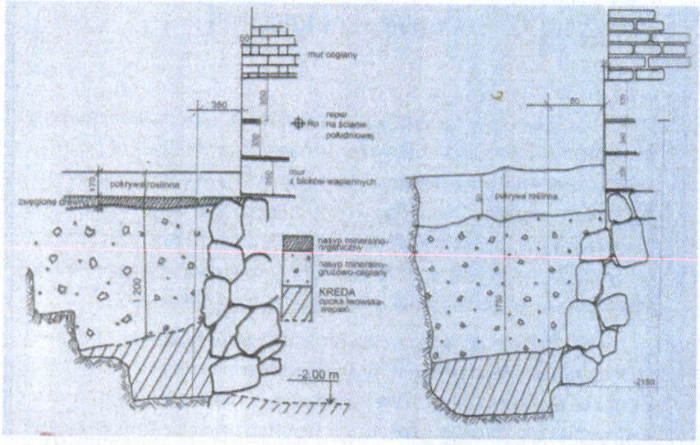

Fig. 7. The photo of archival drawings of the arrangement of the foundation of the church. The archival drawings and field studies showed a good convergence of the results of foundation studies in open pits.

into account the maximum compressive loads on the foundations, we will get significant safety margins for pit No 1 $N_1 = 325–136 = 189$t and for pit No 2 $N_2 = 217–97 = 120$t, respectively.

The reasons for significant deformations of the building, in our opinion, are caused by the following factors:

(1) the phenomenon of horizontal expansion of the vaults under the action of the forces of frost heaving of the brickwork of the walls due to the absence of any roof structure over the entire nave of axes "2"–"5" "A"–"F"; (2) uneven settlement of the foundations under the sole of the foundations as a result of their atmospheric soaking (there is no paving around the perimeter of the building, the roof of axes "2"–"5" "A"–"F" has been completely destroyed); (3) sliding of the slope due to its loading by church structures and rheological phenomena in the soils under the sole of the foundations.

The nature of the development of cracks in the walls of the church indicates their vertical orientation, slightly inclined in places, with an increase in their opening from bottom to top. We consider the order of magnitudes of strut forces in vault structures and a simplified static diagram of vault deformation in the form of statically defined arched cantilevers without braces. We estimate the stress state of the walls in the first approximation, using the assumption that the frost heave pressure is 0.2–0.3 MPa. Taking into account the thickness of the wall 0.75 m, the height of the vault 3 m, the bending moment in the wall at the place of occurrence of cracks (in the middle of their height) is equal to M = 168 tm. Taking into account the moment, the tensile force in the wall is:

$$\sigma_p = M/W + 20\,m/M^2 \tag{1}$$

$$W = (0.75 * 32)/6 = 1.13M^3 \tag{2}$$

$$\sigma_p = 168/1.13 + 20\,m/M^2 = 170\,m/M^2 \tag{3}$$

For brick brand M150, when laying in the horizontal direction, the value of tensile stress does not exceed σp = 100 t/m^2. As we can see, the available calculated value of tensile stress significantly exceeds the tensile strength of the masonry used in the construction of the church. Therefore, we can state with high probability that the possible reason for the appearance of significant cracks in the brick walls of the church is frost heaving of the vaults.

The uneven settlement of foundations under the soles of the foundations as a result of their atmospheric soaking is also one of the main reasons for the formation of significant cracks in brick walls over the past thirty years (we assume that the cracks did not immediately appear after the destruction of the supporting structures of the roof and vaults).

Fig. 8. A general view of the wall of the transept of the main nave of axes "A" "4"–"3".

By carrying out geodetic observations of the beacons that were installed in the walls of the church, the authors of the article measured the magnitude of the movements for

Fig. 9. A general view of the wall of the presbytery (altar) axes "2" "B"–"E".

a period of only half a month. The values of these movements are insignificant (0.35–0.75 mm), but they indicate some uneven settlement of the foundations. Due to the determination of the deformation modulus of the soil under the sole of the foundations, larger subsidence was recorded in the western part of the church (axes "4"–"6"), and smaller in the eastern part (axes "1"–"3"). In the engineering and geological study, it is emphasized that the cracks which are present in the load-bearing structures can be caused by both technical reasons (destruction of the vault and roof) and the possible development of the process of mechanical suffusion in the soil layer No 3. Medium-grained sands lie at the depth of more than 8 m and 12 m, respectively, that is, up to the maximum explored depths that geologists have performed. At a distance of about six meters from the wall of axis "A" (the northern part of the church), a slope is fixed, which is more than 3 m high and is composed of sands, in the thickness of which, under the conditions of their soaking (disordered surface runoff), a suffusion process can develop. We want to note that no signs of the development of shear processes were found within the slope. According to the composition and condition as well as both physical and mechanical properties of soils, fine-grained sands are characterized by high indicators of mechanical properties. Evaluating the possibility of sliding of the slope, we want to emphasise that the plane of possible sliding is outside the sand, taking into account that the natural slope angle does not exceed 11^0 and the normative value of the angle of internal friction of the IGE-3 soil layer is large and is $\varphi = 35^0$. Therefore, the slope is stable, given that $\varphi > \alpha$, $35^0 > 11^0$.

The walls of the examined church are made of full-bodied well-burnt clay bricks with the geometric dimensions of the element 285*140*65 mm on a lime-sand mortar with a seam thickness of at least 10 mm. The thickness of the walls is significant due to their great height without disconnection (vaults begin to rest on the walls from a height of more than 15 m) and for the possibility of absorbing the strut forces from the brick structures of the attic floor. The minimum value of the thickness of two bricks is fixed for walls within the axes "B" "2"–"3", "E" "2"–"3", "B" "4"–"5", "E" "4"–"5", located

Fig. 10. A general view of the wall of the transept of the main nave of axes "E" "3"–"4".

Fig. 11. A general view of the wall of the narthex of axes "5" "E"–"B"

between the brick pillars of axes "2", "3", "4", "5". In places where the wide arched vaults rest, the brick pillars reach a height of more than 2 m in one of the directions (see Fig. 4). Binding of all rows of existing brickwork is at least 70 mm. The brickwork is made for the entire height of the building, the socle is lined with natural stone blocks outside, the socle height is not less than 1500 mm. All the walls of the examined church are load-bearing due to the support of various types of attic vaults: cross, sail and barrel vaults. Window and door lintels are arched, all of them are load-bearing.

The brand of bricks and mortar was determined by the destructive method of compression testing. According to the results of eight samples, the value of the brand M75–150, the brand of undamaged mortar within the limits of M75, was established. The lower grade of brick is located in places where it is significantly damaged due to alternating wetting-freezing-thawing. As a result of the lack of the roof and the immediate impact of precipitation, the brickwork of the examined church is thawing and weathering, especially in its upper part between the 9.500 and 15.900 marks (see Figs. 1, 2, 3, 10, 11, 12 and 13). Such destruction to a greater or lesser degree is present along the entire perimeter of the building, but to the depth of no more than 10%, in some places 15% of the thickness of the church wall. Given the large thickness of the walls and pillars, such destruction at the time of inspection of the building is not critical. This does not apply to the upper rows of masonry (between grade 16,000 ÷ 16,970, in the places where the brick vaults of the floor and parapet slabs of natural stone rest), which are completely destroyed by cycles of alternate moistening-freezing-thawing (see Figs. 8, 9, 10 and 11). These parts of the masonry must be disassembled when performing emergency work.

Fig. 12. A fragment of the wall of the transept of the main nave of axes "F" "3"–"4". A vertical crack is between the 7500 ÷ 5100 with a maximum opening width of 10–14 mm.

Fig. 13. Vertical cracks of the wall of the presbytery (axis of the wall "2" "C"–"D") between marks 9.500 and 2.500. The maximum opening width is 12 mm.

The complete destruction of the upper rows of masonry is also evidenced by the fact that individual parapet slabs collapsed to the bottom. Basement internal and above-basement external parts of the walls are also shocked by cycles of alternate moistening-freezing-thawing. As a result of these destructions, the cross-sectional area of individual sections of the walls decreased by one brick (half a brick on each side).

Therefore, in order not to carry out load-bearing reinforcements of individual sections of the walls (reinforced concrete or steel brackets), we recommend the installation of a light construction of the floor and roof. In this case, the damaged lower sections of the walls will only need to be treated with a cement-sand mortar of the brand M150 on the RABITSA grid.

The nature of the appearance of vertical cracks of different opening widths was analyzed in detail above. The appearance of all the cracks along the perimeter of the building, both from the outside and from the inside, is connected with the destruction of the attic floor and roof, which occurred after the fire. Multiple frost heaving of the brickwork led to the horizontal gap of all the brick vaults, and therefore to the appearance of a large number of cracks with their placement in the middle of the blocks under the vault. The increase in the opening width of the cracks from the bottom to the top (point 16,000 $\div$ 1500) also confirms the fact of the influence of the horizontal spacer after the frost heaving of the brickwork. The appearance of cracks is also due to the uneven settlement of the foundations, due to their constant soaking because of the lack of the roof, attic floor and paving around the perimeter of the building.

4 Practical Expediency of Research

Visual and instrumental studies of the preserved supporting structures of the All Saints Church (an architectural monument of the 17th century) allowed us to establish their technical condition and assess the degree of damage and destruction of the foundations, brick walls, arches and vaults. The conducted engineering-geological and geodetic studies, together with the necessary calculations of the bearing capacity of the foundations, made it possible to determine as accurately as possible the ways of the phased removal of the preserved structures from the emergency state.

The proposed constructive solutions based on the survey results (see the next section, Figs. 4, 14, 15 and 16) will ensure the necessary load-bearing capacity of the preserved foundations and walls, prevent their further destruction and help to recreate the original appearance of the sacred building.

5 Conclusion

The causes of significant deformations of the examined church, which manifested themselves in the form of cracks of an exceedingly wide opening of the brick walls, are the phenomenon of horizontal expansion of the brick vaults under the influence of the forces of multiple frost heaving, as well as uneven settlement of the foundations due to the ingress of atmospheric water into the foundations because of the lack of the roof as well as attic floor and paving. There are no structural defects or damage to the foundations, their load-bearing capacity will be ensured when installing any type of attic floor

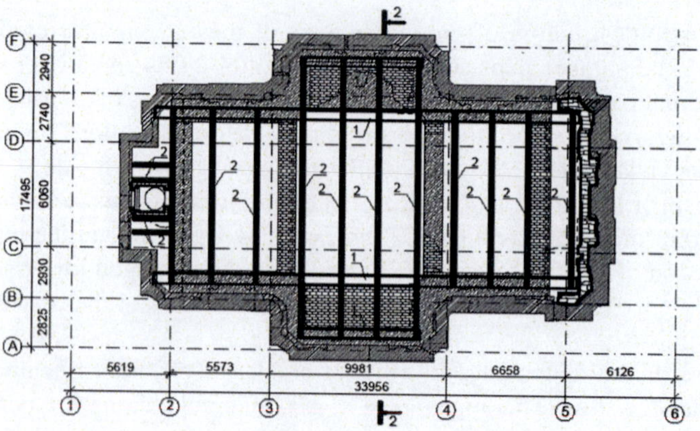

Fig. 14. A scheme of the floor and monolithic reinforced concrete belts. 1—installation of a monolithic reinforced concrete belt along the coronets under the roof and floor structures of the building; 2—metal ceiling beams.

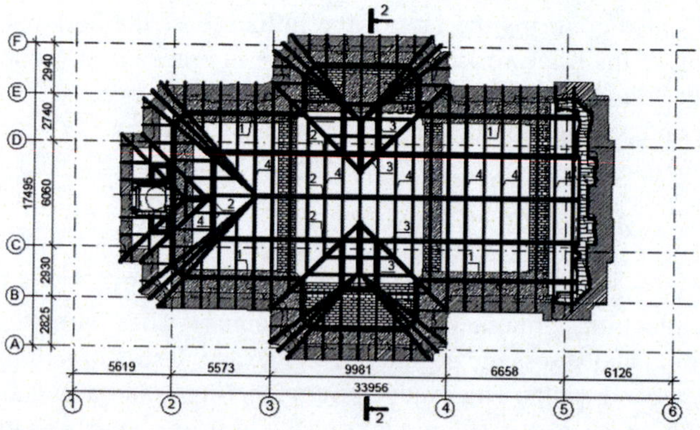

Fig. 15. A scheme of the roof arrangement. 1—mauerlats from a wooden beam; 2—purlins made of a wooden beam; 3—diagonal rafters; 4—rafters;

in the form of vaults (brick, reinforced concrete, steel) with load-bearing roof structures (see Fig. 14, 15 and 16). When carrying out emergency work on the arrangement of the roof, the strip foundations of the church do not need reinforcement. Therefore, we believe that the technical condition of the foundations is satisfactory (category "2").

The technical condition of the walls is emergency (category "4") due to the presence of significant strength defects and damage in the form of a large number of cracks along the perimeter of the building, excessive opening width and destruction of significant sections of the walls along the entire height associated with cycles alternating moistening-freezing-thawing (see Figs. 8, 9, 10, 11, 12 and 13). The upper rows of masonry between the Sect. 15,000 ÷ 16,000 suffered the greatest destruction in places where the brick vaults of the ceiling and parapet slabs of natural stone rest. These parts of the walls are completely destroyed by cycles of alternate moistening-freezing-thawing, which

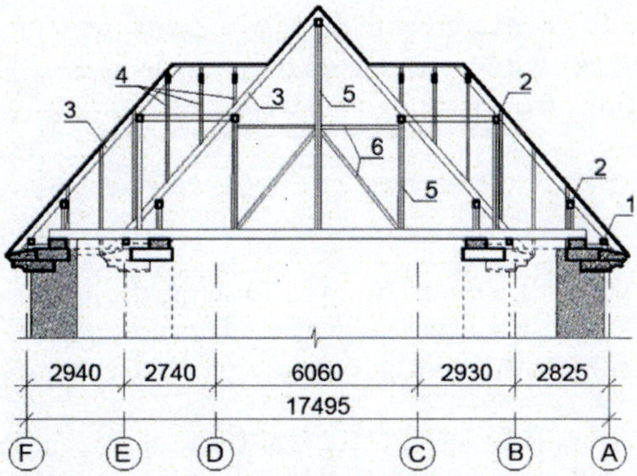

Fig. 16. A section diagram of the 2–2 roof with reproduction of the historical configuration of the destroyed roof 1—mauerlats from a wooden beam; 2—purlins made of a wooden beam; 3—diagonal rafters; 4—rafters; 5—metal racks from the channel into the box; 6 – connections;

leads to the fall of individual parapet slabs to the bottom and individual bricks. We want to note that, with the exception of the mentioned upper sections of the walls along the perimeter of the building, the entire surface of the walls is a subject to repair without significant sections of their dismantling. There is also no need to strengthen the load-bearing elements with the help of steel clips.

Urgent emergency works with the necessary measures to eliminate the consequences of significant deformations of the church building include:

(a) the upper rows of masonry (between marks 15,000 ÷ 16,000, see Fig. 5), which are completely destroyed by cycles of alternating wetting-freezing-thawing, must be urgently dismantled to prevent parapet stone slabs and bricks from falling to the bottom. Instead of damaged (destroyed) bricks along the contour of the walls of the church building, arrange a monolithic reinforced concrete belt, which will increase the rigidity of the building and ensure its even settlement (see Fig. 14);
(b) to increase the rigidity of the walls, we recommend arranging a flat attic floor on steel beams, anchoring them in a monolithic reinforced concrete belt. The support of the roof structures should be put on these load-bearing steel beams, preserving its geometry which had been before the fire started (see Figs. 5, 15 and 16);
(c) as a result of the destruction of the basement inner and upper basement outer parts of the walls, their cross-sectional area decreased by one brick (half a brick on each side). Therefore, in order not to carry out load-bearing reinforcement of individual sections of the walls (reinforced concrete or steel brackets), we recommend the installation of a light structure of the floor and roof (steel structures of the attic floor with a pitched reconstructed wooden roof). The structures of the former brick vaults of the church of axes "2"–"5" "A-"F" will be restored due to the installation of foam concrete vault shells suspended from the metal beams of the floor. Accordingly, the damaged lower sections of the walls will only need to be treated with a cement-sand mortar of the M150 grade according to the RABITSA grid;

(d) after installing the floor covered with the load-bearing structures of the roof and roof itself, inject the cracks with a cement-sand mortar of at least M150 grade. Carry out geodetic observations for possible uneven settlement of the church within one year.

References

1. Kramarchuk A, Ilnytskyy B, Kopiika N (2022) Ensuring the load-bearing capacity of monolithic reinforced concrete slab damaged by cracks in the compressed zone. Lecture Notes in Civil Engineering, vol 290. Proceedings of Ecocomfort 2022. https://doi.org/10.1007/978-3-031-14141-6_21
2. Kramarchuk A, Ilnytskyy B, Lytvyniak O (2022) The features of preservation of architectural sights of national significance in modern urban space. Lecture Notes in Civil Engineering, vol 290. Proceedings of Ecocomfort 2022. https://doi.org/10.1007/978-3-031-14141-6_20
3. Blikharskyy Y, Selejdak J, Kopiika N (2016b) Corrosion fatigue damages of rebars under loading in time. Materials 14(12):3416. https://doi.org/10.3390/ma14123416
4. Blikharskyy Y, Selejdak J, Kopiika N (2021d) Specifics of corrosion processes in thermally strengthened rebar. Case Stud Construct Mater 15:e00646. https://doi.org/10.1016/j.cscm.2021.e00646
5. Blikharskyy Y, Selejdak J, Kopiika N, Vashkevych R (2021c) Study of concrete under combined action of aggressive environment and long-term loading. Materials 14(21):6612. https://doi.org/10.3390/ma14216612
6. Blikharskyy Z, Karchut I, Struk R (2014) Calculation and design of normal and inclined sections of reinforced concrete elements. Tutorial Lviv 141
7. DBN B.1.1–12: 2014 (2019) Construction in seismic areas of Ukraine (with changes No 1 from 01.05.2019). Kyiv, State Enterprise "State Research Institute of Building Structures" (NDIBK), p 116
8. DBN B.2.6–98: 2009 (2011) Concrete and reinforced concrete structures. Ministry of Regional Development of Ukraine, Kyiv, 73
9. DBN V.1.2–14:2018 (2018) General principles for reliability and constructive safety of buildings and civil engineering works. UkrNDIproektstalkonstruktsiya, Ministry of construction of Ukraine, Kyiv, 33
10. DBN B.1.2–2:2006 (2019) Reliability and safety system for construction projects. Loads and effects. Design standards (with change No 1). UkrNDIproektstalkonstruktsiya, Ministry of construction of Ukraine, Kyiv, 77
11. DSTU B B.1.2–3: 2006 (2006) Reliability and safety system for construction projects. Deflections and displacements. Ministry of construction of Ukraine, Kyiv, 15
12. DSTU B B.2.6–156: 2010 (2011) Concrete and reinforced concrete structures made of heavy concrete. Ministry of Regional Development of Ukraine, Kyiv, 123
13. DSTU-N B B.1.1–18: 2016 (2017) Guidelines for the inspection of buildings and structures to determine and assess their technical condition. Kyiv, State Enterprise "State Research Institute of Building Structures" (NDIBK), 47
14. Jelenski T (2018) Practices of built heritage post-disaster reconstruction for resilient cities. Buildings 8(4):53. https://doi.org/10.3390/buildings8040053
15. Ivanyk I, Vikhot S, Vybranets Y, Ivanyk Y (2018) Theoretical research into spatial work of a steel-reinforcedconcrete statically indeterminate combined structure. Eastern-Euro J Enterprise Technol 5(7–95):13–22. https://doi.org/10.15587/1729-4061.2018.143023
16. de Rubeis T, Nardi L, Muttill M, Paoletti D (2020) The restoration of severely damaged churches—implications and opportunities on cultural heritage conservation, thermal comfort and energy efficiency. J Cult Heritage 43:186–203. https://doi.org/10.1016/j.culher.2019.11.008

17. Kramarchuk A, Ilnytskyy B, Lytvyniak O (2018) Arrangement of the foundations under the new hotel in Lviv. In: Ulewicz R, Hadzima B (eds) 12th international conference quality production improvement, QPI,-2018, MATEC Web of Conferences 183:02007. https://doi.org/10.1051/matecconf/201818302007
18. Kramarchuk A, Ilnytskyy B, Bobalo T (2016) Building structures. Lviv, Lviv Polytechnic Publisher, 199
19. Kramarchuk A, Ilnytskyy B, Lytvyniak O, Grabowski A (2019b) The increase of seismic stability for existing industrial buildings. In: Piatkowski T (ed) Scientific session on applied mechanics X: proceedings of the 10th international conference on applied mechanics, AIP Conference Proceedings 2077, 1, 020029. https://doi.org/10.1063/1.5091890
20. Małachowicz E (2007) Konserwacja i rewaloryzacja architektury w środowisku kulturowym. Oficyna Wydawnicza Politechniki Wrocławskiej, Wrocław

Bibliometric Analysis of Renewable Energy Research and Industrial Assets in Poland and Slovakia

Paweł Kut[1] (✉), Katarzyna Pietrucha-Urbanik[2], and Martina Zeleňáková[3,3]

[1] Department of Heat Engineering and Air Conditioning, Rzeszow University of Technology, Al. Powstancow Warszawy 6, 35-959 Rzeszow, Poland
p.kut@prz.edu.pl

[2] Department of Water Supply and Sewerage Systems, Rzeszow University of Technology, Al. Powstancow Warszawy 6, 35-959 Rzeszow, Poland

[3] Department of Environmental Engineering, Faculty of Civil Engineering, Technical University of Kosice, 04200 Kosice, Slovakia

Abstract. Renewable energy sources, such as wind, solar, hydro, and geothermal, play an increasingly important role in the global energy mix. They offer a sustainable alternative to fossil fuels, reducing greenhouse gas emissions and contributing to the fight against climate change. The development and deployment of renewable energy technologies have been growing rapidly worldwide, driven by government policies, technological advancements, and decreasing costs. Despite their advantages, renewable energy sources still face challenges, such as intermittency, storage, and grid integration. However, with continued investment and innovation, renewable energy sources have the potential to become the dominant source of energy in the future. The article presents a bibliometric analysis of renewable energy sources research in Poland and Slovakia. The study was based on publications indexed in the Scopus database. The analysis covered the period from 1990 to 2023. The study investigated the research topics related to renewable energy sources and analysis the most cited publications written by authors from the analyzed countries. Bibliometric analysis demonstrates the growing interest in renewable energy sources research in Poland and Slovakia. However, there is still a need for further research in certain areas, such as energy storage, smart grid technologies, and the use of renewable energy in transport. The results of this study can serve as a valuable resource for policymakers, researchers, and other stake-holders interested in promoting the development and use of renewable energy sources.

Keywords: Renewable energy · Bibliometric · Citespace

1 Introduction

In recent decades, there has been growing public awareness of the impact of greenhouse gas emissions on the Earth's climate and the need to reduce dependence on fossil fuels [1–5]. As a result, renewable energy sources have become one of the most important elements of the energy policy of many countries, including Poland and Slovakia [6–12].

Z. Blikharskyy et al. (Eds.): CEE 2023, LNCE 438, pp. 214–223, 2024.
https://doi.org/10.1007/978-3-031-44955-0_22

Renewable energy sources, such as solar, wind and hydropower, are gaining increasing importance worldwide as an alternative to fossil fuels [13–16]. The European Union's energy policy imposes an obligation on member countries to increase the share of energy from renewable sources in the overall energy mix [17–19]. Poland and Slovakia, as countries with a similar energy structure, are following this trend and intensifying investments in renewable energy sources.

Figure 1 shows the share of renewable energy in Poland and Slovakia from 2004 to 2021, according to Eurostat data [20].

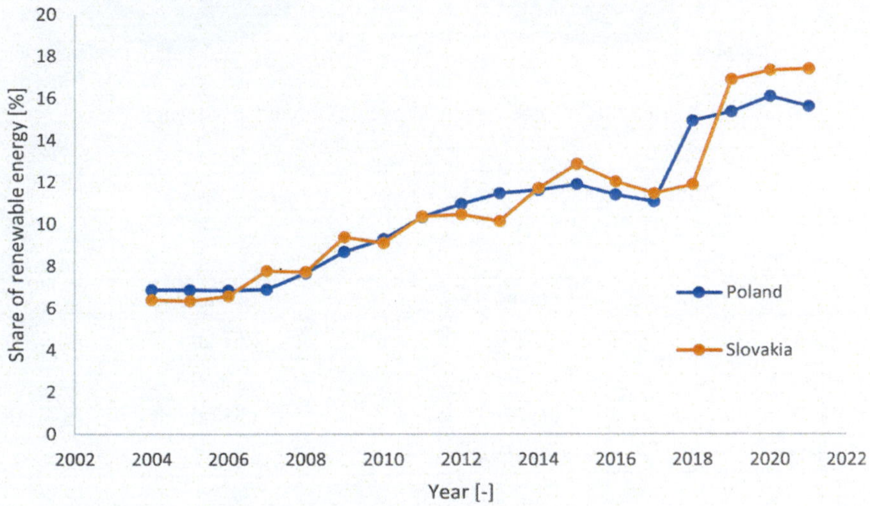

Fig. 1. Share of renewable energy in Poland and Slovakia according to Eurostat data [20].

Analyzing the chart above, it can be seen that Poland and Slovakia are characterized by very similar growth in the share of renewable energy. In 2021, Poland's renewable energy sources share was 15.6%, while Slovakia's was 17.4%. In 2020, both countries reached the target set by the European Union. Poland achieved the required 15%, while Slovakia achieved 14% of the renewable energy sources share in the energy mix.

Figure 2 shows the percentage share of each renewable energy source in Poland and Slovakia.

Despite the similar rate of development of renewable energy sources in Poland and Slovakia, the structure of energy generation from individual renewable sources is different (Fig. 2). The main renewable energy source for Poland is photovoltaics, which account for 54% of installed capacity. In Slovakia, the main source of renewable energy is hydroelectric power, which accounts for 77%. Table 1 shows the installed capacity of each renewable source in 2022.

When comparing the installed capacity of renewable energy sources, it is important to consider the total installed capacity of all power plants. In Poland, the total installed capacity is about 60 GW, while in Slovakia it is about 8 GW. This means that the total installed capacity of renewable energy sources in Poland is about 37% of the total installed capacity, while in Slovakia it is 41%. It should be noted, however, that energy

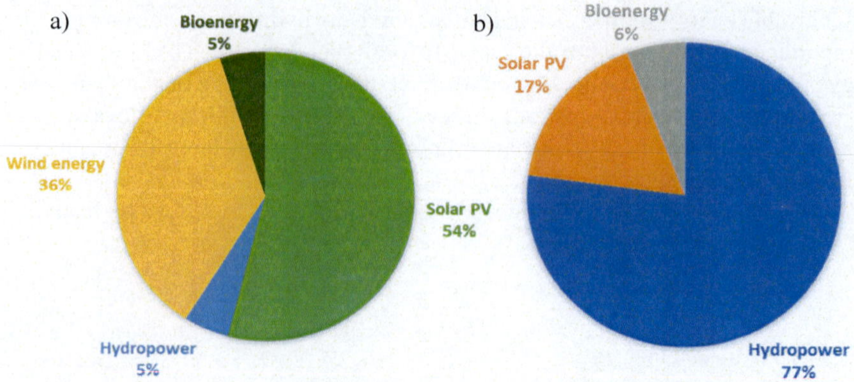

Fig. 2. Structure of renewable energy in a) Poland, b) Slovakia.

Table 1. Installed capacity in renewable energy sources

Installed capacity (MW)		
	Poland	Slovakia
Hydropower	978.0	2574.0
Bioenergy	1248.1	201.0
Solar PV	11, 924.0	573.0
Wind Energy	7864.8	3.0

from renewable sources such as photovoltaics and wind power is characterized by high variability and largely depends on weather conditions. Hydroelectric power plants, on the other hand, very often operate as peaking power plants, i.e. they work only during periods of peak energy demand. As a result, the actual energy obtained from renewable sources is lower than the installed capacity value would suggest.

The article presents a bibliometric analysis of research on renewable energy sources in Poland and Slovakia. The study was based on publications indexed in the Scopus database. The analysis covered the period from 1990 to 2023. The paper analyzes research topics related to renewable energy sources and examines the most cited publications by authors from the analyzed countries and the main research centers.

2 Bibliometric Analysis of the Literature

Bibliometric analysis is a method of studying scientific publications that is based on the use of statistical techniques to collect, process and analyze bibliographic data. The purpose of bibliometric analysis is to assess the impact of scientific publications on the development of a particular scientific field and to identify the most important authors, journals, topics and research centers [21–23].

For the analysis, CiteSpace software was used, which allows analysis of publications and citations and creates visualizations showing the network of connections between publications, authors, institutions and keywords. CiteSpace allows the generation of maps and graphs that show connections between scientific publications in a given research area. The program can detect trends and patterns in scientific publications, as well as identify key authors, institutions, research topics and scientific journals. CiteSpace allows analysis of bibliometric data from a variety of sources, including databases such as Web of Science, Scopus, PubMed and others.

This publication uses data from the Scopus database. "Renewable Energy" was used as the search criterion. Then, through filters, the country was set to Poland and then to Slovakia.

2.1 Bibliometric Analysis for Poland

Figure 3 shows an analysis of publications by authors' affiliations. Through this analysis, it is possible to determine which institutes are most involved in renewable energy research.

Based on Fig. 3, Table 2 was made, which shows the 15 centers with the highest number of publications and the corresponding centrality coefficient. The centrality coefficient in bibliometric analysis signifies how important a node is in the network and its role in connecting other nodes. In the case of bibliometric analysis, this coefficient can help identify authors, universities or publications that play a central role in a particular field of scientific research. This means that publications from universities with the highest centrality index, characterize the greatest innovation and are the most frequently cited.

Based on the analysis performed, it can be concluded that for the adopted criteria, the highest number of publications is shown by AGH University of Krakow, while the highest centrality coefficient, which indicates the key role in research on renewable energy sources, has the Polish Academy of Sciences, which means that the research done by its scientists is most often cited.

Figure 4 shows an analysis of citations by major areas of interest to researchers.

Based on the analysis seen in Fig. 4, the CiteSpace program can analyze publications for the citation burst indicator. This indicator is used to identify articles that have gained a significant number of citations in a short period of time. Citation burst can be used to identify articles with high scientific value and those that have influenced the further development of the field. Figure 5 shows the 10 publications with the highest citation burst.

Analyzing the publications with the highest citation burst, it can be seen that the highest number of citations are articles relating to the issues of Poland's energy transformation, which currently obtains most of its energy from coal [24–28]. An important aspect of these studies is the impact of renewable energy sources on Poland's energy security and the problems of implementing renewable energy sources in Poland, such as the law defining minimum distances of wind turbines from buildings. Attention was also given to the problem of implementing low-carbon technologies in rural areas [29]. This problem is particularly evident during the heating season, when many households use solid fuel stoves to heat their homes. However, due to the lack of access to the gas

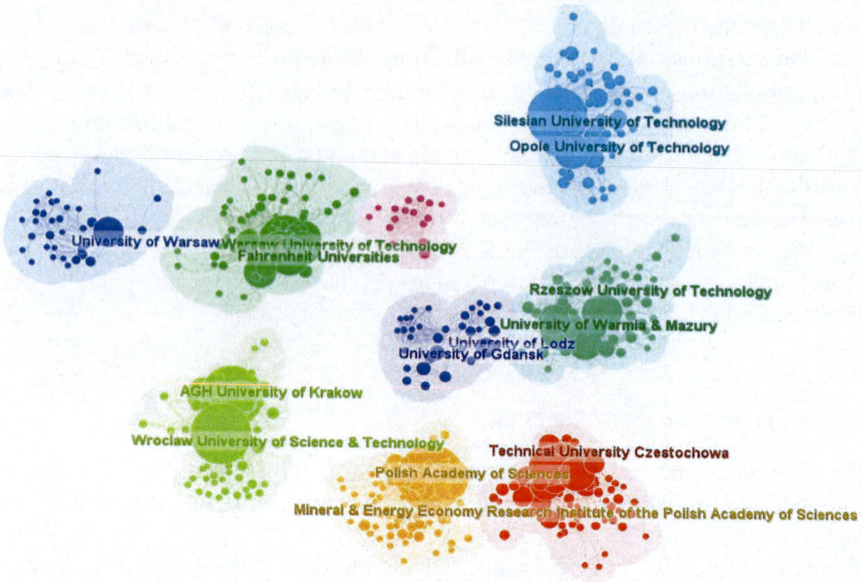

Fig. 3. Cooperation between universities.

Table 2. Number of publications and centrality by university

Institution	Number of publications	Centrality
AGH University of Krakow	412	0.08
Polish Academy of Sciences	390	0.21
Wroclaw University of Science and Technology	368	0.14
Warsaw University of Technology	343	0.13
Silesian University of Technology	339	0.06
Fahrenheit Universities	212	0.02
Lublin University of Technology	206	0.05
Technical University of Czestochowa	190	0.06
Gdansk University of Technology	188	0.02
University of Warmia and Mazury	168	0.01
Warsaw University of Life Sciences	157	0.03
Lodz University of Technology	140	0.01
Poznan University of Technology	133	0.02
Bialystok University of Technology	120	0.02
Krakow University of Technology	115	0.01

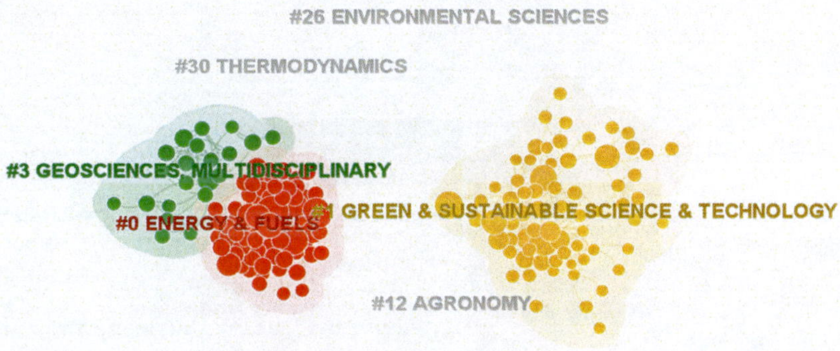

Fig. 4. Citation analysis.

References	Year	Strength	Begin	End	1993 - 2023
Budzianowski WM, 2012, RENEW SUST ENERG REV, V16, P342, DOI 10.1016/j.rser.2011.07.161, DOI	2012	4.32	**2012**	2017	
Szargut J, 2005, EXERGY METHOD TECHNI, V0, P0	2005	4.72	**2014**	2018	
Szargut J, 2002, ENERG CONVERS MANAGE, V43, P1149, DOI 10.1016/S0196-8904(02)00005-5, DOI	2002	4.92	**2015**	2018	
Parasyuk OV, 2012, OPT MATER, V35, P65, DOI 10.1016/j.optmat.2012.07.002, DOI	2012	4.86	**2017**	2018	
Piwowar A, 2019, ENERGIES, V12, P0, DOI 10.3390/en12183558, DOI	2019	5.45	**2020**	2021	
Marks-Bielska R, 2020, ENERGIES, V13, P0, DOI 10.3390/en13184624, DOI	2020	9.71	**2021**	2023	
Pietrzak MB, 2021, ENERGIES, V14, P0, DOI 10.3390/en14082046, DOI	2021	5.81	**2021**	2023	
Brauers H, 2020, ENERG POLICY, V144, P0, DOI 10.1016/j.enpol.2020.111621, DOI	2020	5.04	**2021**	2023	
Gielen D, 2019, ENERGY STRATEG REV, V24, P38, DOI 10.1016/j.esr.2019.01.006, DOI	2019	4.78	**2021**	2023	
Owusu PA, 2016, COGENT ENG, V3, P0, DOI 10.1080/23311916.2016.1167990, DOI	2016	4.53	**2021**	2023	

Fig. 5. Top 10 references with the strongest citation burst.

grid and the costliness of installing heating systems based on renewable energy sources, rural residents continue to use old and environmentally unfriendly solutions.

2.2 Bibliometric Analysis for Slovakia

Figure 5 shows the analysis in terms of authors' affiliations performed for publications from Slovakia (Fig. 6).

Table 3 shows the 10 institutes with the highest number of publications and their corresponding centrality coefficient. The analysis shows that the largest number of publications on renewable energy sources is published by scientists from the Technical University of Košice. Based on the centrality coefficient, it can be concluded that the highest number of citations and the greatest influence on other scientists are characterized by the works of scientists from the Slovak Academy of Sciences.

Figure 7 shows an analysis of citations by the main areas of interest for scientists from Slovakia.

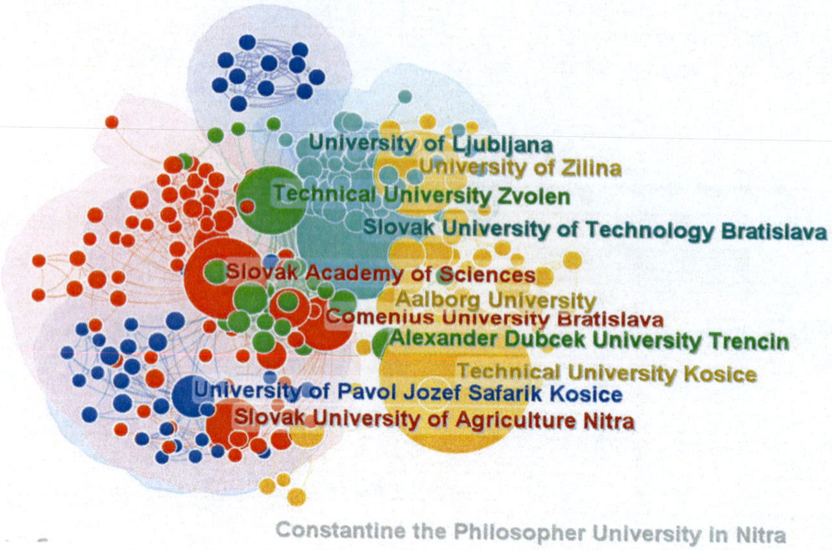

Fig. 6. Cooperation between universities.

Table 3. Number of publications and centrality by university

Institution	Number of publications	Centrality
Technical University of Kosice	323	0.22
Slovak University of Technology Bratislava	239	0.20
University of Zilina	103	0.08
Slovak Academy of Sciences	94	0.57
Technical University of Zvolen	62	0.20
Comenius University Bratislava	42	0.13
Slovak University of Agriculture Nitra	35	0.05
University of Pavol Jozef Safarik Kosice	23	0.10
University of Presov	15	0.04
University of Economics Bratislava	14	0.00

Figure 8 shows the 5 publications with the highest citation burst rate.

Analyzing the results obtained, it can be concluded that the highest number of citations in a short period of time was obtained by research on photovoltaic installations. The authors of the research present maximum power point tracking (MPPT) algorithms that will increase the efficiency of photovoltaic panels [30, 31]. The authors also discuss problems related to DC-DC converters and their impact over the performance of renewable energy systems.

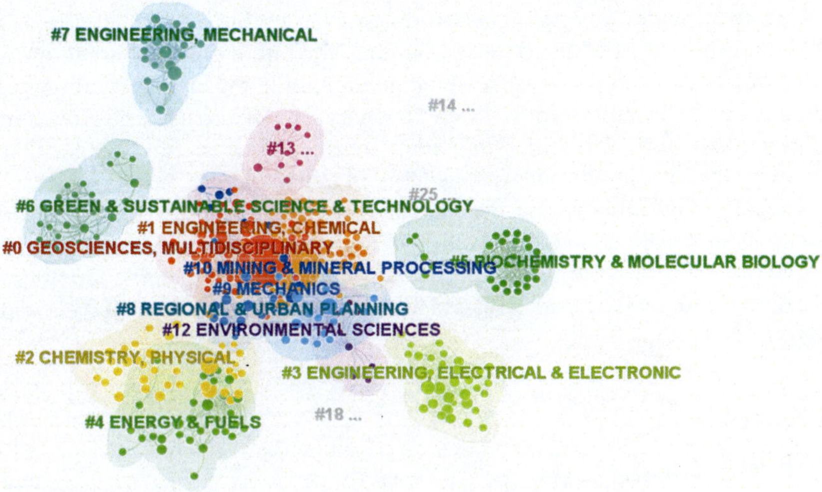

Fig. 7. Citation analysis

References	Year	Strength	Begin	End	1993 - 2023
Sanjeevikumar P, 2015, PROC IEEE INT SYMP, V0, PP214, DOI 10.1109/ISIE.2015.7281471, DOI	2015	5.82	2017	2018	
Kuric I, 2021, PROCESSES, V9, P0, DOI 10.3390/pr9050843, DOI	2021	3.58	2022	2023	
Jain S, 2016, ENG SCI TECHNOL, V19, P1771, DOI 10.1016/j.jestch.2016.07.008, DOI	2016	3.56	2018	2018	
Padmanaban S, 2017, INT J COMPUT INT SYS, V10, P140, DOI 10.2991/ijcis.2017.10.1.10, DOI	2017	3.52	2017	2017	
Blaabjerg F., 2016, J ELECT POWER COMPON, V0, P1	2016	3.52	2017	2017	

Fig. 8. Top 5 references with the strongest citation burst.

3 Conclusion

The article presents a bibliometric analysis of research on renewable energy sources in Poland and Slovakia. Thanks to the use of programs such as CiteSpace, it is possible to identify the scientific centers that constitute the most important links in research on renewable energy sources. The analysis shows that in Poland and Slovakia, the key centers in terms of innovation and number of citations are the Polish and Slovak Academy of Sciences. On the other hand, the number of publications in a given field largely depends on the size of the academic center and the number of scientists working in them, and does not indicate the innovativeness of a given scientific center.

Analyzing specific publications and links between them, it can be concluded that in the case of Poland, publications on Poland's energy transformation have the highest number of citations. Currently, most of Poland's energy is obtained from coal, and the use of renewable energy sources is critical to Poland's energy security. One of the main problems is the implementation of renewable energy sources, including the problem of minimum distances of wind turbines from buildings, as specified in the law. In addition, the problem of implementing low-carbon technologies in rural areas was also studied.

This analysis provides a better understanding of the problems of energy transition in Poland and can help develop effective solutions for future generations. In the case of Slovakia, the most cited publications relate to increasing the efficiency of photovoltaic installations through improved hardware, such as DC-DC converters. However, there is still a need for further research in certain areas, such as energy storage, smart grid technologies, and the use of renewable energy in transport.

The results of the bibliometric analysis can guide researchers and investors who make decisions about the direction of renewable energy research and investment. The analysis can also be used to assess the performance and effectiveness of scientific activities, which is important for making decisions related to their funding and assessing their impact on society.

References

1. Amponsah NY, Troldborg M, Kington B, Aalders I, Hough RL (2014) Greenhouse gas emissions from renewable energy sources: a review of lifecycle considerations. Renew Sustain Energy Rev 39:461–475
2. Wang Q, Guo J, Li R, Jiang X (2022) Economic growth and carbon emission in the organization for economic cooperation and development countries: the effects of oil, gas, and renewable energy. Energy Environ 0958305X2211388
3. Elshafei G, Vilcekova S, Zelenakova M, Negm AM (2021) Towards an adaptation of efficient passive design for thermal comfort buildings. Sustainability 13:9570
4. Telly Y, Liu X, Gbenou TRS, Investigating the Growth effect of carbon-intensive economic activities on economic growth: evidence from Angola. Energies 16:3487
5. Yang H, He Q, Cui L, Mohamed Taha AM (2023) Exploring the spatial relationship between urban vitality and urban carbon emissions. Remote Sensing 15:2173
6. Paska J, Surma T (2014) Electricity generation from renewable energy sources in Poland. Renewable Energy 71:286–294
7. Serowaniec M (2021) Sustainable development policy and renewable energy in Poland. Energies 14:2244
8. Wohlgemuth N, Wojtkowska-Łodej G (2003) Policies for the promotion of renewable energy in Poland. Appl Energy 76:111–121
9. Brożyna J, Strielkowski W, Fomina A, Nikitina N (2020) Renewable energy and EU 2020 target for energy efficiency in the Czech Republic and Slovakia. Energies 13:965
10. Csikosova A, Antosova M, Culkova K (2014) Trading with renewable energy sources in Slovakia. AASRI Procedia. 7:20–25
11. Ondrejka Harbulakova V, Zelenakova M, Purcz P, Olejnik A (2018) Selection of the best alternative of heating system by environmental impact assessment—case study. Environments 5:19
12. Kut P, Pietrucha-Urbanik K, Tchórzewska-Cieślak B (2021) Reliability-oriented design of a solar-PV deployments. Energies 14:6535
13. Szabó G-S, Szabó R, Szabó L (2022) A review of the mitigating methods against the energy conversion decrease in solar panels. Energies 15:6558
14. Li B, Li M, Yan S, Zhang Y, Shi B, Ye J (2023) An optimal energy storage system sizing determination for improving the utilization and forecasting accuracy of photovoltaic (PV) power stations. Front Energy Res 10:1074916
15. Pimentel Da Silva GD, Magrini A, Branco DAC (2020) A multicriteria proposal for large-scale solar photovoltaic impact assessment. Impact Assess Project Appraisal 38:3–15

16. Rabczak S, Nowak K (2022) Possibilities of adapting a free-cooling system in an existing commercial building. Energies 15:3350
17. Can M, Ahmed Z (2023) Towards sustainable development in the European Union countries: does economic complexity affect renewable and non-renewable energy consumption? Sustain Dev 31:439–451
18. Ollier L, Metz F, Nuñez-Jimenez A, Späth L, Lilliestam J (2022) The European 2030 climate and energy package: do domestic strategy adaptations precede EU policy change? Policy Sci 55:161–184
19. Bórawski P et al (2022) Development of renewable energy sources in the European union in the context of sustainable development policy. Energies 15:1545
20. https://ec.europa.eu/eurostat/. Last accessed 15 May 2023
21. Kut P, Pietrucha-Urbanik K (2022) Most searched topics in the scientific literature on failures in photovoltaic installations. Energies 15:8108
22. Özköse H (2023) Bibliometric analysis and scientific mapping of IoT. J Comput Inf Syst 1–22
23. Al Horr Y, Arif M, Kaushik AK, Arsalan H, Mazroei A, Rana MQ (2023) Thermal comfort in buildings: scientometric analysis and systematic review. J Archit Eng 29:03123001
24. Gielen D, Boshell F, Saygin D, Bazilian MD, Wagner N, Gorini R (2019) The role of renewable energy in the global energy transformation. Energ Strat Rev 24:38–50
25. Pietrzak M, Igliński B, Kujawski W, Iwański P (2021) Energy transition in Poland—assessment of the renewable energy sector. Energies 14:2046
26. Marks-Bielska R, Bielski S, Pik K, Kurowska K (2020) The importance of renewable energy sources in Poland's energy mix. Energies 13:4624
27. Brauers H, Oei P-Y (2020) The political economy of coal in Poland: drivers and barriers for a shift away from fossil fuels. Energy Policy 144:111621
28. Budzianowski WM (2012) Sustainable biogas energy in Poland: Prospects and challenges. Renew Sustain Energy Rev 16:342–349
29. Piwowar A, Dzikuć M (2019) Development of renewable energy sources in the context of threats resulting from low-altitude emissions in rural areas in Poland: a review. Energies 12:3558
30. Jain S, Ramulu C, Padmanaban S, Ojo JO, Ertas AH (2016) Dual MPPT algorithm for dual PV source fed open-end winding induction motor drive for pumping application. Eng Sci Technol Int J 19:1771–1780
31. Sanjeevikumar P, Grandi G, Wheeler PW, Blaabjerg F, Loncarski J (2015) A simple MPPT algorithm for novel PV power generation system by high output voltage DC-DC boost converter. In: 2015 IEEE 24th international symposium on industrial electronics (ISIE). IEEE, Buzios, Rio de Janeiro, Brazil, pp 214–220

Vibro-pressed Concrete in Ukraine: Problems, Trends and Prospects for Development

Ihor Lichnov(✉) and Vadym Kahanov

Lviv Polytechnic National University, Lviv, Ukraine
ihor.s.lichnov@lpnu.ua

Abstract. In this publication, the authors' team from the Department of Construction Production at Lviv Polytechnic National University analyzes the current state of implementing vibro-compressed technology for ultra-hard concrete mixtures using high-productivity concrete forming equipment in Ukraine at the end of the XX and the beginning of the XXI century. The article examines the existing industrial capacities of industrial productions for mass production of a wide range of vibro-compressed concrete products in different regions of Ukraine and identifies the specific potential and development prospects of each of these manufacturers. The focus of the authors' publication is on the experience of manufacturing a diverse range of building products using vibro-compressed concrete production facilities under the trademark "Mii Dvir" (Lviv). The study investigates the test results of vibro-compressed concrete paving stones that were produced using the automated block making machine model MASA XL 9.1 (Germany) under real production conditions. The article formulates ways to enhance the operational and physical-mechanical characteristics of both the face-colored layer and the main (base) layer of vibro-compressed concrete paving stones, which are manufactured based on ultra-hard concrete mixtures with a W/C ratio of up to 0.36. In the study, in addition to describing the general trends in the development of vibro-compression for industrial concrete based on ultra-hard mixtures, the authors analyze the results of current testing of actual vibro-compressed concrete specimens obtained in the accredited measurement laboratory of the "Mii Dvir" trademark and the specialized laboratory of construction materials at the Department of Construction Production, Lviv Polytechnic National University.

Keywords: Vibro-compression · Ultra-hard concrete mixtures · Concrete block making machine · Compressive strength of concrete · Water absorption · Tensile splitting strength of concrete

1 Introduction

At the end of the XX and the beginning of the XXI century, Ukraine witnessed a significant growth in the production of vibro-compressed concrete products based on ultra-hard concrete mixtures, manufactured using the semi-dry compacting method and specialized equipment.

Z. Blikharskyy et al. (Eds.): CEE 2023, LNCE 438, pp. 224–231, 2024.
https://doi.org/10.1007/978-3-031-44955-0_23

In most regions of Ukraine, manufacturers of concrete construction products started adopting the semi-dry vibro-compression method for producing ultra-hard concrete mixtures on a mass scale. This was done with the aim of creating modern pavements using small-sized colored paving stones, contributing to the improvement of urban areas and pedestrian zones in populated areas.

The first company to introduce vibro-compressed concrete made from ultra-hard and rigid mixtures in Ukraine was the LLC "Magik" based in Lviv. Since 1997, they have been engaged in large-scale production of paving stones using modern professional block making machines HB 828 MAZ and 940 MAZ, manufactured and engineered by Zenith (Germany) company. The technological equipment from the Zenith company was designed and produced for vibro-compressing ultra-hard and hard concrete mixtures with a W/C ratio ≤ 0.36 in a closed circuit, under the load exerted by punches in individual compartments of the steel molds.

In the year 2000, the Kovalska Plant, the flagship of the construction industry in Eastern European countries, made its choice and installed German block making machines of the "Topwerk Group" corporation, HESS RH 1500-SA model. The concrete products produced by these machines actively met the demand for high-quality vibro-compressed elements with vibrant and architecturally appealing textured surfaces in the beautification projects of Kyiv and the Kyiv region, catering to both urban and rural landscapes.

Continuing—In the city of Dnipro, the local concrete product manufacturer, LLC "Gang", chose the concrete forming equipment of the German conglomerate MASA-HENKE as its primary technological equipment. This equipment was dismantled in Germany during the replacement of existing technological production lines for vibro-compressed concrete elements with a more modern and economically viable forming equipment from industrial premises in the suburbs of Cologne.

In Rivne region (Kostopil) and Rivne city, the manufacturers of paving stones opted for technological lines manufactured by the American company "Columbia," which are designed to leverage the powerful effect of pressing artificial composite products based on mineral binders.

All the aforementioned concrete compacting equipment can be classified as professional equipment capable of producing between 800 and 2000 m^2 of paving stones in a single shift [1].

2 The Purpose and Objectives of Scientific Research

The authors of the article aimed to analyze the production experience of the trademark "Mii Dvir" (Lviv) in the overall process of implementing vibro-compressed products based on ultra-hard and hard fine-grained mixtures for the production of high-quality landscaping elements for populated areas in Ukraine.

3 Methodology of scientific research

Normative laboratory tests were conducted as part of the scientific research, following the requirements of the current standards DSTU B EN 1338:2016 and DSTU B EN 1339:2016, in the certified measurement laboratory of the operating manufacturing enterprise of TM "Mii Dvir," under the Ministry of Economic Development and Trade of Ukraine.

In the analysis of the research results of the natural samples of paving stones conducted by the authors, the parameters obtained during the certification and current acceptance tests conducted by the measurement laboratory of the trademark "Mii Dvir" were used. These tests were carried out during the period from 2018 to 2023.

4 Analysis of recent scientific research based on literary sources.

Significant contributions have been made by researchers in the field of scientific investigations into the prospects and characteristics of the technology of vibro-compressed of ultra-hard and hard concrete mixtures. Polish researcher W. Brylicki thoroughly studied the nature of form formation and the algorithm for constructing a high-quality and dense concrete structure for vibratory-pressed concrete paving stones [2–4].

Researchers from the United States, Ramachandran [5], and Wittmann from the Netherlands [6], dedicated their research to studying the mechanism of structure formation and the dynamics of internal bonding between components in vibratory-pressed rigid concrete.

It is worth noting that studies on the topic of vibratory pressing of rigid concrete were not limited to Europe and the United States. Concrete experts from Australia, D. McCarthy and M. Power, also extensively studied the properties of vibratory-pressed concrete for industrial formation of concrete elements for landscaping [7, 8].

In recent years, significant scientific interest in the properties of vibratory-pressed concrete has been demonstrated by researcher Eduard Batiyanovsky from Belarus [9], and Usachev from the Russian Federation [10], who meticulously investigated the properties and technological aspects of this type of high-quality concrete.

It is important to highlight that researchers from Ukraine have actively participated in the global research process on vibratory-pressed concrete mixtures. Professor Leonid Dvorkin from Rivne, together with his followers Vadim Zhytkovskyi and Vadim Kahanov from Lviv, devoted their scientific works to this issue, focusing on both conventional building materials and the use of screenings and durable rock formations [11–13].

Special role in the study of properties of fine-grained concrete based on high-quality modified mixtures and investigating the influence of particle size distribution on the fundamental characteristics of hard concrete is played by the scientific works of experts from the powerful Ukrainian school led by Professor Myroslav Sanytskyi. This research is being successfully continued by representatives of the Department of Construction Production at the Lviv Polytechnic National University, Doctors of Technical Sciences Uliana Marushchak and Tetiana Kropyvnytska [14–17].

However, despite the extensive research conducted on the properties of vibro-compressed concrete, there are still significant problematic issues related to the formation of concrete structures based on ultra-hard mixtures and the necessity of studying the processes of creating a dense structure in vibro-compressed concrete products manufactured in factories using modern technological equipment. The article prepared by the team of authors from the Department of Construction Production at the Lviv Polytechnic National University aims to identify factors for enhancing the operational and physical-mechanical properties of vibro-compressed concrete products and exploring the prospects of this compacting method in the conditions of modern Ukraine.

5 The main material of the article

From 1997 to 2023, the industrial technology of vibro-compression of ultra-hard concrete mixtures has been significantly developed at the production facilities of the "Mii Dvir". This development involves the use of professional German-made concrete compacting complexes. The semi-dry vibro-compression technology was chosen as the basis, ensuring high operational characteristics of a wide range of concrete products (paving stones, concrete curb stones, small architectural forms, hollow wall blocks). The production principles of the "Mii Dvir" in Lviv are based on the following approaches to the vibro-compression technological process:

- two layers of mixtures are used in the production process of concrete products. The base concrete layer consists of fractionated gravel, with a water-cement ratio of up to 0.36, consisting of 40% of 2/5 mm-sized gravel and 60% of 5/10 mm-sized gravel as the main coarse aggregate. The face textured layer of concrete is made using pre-washed construction sand with a fineness modulus of at least 2.1 (40% of the total aggregate mass) and granite gravel sized 0.5/2 mm (60% of the total aggregate mass).
- Portland cements CEM I 42.5 and CEM II 42.5 A/S, produced by Ivano-Frankivsk Cement Plant, are used as binders in the technological process. The main physical-technical characteristics of these cements are provided in Table 1 and Fig. 1.
- to ensure ease of placement of hard and ultra-hard concrete mixtures in both layers of concrete, plasticizing additives of type I based on lignosulfonates, produced by "Sika-Ukraine," are used.
- to enhance the architectural expressiveness of the textured face layer of concrete, colored inorganic pigments from LANXESS concern (Germany) are used.

The use of semi-dry vibro-compression technology for concrete mixtures, which are sequentially placed in steel moulds with a loading force on the tamperhead of up to 1070 kN per square meter of production, and further vibration compaction of the freshly formed mixture on the work table for 8 to 12 s, has created conditions for obtaining highly durable paving stones with increased density in the early days of curing in the thermo-humidity chamber at an average temperature of +36 °C and humidity up to 90%.

Figure 2 and the results presented in Table 2 show the main physical-mechanical characteristics (compressive strength, tensile strength, water absorption) of the natural samples of concrete products obtained by the certified measurement laboratory of the "Mii Dvir" brand from 2018 to 2023. Particularly interesting is the presence of high strength in the natural concrete samples on the 3rd and 7th day after manufacturing, indicating the appropriate selection of working compositions for concrete mixtures and optimal vibro-compression regimes.

Thus, the implementation of the main principles of organizing the vibro-compression technological process on modern block making machines and the methods of selecting optimal working compositions for both layers of the finished product in the production of paving stones used in the manufacturing process of the "Mii Dvir" (Lviv) has been achieved.

Table 1. Physical and mechanical properties of cement according to the results of laboratory tests of the Mii Dvir

Year	Cement type	Physical-mechanical properties of cement			
		Bulk density, kg/m³	Compressive strenght, Mpa		Normal thickness of the cement paste, %
			48 h	28 days	
2018	IF CEM I 42.5 R	0.925	24.94	51.53	28.3
2019	IF CEM I 42.5 R	0.912	26.24	53.91	28.7
2019	IF CEM II/A-S 42.5 R	0.928	24.08	51.30	27.4
2020	IF CEM I 42.5 R	0.904	27.12	55.76	28.5
2020	IF CEM II/A-S 42.5 R	0.932	24.79	52.58	28.1
2021	IF CEM II/A-S 42.5 R	0.919	25.33	58.86	28.8
2022	IF CEM II/A-S 42.5 R	0.916	24.43	51.96	29.4
2023	IF CEM II/A-S 42.5 R	0.929	20.85	48.11	28.8

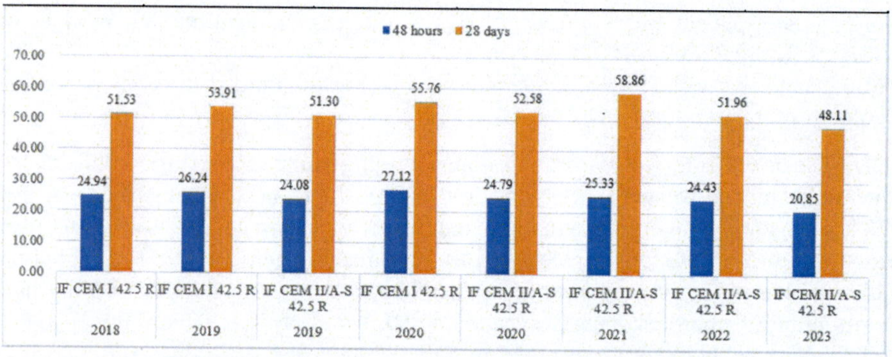

Fig. 1. Diagram of the compressive strength of cement

There are significant reserves for improving the main operational properties of vibro-pressed ultra-hard and hard concrete mixtures in the process of serial production for the manufacture of road and municipal elements for landscaping. These reserves include the following aspects:

- Application of nano-modified components for high-strength and dense fine-grained hard and ultra-hard concrete mixtures.

- Use of special technologies for preparing aggregates from construction sands and cubical shaped aggregates from strong igneous rocks (granite, diabase, basalt, quartzite, and other magmatic rocks).
- Introduction of ready-made building mixtures containing dispersed fibers from polymer materials and steel wire parts with lengths ranging from 5 to 10 mm into the composition of hard and ultra-hard concrete mixtures for vibro-pressing.
- Replacement of Group I plasticize additives and mineral admixtures with high-performance chemical additives of a new generation that improve the process of forming a dense matrix structure of concrete and enhance the rheological properties of mixtures by changing the algorithm of bond formation between components at the meso- and micro-levels of concrete structure.

Table 2. Physical and mechanical properties of paving stones according to the results of laboratory tests of the Mii Dvir.

Physical and mechanical properties					
Year	Compressive strength, MPa			Total water absorption, %	Tensile splitting test, MPa
	24 h	7 days	28 days		
2018	22.9	40.2	50.5	4.9	3.49
2019	22.9	43.4	51.3	4.8	3.55
2020	23.5	44.4	52.5	4.4	3.72
2021	17.4	39.6	52.6	4.8	3.70
2022	24.9	42.8	52.4	4.7	3.80
2023	17.8	40.3	51.8	4.8	3.90

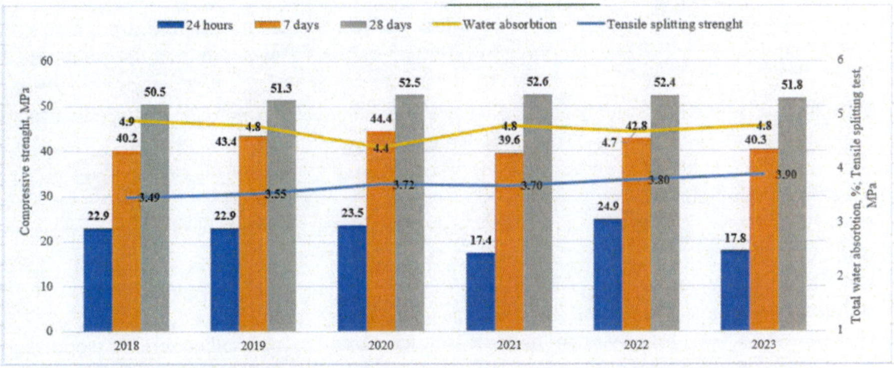

Fig. 2. Diagram of the physical and mechanical properties of the paving stones

The aforementioned directions for enhancing the efficiency of vibro-pressing ultra-hard concrete mixtures are planned to be implemented in the coming years during the mass production of high-quality concrete products for the post-war road and transportation infrastructure of Ukraine at the facilities of the "Mii Dvir" in Lviv.

6 Conclusion

The presented features of producing vibro-pressed hand and ultra-hard concrete mixtures, based on the experience of modern German block making machines at the "Mii Dvir" facilities, lay the foundation for innovative approaches in the process of vibro-pressing high-quality concrete road and municipal products for urban improvement in Ukraine. The development directions for vibro-pressing concrete structural elements discussed in this publication define the immediate prospects for implementing modern manufacturing technologies in Ukraine for the production of high-quality concrete products, aiming to establish road networks and improve urban, rural, and residential areas.

The significant practical experience gained in the production of vibro-pressed hard and ultra-hard concrete mixtures holds considerable interest for Ukrainian manufacturers of high-quality construction products and concrete researchers involved in studying the characteristic properties of structure formation in vibro-pressed concrete and conducting in-depth scientific research on the bond formation between components that constitute the face layer and the base layer of concrete.

References

1. Kahanov VO (2022) Trends in the implementation of modern technology for vibro-compaction of rigid and stiff concrete mixtures in the industrial production of Ukraine. NUWEE Rivne 2(98):49–58
2. Brylicki W (1988) Paving stones made of vibropressed concrete, Polski Cement, Krakow 147
3. Brylicki W, Malolepszy J (1996) Cement-Wapno-Concrete 1
4. Sliwinski J, Brylicki W (2002) In: 3rd International conference on concrete structures, Zilina
5. Ramachandran VS (1982) Concrete admixtures handbook. Noyes Publication, Park Ridge, NJ
6. Wittmann FH (1989) Structure and fracture mechanics of composite materials: fracture toughness and fracture energy. Balkema, Rotterdam
7. McCarthy DJ (1986) Proceedings of the international workshop on interlocking concrete pavements, Melbourne
8. Powers TC (1958) Structure and physical properties of Portland hardened cement paste. J Am Ceram Soc 41(1):1–6
9. Batiyanovsky EI, Bondarovich AN (2018) Vibro-pressed concrete: technology and properties. Belarusian National Technical University., BNTU, Minsk, p 262
10. Usachev SM (2006) Improvement of the technology of vibro-pressed concrete by optimizing the balance of internal and external forces. (Doctoral dissertation, 05.23.05). Voronezh 216
11. Dvorkin LY, Zhytkovskyi VV, Kahanov VO (2006) Concretes based on rigid mixtures. Rivne, RDTsNTEI, 179
12. Zhytkovskyi VV (2003) Vibro-pressed fine-grained concrete using granite crushing screenings. (Doctoral dissertation abstract). Odessa 14

13. Zhytkovskyi VV, Dvorkin LY (2021) Influence of dispersed reinforcement on the properties of road concretes obtained by vibro-compaction of rigid mixtures. KHNADU 92(2):24–31
14. Sanitskyi MA, Kropyvnytska TP, Teviuk IM (2020) Modified rapid-hardening clinker-efficient concretes. Concrete Mater Prod 1:22–28
15. Maruchak U, Sanitsky M, Sydor N, Margal I (2021) Designing of alkaline activated cementing matrix of engineering cementations composites. Theory Build Pract 3(2):52–57
16. Marushchak UD, Sanitsky MA, Gogol MV, Pozniak OR, Mazurak OT (2021) Investigation of nano-modified Portland cement systems. In: 9th international scientific and practical conference on the problems of reliability and durability of engineering structures and buildings in railway transport. Kharkiv, 427–430
17. Sanitsky M, Kropyvnytska M, Fisher H, Kondratieva N (2021) Effect of composite gypsum binders. In: 4. Weimarer Gips stagung, Weimar, 161–168

Influence of Shear Span-to-Depth Ratio (a/d) on the Mechanics of Crack Formation and Failure of Reinforced Concrete Beams on Inclined Sections

Solomiya Maksymovych(✉), Ihor Karkhut, and Olha Krochak

Lviv Polytechnic National University, Lviv 79013, Ukraine
Solomiia.B.Maksymovych@lpnu.ua

Abstract. One of the key points when calculating the strength of inclined sections (displacement calculation) is the issue of the design model. To create it, it is necessary to know not only the failure pattern, but also the order of crack formation and propagation during loading of structures. After the formation of cracks, a reinforced concrete beam has a complex structure that does not correspond to such systems as the arch or truss. The pattern of formation and propagation of diagonal cracks differs significantly from that of vertical cracks, which are the reason for the stretching stress or moment. Diagonal cracks are more dangerous, as they can be formed suddenly and the time interval between the crack formation and failure is considerably shorter than for pure bending. In this case, the strength of the vertical section is not achieved. This paper discusses the features of crack formation and failure of reinforced concrete beams on inclined sections with the shear span to effective depth ratio a/d = 0.5 and a/d = 1. The dynamics of propagation and width of opening of destructive diagonal cracks depending on the load is shown. The analysis of the studies revealed the specifics of formation and propagation of diagonal cracks in the zone of impact of the shear force and moment. The various loading patterns used for experimental samples revealed the specific pattern of the influence of positive and positive/negative moment diagrams on the crack formation and propagation.

Keywords: Reinforced concrete beams · Crack resistance · Shear strength · Mechanisms of shear failure · Critical shear crack · Experimental results

1 Introduction

The methods of calculating the strength of inclined sections are constantly being improved. However, the calculation method recommended by the current design standards [1] and European standards [2] does not allow fully taking into account the numerous factors that affect the nature of the stress-strain behaviour in these elements, causing in many cases an excessive consumption of materials, and sometimes insufficient reliability of structures.

Z. Blikharskyy et al. (Eds.): CEE 2023, LNCE 438, pp. 232–241, 2024.
https://doi.org/10.1007/978-3-031-44955-0_24

Researchers believe that the reason for the lag in solving the problem of "chipping" in reinforced concrete is the complexity of the state of stress in a beam in the areas of simultaneous impact of the shear force and moment, as well as the lack of a reliable theory of concrete strength. The above indicates the need to search for new approaches to assessing the strength of inclined sections through analysis and use of experimental studies.

At this stage of research, the issue of dynamics and opening width of diagonal cracks is one of the least studied. It is considered only in a few special works, and the number of studies in which the results of measuring the width of the opening of diagonal cracks are provided is even smaller. This is especially true for elements with a positive/negative bending moment diagram.

The experimental studies have established that diagonal cracks open much wider than the vertical ones. Often, due to the excessive opening of diagonal cracks in real building structures, it is necessary to reinforce them.

This problem has acquired particular relevance with the use of steel with enhanced mechanical characteristics for reinforced concrete structures, as a result of which the saturation of their cross-sections with longitudinal and transverse reinforcement decreases, the deformation of this reinforcement increases, and hence the width of the vertical and diagonal cracks.

It has been experimentally established that transverse and longitudinal reinforcement, bend, concrete strength, a/d and l/d ratios, and cross-sectional dimensions have the most significant influence on the opening width of diagonal cracks.

The basic studies, which are still referred to by modern researchers of shear reinforcement, are the experiments of Leonhardt and Walther [3] and Kani [4]. They have not lost their relevance and their conclusions are important for us. The authors comprehensively studied single-span and double-span beams of rectangular and T-shaped cross-sections under symmetrical loading with concentrated load and analyzed the dynamics of the formation and propagation of cracks at the stages of loading.

A characteristic pattern of formation and propagation of cracks in the area of constant shear force and a diagram of bending moments passing through the zero point can be observed in the Leonhardt F. HH beams. It is clearly seen in all the photos that there are no inclined cracks in the areas to the right and to the left of the intermediate support in the zones of impact of the constant shear force where the bending moment changes its sign.

The experimental studies of the school from École Polytechnique Fédérale de Lausanne, presented in the articles [5–7], include the testing of 20 reinforced concrete beams with a detailed description and analysis of crack formation and failure of beams. The digital image correlation was used in the work for deformation measurement. The beams with cross-sectional dimensions of 250×600 mm were tested according to different patterns (single-span, continuous, cantilever) under a concentrated or distributed load. The authors noted a large influence of shear span to ratio a/d and l/d on a critical crack formation and on the shear strength. The assumed ratio is $a/d = 1.87 - 6.87$ and $l/d = 3.75 - 10$.

In the article [8], six single-span 100 × 200 mm beams without transverse reinforcement are studied by detailed tracking of crack kinematics with photogrammetry techniques, the ratio a/d = 3.7 is assumed.

The authors [9] after testing five reinforced concrete beams with cross-sectional dimensions of 200 × 400 mm (a/d = 3 – 3.5) concluded that the mechanism of failure is closely associated with the loss of the dowel force. The width of the diagonal crack opening was measured at three levels along the height of the beam, and the maximum value before failure was 0.25 mm.

The total of 381 samples of beams was analysed in the research [10]. Based on statistical analysis, both ACI and CSA STM models were found to underestimate the experimental shear strength. The analysis showed that the ratio (a/d) and horizontal and vertical shear reinforcement ratio are the important parameters affecting the shear strength.

Smith and Vantsiotis [11] found out that the contribution of the vertical reinforcement decreases as a/d drops (a/d < 1.0).

In [12] Ashour A.F. concluded that the horizontal transverse reinforcement is more effective than the vertical transverse reinforcement for a/d < 0.75.

Based on the results of experimental studies in the dissertation [13], Yuguang Yang presented his vision of the mechanism of failure of reinforced concrete structural elements without shear reinforcement due to the opening of a diagonal crack in the critical shear span. A critical shear displacement is used as a failure criterion, and based on it a new method for shear estimation is proposed.

The program of our experimental studies included testing of three series of cantilever beams (19 beams), single-span and double-span ones, with various schemes of inclined planes reinforcement, at a/d = 0.5 and a/d = 1 [14–16].

2 Aim

The aim is, based on our own experimental studies and research carried out by other authors, to analyse the influence of the shear ratio a/d on the mechanics of crack formation and failure of reinforced concrete beams, without shear reinforcement, on inclined planes.

The research also aims at comparing the dynamics of crack formation and width of opening of the destructive diagonal crack of single-span beams for a/d = 0.5, a/d = 1 with the testing results for the cantilever beam with a positive/negative bending moment diagram for a/d = 1 in the zone of simultaneous impact of the moment and shear force.

3 Testing Method

The experimental samples had two loading patterns: single-span and single-span with cantilevers (Fig. 1). The cross-sectional dimensions of beams were b × h = 120 × 400 mm. The longitudinal reinforcements were assumed 4Ø12 A400C (Fig. 2). The beams without transverse reinforcement were chosen for our analysis. The characteristics of test samples of the beams are presented in Table. 1.

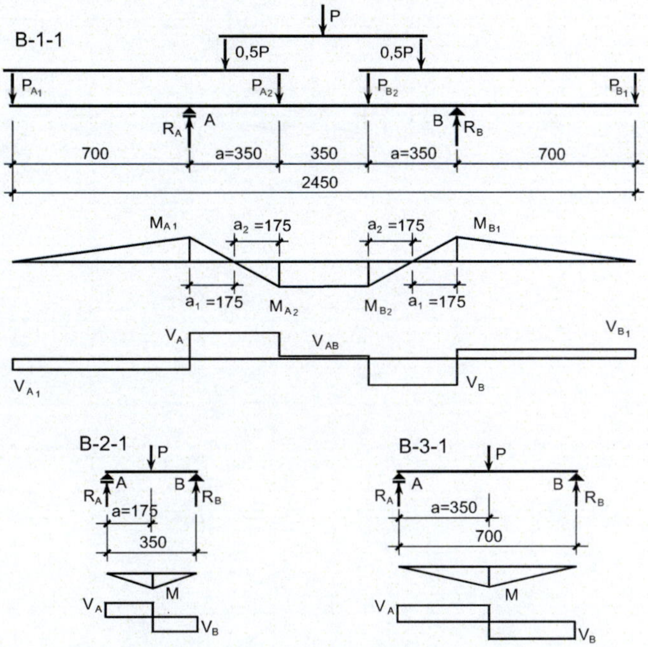

Fig. 1. Schematic diagram of beam testing.

4 Experimental Results

The patterns of formation and propagation of cracks and failure of beams are shown in Fig. 3. The results of experimental studies of beams are shown in Table 2 using the schematic symbols in Fig. 1.

All beams failed on the inclined sections. The total destructive force P, at which each beam failed, is shown in Table 2, column No. 4.

The figures show all the cracks that were recorded at the time of failure of the beams. In addition, the loads under which a crack was formed and propagated were recorded for each crack. Each crack has its own serial number, under which its opening width was fixed at different levels along the height of the beam at each stage of loading.

The first vertical cracks in the B-1-1 beam appeared above the supports A and B at the load $P = 300$ kN. Later, at $P = 400$ kN, span vertical cracks and diagonal cracks formed almost simultaneously. The diagonal cracks formed at supports A and B in the mid-height of the beam.

As the load rose to $P = 500$ kN, the formation of new vertical cracks was observed, which were getting closer to the supports, and the diagonal cracks continued to develop and increase the width of opening. At $P = 600$ kN, the diagonal cracks propagated intensively towards the applied force and towards the support, while the vertical cracks remained unchanged. At $P = 700$ kN, the diagonal cracks reached the bottom and top sides of the beam. A further increase in the load $P = 800 \div 1000$ kN led to an intensive increase in the opening width of diagonal cracks and their merging with other cracks, which ultimately led to the failure of the beam along one of these diagonal cracks.

Table 1. Characteristics of test beams.

Beam grade	l, mm	l_c, mm	b, mm	h, mm	d, mm (support/span)	Class of concrete	Longitudinal reinforcement				Transverse reinforcement		
							Reinforcement (upper and lower)	A_s, cm2	f_y, MPa	ρ_l, % (upper/lower)	Reinforcement	A_{sw}, cm^2	f_y, MPa
1	2	3	4	5	6	7	8	9	10	11	12	13	14
B-1-1	1050	700	120	400	347/350	C45/55	4Ø12 A400C	4.52	410.4	1.09/1.08	Ø10 A240	1.57	258
B-2-1	350	–	122	399	–/355	C45/55	4Ø12 A400C	4.52	410.4	–/1.04	Ø10 A240	1.57	258
B-3-1	700	–	123	399	–/363.	C45/55	4Ø12 A400C	4.52	410.4	–/1.01	Ø10 A240	1.57	258

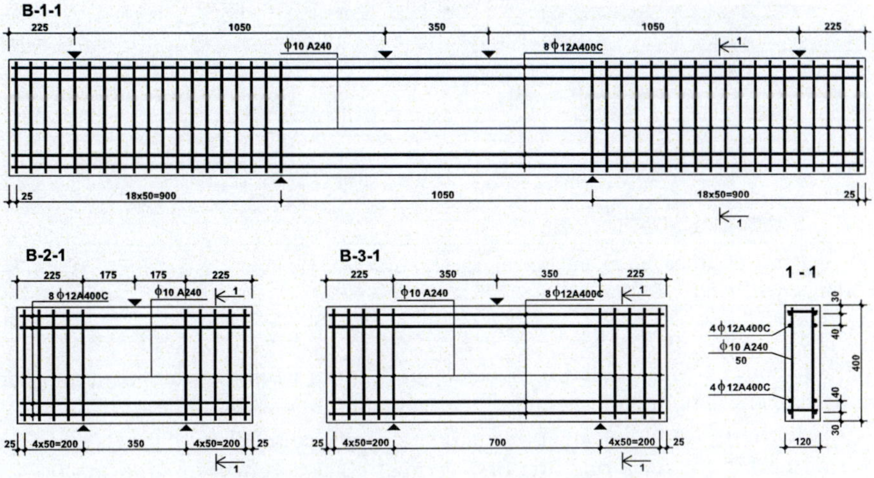

Fig. 2. Structures of the experimental beams.

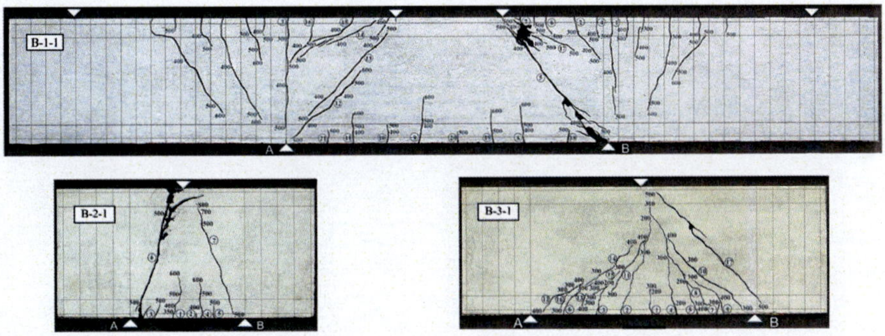

Fig. 3. Patterns of formation and propagation of cracks and failure of the B-1-1, B-2-1, B-3-1 beams.

Table 2. Results of experimental testing of beams.

Beam Grade	a, mm	a/d	P, kN	R_A, kN	R_B, kN	P_{A1}, kN	P_{A2}, kN	P_{B1}, kN	P_{B2}, kN	M_A, kNm	M_B, kNm	V_A, kN	V_B, kN
1	2	3	4	5	6	7	8	9	10	11	12	13	14
B-1-1	350	1.0	1317	655	662.1	131	521	132.4	532.6	91.7	92.7	524	489.7
B-2-1	175	0.49	1090	545	545.0	–	–	–	–	95.4	–	545	–
B-3-1	350	0.96	530	265	265.0	–	–	–	–	92.8	–	265	–

The dynamics of propagation and opening of diagonal cracks, which led to the failure of the B-1-1 beam, is shown in Fig. 4.

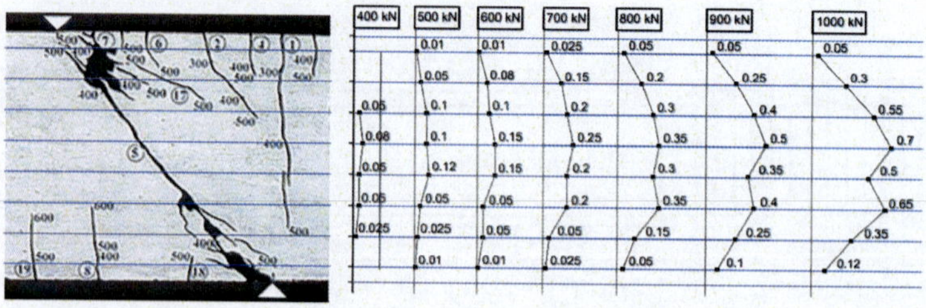

Fig. 4. Patterns of propagation and opening width of a diagonal crack No. 5 in the B-1-1 beam depending on the load.

The diagonal cracks started to nucleate in the mid-height of the beam and had a maximum opening in the mid-height of the beam until the beam failure.

The patterns of formation and propagation of cracks and failure of the B-2-1 beam is shown in the Fig. 5. As a rule, the first vertical cracks were formed inside the span at the load P = 300 kN. Later, at P = 400 kN, other longitudinal vertical cracks developed almost simultaneously with the diagonal cracks. The diagonal cracks formed simultaneously near two supports A and B.

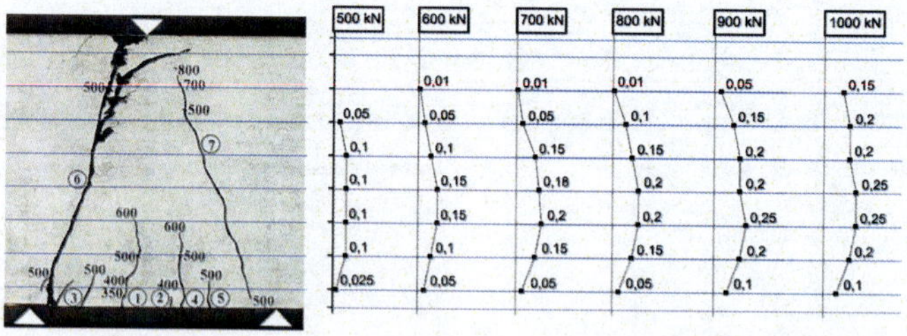

Fig. 5. Dynamics of propagation and opening width of a diagonal crack No. 6 in the B-2-1 beam depending on the load.

The formation of diagonal cracks in the B-2-1 beam occurred at approximately 1/3h off the bottom side (Fig. 5), and they did not reach the lower and upper sides. As the load rose to 500kN, the formation of new vertical cracks was observed, which were getting closer to the supports, and the diagonal cracks continued to propagate and increase their opening width.

The diagonal cracks at the load P = 600 kN started to propagate more intensively towards the supports and the applied force, while the vertical cracks slowed down their propagation.

At P = 700 kN, the diagonal cracks reached the bottom side of the beam, while the vertical cracks almost completely stopped their propagation. An increase in the load up to P = 900, 1000 kN and more led to the merging of diagonal cracks and an intensive increase in the width of their opening, up to the failure of beams along one of these diagonal cracks.

The dynamics of propagation and opening width of a diagonal crack, along which the failure of B-3-1 beam occurred, is shown in Fig. 6.

The first vertical cracks in B-3-1 beam appeared at P = 200 kN earlier than in the B-1-1 and B-2-1 beams. At the load P = 300 kN, new vertical cracks appeared closer to the supports, which started to transform into the diagonal ones, deviating to the applied force, and the diagonal cracks started to form in the middle of the beam at approximately 1/3 h off the bottom side. As the load P = 400 kN, the propagation of existing diagonal and vertical cracks, as well as the formation of new ones, was observed. In this beam, a new diagonal crack appeared at P = 500 kN in the mid-height of the beam along the direct line between the applied force and the support, along which the beam failed at P = 590 kN.

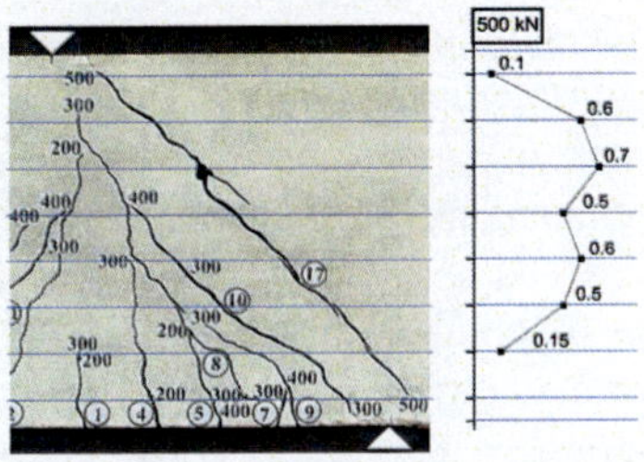

Fig. 6. Dynamics of propagation and opening width of a diagonal crack No. 17 (mm) in the B-3-1 beam depending on the load.

As seen from Fig. 6, the diagonal cracks in the B-3-1 beam started to nucleate in the middle of the beam and at the bottom side. The failure proceeded along the diagonal crack formed in the mid-beam almost immediately before the failure.

In the B-3-1 beam, the diagonal cracks were formed at a lower load and the difference between the crack formation load and the beam failure load is smaller than for the B-2-1 beam. At the load P = 500 kN, the maximum opening width of the diagonal crack along which the failure occurred was: B-3-1 Crack No. 17–0.7 mm.

5 Scientific Novelty

The analysis of the experimental tests of the beams revealed the features of the formation and opening of cracks in single-span beams and in beams with a sign-changing moment, when the force is applied near a support (at a/d = 0.5 and a/d = 1.0).

The comparison of the results of the tests showed the differences in the patterns of crack formation and failure of the B-3-1 single-span and the B-1-1 cantilever beams for a/d = 1 ratio, the moment being positive and positive-negative.

Comparing the ratio of bending moments and shear forces of the B-1-1, B-2-1 and B-3-1 beams without transverse reinforcement (Table 2):

$$\frac{M_{u,inc(\text{B-1-1})}}{M_{u,inc(\text{B-2-1})}} = \frac{92.7}{95.4} = 0.97; \qquad \frac{V_{u,inc(\text{B-1-1})}}{V_{u,inc(\text{B-2-1})}} = \frac{529.7}{545.0} = 0.97;$$

$$\frac{M_{u,inc(\text{B-1-1})}}{M_{u,inc(\text{B-3-1})}} = \frac{92.7}{92.8} = 0.998 \approx 1.0; \qquad \frac{V_{u,inc(\text{B-1-1})}}{V_{u,inc(\text{B-3-1})}} = \frac{529.7}{265.0} = 1.99;$$

$$\frac{M_{u,inc(\text{B-2-1})}}{M_{u,inc(\text{B-3-1})}} = \frac{95.4}{92.8} = 1.028 \approx 1.03; \qquad \frac{V_{u,inc(\text{B-2-1})}}{V_{u,inc(\text{B-3-1})}} = \frac{545.0}{265.0} = 2.06.$$

Shows that the value of the ratio of the moments at which the beams failed on inclined sections varied from 0.97 to 1.03, and the value of the ratio of shear forces varied from 0.97 to 2.06. Perhaps that is why the assessment of the bearing capacity of inclined sections on the impact of the shearing moment M_{su} (theory of Walther R.) is more reliable and logical than against the impact of the shear force.

6 Practical Value

The detailed tests of reinforced concrete beams made it possible to increase the experimental basis for further analysis of the strength of inclined sections. The obtained results provide the rationale for understanding the shear phenomenon in reinforced concrete beams without transverse reinforcement.

According to the current design regulations, the inclined sections are calculated based on the shear force that is constant in the segment from the force to the intermediate support, whereas the moment changes the sign. Although the behaviour of inclined sections of such elements has much in common with the behaviour of inclined sections in simply supported beams, it also differs significantly from them. A change in the sign of the bending moment diagram in the area of constant shear force indicates that the compressed zone of the beam changes its position. Therefore, this should be taken into account when calculating the strength of inclined sections.

7 Conclusions

1. The formation of diagonal cracks in the beams in the areas with a positive/negative moment diagram at $a/d = 1$ occurs in the mid-height of the beam. In the beams with a positive bending moment diagram or negative one at $a/d = 0.5$, diagonal cracks also start to form in the middle of the beam.
2. In the single-span beams at $a/d \geq 1$, the formation of diagonal cracks starts on the bottom side of the beam from the vertical cracks that developed earlier and the maximum width of opening of diagonal cracks is observed on the bottom stretched side of the beam.
3. In the beams with a positive/negative bending moment diagram at $a/d > 2$ (HH beams of Leonhardt F.) no cracks were observed in the areas on both sides from the zero point, though the transverse force is constant in this area.
4. The maximum opening width of the diagonal cracks, beginning from their formation and until the beam failure, is observed at the sites of their formation and equals to 0.7 mm for B-1-1, 0.25 mm for B-2-1 and 0.7 mm for B-3-1.
5. The comparison of the ratio of bending moments and shear forces of the B-1-1, B-2-1 and B-3-1 beams without transverse reinforcement shows that it is more reliable and logical to assess the bearing capacity of inclined planes against the impact of the shearing moment M_{su} than against the impact of the shear force.

References

1. DSTU B V.2.6-156:2010 (2011) Konstruktsii budynkiv i sporud. Betonni ta zalizobetonni kon-struktsii z vazhkoho betonu. Pravyla proektuvannia.—K. Minrehionbud Ukrainy. 118 s
2. Eurocode 2 (2004) Design of concrete structures. EN 1992–1.1: general rules and rules for buildings. Brussels, CEN, 226 p
3. Leonhardt F, Walther R (1962) Shear tests in single span reinforced concrete beams with and without shear reinforcement (in German: Schubversuche an einfeldrigen Stahlbetonbalken mit und ohne Schubbewehrung), Deutscher Ausschuss für Stahlbeton vol 151, p 83
4. Kani GNJ (1964) The riddle of shear failure and its solution. ACI J Proc 61(4):441–68. USA. https://doi.org/10.14359/7791
5. Cavagnis F, Fernández Ruiz M, Muttoni A (2015) Shear failures in reinforced concrete members without transverse reinforcement: an analysis of the critical shear crack development on the basis of test results. Eng Struct 103(1):157–173. https://doi.org/10.1016/j.engstruct.2015.09.015
6. Cavagnis F, Fernández Ruiz M, Muttoni A (2017) An analysis of the shear-transfer actions in reinforced concrete members without transverse reinforcement based on refined experimental measurements. Struct Concr 19(1):49–64. https://doi.org/10.1002/suco.201700145
7. Cavagnis F, Simões JT, Fernández Ruiz M, Muttoni A (2020) Shear strength of members without transverse reinforcement based on development of critical shear crack. ACI Struct J 117(1):103–118. https://doi.org/10.14359/51718012
8. Montoya-Coronado LA, Ribas C, Ruiz-Pinilla JG, Cladera A (2023) Time-history analysis of aggregate interlock in reinforced concrete beams without stirrups. Eng Struct 283(1):1–14. https://doi.org/10.1016/j.engstruct.2023.115912
9. Chana PS (1987) Investigation of the mechanism of shear failure of reinforced concrete beams. Mag Concr Res 39(141):196–204. https://doi.org/10.1680/macr.1987.39.141.196
10. Shahnewaz M, Rteil A, Alam MS (2020) Shear strength of reinforced concrete deep beams—a review with improved model by genetic algorithm and reliability analysis. Structures 23(2020):494–508. https://doi.org/10.1016/j.istruc.2019.09.006
11. Smith KN, Vantsiotis AS (1982) Shear strength of deep beams. J Am Concr Inst 79(3):201–13. https://doi.org/10.14359/10899
12. Ashour AF (2000) Shear capacity of reinforced concrete deep beams. J Struct Eng 126(9):1045–1052. https://doi.org/10.1061/(ASCE)0733-9445(2000)126:9(1045)
13. Yang Y (2014) Shear behavior of reinforced concrete members without shear reinforcement, a new look at an old problem. PhD thesis, Delft University of Technology, Delft, Netherlands, p 344. https://doi.org/10.4233/uuid:ac776cf0-4412-4079-968f-9eacb67e8846
14. Maksymovych SB (2015) Experimental study of crack resistance and strength of oblique planes of reinforced concrete point-force loaded beams with positive and negative bending moment diagram. J Theory Pract Build 823:334–342
15. Maksymovych SB, Krochak OV (2019) Eksperementalni doslidzhennia trishchynostiikosti i mitsnosti pokhylykh pereriziv zalizobetonnykh balok, zavantazhenykh zoseredzhenymy sylamy. J Resour Saving Mater Build Struct 37:164–174. Rivne
16. Maksymovych S, Krochak O, Karkhut I, Vashkevych R (2020) Experimental study of crack resistance and shear strength of single-span reinforced concrete beams under a concentrated load at a/d = 1. Lecture notes in civil engineering, vol 100. Proceedings of 2nd international scientific conference on ecocomfort and current issues of civil engineering ecocomfort, Lviv; Ukraine, 16–18 September 2020, pp 277–285. https://doi.org/10.1007/978-3-030-57340-9_34

BIM Technology and Impact on Safety KPIs in Construction Projects in Slovakia, Slovenia, and Croatia

Tomáš Mandičák(✉) and Marcela Spišáková

Technical University of Košice, Košice 042 00, Slovakia
tomas.mandicak@tuke.sk

Abstract. Key performance indicators represent the basis for effective management. At the same time, they can also be an indicator of the efficiency of using. BIM technologies represent a tool that increases information value in managing not only businesses, but also projects. Project management using BIM technology can bring benefits in several areas. The area of health and safety is also a much-discussed topic in the construction industry. Monitoring and increasing security includes search for suitable construction materials and technological procedures that consider a high degree of protection and safety of workers but also simulations and scenarios of eventual accidents. Another point of view is to aim at malfunctions caused by inadvertence, error rate, and human damage. The costs of occupational safety and health administration (OSHA) and average solution time as safety-related indicators cannot be neglected. The research aims to analyze the use level of BIM technology and the impact on the so-called Safety KPI in construction projects designed and implemented in Slovakia, Croatia and Slovenia. The research results show the increasing trend of using BIM technology in construction projects. BIM technology's use rate for risk and risk purposes is relatively low. However, measurement indicators show that it has a positive impact on safety and reduction of occupational safety and health administration costs in construction projects.

Keywords: BIM technology · Key performance indicators · Safety · Construction management

1 Introduction

The use of key performance indicators (KPIs) is a prerequisite for effective management of not only businesses, but also construction projects. The issue of implementation and use of BIM technology is also connected with safety issues in the design and subsequent implementation of construction projects [1]. Several studies point to BIM as a tool [2]. As part of the efforts to achieve KPIs, as measurable results of the success of construction projects, it is important to examine the impact of the use of BIM technology on the field of safety in the construction industry [3, 4]. This includes not only the search for suitable construction materials and technological procedures that consider a high degree

Z. Blikharskyy et al. (Eds.): CEE 2023, LNCE 438, pp. 242–250, 2024.
https://doi.org/10.1007/978-3-031-44955-0_25

of protection and safety of workers, but also simulations and scenarios of eventual accidents [5]. Another point of view is to aim at malfunctions that can be caused by inadvertence, or error rate and human damage. The costs of OSH prevention and Average solution time as safety-related indicators cannot be neglected. The aim of the research is to analyze the level of use of BIM technology and the impact on the so-called Safety KPI in construction projects designed and implemented in Slovakia, Croatia and Slovenia. The research results point to the increasing trend of using BIM technology in building design. The rate of use of BIM technology for risk and risk purposes is relatively low, but indicators of measurement that it has a positive impact on safety and reduction in construction projects.

2 Theoretical Background

2.1 Safety KPIs

Considering the critical KPIs and the areas they are devoted to, it is necessary to pay attention to the selected KPIs that reflect the investigated state. Several KPIs in the security package reflect the results in the researched area. The Indeed Career Guide points to indicators such as the frequency of lost time injuries, the average overtime hours of an employee, and the costs of health and safety prevention [6]. Other international labor organizations point to the measurement of performance indicators in the field of health and safety at work, among others [7]: The number of trained workers from organizations working on OSH reporting and the Number and percentage of stakeholders who rate involvement in program activities as applicable to improve health and safety outcomes for workers.

When trying to understand KPIs in safety (so-called Safety KPIs), it is essential to look at them from several perspectives. The first view is from the point of view of worker safety and health, which is closely related to compliance with safety regulations, legislation, and standards. Another point of view talks about the costs necessary to ensure these measures, respectively solving the problems that have arisen. Another point of view is the number of incidents that can be classified as those that impact the safety of employees in the form of injuries, defects on construction sites, repairs, etc. However, as AISeL (2009) notes, the incident itself needs to provide information on the quality of management [8]. This is explained by the fact that a minimal level of risk can occur even with absolute compliance with the rules and every activity. In other words, zero risk does not exist. The Cascade team defined several health and safety KPIs that are recommended to be monitored and measured for project management [9].

The first group includes direct safety risks and health impacts [10]. They are described in more detail in Table 1.

Another source lists these ten safety KPIs. First, economic impacts such as Health and safety prevention costs. Furthermore, related to time impacts, namely Average resolution time (ART), Total recordable injury frequency rate (TRIFR), Employee attendance rates, Average employee overtime hours. Others also look at repairs and corrections such as Corrective actions, Equipment breakdowns and general ones such as Reported incidents, Lost time injury frequency rate (LTIFR).

Table 1. Key performance indicator groups in safety construction

Safety risk and health KPIs	Equipment safety KPIs	Economical safety KPIs
• Fatalities • Total recordable incident rate (TRIR) • Lost time injury frequency rate (LTIFR) • Lost time injury incidence rate (LTIIR)	• Near misses rate • Equipment breakdowns • Number of safety violations • Fleet safety	• Penalties paid • Prevention cost • Average time to resolution of risks and issues

2.2 BIM, BEP, and Safety KPIs

A building execution plan (BEP) represents a plan or activity, or document based on which it informs about the implementation steps of BIM and provides information on how the project team will use the workflows, tools, and techniques of BIM to achieve the desired project results.

According to planners, making changes sooner means making changes better and cheaper. This is precisely what BIM [10] makes possible. The following picture shows this fact. The ability to change and impact the project decreases over time.

Conversely, each requested change over time is more costly. This proves that making these changes at the initial idea concept planning stage is much easier and cheaper. On the contrary, changes can be made faster and more efficiently now.

One of the reasons for the BEP is the setting of project goals and the setting and coordination of all aspects of the project. In other words, plan how to implement and use BIM to meet project goals. Part of this process is also the establishment of key performance indicators (KPIs) that will be used.

Developing a BIM execution plan (BEP) aims to facilitate information management in a BIM project. It can be defined as a plan prepared to streamline how the "Information Modeling" (United BIM) part of the project will be carried out [11].

There are many participants in construction projects in this process. For management, even in the construction industry, it is essential to have set goals and know how to have clearly defined competencies [12]. (Lucia). BEP is a comprehensive and powerful project accountability tool that moves work forward in the various planning and construction phases [13].

A BIM execution plan is often called a BIM management plan (BMP), i.e. it is a formal document that defines how the project will be implemented, monitored and controlled in relation to BIM [14]. Another group of researchers also defines BEP as a guide on how and when to use BIM, specifying in more detail the processes associated with the use of BIM in a construction project.

Nowadays, the goal of sustainable performance is increasingly mentioned in construction projects. This means that the performance needs to be managed, or the construction project needs to be managed so that the project performance can be achieved during the entire phase of the life cycle. This includes several activities, and one of them is the establishment of key performance indicators.

Safety has been emphasized several times as one of the key performance indicators [15]. Overall, the field of safety was also investigated in another study, which also points to the relationship between solving this problem and safety as a key performance indicator.

3 Methodology and Research Aim

The research aims to analyze the use level of BIM technology and the impact on the so-called Safety KPI in construction projects designed and implemented in Slovakia, Croatia, and Slovenia. The research results show the increasing trend of using BIM technology in construction projects. Based on theoretical analysis, the research determined the key performance indicators in the field of safety. These are the following indicators:

- Prevention cost level
- Defect Costs level
- Average time to resolution of risks and Issues level.

Project managers should have answered based on obtaining information from all people involved in the management of construction projects (including accountants, financial directors, etc.) about the extent of the use of BIM technology on these indicators and the impact on KPIs.

These values, for the need of comparability of research data, were set on a Likert scale of 1 to 5, where the values and description were clearly defined. These results were subsequently processed and evaluated in the relevant software.

The research sample consisted of three countries that have roughly the same market size. From an economic point of view, however, there are small differences in these construction markets. In terms of size, the Slovak and Croatian construction markets are approximately the same. Vocabulary represents half the volume as the mentioned two. On the other hand, as regards the involvement and type of construction projects, many infrastructure projects were in Slovakia and Croatia. In Slovakia, residential buildings were more represented. From the point of view of the share of GDP, there were also differences. The construction industry has the largest share of GDP in Slovakia. On the contrary, it is the smallest and most different in Croatia, where, for example, tourism clearly exceeds it. However, it is necessary to mention that many construction projects in Croatia are implemented in the field of tourism, or these projects are in some way closely related to this field.

The research results are processed by descriptive statistics. The research uses both quantitative and qualitative methods. Average values and follow-up dependence were subject to statistical testing. The results thus point to trends in the monitored areas and selected KPIs.

4 Results and Discussion

Solving the issue of security KPIs is important from the point of view of the perception of the effectiveness of investments in security within the management and implementation of construction projects. Quantifying the level of need to invest in security is often difficult decision from an ethical point of view. However, since the decision-making process regarding safety includes factories that limit this decision-making in a certain way (for example, legislation, limited financial, material, and human resources, the nature of the construction process, natural conditions, etc.), it is important to know the results and the impact on the performance of these decisions even on an economic scale. Also, investing in BIM technologies is not a small investment, therefore monitoring the performance of these parameters is important not only from the performance point of view.

The Fig. 3 shows the level of use of BIM technology and the results in the field of safety KPIs, such as prevention cost level, defect cost level and average time to resolution of risks and issues level.

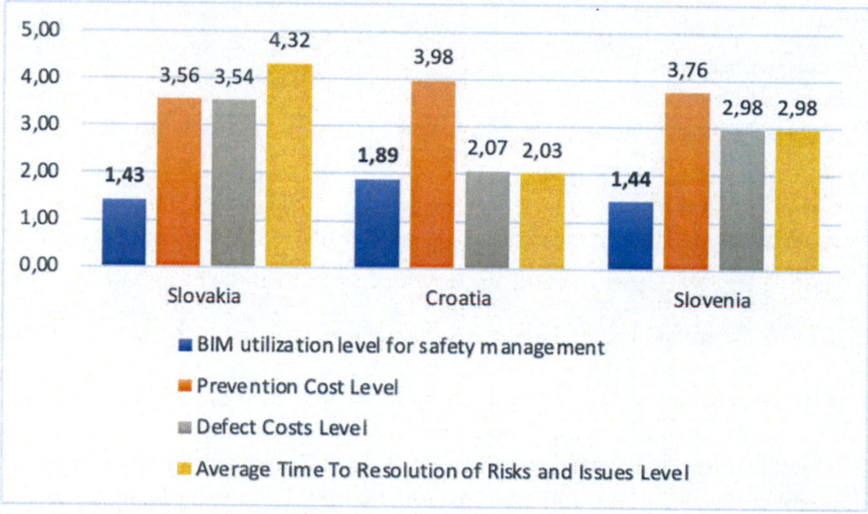

Fig. 1. BIM technology utilization level in construction projects in individual countries and the results of safety KPIs

In Fig. 1, it can be seen that the rate of use of BIM technologies in connection with the identification of safety risks in construction projects in the countries is relatively low. However, there are also differences between countries. Slovakia achieved the lowest rate of BIM use for these purposes, only 1.43 compared to the best result in Croatia, which represents an average value of 1.89. There are even smaller differences when comparing the already achieved value of prevention cost level. This means that, from the given data, the rate of use of BIM technology for safety management does not have a significant impact on the level of prevention cost.

On the contrary, the results were more differentiated for indicators such as defect costs level and average time to resolution of risks and issues level. The best values, i.e., the lowest, were achieved by projects in Croatia, where the rate of use of BIM technology for this purpose was the highest. The same trend in the relationship between the use of BIM technology and defect costs level was observed in Slovenia, where the rate of use and this indicator was the second highest in the comparison. The trend can be deduced from this, that if the rate of use of BIM technology is higher, the costs of defects are lower. A similar trend was also recorded in the average time to resolution of risks and issues level.

Figure 2 offers information regarding the difference in results in projects where BIM technologies were used and without BIM. These results indicate differences in use, as well as in impacts and thus safety KPIs. A more detailed description of all monitored safety KPIs can be seen on separate graphs (Figs. 3, 4 and 5).

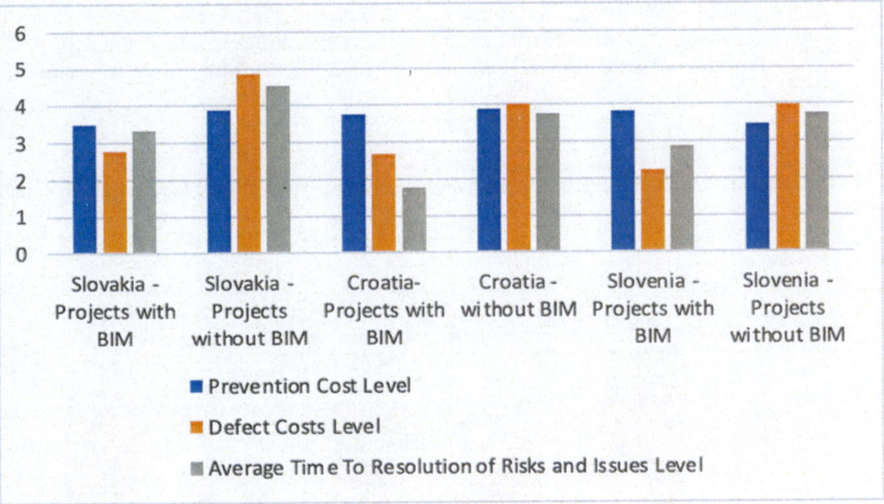

Fig. 2. Selected indicators of safety KPIs in individual countries for projects with and without the use of BIM technology

The average prevention cost level was lower for projects that used BIM technology in each country than for projects that did not use BIM technology (Fig. 3). The most significant difference was in the projects in Croatia, where the difference in the use of BIM technology and the impact on this indicator was demonstrated. However, the difference in the results is so significant that it cannot be clearly stated that the results are significantly differentiated based on the use of BIM technology.

Figure 4 describes the difference in defect costs level results for projects that used BIM technology and projects without using BIM. Here, the trend is clearer, and projects where BIM technologies were used recorded better results in this indicator, regardless of the countries in which they were implemented.

There were also significant differences in the indicator average time to resolution of risks and issues level (Fig. 5). Probably the most visible differences were in Croatia, where a significant difference in the use of BIM technology was demonstrated. These projects achieved the lowest rate at the level of 1.76.

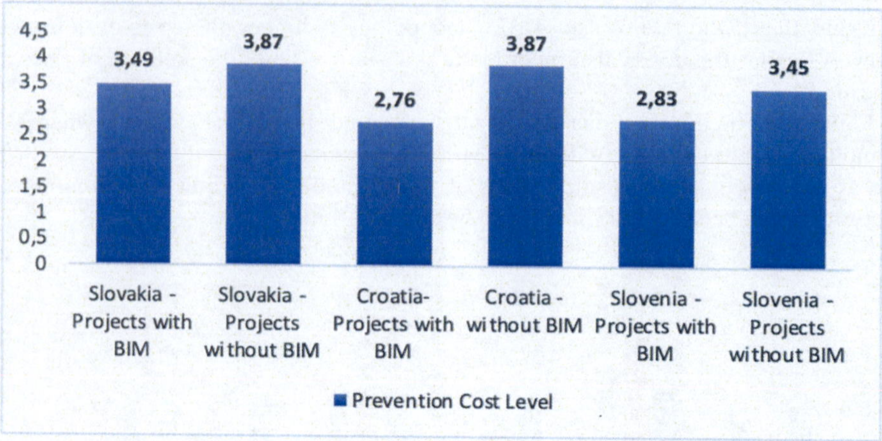

Fig. 3. Prevention cost level for projects with and without the use of BIM technology in selected countries

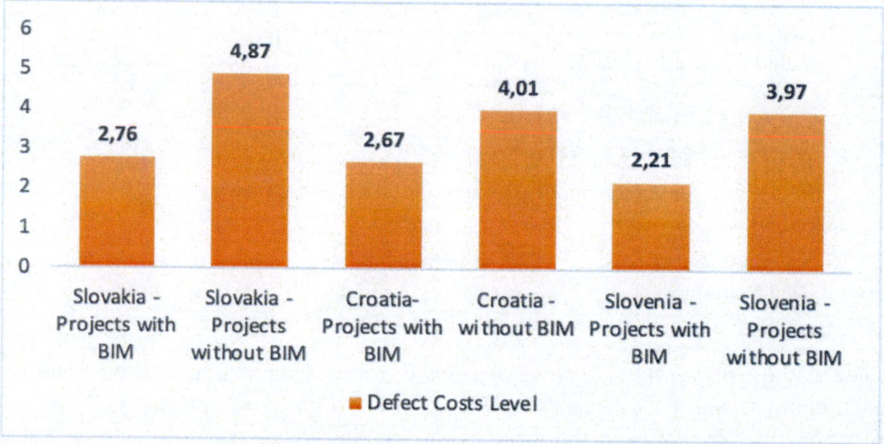

Fig. 4. Defect costs level for projects with and without the use of BIM technology in selected countries

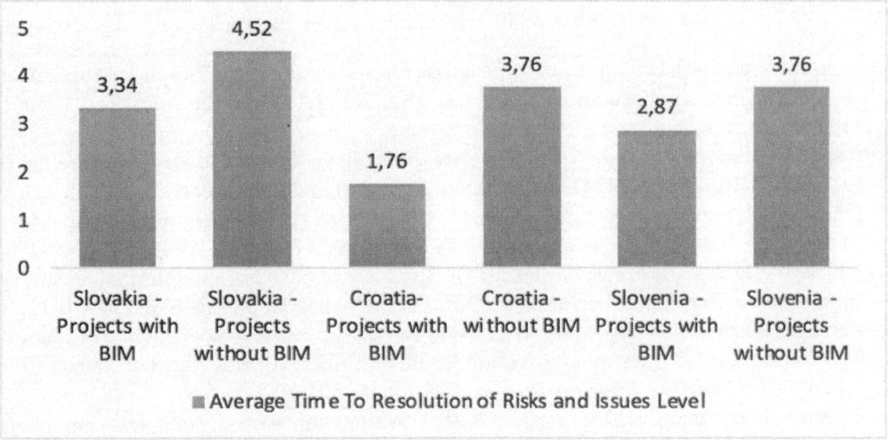

Fig. 5. Average time to resolution of risks and issues level in projects with and without the use of BIM technology in selected countries

5 Conclusion

Research focused on the use of BIM technologies and the monitoring of their impact on selected performance indicators around safety brought interesting findings. First of all, the rate of use of BIM technology for these purposes, i.e., safety management in construction projects, is relatively low. The potential brought by BIM technologies represents much greater expectations and opportunities for involvement. Within the framework of the use of BIM technology, Croatia recorded the best results. However, it must be said that the differences were not significant.

When it came to monitoring KPIs in safety, here the results were clearer and more pointed to the trend of BIM impacts in this area. The trend for the prevention cost indicator was the least significant, where there were no significant and statistically significant impacts of BIM, it only pointed to a trend and difference in that this indicator was little different in projects where BIM technologies were used. However, more significant results were obtained with indicators such as defect costs level and average time to resolution of risks above issues level. The differences were significant. Projects where BIM was used, these indicators in each country were significantly more positive.

This research showed the situation this problem in three markets. The results were comparable in many cases, but small differences were noted. For further research direction, it is important to look at the focus of the projects and explore the differences.

Acknowledgements. Paper presents a partial research result of projects: project KEGA 009TUKE-4/2022 An interactive tool for designing a safe construction site in an immersive virtual reality; project VEGA 1/0336/22 Research on the effects of Lean Production/Lean Construction methods on increasing the efficiency of on-site and off-site construction technologies.

References

1. Kravanja S, Klanšek U, Žula T (2021) Mass, direct cost and energy life-cycle cost optimization of steel-concrete composite floor structures. Appl Sci 11:10316. https://doi.org/10.3390/app112110316
2. Begić H, Galić M (2021) A systematic review of construction 4.0 in the context of the BIM 4.0 premise. Buildings 11:337. https://doi.org/10.3390/buildings11080337
3. Johansson B, Sudzina F, Newman M (2011) ERP system implementation costs and selection factors of an implementation approach. Int J Bus Inf Syst 8(1):87–105. ISSN 1746-0972
4. Johansson B, Sudzina F, Pucihar P (2014) Alignment of business and information strategies and its impact on business performance. J Bus Econ Manag 15(5):886–898. ISSN 1611-1699
5. Mihić M, Vukomanović M, Završki I (2019) Review of previous applications of innovative information technologies in construction health and safety. Organ, Technol Manag Constr: Int J 11(1):1952–1967. https://doi.org/10.2478/otmcj-2019-0004
6. Indeed editorial team: what are safety KPIs? (With definition and examples). https://www.indeed.com/career-advice/career-development/what-are-safety-kpis
7. International Labour Organization (2022) Collecting data and measuring key performance indicators on occupational safety and health interventions—a guidebook for implementers. https://www.ilo.org/wcmsp5/groups/public/---ed_protect/---protrav/---safework/documents/publication/wcms_864235.pdf
8. Association for information systems electric library: key performance indicators framework—a method to track business objectives, link business strategy to processes and detail importance of key performance indicators in enterprise business architecture (2009)
9. Cascade team, 18 health and safety KPIs to track in 2023 (+ Template). https://www.cascade.app/blog/kpis-health-and-safety
10. Plannerly, 7 things you should consider for your BEP—includes 16 point checklist of what should go into your BIM execution plan (BEP). https://plannerly.com/bep-bim-execution-plan-guide/
11. United BIM, BIM execution plan (BXP)—what, why, when and how (2023). https://www.united-bim.com/bim-execution-plan-bep-guide-for-successful-bep-design-and-execution/
12. Behún M, Knežo D, Cehlár M, Knapčíková L, Behúnová A (2022) Recent application of Dijkstra's algorithm in the process of production planning. Appl Sci 12:7088. https://doi.org/10.3390/app12147088
13. Remage M. What is a BIM execution plan and what should it include? https://constructible.trimble.com/construction-industry/what-is-a-bim-execution-plan-and-what-should-it-include
14. Bakar ARA, HAron AT, Rahman RI (2020) Building information modelling execution plan (BEP): a comparison of global practice. Int J Eng Technol Sci 7(2):63–73
15. Bapat H, Sarkar D, Gujar R (2023) Application of multi-criteria decision making for evaluation of key performance indicators of integrated project delivery and BIM model for an infrastructure transportation project in Western India. Int J Constr Manag 23(12):2077–2086

Development Directions of Small Historic Cities in the Subcarpathia

Anna Maria Martyka(✉)

Faculty of Civil and Environmental Engineering and Architecture, Rzeszow University of Technology, Rzeszow, Poland
amartyka@prz.edu.pl

Abstract. The study of urban forms takes into account historical, spatial and functional perspectives in order to better understand contemporary trends and effectively manage future development. Poland's medieval cities, with their unique heritage, are of particular value for research. Small cities with historical features are often underestimated elements of the European cultural heritage, with preserved monuments being of key importance for them, shaping their urban character and identity. The study focuses on the integration of historic centers with the rest of the urban landscape, which can create a coherent, sustainable environment that balances heritage and modernity. The practical aim is to provide general guidelines for urban planning. The examples of Brzostek, Kołaczyce and Pilzno, small cities in the Subcarpathian Voivodeship with a rich heritage of medieval urban planning, serve as a model for the research objective. Classical tools and methods of urban planning, together with qualitative research methods, are used to assess the Urban Integration Index of these historic centres. Research has shown that, despite changes, they have retained their characteristic *genius loci* and key elements of spatial structure. Although traditional functions such as housing, commerce, services and administration still exist, the research suggests that contemporary functions such as the concentration of social, tourist and recreational life are lacking. Recommendations include improving the integration of public spaces, enhancing residential functions, reorganizing pedestrian and cycle transport, controlling advertising and improving the quality of architecture. All of this can be achieved through effective spatial and urban planning, which will contribute to the sustainable use of resources and improve the quality of life of residents.

Keywords: Material heritage · Urban integration · Historic cities · Urban planning

1 Introduction

Cities of medieval origin throughout Europe deserve unique protection because of their profound significance for European history and culture, serving as universal symbols of Western civilization [1]. The distinguishing features of these medieval cities have always been their intimate urban and architectural structures and their close social ties [2, p. 358]. The study of urban forms requires a careful examination of their historical,

Z. Blikharskyy et al. (Eds.): CEE 2023, LNCE 438, pp. 251–264, 2024.
https://doi.org/10.1007/978-3-031-44955-0_26

compositional, spatial and functional characteristics [3, 4]. The analysis of historical urban patterns promotes a more comprehensive understanding of modern development trends and enables more effective conservation of spatial structures and thoughtful, ethical guidance for their future development [5]. Medieval architectural and urban monuments in Poland are particularly valued as historical and cultural heritage, with a comparatively small number surviving to the present day [6].

In small cities, historical structures are of immense value, often serving as the only indicator of their urban essence and contributing significantly to local identity. Studies of many medieval city plans reveal the existence of sites of exceptional importance that have a profound impact on the overall spatial arrangement of cities, making their integration into modern urban forms a key determinant of urban evolution. Urban integration, in the context of urban planning and development, is seen as a method of creating harmonious links within the urban environment, resulting in an area with a compact spatial layout and a high level of integration. A well-integrated city provides a better quality of life for its inhabitants, promotes sustainability, encourages equitable practices and strengthens social solidarity. This is also true for smaller cities with historic centres, where both heritage and modernity have their place. In essence, urban integration is about bringing together the spatial configuration of a city by creating tangible links between different urban areas and strengthening those links that are particularly important for the dynamism of urban spaces [7].

Historically rooted small settlements are an important part of Europe's cultural heritage. However, they are often overlooked in the same way that large cities are sometimes undervalued. Research focusing on the urban heritage of smaller provincial towns often takes a back seat in mainstream studies, but has gained traction in recent European initiatives on spatial culture. Notable among these initiatives are the New European Bauhaus (NEB) and Baukultur programmes. The main aim of these programmes is to inspire Europeans to participate in the creation of sustainable, aesthetically attractive and inclusive living spaces [8].

The aim of the analyses is to identify the level of integration of the historic centre, which can be used as an information tool for zoning the development of a small city according to sustainable development principles. The focus is on the process of integration with the rest of the city, which is important for historical spatial arrangements. The integration of these two different areas can create a coherent and harmonious environment in which both culture and modernity have their place. The research has made it possible to achieve the practical objective of drawing up general guidelines for urban planning in order to strengthen physical and functional links. Three small cities in the Subcarpathian Voivodeship—Brzostek, Kołaczyce and Pilzno—were used as examples. They were founded in the 14th century under the Magdeburg Law, during the important urbanisation campaign of Casimir the Great between 1333 and 1370. These cities have a high quality heritage of medieval urban layouts, which should be protected on the one hand and used as a development potential on the other. They are therefore very good examples of model cities for the realisation of the main objective of the research. This paper presents a multidimensional study of small urban layouts using classical tools and methods of urban planning through architecture. Qualitative methods were used to calculate the Urban Integration Index of the historic centre. A description of the methodology used to determine the urban integration index is given in Sect. 3.

The novelty of the research lies in several aspects. The presented study fills a gap in research on small historic towns in the Subcarpathian Voivodeship, with an emphasis on their urban layouts. The use of juxtapositions and comparisons between small historic towns will allow for a better understanding of their uniqueness, as well as the transformations they have undergone over the years. The search for and identification of information tools that can be used in spatial policy. Finally, research can provide valuable guidance for the protection and conservation of urban and cultural heritage, emphasising the need to understand the history and character of these places in order to preserve their uniqueness and authenticity.

2 Medieval Urban Construction of Small Cities in the Subcarpathian

2.1 Brzostek

Brzostek lies north of Jasło on the Słony stream. Originally, Brzostek was a rural settlement belonging to the Benedictine monastery in Tyniec. The settlement was mentioned in a document of the papal legate Idzi from 1105 as an antiqua villa village [9, p. 12]. The village of Brzostek probably had a market function, as it lay on the route of the Hungarian road along the Wisłok valley from Sandomierz to Hungary. This is the main reason why the city was founded on the site of a prosperous village that did not have its own church and belonged to the parish of neighbouring Kleć [9, p. 45].

Before 1367 Brzostek was granted the status of a town, which was recorded in a document of the Abbot of Tyniec [10, p. 77]. The city was built on a promontory bordered on three sides by the slopes of the Słony stream and the Wisłoka valley. The medieval urban plan of Brzostek is characterised by a simple spatial and functional structure, similar to those of Kołaczyce and Pilzno. The city was planned in the central part of the village. The market square in Brzostek was created by transforming the buildings of the former village according to the directions of the three trade routes (Figs. 1 and 2).

The market was laid out at a ratio of 2:1 and surrounded by individual blocks of buildings. The proportions of the market are somewhat atypical compared to other such medieval layouts, as it has an elongated rectangular shape, but its area is similar to other markets in cities founded in this period, ranging from 0.8 to 1.0 ha [11, p. 91]. A city hall was built in the centre of the market some time after the city was founded. The city hall was first mentioned in the 16th century [9, p. 51].

2.2 Kołaczyce

The city was founded on the main road running through the valley of the Wisłoka River at the junction with the road leading east. To the north there was a road to Pilzno, which joined the Krakow-Lviv Ruthenian route. Another road ran through the city to the east, towards Frysztak. The first mention of settlements in this area comes from a document of

Fig. 1. The market square in Brzostek. *Photo* by Martyka 2023.

Leszek the Black from 1288, in which he granted the monastery of Tyniec near Kraków the right to found cities under German law and freed the monastic villages from secular jurisdiction [9, p. 61].

The foundation document of Kołaczyce has not been preserved, so the exact date of its foundation is unknown. However, it is known that it was founded before 1339, when the deed of foundation of the parish church in Kołaczyce was issued [10, p. 78]. The city was founded on the bank of the river, in an area favourable for the defence of the centre, on the site of a functioning market settlement of the rural type [9, p. 73]. The market character of the square is also evidenced by the irregularly shaped market frontages, especially the southern and eastern ones, where the location of the buildings follows the line of the streets in the square. The aim of the urban design was to create a checkerboard plan with individual blocks of buildings surrounding the square. The square is rectangular, with proportions of 3 × 4 (Figs. 3 and 4).

Two streets led from each corner of the square. The rectangle formed by the blocks around the square was extended to the south by the church square, and two additional small blocks completed the rectangle at the level of the church square. Originally the main road entered the city from the south near the church. There was a gate on the line of the defensive fortifications. The main road crosses the square diagonally and cuts through its north-western corner in the direction of Pilsen. The oriented parish church is situated outside the square, behind the south-eastern corner. Crafts, organised in guilds, developed mainly in Kołaczyce. The most important were the shoemakers' and potters' guilds. The products made in the city satisfied not only the local market, but were also exported to other regions of the country and abroad 2. The area of the city was surrounded by earth and wooden fortifications, along which an elevated road ran from the centre of the city, which can still be seen on the cadastral map and in the field.

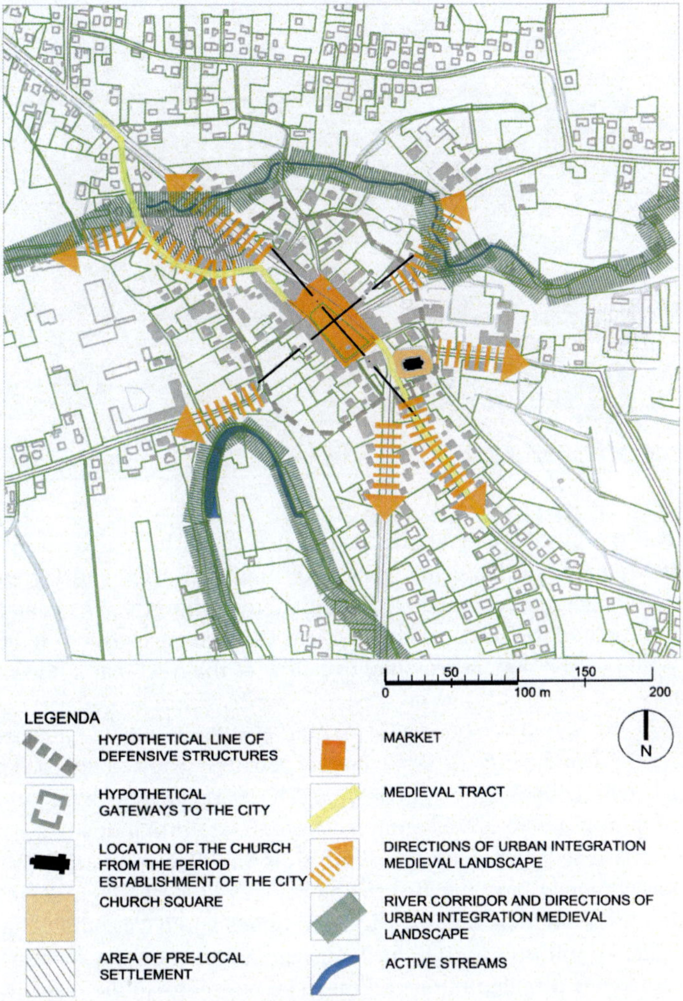

Fig. 2. Spatial structure and main directions of urban integration in Brzostek. By Martyka

2.3 Pilzno

Pilzno is situated on the edge of the Ciężkowice Foothills, a part of the Carpathian Foothills, on the Dulcza River—a left-bank tributary of the Wisłoka River. In the Middle Ages international roads passed through Pilzno. The city was founded in 1354 under the Magdeburg Law by King Casimir the Great on the site of an older settlement, which he had previously confiscated from the Benedictines of Tyniec. The flourishing economic development of the city is evidenced by the statutes of the following guilds: clothiers, furriers, tailors, butchers, weavers, potters, bakers, wheelwrights, stonemasons, carpenters, saddlers, turners, rope makers and locksmiths [12, p. 219].

The centre of the town was situated on a hill overlooking the floodplain of the Wisłoka valley. It was also the site of a road junction, which facilitated the development of the

Fig. 3. The market square in Kołaczycach. *Photo* by Martyka 2023.

village and then the city [13]. The size and shape of the hill, created by the erosive activity of the Wisłoka and Dulcza Rivers, determined the size and layout of the site city, which was designed according to a theoretical modular grid. The centre of the city, according to custom and for practical reasons, was situated on a terrace on the river bank, protected by a high embankment from the overflow of the river, but at the same time by natural defences.

The centre of the layout is a quadrilateral market square with an almost square plan, from the corners of which pairs of parallel streets emerge (Figs. 5 and 6). The market is very close to a square with a side length of 90 m. The market blocks have largely retained their original depth, if we do not take into account the deformations that occurred later for various reasons. The side boundaries of the plots have been obliterated, so it can only be assumed that they may have had dimensions of 45 m long and 9 m wide [14, p. 23].

The spatial ensemble includes two parish churches and a Carmelite church—originally Augustinian—built on the edge of the urban complex on the eastern side, which is not typical for cities founded in this region. The buildings in the historic quarter are almost entirely of brick, with a few exceptions of one- and two-storey houses, historically located on the frontages of the square and the compact streets. The city was surrounded by fortifications. Communication with the outside world took place through three gates: Kraków, Lwów and Strzegom. The Lwów Gate was first mentioned in 1403, so we can conclude that the earthen fortifications were built in the second half of the 14th century. However, the existence of an exclusively wooden fence cannot be excluded [11, p. 103]. The Podwalna Street has been preserved in the field as a proof of the existence of the fortification. This type of fortification was only partially effective against invasions and required constant maintenance. It is interesting to note that the Sokolnia in Pilzno was established in 1907, together with a park complex, which has been preserved in its original state to this day.

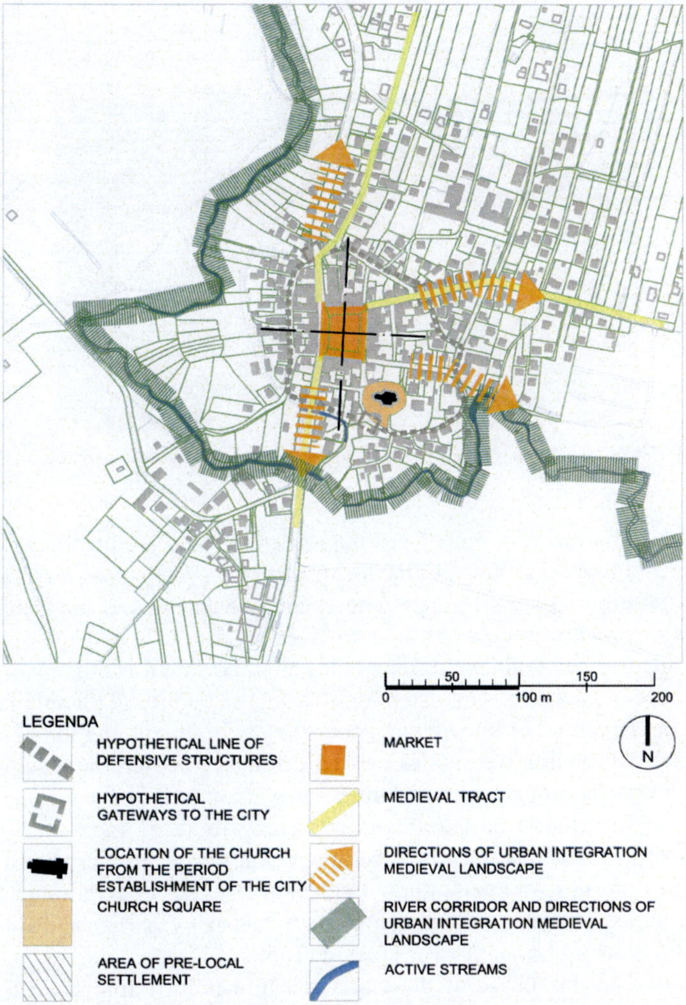

Fig. 4. Spatial structure and main directions of urban integration in Kołaczyce. By Martyka.

3 Urban Integration Index of the Historic Layout

Integration usually appears as one of the many threads of different types of urban processes such as revitalisation, synergy, transformation, operational urbanism, programming and steering urban development [15]. Zuziak states that integration is one of the most important processes initiated in a city. He understands urban integration as a purposeful and desirable action that creates a synergistic effect in response to the manifestations of the disintegration of urban structures, which are particularly visible in Polish cities [16].

Fig. 5. The market square in Pilzo. *Photo* by Martyka 2023.

The tools of qualitative research in urban planning and architecture were used to assess the Urban Integration Index of the historic spatial system. The qualitative assessment made it possible to identify the strengths and weaknesses of the area in question and to rationally plan the process of transformation in order to improve its quality. Qualitative assessments of the built environment play an important role not only in improving the quality of life of residents, but also in improving the quality of planning and design, allowing the development of knowledge about the environment and the comparison of different areas. The qualitative studies carried out have focused on specific problems and evaluation criteria, in this case the Urban Integration Index.

The level of integration is determined according to seven main categories. Each category was subdivided into a set of several specific characteristics (28 in total). The resulting set of criteria systematises how the level of spatial integration of each of the small historic cities analysed was determined. A matrix of partial and total levels of integration of the historic systems was created (Table 1).

The tool used in this phase of the work is a four-point rating scale from 0 to 3, modelled on the scale used in Landscape and Visual Impact Assessment (LVIA). LVIA is an informative procedure used in spatial planning and environmental management to assess how a proposed development or spatial change will affect the character and appearance of the landscape and the views that characterise it [17, 18; pp. 124–127]. Points were then assigned for each of the detailed characteristics identified in each category, which could range from 0 to 3. The general rationale for assigning a value to a particular detailed characteristic was as follows:

- 0 points (Negligible)—Lack of or insignificant level of features that characterise the area in the criterion considered,
- 1 point (Moderate)—a sufficient level of specific features that determine the level of integration, but requiring significant strengthening,
- points (Substantial)—a good level of a set of characteristics that influence the level of integration to a fairly good extent, but it is possible and even advisable to strengthen or develop them,

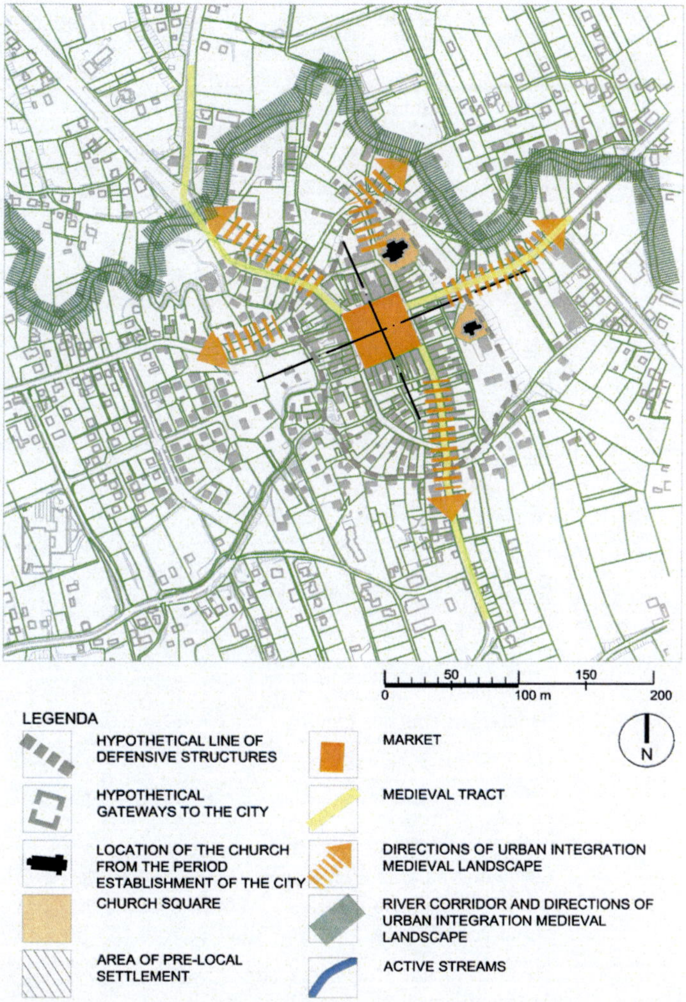

Fig. 6. Spatial structure and main directions of urban integration in Pilzno. By Martyka.

– points (High)—characteristics which, in a given category, indicate a very good level of integration that does not require intervention in the short and medium term.

On the basis of the earlier stages of the research process and expert knowledge, the scores for each feature have been adopted and grouped into seven categories. For each category, a partial level of integration was calculated as the arithmetic mean. The final level of integration was also calculated as the arithmetic mean of all characteristics. The value of the urban integration index can range from 0 to 3 points. The proposed method allows the use and analysis of qualitative data collected from different sources. The advantage of this method is its simplicity of calculation.

Table 1. Assessment of the urban integration index of medieval spatial systems by category and specific characteristics

Criteria	Specific features	City names		
		Brzostek	Kołaczyce	Pilzno
Urban composition	Readability of the historical layout	3	3	3
	Readability of the market's urban design	3	3	3
	Morphological homogeneity of the fabric around the historic layout	2	2	3
	No adverse architectural features	2	1	2
	Preservation of a strong vertical dominant in the form of the parish church from the time of the founding of the city	3	3	3
	Roofscape in historic layout	2	2	2
	Arithmetic average	2,50	2,33	2,67
Aesthetics architecture	Elements that give the centre individual characteristics with which residents can identify	1	3	3
	Colour consistency of facades	1	1	2
	Advertising aesthetics	0	1	2
	State of building maintenance	1	1	3
	Surfacing materials for squares and pavements, for the creation of individual character within the square itself	0	2	3
	Arithmetic average	0,60	1,60	2,60
Public space	Attractiveness of public spaces and facilities, especially the city square	2	2	3
	Appropriate hierarchy of spaces (private, social public)	3	3	3
	Accessibility for all—inclusiveness and secu	2	3	3
	Arithmetic average	2,33	2,67	3,00
Communication	Continuity of footpaths	2	2	2

(*continued*)

Table 1. (*continued*)

Criteria	Specific features	City names		
		Brzostek	Kołaczyce	Pilzno
	Cycle path continuity	0	0	1
	Cycling infrastructure	0	0	2
	No inconvenience to transit traffic	0	0	2
	Slowing down and reducing car traffic in the area of the historic layout	0	0	2
	Arithmetic average	0,40	0,40	1,80
Greenery	Attractiveness of green spaces in the market area	1	3	2
	Green links in the historic area with neighbouring areas	2	2	3
	Formal green diversity	3	2	2
	Pedestrian and cycle access to the river	0	2	0
	Arithmetic average	1,50	2,25	1,75
Diversity of functions	Concentration of residential functions	3	3	2
	Concentration of service and retail functions	3	3	3
	The presence of functions that integrate and activate social life	2	1	3
	Arithmetic average	2,67	2,33	2,67
Legal protection	Registration of the spatial layout in the register of monuments and historic buildings	3	0	3
	Number of monuments protected by an entry in the Register or in the Monument Records	2	1	2
	Arithmetic average	2,5	0,5	2,5
Urban integration index		**1,64**	**1,75**	**2,39**

4 Results

The results of the study show that the highest Urban Integration Index was achieved in Pilzno (2.39 points), followed by Kołaczyce (1.75 points). Brzostek achieved a slightly lower level of integration (1.64 points) (see Fig. 7). These results show that cities can increase the level of integration by creating stronger structural links between the layouts with the highest values and the remaining urbanised areas.

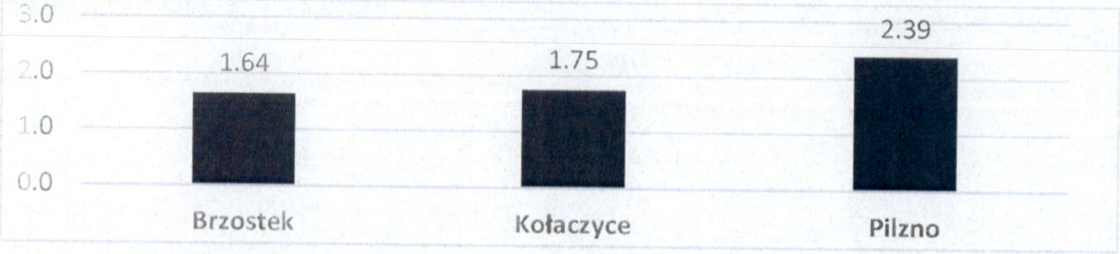

Fig. 7. Urban integration index of the historical spatial systems in the cities analysed.

The construction of the Urban Integration Index, which is made up of detailed characteristics grouped into several categories, and its evaluation make it possible to indicate the specific areas in which it is advisable to direct the city's spatial policy in order to improve the quality of life of the inhabitants, attract new residents and attract more tourists (see Fig. 8). For Brzostek and Kołaczyce, for example, transformations in terms of improving architectural aesthetics and making mobility more attractive and accessible are key. These cities lack safe cycle paths and cycling infrastructure. Walking also needs to be reorganised. It is worth planning these measures in parallel with the development of a network of connections to publicly accessible green spaces. In particular, it is important to exploit the lost potential of picturesque meandering rivers, which were a key natural feature in the formation of the settlement and the city in general. The river provided not only access to drinking water, but also security, transport and economic development [19].

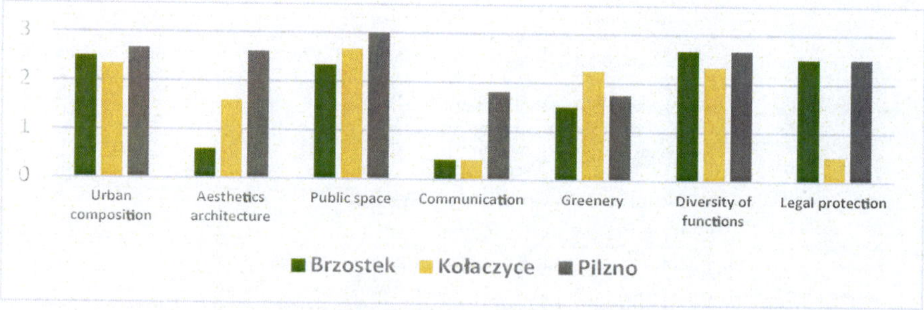

Fig. 8. Urban integration index of historic spatial systems in each category.

Apart from the low level of communication and mobility, Kołaczyce is also characterised by a lack of legal protection of its historical layout. It is puzzling that the area of the medieval city has not yet been included in the Register of Historical Monuments, especially as the arrangement of the building blocks around the market square and the street layout at the time of the city's foundation have not been significantly obliterated and remain legible. Although Pilsen is the best performing of the cities surveyed, there is still room for improvement in terms of cycling infrastructure, cycle lanes and links to green spaces.

5 Summary and Conclusions

The article defines the Urban Integration Index of historical layouts on the example of three small cities in the Subcarpathian Voivodeship. In order to determine this level, a matrix of detailed characteristics influencing the integration and the establishment of connections between the layers of the functional-spatial structure was created. A brief history of the origin of the cities was also outlined. In their more than 650-year history, the layout of the cities has undergone changes, but above all the genius locus of the place and the characteristic elements of the spatial structure in the form of the urban morphology, the market square, the monuments and the main traffic system have been preserved. The level of urban integration has contributed to the creation of the following design guidelines.

Today, traditional functions such as housing, retail, services and administration continue in the historic areas. Unfortunately, surveys show that these cities lack contemporary functions such as those focused on social life, tourism and recreation. The preservation and authentication of urban interiors and the creation of a coherent network of public spaces, both in the historic centre and integrated with the rest of the urban fabric. It is worth creating facilities and incentives to strengthen the residential functions within the development of the market, so that this main space of the city is vibrant with social life at all times of the day and year.

A basic guideline for improving the quality and attractiveness of the public space is the reorganisation of communication in the markets and within the historical spatial complex. This is particularly true for cycling and walking, which are currently completely neglected in the spatial policies of the cities studied, as evidenced by the lack of cycling infrastructure and paths. Another issue in terms of creating a high quality fabric is the way in which the facades of buildings along the main streets and market squares are renovated. At present, the chaos of advertising and its quantity and quality are not in keeping with the noble surroundings. The way in which advertising and visual identity are shaped in the centres should respect the context of the material heritage. Compact development, appropriate functional diversity and good quality architecture, as well as easy access to green and open spaces, form the basis for efficient and sustainable use of resources in a city. The right design of public spaces through appropriate landscaping of green areas mitigates the negative effects of climate change and improves the quality of urban life [20].

The aforementioned design guidelines in the historic zones do not require too much money and can be implemented in phases through small smart investments. What is important, however, is a defined plan and vision that is consistently implemented by city authorities with the involvement of local communities. A sense of continuity stemming from the city's historical planning and development is a valuable element in maintaining a high quality of life, including psychological comfort and an individual's sense of belonging and security. All this can be achieved through spatial and urban planning.

References

1. Weber M (2002) Gospodarka i społeczeństwo. Zarys socjologii rozumiejącej. Wydawnictwo Naukowe PWN, Waeszawa
2. Mumford L (1979) The city in history. Its origins, its transformations, and its prospects, Penguin Books
3. Sikora A (2019) Heritage of the polish central industrial district—New cities. In: IOP conference series: earth and environmental science, vol 362
4. Sikora A (2018) The development of small settlements in the industrial plant-city model-based on the examples of Nowa Dęba and Nowa Sarzyna. In: E3S web of conferences, vol 45
5. Zuziak Z (2021) Idea piękna i architektura miasta. O estetyce formy miejskiej i etyce rozwoju. Teka komisji urbanistyki i architektury XLIX:185–202
6. Martyka A (2020) Ruins of Czchów castle—analysis of conservation and construction work. Wiadomości Konserwatorskie-J Heritage Conserv 2020:28–44
7. Zuziak Z (2008) O tożsamości urbanistyki. Wydawnictwo Politechniki Krakowskiej Kraków
8. Domagalska K (ed) (2022) Notes Archikultury. Praktyki i inspiracje. Nadrenia Północna-Westfalia. Meklemburgia-Pomorze Przednie. Narodowy Instytut Architektury i Urbanistyki, Warszawa
9. Kuśnierz-Krupa D (2014) Fundacje miejskie Benedyktynów Tynieckich w Małopolsce południowej w okresie Średniowiecza. Wydawnictwo Politechniki Krakowskiej, Kraków
10. Wyrozumski J (1964) Początki miast w regionie jasielskim. In: Garbacik J (ed) Studia z dziejów Jasła i powiatu jasielskiego
11. Berdecka A (1982) Lokacje i zagospodarowanie miast królewskich w Małopolsce za Kazimierza Wielkiego. Zakład Narodowy im. Ossolińskich, Warszawa, pp 1333–1370
12. Kumor B (1964) Archidiakonat sądecki. Opracowanie materiałów źródłowych do atlasu historycznego kościoła w Polsce, vol 8
13. Żaki A (1964) Najdawniejsze dzieje Jasła i okolicy (od starożytności po wczesne średniowiecze). In: Garbacik J (ed) Studia z dziejów Jasła i powiatu jasielskiego
14. Książek M (1973) Układy urbanistyczne niektórych zabytkowych miasteczek w województwie rzeszowskim. in Materiały Muzeum Budownictwa Ludowego w Sanoku, vols 17–18
15. Martyka A (2022) Spatial integration level of housing complex with the city structure. Case of Rzeszów. Civil Environ Eng Rep 3:36–61
16. Zuziak Z (2017) O synergii planu w urbanistyce. Budownictwo i Architektura 16
17. Niezabitowska E, Masły D (2007) Oceny jakości środowiska zbudowanego i ich znaczenie dla rozwoju koncepcji budynku zrównoważonego. (Wydawnictwo Politechniki Śląskiej, Gliwice
18. Diego JC, Bonete S, Chías P (2022) VIA-7 method: a seven perceptual parameters methodology for the assessment of visual impact caused by wind and solar facilities on the landscape in cultural heritage sites. Renew Sustain Energy Rev 165:112528
19. Jopek D, Martyka A (2018) Miejski krajobraz nabrzeży rzecznych. In: Architektura w krajobrazie. Harmonia, kompromis, piękno, konflikt, pp 187–202
20. Przesmycka N, Kwiatkowski B, Kozak M (2022) The thermal comfort problem in public space during the climate change era based on the case study of selected area in Lublin city in Poland. Energies (Basel), p 15

Effect of Dry–Wet Cycles on Properties of High Strength Fiber-Reinforced Concrete

Uliana Marushchak[1], Nazar Sydor[1](✉), Serhii Braichenko[1], and Marko Hohol[2]

[1] Department of Building Production, Lviv Polytechnic National University, Lviv, Ukraine
nazar.i.sydor@lpnu.ua

[2] Department of Highways and Bridges, Lviv Polytechnic National University, Lviv, Ukraine

Abstract. The development of modern civil construction has created a significant demand for concrete, which must characterize increased qualities, such as high compressive strength, flexural strength, impact resistance. While concrete is designed primarily to withstand structural loads, it must also contend an array of environmental effects, which determines its durability. The paper is presented the investigation results of the durability of modified high strength fiber-reinforced concrete under the dry-wet cycles effect, which can occur both under the weathering and human activities influence (car washes, industrial floors etc.). The analysis criteria were deterioration in compressive strength and impact resistance of concrete. The compressive strength of developed concrete after 100 dry-wet cycles decreases by 1.9–2.8%. Dispersed reinforcement with polypropylene fiber provided an effective increase in resistance to impact loads under the combined effect of dry-wet cycles. The impact strength of modified high strength fiber-reinforced concrete after 50 and 100 dry-wet cycles decreases by 2.2–3.4% and 3.1–5.4% respectively.

Keywords: High strength concrete · Polypropylene fiber · Complex modifier · Wet-dry cycle · Impact resistance · Compressive strength

1 Introduction

Concrete has become the most economical and most used structural material in construction. However, its environmental impact in the world is significant, as it requires a significant amount of natural resources and energy, and generates an enormous volume of waste [1]. In this regard, concrete technology has evolved from conventional to high strength concrete and in recent decades—to high strength concrete with high durability. Therefore, when design structures, it is necessary to take into account not only the influence of loads, but also the influence of environmental conditions, which plays an important role in effective forecasting of the life cycle of structures. Factors which affect the durability of concrete or damage to structures can be related to changes in internal pore hymidity or temperature (carbonation, freeze–thaw cycles, wet–dry cycles, effect of chloride salts, sulfates), as well as with the influence of various forms of loading

Z. Blikharskyy et al. (Eds.): CEE 2023, LNCE 438, pp. 265–272, 2024.
https://doi.org/10.1007/978-3-031-44955-0_27

[2–4]. Considerable attention of researchers has been focused on the problem of the behavior of concrete under the influence of dry-wet cycles [4–7]. In practical construction, taking into account the single factor of the wet-dry cycle makes the prediction of concrete durability somewhat limited, and it does not contribute to the accurate determination of the service life of buildings. The operating conditions of building structures are often associated with cyclic changes of moisture, which are caused by natural precipitation or technological processes and impact loads (hydraulic structures and transport infrastructure, industrial floors, car washes, production shops) [7, 8]. Multiple change of moisture causes the development of swelling and shrinkage deformations, which causes the breaking of material structure and leads to irreversible changes and a premature decrease in operational characteristics [5]. Therefore, the development of concretes with high resistant to dynamic loads and wet–dry cycles is of practical importance.

First of all, the load-bearing capacity and resistance to destruction of building structural materials is determined by the compressive strength [9]. The development of high strength concrete is based on the use of high-quality cements, fine aggregates, superplasticizers and supplementary cementitious materials, including ultrafine mineral additives [10–14]. The use of organic and mineral nanocomposites—materials of mineral and polymeric components based on silica additives of different origin and polycarboxylate superplasticizers is effective for obtaining of high strength concrete [15].

The high strength concrete is a brittle material in which deformation is localized at the site of manifestation of the first crack after the ultimate loading. Fiber-reinforced concrete has been developed to effectively address this problem, with high tensile strength, crack resistance, durability and impact resistance [16–18]. Dispersion fiber over-comes the opening of surface cracks at ultimate loading and disperses deformation. The effect of disperse reinforcement depends on the structure, strength of the matrix, as well as on the fiber parameters and their properties [19–23]. With the introduction of the disperse fiber the part of it works in tension, while the part prevents the displacement of the particles, ensuring the integrity of the system [24].

The purpose of the study is to research the influence of dry-wet cycles on the strength and impact resistance of high strength concrete and fiber-reinforced concrete.

2 Experimental Program

2.1 Materials

Ordinary Portland cement CEM I 42.5R according to EN 197-1 was used in the investigations. Natural sand of Zhovkva quarry (MF = 2.1) and coarse aggregate of 5–20 mm fraction were used for concrete production. For the research of the effect of dry-wet cycles five concrete mixes were produced containing varying amounts of Portland cement (350 and 400 kg per 1 m^3) and fiber. The mix proportions of concretes are given in Table 1, which were designed to have an average slump between 160 and 210 mm under EN 206-1 for S4 concrete workability class.

Microsilica and Aerosil-380 were used as supplementary cementitious materials for modifying of concrete. The specific surface area of microsilica and Aerosil-380 is 18600 cm^2/g and 300 ± 30 m^2/g respectively. A lignosulphonate plasticizer Centrament N3 (LS) was used as a water reducing agent for concrete mix C1 and polycarboxylate

Table 1. Composition of concrete mixtures (kg/m^3).

Component	Sample codes				
	C1	C2	C3	C4	C5
Portland cement	350	350	350	400	400
Sand	700	710	710	670	670
Coarse aggregate 5/20	1200	1150	1150	1120	1120
Microsilica	–	17.5	17.5	20	20
Aerosil-380	–	1.75	1.75	2	2
Water (W/C)	142 (0.41)	123 (0.35)	137 (0.39)	144 (0.36)	160 (0.40)
Polypropylene fiber	–	–	3.5	–	4
Superplasticizer PCE		3.5	3.5	4	4
Plasticizer LS	1.75	–	–	–	–

superplasticizer Master GLENIUM ACE 430–PCE (BASF Construction Chemicals) was used as a water reducing agent for concrete mixes C2-C5. Concrete mixtures C3 and C5 were proportioned with polypropylene fibers measuring 12 mm in length and 18 μm in diameter to increase an impact resistance of modified high strength concrete after effect of dry-wet cycles.

2.2 Experimental Process and Methods

A compulsory mixer was used to prepare concrete specimens with dimensions of 100 × 100 × 100 and 70.7 × 70.7 × 70.7 mm. For this purpose, the coarse and fine aggregates, Portland cement and mineral additives were firstly added to the mixer separately and mixed for about 2 min; then, mixing water with plasticizing admixtures was added and mixed. Then the polypropylene fibers were added in the case of preparing of fiber reinforced concrete. Mixing was continued for further five minutes to achieve uniform distribution of the fiber. The concrete is mixed according to the mix ratio shown in Table 1. Workability of fresh concrete was determined by the slump test according to EN 12350-2. Finally, the mixture was put into the mold. The specimens were molded for 24 h and then cured under standard conditions (temperature = 20 ± 2 °C; RH = 90–100%) for 28 days. The samples100 × 100 × 100 mm were testing for compressive strength and water absorption (open porosity). The concrete specimens with dimensions of 70.7 × 70.7 × 70.7 mm were subjected to wet-dry cycles test.

The accelerated application of wetting and drying cycles in the laboratory was simulated to intensify the weathering or process. Initially, the strength and impact resistance of the main samples each composition were determined. Other samples were dried at a temperature of 100 ± 5 °C in a laboratory oven for 10 h and then saturated in water for 14 h (t = 20 ± 2 °C). Some properties of the specimens were tested, such as the compressive strength and impact resistance after 50 and 100 dry–wet cycles.

The method of estimating the value of the specific impact viscosity recommended by ACI Committee 544 was used to determine the impact resistance of the developed high strength concrete and fiber-reinforced concrete. The resistance of concrete to the impact load was established as the ability to withstand repeated impacts, taking into account the absorbed energy, until the moment of sample destruction.

After effect of dry-wet cycles, the concrete samples were tested on compressive strength and sprayed with a 0.1% alcoholic solution of phenolphthalein (a colorimetric indicator for calcium hydroxide) to determine the carbonization depth [2]. Spraying the phenolphthalein solution onto the freshly chipped surface of a concrete sample can create a clear color change boundary between carbonated zone (gray) and noncarbonization zone (crimson).

3 Results and Discussion

During the wetting period, the open pores are filled with water, and during the drying period, only the surface area of the element is completely dried. This phenomenon causes the outer layers to shrink, while the inner core (wetter than the outer surface) blocks the volume change [4]. It leads to the formation of microcracks in concrete and a decrease in physical and mechanical properties. As can be seen from Fig. 1, modification with an organic and mineral additive allows to adjust the porosity parameters of concrete. Thus, the volumetric water absorption (open capillary porosity) of modified concretes C2 and C4, as well as fiber-reinforced concretes C3 and C5 decreases by 2.3–2.7 times compared to concrete C1. Due to the density of the surface layer of concrete, a smaller amount of water will be absorbed and such concrete will be characterized by better resistance to dry-wet degradation.

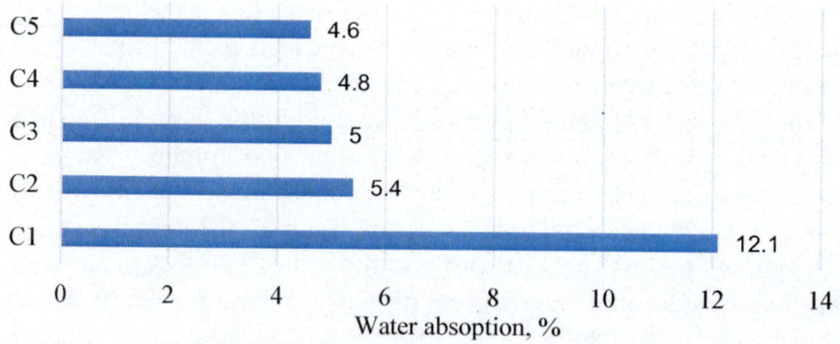

Fig. 1. Water absorption of concrete.

The analysis of the test results showed that compressive strength of concretes C2-C5 slightly increases after 50 cycles (Fig. 2). In this case, the stability coefficient Kst, defined as the ratio of strength after cycles to the strength of concrete after 28 days, is 1.01. It is explained by the activation of pozzolanic reactions in the non-clinker part of the cement matrix of concrete modified with a complex additive based on microsilica, aerosil and polycarboxylate superplasticizer, as well as lower porosity of the surface layer of concrete. At the same time, the strength of concrete C1 after 50 cycles increases by 20% (Kst = 1.2). The authors [25] explain this growth by the emergence of new interface surfaces, which contributes to an increase in the degree of hydration of Portland cement and restores the functional properties of the material.

The compressive strength of modified high strength concretes after 100 dry-wet cycles decreases by 1.9–2.8% compared to the strength of concretes after 50 cycles. When polypropylene fiber, which perceives tensile stresses during drying, was added, the compressive strength of modified fiber-reinforced concrete does not change after 100 cycles, regardless of the consumption of Portland cement. While the strength of modified concrete C1 after 100 cycles decreases by 8.5%.

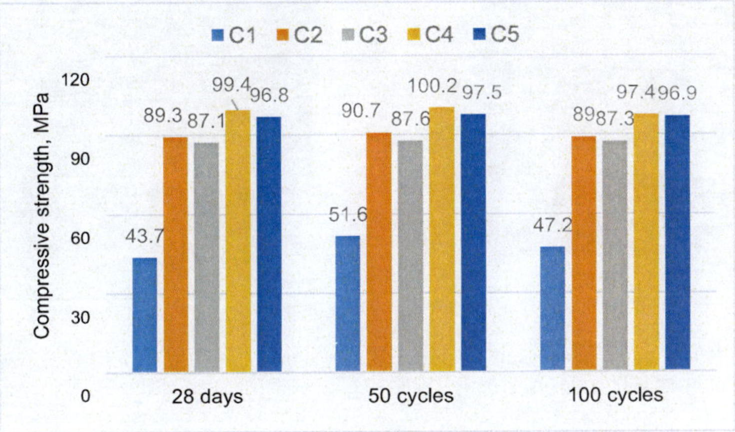

Fig. 2. Compressive strength after wet-dry test.

An increasing of compressive strength can also be associated with partial carbonization of the surface layers of concrete samples. When carbon dioxide acts on cement concrete under normal conditions, not only calcium hydroxide is carbonized, but also other hydrates, in particular calcium hydrosilicates, calcium hydroaluminates, etc. [2]. The rate of carbonization increases with alternating wetting and drying and can cause some increase in strength, but negatively affects the stability of reinforcement in concrete, reducing its hydrogen index. The carbonation depth of concrete C1 is 3 mm after 28 days (Fig. 3a) and after 100 dry-wet cycles carbonation depth increases to 6 mm (Fig. 3b), which is due to the increased porosity of concrete C1. The depth of carbonization of the modified high strength fiber-reinforced concrete C5 before the dry-wet test is 1 mm (Fig. 3c), after it −1.5 mm (Fig. 3d).

The combination of impact loads with multiple dry-wet cycles leads to irreversible damage of the concrete structure and a decreasing of its stability. The fatigue process of concrete under the action of alternating wetting and drying, during which sign-changing deformations were created, led to a decreasing of the impact strength of concrete C1 by 19.1% after 50 cycles (Fig. 4). After 100 dry-wet cycles the reduction of impact strength was 27.4%.

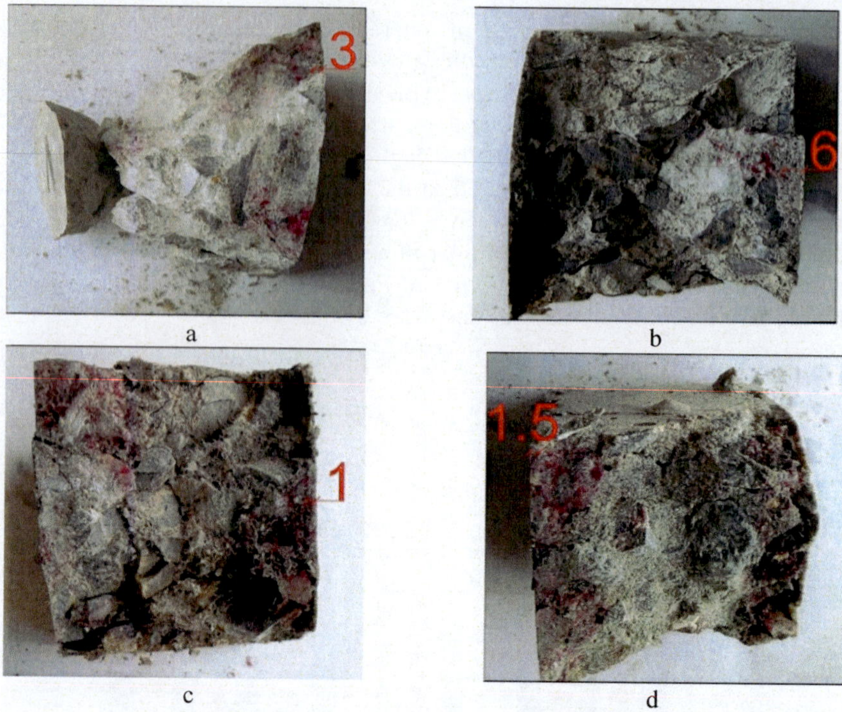

Fig. 3. Carbonation depth: concrete C1 after 28 days **a** and after 100 cycles **b**; concrete C5 after 28 days **c** and after 100 cycles **d**.

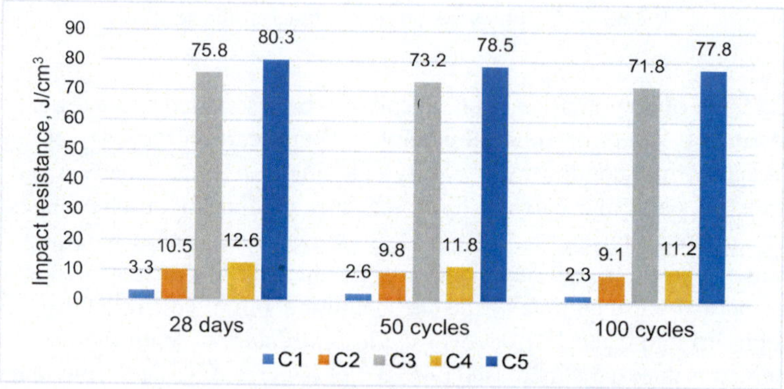

Fig. 4. Impact resistance of concrete after wet-dry test.

The adding of a complex organic and mineral modifier into concrete led to an increasing of strength to 89.3 and 99.4 MPa, respectively, for concrete with a Portland cement consumption of 350 and 400 kg per m^3. In this case, the impact resistance also increased to 10.5 and 12.6 J/cm^3, respectively. The creating of optimal microstructure of the modified concrete ensured a decreasing of the impact resistance of concrete C2 and C4 by 11.4

and 12.9%, respectively, after 100 cycles. Dispersed reinforcement with polypropylene fiber provided an effective increasing of resistance to impact loads with the combined dry-wet effect. Thus, impact resistance of modified fiber-reinforced concrete decreases by 2.2–3.4% and 3.1–5.4% after 50 and 100 dry-wet cycles.

4 Conclusions

It has been established that complex modification of the structure of the cement matrix of concrete and dispersed reinforcement with fibers provide an increasing of the compressive strength of concrete and resistance to dynamic impact (impact resistance is 75–80 J/cm^3). Significant impact energy absorption of the modified matrix when using dispersed fiber is determined by the reduction of defects in the surface layer, compaction of the structure and creation of three-dimensional cross-linking of the structure.

The loss of compressive strength of developed concretes is 1.9–2.8% after dry-wet cycles. Disperse reinforcement with polypropylene fiber provided an effective increasing of resistance to impact loads with the combined dry-wet effect. The impact strength of modified dispersed reinforced concrete after 50 and 100 dry-wet cycles decreases by 2.2–3.4% and 3.1–5.4%, respectively.

References

1. Miller SA, Moore FC (2020) Climate and health damages from global concrete production. Nat Clim Chang 10:439–443
2. Van den Heede P, De Belie N (2014) A service life based global warming potential for high-volume fly ash concrete exposed to carbonation. Constr Build Mater 55:183–193
3. Murali G, Santhi AS, Mohan Ganesh G (2014) Impact resistance and strength reliability of fiber-reinforced concrete in bending under drop weight impact load. Int J Technol 5(2):111–120
4. Zhang J, Gao Y, Han Y (2012) Interior humidity of concrete under dry-wet cycles. J Mater Civ Eng 24:289–298
5. Liu F, Zhang T, Luo T, Zhou M, Zhang K, Ma W (2020) Study on the deterioration of concrete under dry-wet cycle and sulfate attack. Materials 13:4095
6. Wang K, Guo J, Liu X (2021) Effect of dry–wet ratio on pore-structure characteristics of fly ash concrete under sulfate attack. Mater Struct 54:100
7. Guo J-J, Liu P-Q, Wu C-L, Wang K (2021) Effect of dry-wet cycle periods on properties of concrete under sulfate attack. Appl Sci 11:888
8. Marushchak U, Sydor N, Braichenko S, Margal I, Soltysik R (2019) Modified fiber reinforced concrete for industrial floors. In: IOP conference series: materials science and engineering vol 708/1, p 012094
9. Vybranets Y, Vikhot S (2023) Spatial calculation of metal truss structure in joint work with reinforced concrete slab. Lect Notes Civil Eng 290:450–458
10. Sanytsky M, Usherov-Marshak A, Marushchak U, Kabus A (2021) The effect of mechanical activation on the properties of hardened Portland cement. Lect Notes Civil Eng 100:378–384
11. Chepurna S, Borziak O, Zubenko S (2019) Concretes, modified by the addition of high-diffused chalk, for small architectural forms. Mater Sci Forum 968:82–88
12. Markiv T, Sobol K, Petrovska N, Hunyak O (2020) The effect of porous pozzolanic polydisperse mineral components on properties of concrete. Lect Notes Civil Eng 47:275–282

13. Hunyak O, Hidei V, Sobol K, Petrovska N (2023) Valorization of wastepaper sludge ash as supplementary cementitious material in concrete. Lect Notes Civil Eng 290:94–100
14. Shamsad A, Baghabra Al-Amoudi OS, Khan SMS, Maslehuddin M (2022) Effect of silica fume inclusion on the strength, shrinkage and durability characteristics of natural pozzolan-based cement concrete. Case Stud Constr Mater 17:01255
15. Sanytsky M, Kropyvnytska T, Heviuk I, Sikora P, Braichenko S (2021) Development of rapid hardening ultra-high strength cementitious composites using superzeolite and N-C-S-H-PCE alkaline nanomodifier. Eastern-Eur J Enterp Technol 5(6):62–72
16. Stechshyn M, Sanytsky M, Poznyak O (2015) Durability properties of high volume fly ash self-compacting fiber reinforced concretes. Eastern-Eur J Enterp Technol 3(11):49–53
17. Krishna A, Mohan Kaliyaperumal SR, Kathirvel P (2022) Compressive strength and impact resistance of hybrid fiber reinforced concrete exposed to elevated temperatures. Struct Concr 23:3
18. Marushchak U, Sanytsky M, Sydor N, Braichenko S (2018) Research of impact resistance of nanomodified fiber reinforced concrete. MATEC Web Conf 230:03012
19. Najaf E, Abbasi H (2022) Impact resistance and mechanical properties of fiber-reinforced concrete using string and fibrillated polypropylene fibers in a hybrid form. Struct Concr 24:10
20. Mohtasham Moein M et al (2022) Investigation of impact resistance of high-strength Portland cement concrete containing steel fibers. Materials 15:7157
21. Marushchak U, Sanytsky M, Sydor N, Braichenko S (2018) Research of nanomodified engineered cementitious composites. In: Proceedings of the 2018 IEEE 8th international conference on nanomaterials: applications and properties, p 8914835
22. Sydor N, Marushchak U, Braichenko S, Rusyn B (2021) Development of component composition of engineered cementitious composites. Lect Notes Civil Eng 100:459–465
23. Marushchak U, Sydor N, Margal I (2023) Impact of polypropylene fibers on the properties of engineered cementitious composites. Lect Notes Civil Eng 290:262–269
24. Turba Y, Solodkyy S (2021) Crack resistance of concretes reinforced with polypropylene fiber. Lect Notes Civil Eng 100:474–481
25. Vyrovoy V, Korobko O, Antoniuk N, Zakorchemny Y (2021) Influence of cracks on the safe functioning of building structures. Mater Sci Forum MSF 1038:383–392

Substantiation of the Method for Studying the Behavior of Enclosing Structures with Glazing Under Conditions of Fire Thermal Influence

Vadim Nizhnyk, Serhii Pozdieiev, Valeriia Nekora(✉), and Oleksandr Teslenko

Institute of Public Administration and Research in Civil Protection, Kyiv 04074, Ukraine
valerianekora@gmail.com

Abstract. The work is devoted to the actual scientific and technical task of researching the behavior of enclosing structures with glazing under the conditions of the thermal influence of fire. For this, a method is proposed, the essence of which is to create a standard fire temperature regime on model samples of glass elements of enclosing building structures using an experimental installation based on a fire furnace, the fuel system of which ensures the creation of a standard fire temperature regime. With the help of measuring and recording equipment, the temperature indicators in the space of the furnace and in the samples are fixed, as well as the onset of the limit states of loss of integrity and heat-insulating capacity. The purpose of these studies is to experimentally determine the patterns of changes in the parameters of thermal processes and the stress-strain state in glass elements of enclosing building structures under the conditions of thermal influence of the standard fire temperature regime as a scientific basis for the development of standardized calculation methods for assessing their fire resistance. The tasks of the research are checking and obtaining additional data on the thermophysical properties of glass and other additional materials and substances that can be used in the enclosing building structure, checking and obtaining additional data on heat exchange in the air cavities between the glazing, checking mathematical models of the stress-strain state of the glass elements of the enclosing structures under the conditions of thermal exposure of the standard fire temperature regime.

Keywords: Enclosing structures with glazing · Thermal influence of fire · Standard fire temperature regime

1 Introduction

An important factor that determines the possibility of enclosing structures with glazing is their ability to resist destruction in the event of fire, which affects the intensity of fire development and its elimination [1–3]. This is due to the fact that when the enclosing structures with glazing are destroyed in the combustion zone, there is a sharp increase in air flow, in which the room quickly burns out. Increasing the fire resistance of enclosing

Z. Blikharskyy et al. (Eds.): CEE 2023, LNCE 438, pp. 273–285, 2024.
https://doi.org/10.1007/978-3-031-44955-0_28

structures with glazing will allow significantly slowing down the intensity of fire and increasing the safety of people in buildings [4, 5]. Fire resistance for reinforced concrete structures, in which due to heating in the reinforcement, fluidity [6–8] and sudden destruction of the building or structure is a relevant issue [9]. This especially applies to damaged building structures that lack a protective layer of concrete [9–12]. When restoring and strengthening reinforced concrete structures, it is necessary to take into account proper fire protection [8, 13].

Research aimed at developing theoretical foundations and obtaining new experimental data on the behavior of enclosing structures with glazing during a fire is relevant [14–16]. There is a developed system of norms and recommendations for the estimated fire resistance of all types of building structures. At the same time, there is a lack of systematic recommendations for evaluating the fire resistance of enclosing structures with glazing. Calculation methods are insufficiently developed and require theoretical justification. Since these methods are a generalization of extensive experimental experience in improving the experimental base of research in this direction, they are relevant.

2 Purpose

The main goal is to develop sequences of experimental procedures for determining the regularities of changes in the parameters of thermal processes and mechanisms of glazing failure in the elements of enclosing building structures with glazing elements under the conditions of thermal influence of the standard fire temperature regime as a scientific basis for expanding the reference base on the thermal properties of glazing elements and bases for the validation of calculated estimates fire resistance of these structures.

To fulfill this goal, the following **research tasks** were set:

- to analyze approaches to the experimental study of fire resistance of enclosing building structures with glazing elements;
- substantiate experimental methods for studying the regularities of changes in the parameters of thermal processes and mechanisms of glazing destruction in the elements of enclosing building structures with glazing elements under the conditions of thermal influence of the standard fire temperature regime.

3 Research Methods

The essence of the experimental research method is to create a standard fire temperature regime on model samples of glass elements of enclosing building structures using an experimental installation based on a fire furnace, the fuel system of which ensures the creation of a standard fire temperature regime. With the help of measuring and recording equipment, the temperature indicators in the space of the furnace and in the samples are fixed, as well as the onset of the limit states of loss of integrity and heat-insulating capacity.

4 Results

The results of the work will be a prerequisite for further research in terms of substantiating the design parameters of glass elements of enclosing building structures on their predicted fire resistance limits and fire resistance limit states under the conditions of thermal influence of the standard fire temperature regime. The main results of the research consist in checking and obtaining additional data on the thermophysical properties of glass, checking and obtaining additional data on heat exchange in air cavities between glazing, as well as checking mathematical models of the stress-strain state of glass elements of enclosing structures under the conditions of thermal exposure of the standard fire temperature regime.

4.1 Experimental Research Method of Fire Tests of Enclosing Structures with Glazing

All tests are proposed to be performed in an installation consisting of a small-sized fire furnace. In this furnace, it is proposed to test reduced fragments of enclosing building structures with glazing. The structural scheme of furnace is shown in Fig. 1.

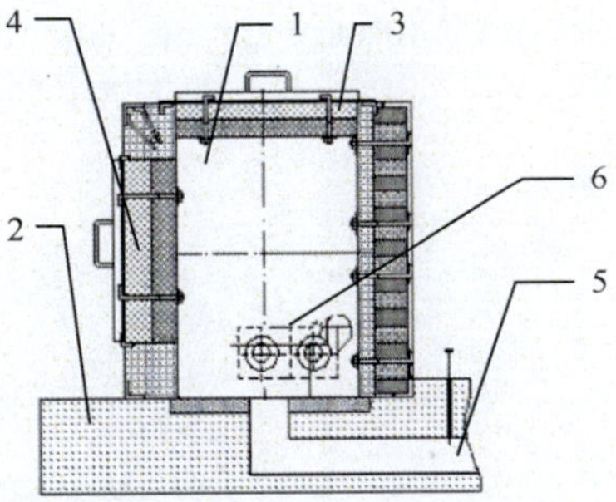

Fig. 1. Test fire furnace: 1—fire chamber; 2—concrete base; 3, 4—thermal insulation covers; 5—outlet channel of combustion products; 6—burners.

During operation of the furnace (Fig. 2), diesel fuel is supplied from tank 6 to burner 5. The temperature in the fire chamber 7 is regulated by the volume of the combustible air-fuel mixture using the valve 10. Combustion products are removed through the smoke channel 4 and chimney 8. The leakage of combustion products is regulated by the valve of the forced gas outlet 9. Diesel fuel is used to heat the furnace.

The furnace 1 is designed for conducting fire resistance tests of fragments of building structures, cable penetrations and tests to check the fire resistance of coatings of building metal structures at a standard temperature regime in accordance to DSTU B V. 1.1-4-98.

The furnace is made of a metal case, which is located on a concrete base 2, which has a channel for the exit of combustion products.

The fire chamber is lined with plates made of material made on the basis of silica fiber (protective thermosquare PTKV-220) and one row of fireproof bricks.

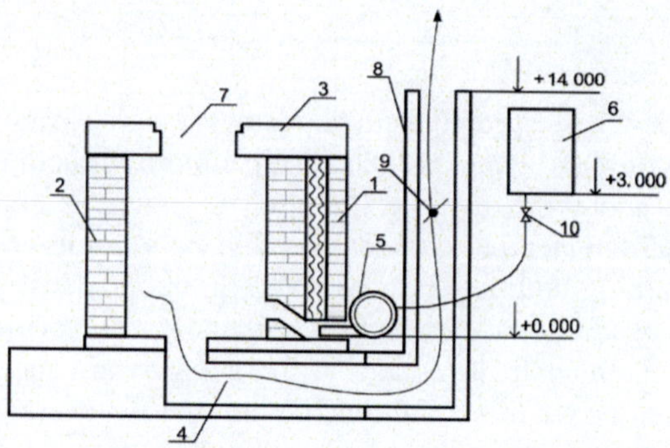

Fig. 2. Functional scheme of the experimental installation: 1, 2—enclosing structures; 3—horizontal heat-insulating cover; 4—outlet channel of combustion products; 5—burner; 6—tank with diesel fuel; 7—fire chamber; 8—chimney; 9—valve for the forced exit of combustion products; 10—faucet.

Lining thickness with PTKV-220 – 60 mm. Thermal inertia of the lining at a temperature of 500 °C–230 $Ws^{0.5}$ m^{-2} K^{-1}.

The experimental study should be carried out in the vertical position of the enclosing building structures with glass elements under the conditions of the standard furnace temperature. The cross-section of the furnace and the conditions for installing samples of enclosing building structures with glass elements in the furnace are presented in Fig. 3.

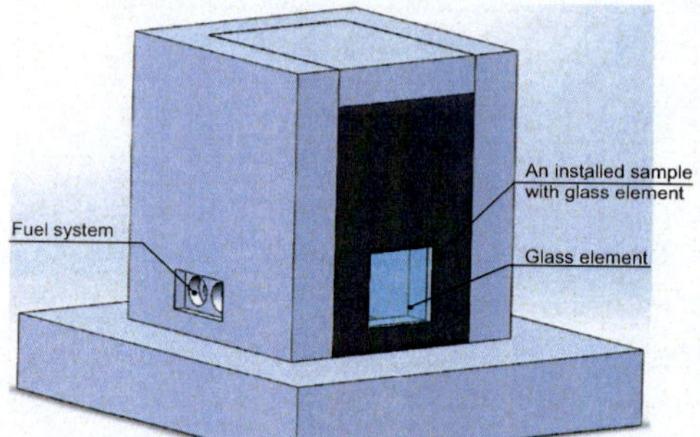

Fig. 3. Conditions of location of samples with glass elements in a fire furnace.

4.2 Research Equipment

During the research, equipment is used, including: an experimental installation based on a fire furnace (Fig. 2), measuring equipment (Table 1), equipment for photo and video shooting.

The measurement and registration of the temperature in the furnace is carried out by the measurement and information system "Thermokont" and five thermoelectric converters (hereinafter thermocouples) of the TCA (thermocouple chromel-alumel) type

(DSTU 2837-94). The diameter of the thermocouple wire is 0.7 mm. The protective screen of the thermocouple has been removed (cut and removed) at a distance of 35 mm ± 5 mm from the measuring junction of the thermocouples. Measurement and registration of temperature values should be carried out at an interval of 60 s.

The temperature in the samples should be measured using five thermocouples with a diameter of 0.7 mm located in the cross-section along the middle of the height of the samples and connected to the Thermokont measuring and information system.

During experimental research, it is necessary to use the test equipment and measuring equipment listed in Table 1.

Table 1. Measuring equipment

Name and type of measuring equipment	Range of measurement	Accuracy class, error of measuring equipment
1	2	3
Measurement and information system "Thermokont"	From 0 to 1200 °C	$\Delta = \pm$ 0.35%
Thermocouples TCA	From 0 to 333 °C From 334 to 1200 °C	$\Delta = \pm$ 2.5 °C $\Delta = \pm$ 0.0075·Tmeas
Receiver of heat flow PAΠ-12 M2	From 0 to 100 kW/m^2	$\Delta = \pm$ 5%
Thermohygrometer Testo 608-H1	From 0 to 50 °C From 2 to 98%	$\Delta = \pm$ 0.5 °C $\Delta = \pm$ 3%
Stopwatch COC пр. 2Б-2-000	From 0 to 3600 s From 0 to 60 s Over 60 s	2 accuracy class; $\Delta = \pm$ (0.4·τ meas./60) s $\Delta = \pm$ (0.4 + 1.5·(τ meas. – 60)/3540 s
Aneroid barometer M 67	From 610 mmHg to 790 mmHg	$\Delta = \pm$ 1 mmHg
Traction gauge TNZ-H	From 0 Pa to 100 Pa	$\Delta = \pm$ 2%
Vernier caliper SHC-1	From 0 to 125 mm	2 accuracy class $\Delta = \pm$ 0.1 mm
Measuring ruler	From 0 to 1000 mm	$\Delta = \pm$ 1 mm
Magnifier	×5	Focal distance 62.5 mm

All measuring equipment must have valid calibration certificates, and test equipment must have verification certificates.

It is allowed to use other measuring equipment during experimental research, the metrological characteristics of which are not worse than those listed in the Table 1.

4.3 Requirements for Making Samples

Four types of samples are used for research. The first type of samples for checking and obtaining additional data on the thermophysical properties of glass and other materials and substances that can be used in enclosing building structures. In Fig. 4 shows the structural scheme of the first type of samples.

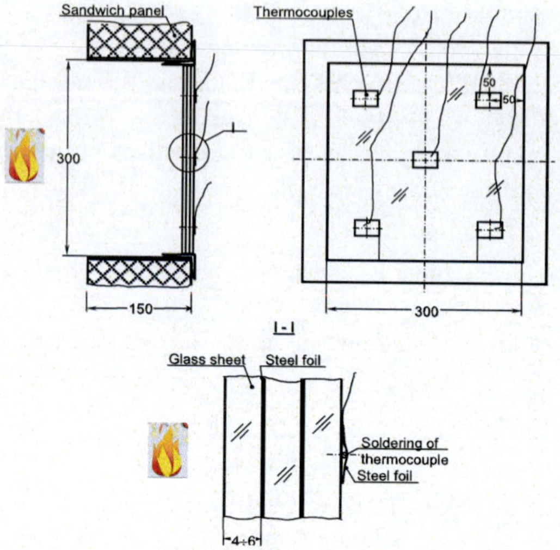

Fig. 4. The structural scheme of the first type of samples and the arrangement scheme of thermocouples.

The peculiarity of this type of samples is that they have a multi-layer construction and consist of three sheets of glass 4 ÷ 6 mm thick. Layers of aluminum or steel foil are placed between sheets of glass to exclude the effect of thermal radiation on the joints of the layers. Thermocouple junctions are protected by layers of aluminum foil for the same purpose.

The second type of samples is designed to check and obtain additional data on the thermophysical parameters of the layer of thermal insulating gel between sheets of glass in fire-resistant enclosing structures. The design of this type of samples is shown in Fig. 5.

The third type of samples is designed to check and obtain additional data on heat exchange in the air cavities between the glazing. The design of this type of samples is shown in Fig. 6.

The fourth type of samples is designed for checking mathematical models of the stress-strain state of glass elements of enclosing structures under the conditions of the thermal effect of a fire. The design of this type of samples is shown in Fig. 7.

The main material of the first two types of samples is ordinary window silicate glass, the properties of which are given in the Table. 2.

The material of the third type of samples is ordinary tempered silicate glass, the properties of which are given in the Table 3.

For the sample of the second type, a fire-resistant double-glazed unit with heat-insulating gel PROMAGLAS manufactured by Promat Inc. is used.

At a distance of 100 mm from the central axis of filling the test sample on the side of its unheated surface, we install a heat flux receiver to measure the heat flux density.

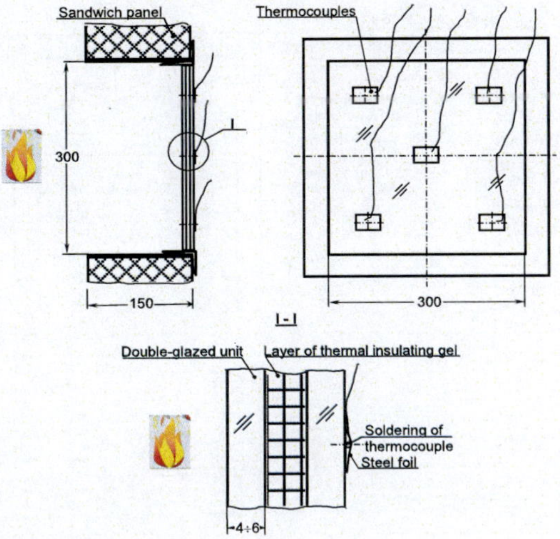

Fig. 5. The structural scheme of the second type of samples and the arrangement scheme of thermocouples.

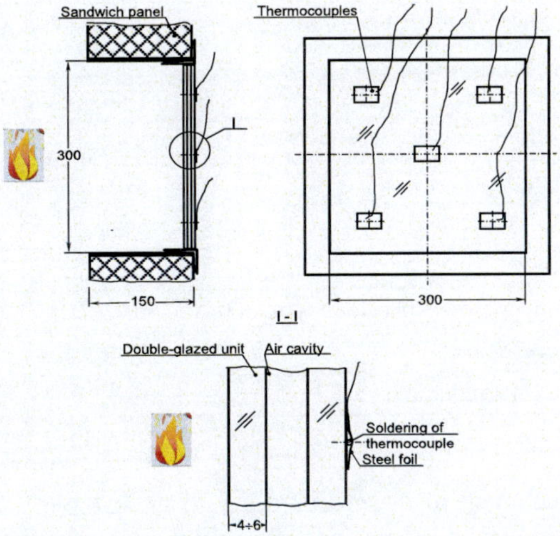

Fig. 6. The structural scheme of the third type of samples and the arrangement scheme of thermocouples.

The installation scheme of the heat flow receiver relative to the test sample is presented in Fig. 8.

4.4 The Program of Experimental Research

Experimental studies should be carried out in the following sequence:

Preparatory procedures:

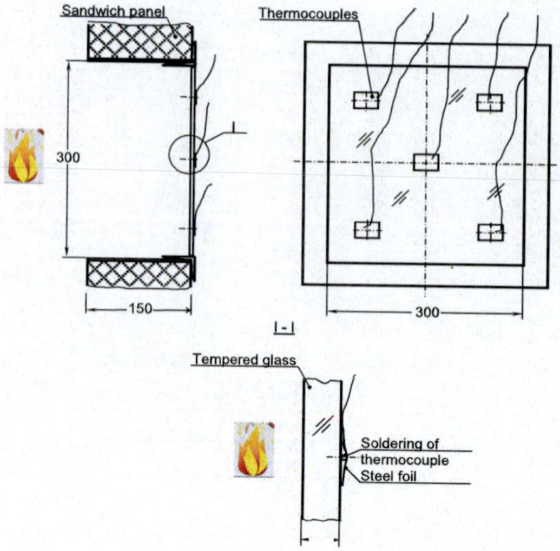

Fig. 7. The structural scheme of the fourth type of samples and the arrangement scheme of thermocouples.

Table 2. Physical properties of ordinary window glass

Parameter	Units of measurement	Value
Density	Kg/m^3	2500
Average bending strength	MPa	40
Elasticity modulus	MPa	6.8 · 104
Temperature of softening	°C	480
Coefficient of linear temperature expansion (up to 300 °C)	°C^{-1}	95 · 10^{-7}
Thermal conductivity coefficient	W/(m·°C)	0.814
Passing capacity for thermal radiation	%	87

- verification of attachment of thermocouples 5 to the sample, which are stationary installed in the furnace, and their connection to the information and measurement system "Thermokont";
- installation of thermocouples at the same time in samples in the amount of 5 pcs. in each (Fig. 9) and connecting them to the information and measurement system "Thermokont";
- inspection and adjustment of the fuel system;
- checking the air conditioning system;
- installation of the tested samples in the furnace, insulation of the edges;
- inspection and adjustment of measuring equipment, devices, PCs.

Table 3. Physical properties of tempered silicate glass

Parameter	Units of measurement	Value
Density	Kg/m^3	2500
Average bending strength	MPa	200
Elasticity modulus	MPa	7.2·104
Temperature of softening	°C	680
Coefficient of linear temperature expansion (up to 300 °C)	°C^{-1}	95·10–7
Thermal conductivity coefficient	W/(m·°C)	0.93
Passing capacity for thermal radiation	%	87

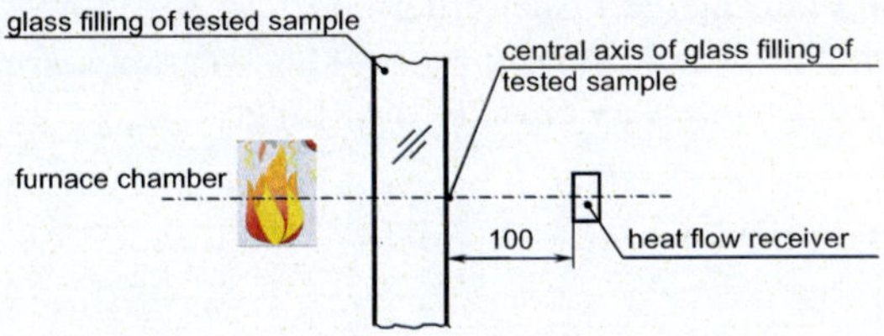

Fig. 8. Scheme of installation of the heat flow receiver relative to the test sample.

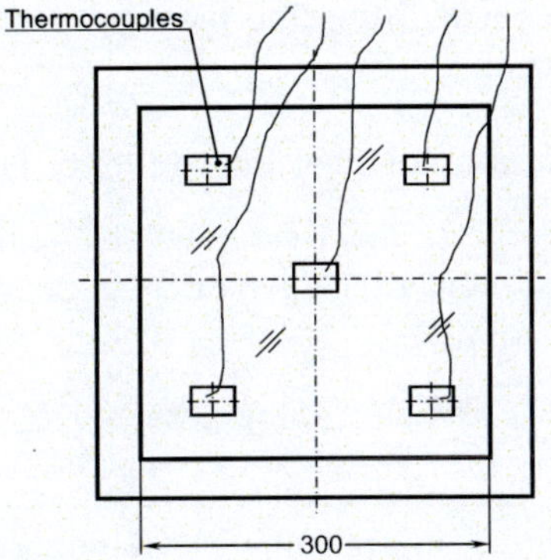

Fig. 9. Thermocouple location scheme

4.5 Conducting an Experiment

- the tanks are filled with diesel fuel;
- the electric motor of the pump is connected to the power grid, it is turned on, the valve is opened, the supply of diesel fuel to the burner with the help of the regulator begins, the diesel fuel is ignited;
- the sensitivity of the secondary devices of the thermocouples located inside the furnace is checked;

- with the help of a regulator (valve), the temperature regime is controlled, as close as possible to the standard one, according to DSTU B V.1.1-4-98;
- readings are taken every minute from all thermocouples (5 thermocouples installed on test samples and 5 thermocouples installed in the furnace), listed into the protocol during 60 min of heating.

4.6 Evaluation of Experimental Research Results

According to the research results, for each moment of measurement time t_i, we determine the temperature Θ_i of the unheated surface of the test sample and the heat flux density q_i from the side of the unheated surface of the test sample. Experimental data are recorded in the journal.

We evaluate the results of experimental studies for the presence of emissions and quasi-emissions using the Grubbs criterion, and also check whether they belong to the same general population using the Fisher criterion.

We present the final results of experimental studies in graphic form as a dependence of temperature and heat flux density changes over time.

4.7 The End of the Experiment

- cooling of the sample;
- photofixation of the sample;
- visual examination of the sample with a description of the state of the sample;
- measurement of charring layers on each area of sample;
- drawing up test protocols and listing the received data for processing using the Mathcad software product.

4.8 Methods of Implementation of Thermal Modes of Heating Samples

According to [1, 3] during the tests, the average volume temperature in the furnace T_f must change according to the standard temperature regime, which is determined by the dependence:

$$T_f = T_0 + 345\log_{10}(8\tau + 1), \tag{1}$$

where t—test time, min, T_0—the initial temperature of the environment, °C (T_0 ~ 20 °C), T_f—the temperature in the fire chamber of the installation for determining the limits of fire resistance of structures depending on the time t of the test.

The standard temperature regime must be maintained in the furnace by adjusting the intensity of fuel supply through the valve by controlling the correspondence of the values found by Eq. (1) and the indicators of the thermocouples located in the fire exposure chamber. The deviation should not exceed the permissible values of DSTU B V.1.1-4-98 and should be calculated according to the Eq. (2):

$$d = (T_f - T_s) \cdot 100\%/T_s, \tag{2}$$

where T_s—real fire temperature, °C; T_f—standard fire temperature, °C.

At the same time, the deviation d of the average temperature in the furnace T_f from the standard temperature regime must not exceed the following values:

(a) ±15% for $0 < t \leq 15$ min
(b) ±15–0,5(t−10) % for $15 < t \leq 30$ min
(c) ±5–0,083 (t−30) for $30 < t \leq 60$ min.

During the tests of the studied samples, on individual thermocouples, after 10 min of testing, the temperature deviation from the standard temperature regime is allowed by no more than 100 °C, which can be seen from Fig. 10.

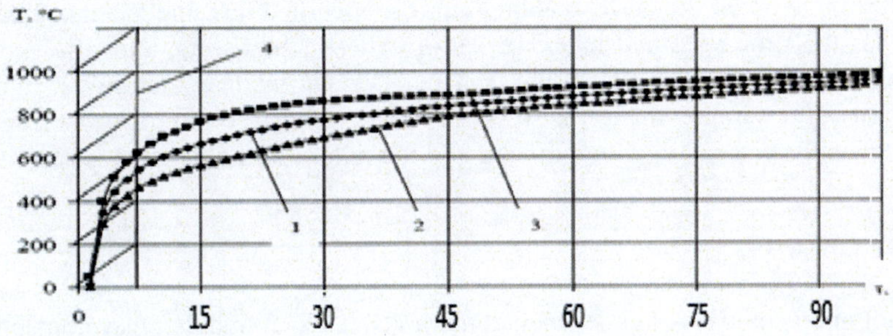

Fig. 10. Standard temperature regime: 1—standard temperature regime; 2—the lower permissible limit of tests; 3—the upper permissible limit of tests; 4—the period of time during which the deviation of the average temperature is uncontrolled.

Fire tests should be carried out in sequence with the program of experimental studies (Table 4).

Table 4. Nomenclature of studied samples for testing

No	Sample number	Amount	Sample type	Test time
1	1.3.1, 1.3.2	2	Type 1 sample with three layers of glass	60 min
2	1.4.1, 1.4.2	2	Type 1 sample with four layers of glass	
3	2.1.1, 2.1.2	2	Type 2 sample (single-chamber double-glazed unit with a layer of thermal protection gel)	
4	3.1.1, 3.1.2	2	Type 3 sample (single-chamber double-glazed unit with an air cavity)	
5	4.1.1, 4.1.2	2	Type 4 sample (one layer of tempered silicate glass)	

The first number in the sample marking:
1—a series of samples of the first type;
2—a series of samples of the second type;
3—a series of samples of the third type;
4—a series of samples of the fourth type.
The second number in the sample marking:

1—one layer of glass element;

3—three layers of glass in a sample of the corresponding series;

4—four layers of glass in a sample of the corresponding series.

The third number in the sample marking:

1, 2—Sample number in the series.

In this way, the marking of sample 1.3.1 is deciphered:

sample of the first type with three layers of glass number 1 in a series of experimental tests.

Each sample under the corresponding number specified in Table 4 is installed in the furnace for the time specified in the series of experimental tests, after the expiration of the set time, it is removed from the furnace. The sample is cooled in air. In case of destruction of the sample, the test is stopped. After cooling the sample or its fragments, they are visually inspected, photographed and the nature of the damage is identified.

5 Conclusions

As a result of the analysis of experimental methods, it was found that thermomechanical processes in the glass of enclosing structures can be investigated with the involvement of small-sized installations when testing reduced fragments.

Experimental methods for studying patterns of changes in parameters of thermal processes and mechanisms of glass destruction in elements of enclosing building structures with translucent elements under the conditions of thermal influence of the standard temperature regime of fire by establishing structural parameters of samples, location of measuring fittings and furnace construction were substantiated.

References

1. Shnal T, Pozdieiev S, Nuianzin O, Sidnei S (2020) Improvement of the assessment method for fire resistance of steel structures in the temperature regime of fire under realistic conditions. Mater Sci Forum MSF 1006:107–116. https://doi.org/10.4028/www.scientific.net/MSF.1006.107
2. Sidnei S, Shnal T, Kholod P, Pozdieiev S (2023) Forecasting the behavior of steel beams with corrugated walls under the thermal exposure of a fire. Lect Notes Civil Eng LNCE 290:388–398. https://doi.org/10.1007/978-3-031-14141-6_39
3. Cacho DV, Sobko Y (2020) Fire situation in case of RC members by sika CFRP strengthening. Lect Notes Civil Eng 47:436–443. https://doi.org/10.1007/978-3-030-27011-7_55
4. Shnal T, Pozdieiev S, Yakovchuk R, Nekora O (2021) Development of a mathematical model of fire spreading in a three-storey building under full-scale fire-response tests. Lect Notes Civil Eng LNCE 100:419–428. https://doi.org/10.1007/978-3-030-57340-9_51
5. Shnal T, Pozdieiev S, Sidnei S, Shvydenko A (2021) Determination of the charring rate of timber to estimate the fire resistance of structures at real temperature modes of fires. Lect Notes Civil Eng LNCE 100:409–418. https://doi.org/10.1007/978-3-030-57340-9_50
6. Blikharskyy Z, Shnal T, Khmil R (2017) The influence of the damaged reinforcing bars on the stress-strain state of the reinforced concrete beams. Prod Eng Arch 14(14):23–26. https://doi.org/10.30657/pea.2017.14.06

7. Blikharskyy Y, Selejdak J, Kopiika N (2021) Specifics of corrosion processes in thermally strengthened rebar. Case Stud Constr Mater 15:e00646. https://doi.org/10.1016/j.cscm.2021.e00646
8. Blikharskyy Y, Selejdak J, Bobalo T, Khmil R, Volynets M (2021) Influence of the percentage of reinforcement by unstressed rebar on the deformability of pre-stressed RC beams. Prod Eng Arch 27(3):212–216. https://doi.org/10.30657/pea.2021.27.28
9. Koteš P, Farbák M, Kotula P, Brodňan M, Čavojcová A (2013) Using CFRP lamellas for strengthening of dynamically loaded beams. Procedia Eng 65:302–310. https://doi.org/10.1016/j.proeng.2013.09.047
10. Blikharskyy Y, Selejdak J, Kopiika N, Vashkevych R (2021) Study of concrete under combined action of aggressive environment and long-term loading. Materials 14(21):6612. https://doi.org/10.3390/ma14216612
11. Guo C, Sun L, Tan Z, Liu G (2019) Numerical analysis of the influence of concrete damage in the compressed zone on the strength of reinforced concrete beams. Adv Struct Eng 22(6):1419–1431. https://doi.org/10.1177/1369433218818156
12. Vegera P, Vashkevych R, Blikharskyy Y, Khmil R (2021) Development methodology of determinating residual carrying capacity of reinforced concrete beams with damages tensile reinforcement which occurred during loading. Eastern-Eur J Enterp Technol 4(7–112):6–17. https://doi.org/10.15587/1729-4061.2021.237954
13. Khmil R, Tytarenko R, Blikharskyy Y, Vegera P (2021) The probabilistic calculation model of RC beams, strengthened by RC jacket. Lect Notes Civil Eng LNCE 100:182–191. https://doi.org/10.1007/978-3-030-57340-9_23
14. Wang S, Zhang C, Zhang S, Zhou Y, Arıcı M (2023) Characteristics of PCMs-filled double glazing unit under fire: a detailed thermal structural analysis. J Build Eng 65:105672
15. Wang Y, Wang Q, Wen JX, Sun J, Liew KM (2017) Investigation of thermal breakage and heat transfer in single, insulated and laminated glazing under fire conditions. Appl Therm Eng 125:662–672
16. Rickard I, Spearpoint M, Lay S (2021) The performance of laminated glass subjected to constant heat fluxes related to building fires. Fire Mater 45(2):283–295

The Study of the Behavior of Reinforced Concrete Structures of Modular Shelter in Conditions of Explosion

Alina Novhorodchenko[1(✉)], Taras Shnal[2], Roman Yakovchuk[3], and Nazarii Tur[3]

[1] Cherkassy Institute of Fire Safety of National University of Civil Defense of Ukraine, 8 Onoprienka Str., Cherkassy 18034, Ukraine
novhorodchenko_alina@chipb.org.in
[2] Lviv Polytechnic National University, 12 Bandera Str., Lviv 79013, Ukraine
[3] Lviv State University of Life Safety, 35 Kleparivska Str., Lviv 79000, Ukraine

Abstract. The article presents the results of mathematical modeling of the impact of the explosion and the corresponding stress-strain state in reinforced concrete blocks of protective structures. The purpose of the work is to establish the safety of the developed protective shelters in the conditions of an explosion during aerial attacks. After conducting the calculations, the results were obtained, which allow to investigate the mechanisms of destruction or loss of integrity of shelter structures and to establish the relationship of these aspects with ensuring the performance of its protective functions under the influence of an explosion. To study the mechanisms of destruction or loss of integrity, the impact of the explosion was investigated. The scientific novelty consists in the development of a new approach to calculations, based on the application of the warhead value of the corresponding projectile in TNT equivalent, the distance of the explosion and the position of the point where the explosion occurs. The determined pressure according to these parameters can be used to study its effect on structures. The practical significance of the research—on the basis of the conducted research, the results of mathematical modeling of the behavior of protective shelters in the conditions of an explosion were obtained, which allow to investigate the mechanisms of destruction or loss of integrity of the structures of the shelter and to establish the relationship of these aspects with ensuring the performance of its protective functions under the influence of an explosion.

Keywords: Reinforced concrete structures · Modular shelter · Stress-Strain state · Protective structures · Mathematical modeling · Impact of explosions

1 Introduction

Since the beginning of February 24, 2022, during the year of the war, Russia has launched almost 5000 missile strikes and about 1000 kamikaze drone attacks on the territory of Ukraine. Unfortunately, the greatest destruction of buildings and structures and a large

Z. Blikharskyy et al. (Eds.): CEE 2023, LNCE 438, pp. 286–294, 2024.
https://doi.org/10.1007/978-3-031-44955-0_29

number of victims are caused by S-200, S-300, X-22, "Iskander" missiles, which cannot be intercepted by the air defense of Ukraine. Therefore, today it remains an urgent issue to establish standardized ground shelters that, in the conditions of hostilities, can block the influence of impact factors (aircraft, missile and artillery fire) and protect people from damage by debris and fragments of building structures. An important aspect that determines the effectiveness of protective structures is the possibility of safe evacuation from them, after the destruction of the building structures of the objects from hitting them with combat shells. This determines the urgency of modernizing and improving existing or setting up new ground defense structures.

2 Purpose

The main goal is to determine the patterns of behavior of reinforced concrete structures of ground protective shelters, which are installed directly on the ground, under the conditions of an explosion during air strikes during hostilities on the territory of Ukraine. The obtained regularities can serve as a scientific basis for methods of establishing their compliance with the current norms of Ukraine [1, 2].

To fulfill this goal, the following research tasks were set:

- to analyze approaches to modeling the behavior of reinforced concrete structures under explosion conditions;
- to substantiate mathematical models of the behavior of reinforced concrete structures under explosion conditions;
- to develop a methodology and, in accordance with the methodology, conduct a numerical experiment on the effect of an explosion on reinforced concrete structures of a ground shelter;
- to establish the main regularities and peculiarities of the behavior of reinforced concrete structures of ground shelter in the conditions of an explosion.

3 Research Methods

To simulate the impact of the explosion and the corresponding stressed and deformed state (SDS) in reinforced concrete blocks under the conditions of the explosion, a generalized engineering approach was applied, which is based on the following basic methods. 1. For mathematical modeling of the SDS of a deformed solid, a generalized theoretical approach is used, which is based on the integration of equations using the finite element method (FEM) [3]. 2. To describe the nonlinear behavior of concrete under active loads, the plastic model of damaged concrete described in the works of Grassl and Jirasek [4] is used. 3. The Drucker-Prager model is used to model the nonlinear behavior of the soil [5]. 4. A model of contact interaction is used to describe the interaction between the support surfaces of reinforced concrete blocks of protective shelter on the ground. The bearing surfaces of the soil layer on the deep layers, according to the accepted assumptions, consist of absolutely undeformed material. 5. The combat charge of shells is assumed to be equivalent to 10 kg in TNT equivalent as the most common. To determine the change in pressure on the walls of reinforced concrete shelter blocks and on the surface of the soil, the Taylor model [6] is used. 6. When studying the behavior of ground protective shelters during an explosion, the LS-DYNA computer system is used, which is designed to simulate impulse effects on structural systems using the dynamics equations in finite-element implementation.

4 Results

The results of the work will become a prerequisite for further research in the part of substantiating the constructive parameters of ground shelters in the conditions of martial law. And the detection of regularities in the dependence of the degree and severity of damage of ground structures on the parameters of the effects caused by the action of combat shells, as well as the scattering of debris and fragments of building structures. The main results of the research consist in the development and improvement of the calculation base for the assessment of storage by the enclosing structures of ground protective structures in the conditions of explosions and impulse actions, as a scientific basis for the improvement of the existing regulatory framework of Ukraine.

The theoretical base of the main approach to calculating the dynamic interaction of mechanical systems, implemented in the code of the software complex LS-DYNA.

4.1 Peculiarities of Mathematical Modeling of the Stress-Strain State

For mathematical modeling of the SDS of a deformed solid body, a generalized theoretical approach is used, which is based on the initiation of movements of the points of the mechanical system of rigid deformed bodies using a system of general dynamics equations and equations of the SDS caused by these movements. The main methods of calculating the dynamic interaction of mechanical systems for ground protective structures are detailed in the Table 1 [7–15].

4.2 Mathematical Model of the Behavior of Reinforcing Steel Under Mechanical Load Conditions

The mathematical description of the phenomenology of the behavior of reinforced steel corresponds to the material model, which describes a simple isotropic material with a deformation diagram that is symmetrical for compression and for tension [7]. When describing concrete, a model of a continuous failure surface with a limiting dome was used, which is described in works Murray, Abu-Odeh i Bligh [8].

The main feature of this model is that the combinations of stresses that the material can withstand are limited by a special surface, which is composed of a yield surface and a brittle failure surface. So this surface covers all cases of destruction of concrete in a complex way and thereby establishes the conditions for either brittle or viscous destruction Fig. 1 shows the general view of the geometry of this surface.

At the same time, parts of the destruction surface are unequal, and therefore this surface is divided into zones—a zone of destruction due to shear deformations and a zone of strengthening during compaction of concrete due to compression, which are interconnected and form a smooth and continuous surface. Figure 2 shows the meridional section of this surface.

To determine the nature of deformation, deformation diagrams are used. They are presented in Fig. 3.

Table 1. A generalized engineering approach to calculation

№	Peculiarities of mechanical processes	Mathematical model or methods
1	The stressed-deformed state of a solid body	Integration of dynamics equations and stress-deformed state equations using the finite element method
2	Modeling of concrete base and soil layer of shelter blocks	Application of three-dimensional massive finite elements of hexahedral shape with eight nodes
3	Modeling of steel plate elements of the anchoring system and door elements	Using two-dimensional shell finite elements of a rectangular shape of the Belychko-Tsai type with four nodes and five integration points
4	Modeling of steel reinforcement in the form of reinforcing bars	The use of one-dimensional linear elements with a defined cross-sectional shape, known as Hughes-Liu rod-beam elements
5	Description of non-linear behavior of concrete under active loads	Application of the plastic model of damaged concrete according to the works of Grassl and Jirasek, which is built on the basis of nonlinear deformation diagrams with descending branches
6	Modeling of nonlinear soil behavior	Application of the Drucker-Prager model
7	Interaction between the support surfaces of reinforced concrete blocks of protective shelter on the ground	Using the model of contact interaction
8	Modeling the explosion and determining the change in pressure on the walls of reinforced concrete shelter blocks	Application of the Taylor model
9	Program code for the sample testing process	The use of the LS-DYNA computer system is part of the complex of engineering calculation systems ANSYS Workbench and is also included as a block in a separate part of this complex ANSYS APDL

4.3 Modeling the Effect of the Explosion on the Structures of Protective Structures

The guidelines for the design of protective structures, recommended in the norms [5] in force in Ukraine, determine the excess pressure of the shock wave as the main influence. This pressure is defined as a reference value and is a force factor applied to the surfaces of the building structures of the bomb shelter. At the same time, appropriate coefficients are used to take into account the dynamics of load application. This approach is somewhat outdated, as it uses a quasi-static approach for calculation. For a more accurate description, it is proposed to use a more physically based approach, which is based on the application of the warhead value of the corresponding projectile in TNT equivalent,

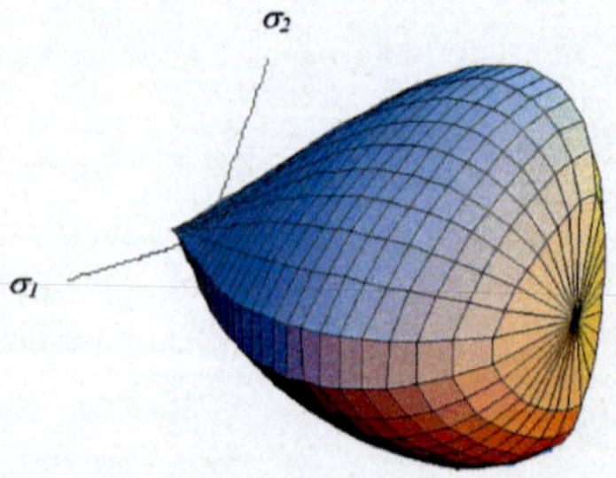

Fig. 1. General view of the geometric configuration of the surface of concrete destruction.

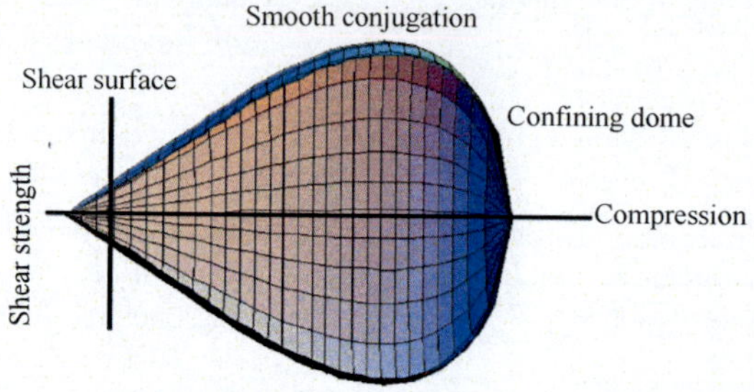

Fig. 2. Meridional section of the surface of concrete destruction

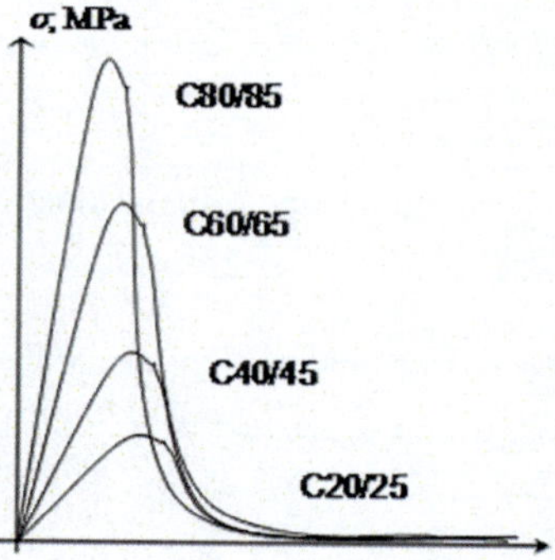

Fig. 3. Diagrams of concrete deformation depending on the strength class.

the distance of the explosion and the position of the point where the explosion occurs. The determined pressure according to these parameters can be used to study its effect on structures.

A special empirical curve according to [9] is used to simulate pressure under the influence of an explosion. This curve is shown in Fig. 4.

The parameters of this curve depend on the scaled distance from the wall to the center of the explosion, which is determined by the formula:

$$Z = R \cdot M^{-\frac{1}{3}} \tag{1}$$

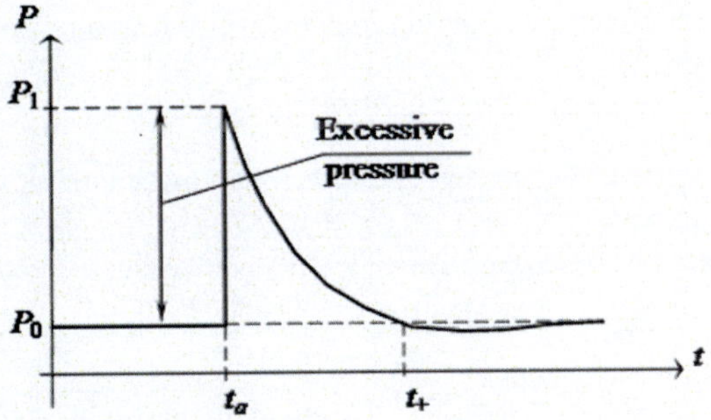

Fig. 4. The curve of pressure on the surface of the walls of shelter structures.

where

R—the distance of the wall to the center of the explosion;
M—TNT mass equivalent.

4.4 Modeling the Impact of the Explosion on the Ground Protective Structures

Two types of load are used to study the impact of an explosion on shelter structures: the load from the structures' own weight and the load from the explosion. The load from its own weight is applied to all components of the model and has a constant effect, however, at the moment of its application, when using the explicit method, unwanted kinetic energy appears, which is manifested, as a rule, in an oscillatory process of a certain duration of its damping. When applying the load from the explosion, the Taylor model [9] and the calculation scheme shown in Fig. 6 are used. According to the diagram in Fig. 5 to simulate the explosion two main parameters—the minimum distance from the epicenter to the surface of the shelter and the equivalent mass of the TNT charge are used.

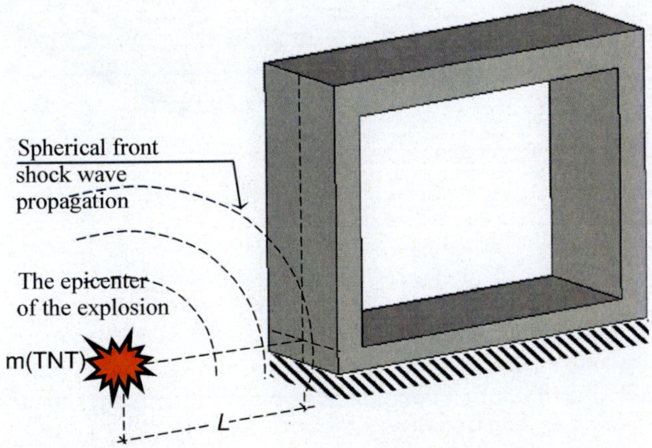

Fig. 5. The scheme of applying the load from the explosion.

In Table 2 sets of such parameters and the corresponding maximum overpressure of the explosion are given.

Table 2. Parameters of the explosion impact model

A variant of the impact of the explosion	Distance, L, m	TNT equivalent, m(TNT), kg	Maximum excess pressure, kPa
1	1	10	4000
2	1	20	8450
3	1	30	13640

Figure 6 shows several options for the location of the epicenter of the explosion: in the middle of the side surface of the shelter, on the side of the entrance to the shelter and above the upper slab of the shelter.

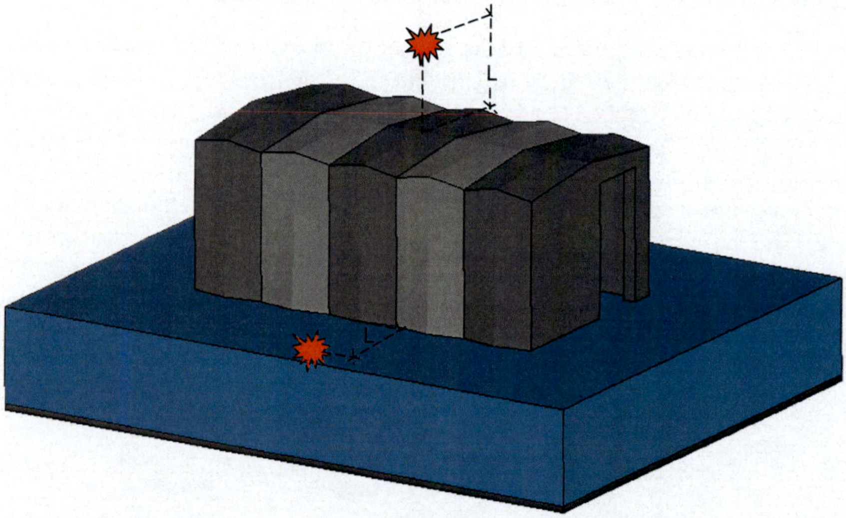

Fig. 6. Schemes of different options for the location of the epicenter of the explosion.

4.5 Finite Element Scheme of the Ground Protective Shelter

After carrying out the above calculations, according to the developed methods, the finite-element scheme of the shelter was obtained, shown in Fig. 7. This diagram shows the finite-element discretization of all components-parts of the mathematical shelter model [8–15].

Thus, the results that will allow in the future to investigate the mechanisms of destruction or loss of integrity of shelter structures and to establish the relationship of these

Fig. 7. Finite element diagram of shelter components.

aspects with ensuring the performance of its protective functions under the influence of an explosion were obtained.

Scientific novelty and practical significance

The results of the work will become a prerequisite for further research in the part of substantiating the structural parameters of ground reinforced concrete shelters during shelling in martial law conditions. They will also be useful for project organizations during the design and evaluation of the effectiveness of the functioning and protective properties of ground defense structures. The above determines the urgency of developing and improving the relevant regulatory framework, taking into account the generalized practical experience gained during Russia's war against Ukraine.

5 Conclusions

So, in the work, scientific results were obtained regarding the determination of the regularities of the behavior of reinforced concrete structures of ground protective shelters, which are installed directly on the ground, under the conditions of an explosion during air strikes during hostilities on the territory of Ukraine. It is worth noting that in the current regulatory documents, the calculation method is expressed by the following criteria: overpressure of the explosion and the protection factor (reduction of ionizing radiation), and such criteria as mechanical impact from military weapons are not defined. Therefore, the developed mathematical models and methods provide a solution to the current problem—this is the improvement of the methodological and regulatory framework for determining the structural parameters of ground protective structures in Ukraine.

References

1. DBN B.2.2-5-97 (2018) Civil defense protective structures. Buildings and structures. With changes. Kyiv, p 81. (in Ukrainian)

2. DSTU-H B EN 1993-1-2:2012 (2012) Projection steel structures of steel structures. Part 1–2. Terms. Calculation of structures for fire resistance. Kyiv, p 77. (in Ukrainian)
3. Perehin A, Nuianzin O, Shnal T, Shchipets S, Myroshnyk O (2023) Improvement of means for assessing fire resistance of fragments of reinforced concrete structures. AIP Conf Proc 2684:030032. https://doi.org/10.1063/5.0120061
4. Murray YD, Abu-Odeh AY, Bligh RP (2007) Evaluation of LS-DYNA concrete material model 159. No. FHWA-HRT-05-063. United States. Federal highway administration. Office of research, Development, and Technology
5. Hallquist JO (2006) LS-DYNA theory manual. Livermore software technology corporation 3:25–31
6. Shnal T, Pozdieiev S, Nuianzin O, Sidnei S (2020) Improvement of the assessment method for fire resistance of steel structures in the temperature regime of fire under realistic conditions. Mater Sci Forum, MSF 1006:107–116. https://doi.org/10.4028/www.scientific.net/MSF.1006.107
7. Sidnei S, Shnal T, Kholod P, Pozdieiev S (2023) Forecasting the behavior of steel beams with corrugated walls under the thermal exposure of a fire. Lect Notes Civil Eng, LNCE 290:388–398. https://doi.org/10.1007/978-3-031-14141-6_39
8. Pozdieiev S, Nekora O, Kryshtal T, Zazhoma V, Sidnei S (2018) Method of the calculated estimation of the possibility of progressive destruction of buildings in result of fire. MATEC Web Conf 230:02026. https://doi.org/10.1051/matecconf/201823002026
9. Shnal T, Horzhevsky V, Pavlyuk, Y., Parhomenko R (2002) Technical diagnostics and further operation of reinforced concrete structures after the fire. Publishing House—National University Lviv Polytechnic, pp 128–135
10. Vegera P, Vashkevych R, Blikharskyy Y, Khmil R (2021) Development methodology of determinating residual carrying capacity of reinforced concrete beams with damages tensile reinforcement which occurred during loading. Eastern-Eur J Enterp Technol 4(7–112):6–17. https://doi.org/10.15587/1729-4061.2021.237954
11. Blikharskyy Z, Shnal T, Khmil R (2017) The influence of the damaged reinforcing bars on the stress-strain state of the reinforced concrete beams. Prod Eng Arch 14(14):23–26. https://doi.org/10.30657/pea.2017.14.06
12. Blikharskyy Y, Selejdak J, Bobalo T, Khmil R, Volynets M (2021) Influence of the percentage of reinforcement by unstressed rebar on the deformability of pre-stressed RC beams. Prod Eng Arch 27(3):212–216. https://doi.org/10.30657/pea.2021.27.28
13. Aagaard BT (2000) Finite-element simulations of earthquakes. Ph.D-Thesis, California Institute of Technology, Pasadena, California
14. ANSYS, Inc. (2004) ANSYS structural analysis guide, ANSYS, Inc., Canonsburg, PA
15. Bakeer T (2009) Collapse analysis of masonry structures under earthquake actions. Publication series of the chair of structural design, TU Dresden, p 229

Rainwater Treatment Technology for the Hygienic and Food Purposes in Households

Dorota Papciak(✉) and Monika Zdeb

Department of Water Purification and Protection, Faculty of Civil, Environmental Engineering and Architecture, Rzeszow University of Technology, 35-959 Rzeszow, Poland
dpapciak@prz.edu.pl

Abstract. The aim of the research was to develop a comprehensive technology for the treatment of rainwater collected from roof surfaces for hygienic and food purposes in households. Optimized technology will allow obtaining water with parameters enabling its widespread use "in situ". Based on the results of the analyses, the quality of rainwater was assessed and critical parameters were selected, determining its suitability for various household purposes. Due to the pH < 6.5, low alkalinity and negligible hardness, rainwater tends to dissolve $CaCO_3$ deposits. Rainwater cannot be considered as an emergency source of water intended for consumption in crisis conditions and used for hygienic purposes without its prior treatment. The proposed rainwater treatment technology is based on the following processes: pre-filtration (net filter 20 microns)—de-acidification filtration (Corosex bed)—main filtration (membrane filter with a filtration threshold of 0.1 micron)—disinfection (UV lamp, UV-LED diodes), improved the microbiological quality and physical stability of water by lowering the turbidity to 0.25 NTU, increased the pH to the normative value, and also minimized the corrosive aggressiveness of rainwater.

Keywords: Rainwater quality · Rainwater treatment · Rainwater stability

1 Introduction

Currently, around 36% of the world's population is affected by the water crisis. In many areas, assurance of the growing demand for this raw material is not possible due to limited access to fresh water, which is why storing and collecting rainwater is one of the most promising methods of water supply for economic and consumption purposes [1–3]. The use of rainwater harvesting techniques and good water management are the means by which water resources can be developed to solutions to the water crisis. The interest in the management of rainwater is influenced by the increasingly widespread pro-ecological awareness and the constantly increasing price of high-quality water. In addition, the analysis of the structure of water consumption in a household indicates that about 50% of tap water can be replaced with rainwater, and in public buildings, this value is almost 65% [4, 5]. Rainwater collected directly from precipitation is considered relatively clean

Z. Blikharskyy et al. (Eds.): CEE 2023, LNCE 438, pp. 295–307, 2024.
https://doi.org/10.1007/978-3-031-44955-0_30

compared to groundwater and surface water. However, many studies show that it is physical parameters and chemical and microbiological quality are not beneficial for humans [6–8]. The source of various types of rainwater pollution are, to a large extent, substances leached from the atmosphere and from roof surfaces and catchment areas (roads, land, storm and combined sewage systems). In the context of the use of rainwater in households, the greatest concern is the presence of microorganisms such as bacteria, viruses and protozoa, including pathogens. Also for this reason, the collected and stored rainwater can be used without treatment and disinfection only for irrigation of crops and for cleaning purposes [9, 10]. For several decades, systems for collecting and storing rainwater have been developed and improved, which, due to their physicochemical and microbiological parameters, cannot be directly used, e.g. for hygienic purposes or contact with food [11–14]. The optimal technology of purification of such waters will allow to obtain its quality per the requirements and allow for widespread use. Purified rainwater can also be used in crisis situations (e.g. contamination of tap water, drought) [15, 16]. The research work described in the article is a response to the globally progressing climate changes and the alarming hydrological situation.

The aim of the research was to develop a comprehensive technology for treatment rainwater collected from roof surfaces for hygienic and food purposes in households. Optimized technology will allow obtaining water with parameters enabling its widespread use "in situ". The development and confirmation of the effectiveness of rainwater treatment technologies is one of the elements of the strategy to enable human access to purification and protect water. Such actions can significantly improve the quality and comfort of life, but also sanitary safety in areas with significant water deficits or crisis situations (droughts, war conflicts, failures of water treatment systems).

2 Methods

2.1 Purpose and Scope of Research

The research presented in the work was aimed at:

- determination of physical and chemical parameters of rainwater collected from the roof surface,
- assessment of the quality of stored rainwater and the selection of critical parameters determining its suitability for various household purposes,
- selection of unit elements of the rainwater treatment technological system,
- testing the effectiveness of rainwater treatment in an experimental technological system and assessing its suitability for drinking.

2.2 Subject of Study

The subject of the research was rainwater from the area of Krasne, located east of Rzeszow in Poland. The suburban area from which the samples for testing were taken is characterized by single-family housing and a significant density of vegetation. Rainwater was collected in the period April–August 2022 from the roof covered a galvanized sheet with anti-corrosion coating. The test samples were taken directly from the roof gutters, with the rejection of the first run-off. Physico-chemical and microbiological analyzes were performed in accordance with the applicable procedures contained in the Table 1. On the basis of the legal acts in force in Poland [Regulation of the Minister of Health on the quality of water intended for human consumption of December 7, 2017, item 2294], critical parameters were selected that determine the suitability of rainwater for drinking.

2.3 Research Procedures

The assessment of the quality of rainwater was made on the basis of the parameters and methods contained in Tables 1 and 2.

Table 1. Methods of assessing the microbiological quality of rainwater.

Parameter	Unit	Analytical method/standard
The total number of bacteria at 37 °C (*mesophilic bacteria*)	CFU/ml	A Agar (P-0231) manufactured by BTL by the department of enzymes and Peptons (Poland) (HPC at 22 °C (incubation 2 days)
The total number of bacteria at 22 °C (*psychrophilic bacteria*)	CFU/ml	A Agar (P-0231) manufactured by BTL by the department of enzymes and Peptons (Poland) (HPC at 22 °C (incubation 3 days)
Escherichia coli	CFU/100 ml	Membrane filtration procedure using Chromocult® Coliform Agar (MERCK, Poland)
Enterococcus	CFU/100 ml	Membrane filtration procedure using Slanetz and Bartley Agar (MERCK, Poland)
Clostridium perfringens	CFU/100 ml	Membrane filtration procedure using Tryptose sulfite Agar perfringens TSC (Argenta)

2.4 Selection of Unit Elements of the Treatment System

A series of preliminary studies were carried out to: select the type of pre-filter, select the alkalizing bed, select the type of filtration membranes and an effective method of rainwater disinfection.

The selection of the type of pre-filter was based on tests of the effectiveness of removing suspensions and compounds causing water turbidity. The choice was made from: reusable nylon net filters with a filtration degree of 20, 50, and 100 microns, yarn filters of 10, and 20 microns, foam sponge filter of 5 microns, and a carbon filter with a filtration threshold of 5 microns.

Table 2. Methods of assessing the physico-chemical quality of rainwater.

Parameter	Unit	Analytical method/Standard
pH	–	ELMETRON CX-505
Conductivity	µs/cm	ELMETRON CX-505
Turbidity	NTU	2100P ISO TURBIDIMETER HACH
Alkalinity	mg CO_2/L	Titration method
Total hardness	mg $CaCO_3$/L	Titration method
TOC	mg C/L	TOC analyzer Sievers 5310 C (SUEZ, Boulder, CO, USA)
UV Absorbance	254 nm	Spectrophotometric method using Hach-Lange DR 500 spectrophotometer
Ammonium nitrogen	mg N-NH_4^+/L	Spectrophotometric method 8155 (sachet tests—ammonia salicylate (1) and cyanurate (2)) using Hach-Lange DR 500 spectrophotometer
Nitrite Nitrogen	mg N-NO_2^-/L	Colorimetric method by Nitrite test Merck 1.14408
Nitrate Nitrogen	mg N-NO_3^-/L	Spectrophotometric method 8039 (sachet tests—NitraVer5) using Hach-Lange DR 500 Spectrophotometer
Phosphates	mg PO_4^{3-}/L	Spectrophotometric method 8048 (sachet tests—PhosVer3) using Hach-Lange DR 500 Spectrophotometer

The next stage is the selection of alkalizing deposits to correct the pH and reduce the corrosive aggressiveness of rainwater. As part of the work, the following neutralizing deposits were tested: dolomite deposit (magnesium-calcium), Akdolit carbonate deposit (calcium), and Corosex deposit (magnesium). The alkalizing bed was selected based on pH and alkalinity. In addition, the values of calcium precipitation potential (CCCP) were analyzed to determine the corrosive aggressiveness of rainwater.

The third stage of the research concerned the selection of the ultrafiltration/microfiltration method. There are 4 variants: ultrafiltration membrane 0.02 micron, ceramic cartridge 0.3 micron, membrane filter 0.1 micron, and carbon cartridge 0.5 micron.

The last element of the technological system—disinfection—was tested for the use of various elements reducing the number of indicator microorganisms, including pathogenic ones. The choice was made from 4 variants: ceramic filtration, membrane filtration, and disinfection using a UV lamp and LED-UV lamp.

3 Results

3.1 Rainwater Quality and Stability

The tested rainwater in terms of physical and chemical parameters does not meet the requirements described in the current Regulation of the Minister of Health of December 7, 2017, on the quality of water intended for human consumption (Journal of Laws of 2017, item 2294), due to the above-norm concentration ammonium nitrogen and turbidity (Table 3).

Table 3. Quality parameters of rainwater.

Parameter	Unit	Drinking water requirements	Max	Min	Medium	SD	VC* [%]
Temperature	°C	–	16.60	11.50	14.75	2.29	15.89
pH	–	6.5–9.5	7.22	6.46	6.75	0.33	4.88
Conductivity	µS/cm	2500	22.20	11.73	15.99	4.32	26.20
Turbidity	NTU	1	5.19	2.35	3.13	1.23	35.62
Total hardness	mg $CaCO_3$/L	60–500	19.20	6.40	9.80	5.70	50.42
Alkalinity	mg CO_2/L	–	4.62	2.20	2.53	1.14	38.49
TOC	mg C/L	No invalid changes	2.61	1.98	2.11	0.28	12.73
$N\text{-}NH_4^+$	mg $N\text{-}NH_4^+$/L	0.39	0.94	0.56	0.83	0.16	20.52
$N\text{-}NO_2^-$	mg $N\text{-}NO_2^-$/L	0.15	0.08	0.01	0.05	0.03	58.16
$N\text{-}NO_3^-$	mg $N\text{-}NO_3^-$/L	11	1.00	0.50	0.90	0.24	28.64
PO_4^{3-}	mg PO_4^{3-}/L	–	0.29	0.06	0.07	0.11	94.53

* VC variation coefficient

On the basis of the quality of rainwater, the following critical parameters were proposed, which will determine the technological system of treatment: pH, turbidity, alkalinity, and indicators of bacteriological contamination.

Turbidity is caused by the presence of colloids and fine suspensions of mineral and organic origin in the water. This indicator is very important in determining the physical stability of rainwater. Turbid water should not be used for economic purposes and is not suitable for consumption. The turbidity of water not only determines the aesthetic value but also may indicate microbiological contamination present in it. The cause of high turbidity in the tested water may be an increased number of microorganisms.

The alkalinity of water and pH determine chemical stability, which is determined on the basis of the Strohecker Index from the formula (1) and (2):

$$J_{st} = pH_n - pH, \tag{1}$$

$$pH_n = 11,39 - 2\log A, \tag{2}$$

where:

J_{st} stability index,
pH_n water pH corresponding to the carbon-calcium balance,
pH reaction value of the tested water,
A amount of bound CO_2 [mg/dm^3] (alkalinity [mval/L] x 22 $mgCO_3^{2-}$).

If:

- $J_{st} < 0$ ($pH_n < pH$)—unstable water, characterized by the ability to precipitate $CaCO_3$,
- $J_{st} = 0$ ($pH_n = pH$)—stable water,
- $J_{st} > 0$ ($pH_n > pH$)—unstable water, characterized by the ability to dissolve $CaCO_3$ deposits.

In the analyzed rainwater, the average value of the Strohecker Stability Index was 3.65. It was found that the tested water is chemically unstable and tends to dissolve $CaCO_3$ deposits.

Significant microbiological contamination, including bacteria of fecal origin (mainly *Escherichia coli* and *Enterococcus*), was found in each sample of stored rainwater (Table 4).

Table 4. The number of indicator bacteria in rainwater.

Microbiological indicator	Rainwater	
	Received value	Required value
The total number of bacteria at 37 °C (*mesophilic bacteria*) [CFU/ ml]	1120	**20**
The total number of bacteria at 22 °C (*psychrophilic bacteria*) [CFU/ ml]	7860	**100**
Escherichia coli [CFU/100 ml]	>50	**0**
Enterococcus [CFU/100 ml]	>50	**0**
Clostridium perfringens [CFU/100 ml]	0	**0**

3.2 Selection of Elements of the Treatment System

The critical parameters selected on the basis of the research allowed to determine the individual processes of the technological treatment system, i.e.:

pre-filtration—de-acidification filtration—main filtration—disinfection.

Pre-filtration—protects subsequent devices of the technological system against larger impurities (leaves, algae, insects, suspension, etc.). The 10 micron yarn filter was the most effective in reducing turbidity (Fig. 1). The lowest turbidity reduction was noted for net filters (Fig. 2.). Based on the obtained results, a decision was made to use all the tested filters in subsequent tests, but in different configurations (Fig. 9). The possibility of rinsing the filter cartridges and their repeated use was taken into account, despite the least favorable turbidity values, often exceeding the value of 1 NTU.

De-acidification filtration (correction of the pH value)—Of the three tested de-acidification deposits, two deposits are proposed (to choose from): Dolomit, and Corosex. Both guarantee a stable pH (Fig. 3) of the treated water and the lowest corrosion index (CCPP) (Table 5). The maximum alkalinity value that was obtained was 0.36 mval/L (Fig. 4). The alkalinity value and trace hardness of rainwater would require a pH greater than 9.5 to minimize the aggressive nature of the water. The higher the alkalinity, the lower the stability pH. Waters are not aggressive and their corrosiveness is lowest when the alkalinity is greater than 1.3 mval/L, regardless of other indicators.

Basic filtration (Microfiltration/Ultrafiltration)—it is recommended to use a 0.1 micron membrane filter due to: availability, scalability, high efficiency of removing micropollutants and microorganisms (Fig. 5). Most of the tested filter cartridges allowed to reduce the turbidity below 1 NTU. The required turbidity value was not obtained only in the case of the ceramic filter (Fig. 6).

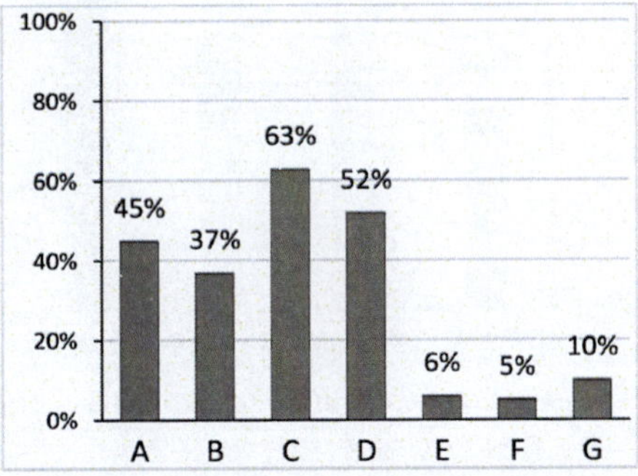

Fig. 1. Efficiency of turbidity removal in the W—rainwater pre-filtration process, A—carbon filter 5 microns, B—foam filter 5 microns, C—yarn filter 10 microns, D—yarn filter 20 microns, E—net filter20 microns, F—net filter 50 microns, G—net filter 100 microns.

Disinfection—in most of the research variants concerning disinfection, the bacteriological quality of water was obtained in accordance with sanitary requirements (Figs. 7 and 8). The greatest effectiveness in reducing microorganisms, including fecal bacteria (coliform bacteria and fecal streptococci) was demonstrated in contact of the tested waters with a UV mercury lamp. Together with the universal availability and relatively low price, this is the basis for recommending this type of devices as an element of the technological system for rainwater treatment for the purpose of their final disinfection. The ceramic filter was also highly effective in removing bacteria from the tested waters, however, with prolonged use, a gradual loss of the ability to remove microorganisms from the water was observed. Besides, as mentioned above, it did not improve other water parameters. UV-LEDs did not provide the sanitary quality required for drinking water, and the latest literature reports indicate their greater effectiveness on surfaces (e.g. food disinfection) [17]. It is worth mentioning that even the best disinfection effects are not permanent, and the bacterial microflora is rapidly multiplied again [18, 19].

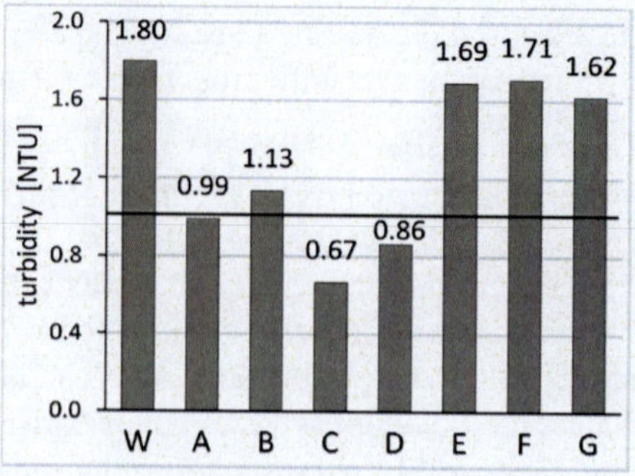

Fig. 2. Rainwater turbidity after pre-filtration process: W—rainwater; A—carbon filter 5 microns, B—foam filter 5 microns, C—yarn filter 10 microns, D—yarn filter 20 microns, E—net filter 20 microns, F—net filter 50 microns, G—net filter 100 microns.

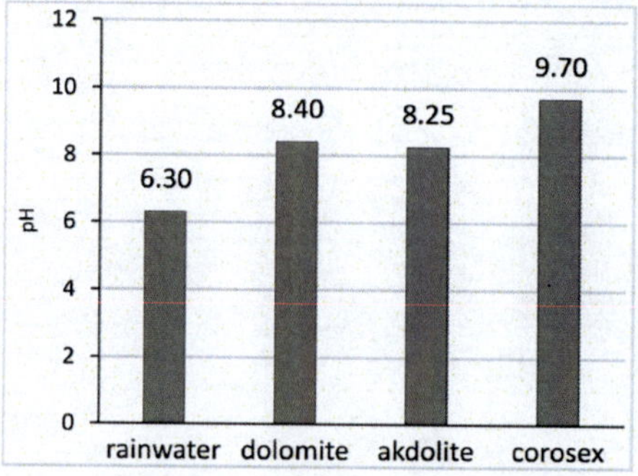

Fig. 3. The pH of rainwater after deacidification filtration.

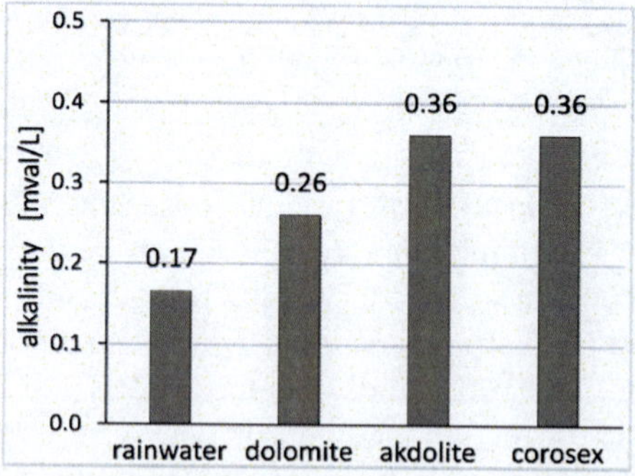

Fig. 4. The alkalinity of rainwater after de-acidification filtration.

Based on the analysis of the results of the research carried out so far, the selection and optimization of individual unit processes of the treatment system was made and 5 treatment systems were proposed (Fig. 9).

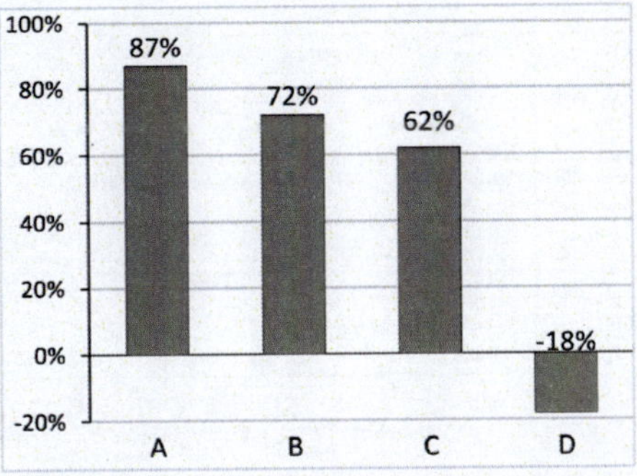

Fig. 5. Primary filtration efficiency (microfiltration/ultrafiltration): A—membrane filter 0.02 micron; B—membrane filter 0.1 micron; C—carbon filter 0.5 micron; D—ceramic filter 0.3 micron

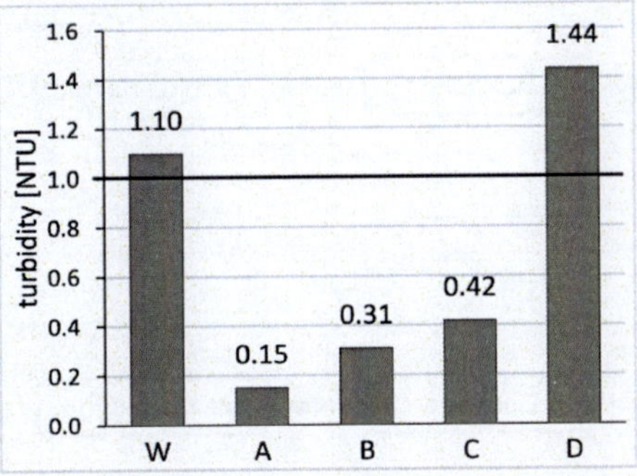

Fig. 6. Rainwater turbidity after basic filtration A—membrane filter 0.02 micron; B—membrane filter 0.1 micron; C—carbon filter 0.5 micron; D—ceramic filter 0.3 micron.

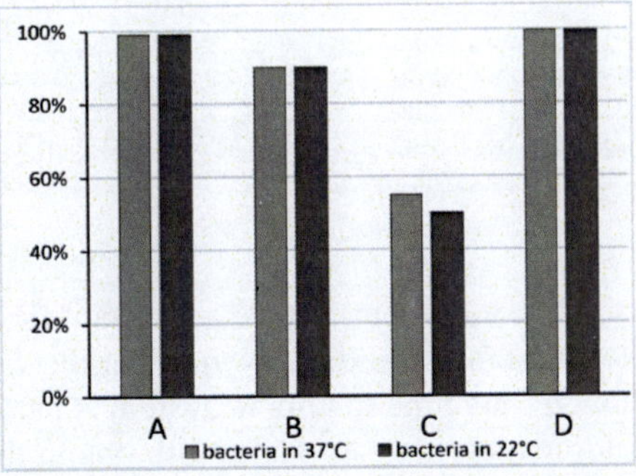

Fig. 7. Disinfection efficiency: A—ceramic filter, B—membrane filter 0.02 micron; C-UV-LED lamp, DUV lamp.

3.3 Optimization and Selection of the Most Effective Treatment System

Based on the results of the tests contained in Table 5, an optimal technological system for rainwater treatment was proposed (Fig. 10). The preliminary tank retains larger

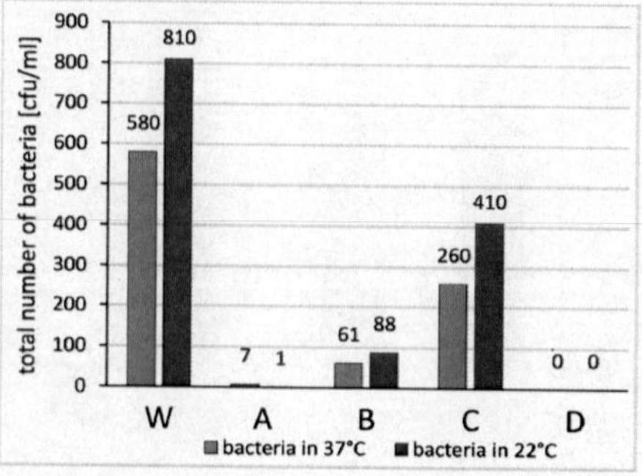

Fig. 8. Total number of bacteria at 37 and 22 °C before and after disinfection of rainwater: W—rainwater, A—ceramic filter, B—membrane filter 0.02 micron, C-UV-LED lamp, D-UV lamp.

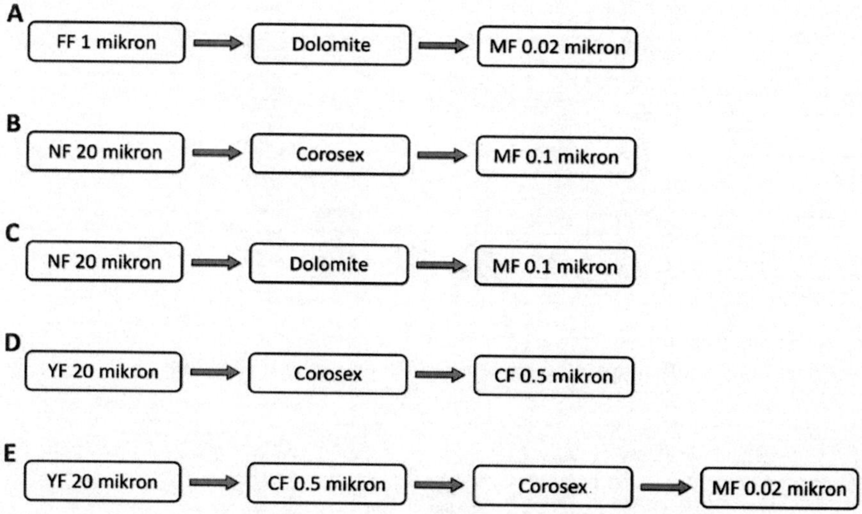

Fig. 9. Rainwater treatment systems: YF–yarn filter, FF–foam filter, CF–carbon filter, MF-membrane filter, NF-net filter.

mechanical impurities and allows to maintenance a constant flow of water in the treatment system. The treated water tank provides coverage for the variability of daily water consumption. Constant flow through the treatment system is particularly important during contact with the deacidification bed, as it allows to obtain the assumed parameters in the pH range of 6.5–9.5.

The proposed treatment system can be scaled and modified depending on the requirements and use of water. It is a system that ensures health safety of water intended for drinking and food purposes. It can also be used after modification for other technological purposes, e.g. cooling purposes or other technical applications. The optimized rainwater treatment system takes into account the physical, chemical, and biological stability of the treated water, which is important when the rainwater will not be used for a long period of

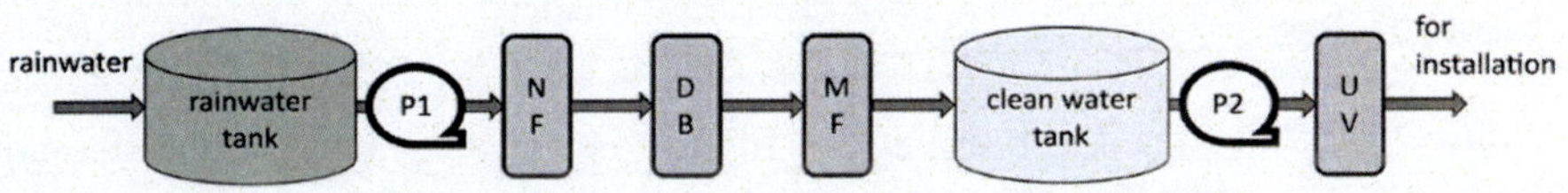

Fig. 10. Optimized rainwater treatment system: P1-first stage pump, P2-second stage pump, NF-net filter, DB-deacidification bed, MF-membrane filter, UV-UV lamp.

Table 5. Parameters of treated water obtained in the analyzed treatment systems.

Parameter	Parameter value for drinking water	Rainwater before treatment	Treatment system*									
			A	A + UV	B	B + UV	C	C + UV	D	D + UV	E	E + UV
pH	6.5–9.5	6.02 ÷ 6.38	9.00	–	8.52	–	8.07	–	7.16	–	9.09	–
Turbidity [NTU]	<1	**0.45 ÷ 1.2**	0.35	–	0.36	–	0.33	–	0.47	–	0.29	–
Parameter CCPP [–]	0	**−33 ÷ −22**	**−4.00**	**–**	**−6.81**	**–**	**−6.82**	**–**	**−7.75**	**–**	**−2.05**	**–**
The total number of bacteria at 37 °C (*mesophilic bacteria*) [CFU/ ml]	20	**5460**	**3270**	**25**	**95**	**38**	12	6	**833**	3	**96**	3
The total number of bacteria at 22°C (*psychrophilic bacteria*) [CFU/ ml]	100	**6930**	**2660**	95	12	80	95	8	68	66	**136**	72
Number of *Escherichia coli* [CFU/100 ml]	0	**>50**	**7**	0	**2**	0	**2**	0	**45**	0	0	0
Number of *Enterococcus* [CFU/100 ml]	0	0	0	0	0	0	0	0	0	0	0	0
Number of *Clostridium perfringens* [CFU/100 ml]	0	0	0	0	0	0	0	0	0	0	0	0

* The values in the table are the average of at least 12 measurements

time. Then it is worth paying attention to maintaining its stability, because stable water will not change parameters during its retention. Biogenic compounds are of great importance for maintaining biological stability, i.e. the multiplication of microorganisms in rainwater. The content of nitrogen and inorganic phosphorus as well as organic substrates create favorable conditions for the growth of pathogenic and heterotrophic organisms. Secondary growth of microorganisms in the stored treated water will be eliminated if the requirements in Table 6 are met.

Table 6. Biological stability of water—recommended values and parameters [20].

Parameter	Unit	The condition of stability
BDOC	mg C/L	≤0.25
$\sum N_{inorg}$	mg N/L	≤0.2
PO_4^{3-}	mg PO_4^{3-}/L	≤0.03

4 Conclusion

Rainwater collected from the roof surface for use in the household as an alternative source of water supply does not meet the requirements for water intended for human consumption (Journal of Laws of 2017, item 2294), due to the excessive content of ammonium nitrogen, turbidity and the total number of microorganisms, the presence of *E. Coli* and *Enterococcus* bacteria. The bacteriological quality of rainwater collected from the roof surface may pose a health risk. The rainwater to be tested is physically, chemically, and biologically unstable. Due to the pH < 6.5, low alkalinity and negligible hardness is characterized by a tendency to dissolve $CaCO_3$ deposits. Rainwater cannot be considered as an emergency source of water intended for consumption in crisis conditions and used for hygienic purposes without its prior treatment. The proposed rainwater treatment technology based on pre-filtration processes (yarn/net filter 20 microns); deacidification filtration (Corosex bed); basic filtration (membrane filter with a filtration threshold of 0.1 μm); disinfection (UV lamp) made it possible to improve the physical stability of the water by lowering the turbidity to 0.25 NTU, increasing the reaction to the normative value, and improving the sanitary quality of the water. Treated water is not chemically and biologically stable, therefore special attention should be paid to the method of storage (tank material, temperature) and effective disinfection, which should be applied before introducing water into the internal system.

Acknowledgment. The research was carried out in the "RainTech" project (N2_022), as part of the Grant Program for R&D works of scientific units under the project entitled "Podkarpackie Center of Innovation" implemented under the Regional Operational Program of the Podkarpackie Voivodeship for 2014–2020, Priority Axis I: Competitive and innovative economy.

References

1. Bertuzzi G, Ghisi E (2021) Potential for potable water savings due to rainwater use in a precast concrete factory. Water 13:448. https://doi.org/10.3390/w13040448
2. Struk-Sokołowska J, Groździej-Mazur J, Jadwiszczak P, Butarewicz A, Ofman P, Wdowikowski M, Kaźmierczak B (2020) The quality of stored rainwater for washing purposes. Water 12:252. https://doi.org/10.3390/w12010252
3. de Sa R, Silva CA, Bimbato AM, Perrella Balestrieri JA, Nogueira Vilanova MR (2022) Exploring environmental, economic and social aspects of rainwater harvesting systems: a review. Sustain Cities Soc 76:103475. https://doi.org/10.1016/j.scs.2021.103475
4. Słyś D, Stec A (2017) Analysis of the possibility of using rainwater in a university campus in Poland. ECOpole'16 conference, 5–8.10.2016; Zakopane, Poland. https://doi.org/10.2429/proc.2017.11(1)031 (in Polish)
5. Słyś D, Stec A, Zeleňáková M (2012) A LCC analysis of rainwater management variants. Ecol Chem Eng S 19(3):359–372. https://doi.org/10.2478/v10216-011-0026-7
6. Zdeb M, Zamorska J, Papciak D, Skwarczyńska-Wojsa A (2021) Investigation of microbiological quality changes of roof-harvested rainwater stored in the tanks. Resources 10(10):103. https://doi.org/10.3390/resources10100103
7. Waso M, Khan S, Khan W (2018) Microbial source tracking markers associated with domestic rainwater harvesting systems: corelation to indicator organisms. Environ Res 161:446–455. https://doi.org/10.1016/j.envres.2017.11.043

8. Zdeb M, Zamorska J, Papciak D (2016) Studying microbiology of rain water for of their use in economy. J Ecol Eng 17(3):203–208. https://doi.org/10.12911/22998993/63314
9. Zdeb M, Zamorska J, Papciak D, Słyś D (2020) The quality of rainwater collected from roofs and the possibility of its economic use. Resources 9(2):12. https://doi.org/10.3390/resources9020012
10. Bae S, Maestre JP, Kinney KA, Kirisits MJ (2019) An examination of the microbial community and occurrence of potential human pathogens in rainwater harvested from different roofing materials. Water Res 159:406–413. https://doi.org/10.1016/j.waters.2019.05.029
11. Dissanayake J, Mooyoung Han M (2020) Effect of inlet/outlet configuration on water quality in a rainwater harvesting tank. Water 12:1970. https://doi.org/10.3390/w12071970
12. Sajeeve L, Alim MA, Rahman A (2022) Disinfection methods for domestic rainwater harvesting systems: a scoping review. J Water Process Eng 46:102542. https://doi.org/10.1016/j.jwpe.2021.102542
13. Pineda E, Guaya D, Rivera G, Garcia-Ruiz MJ, Osorio F (2021) Rainwater treatment: an approach for drinking water provision to indigenous people in Ecuadorian Amazon. Int J Environ Sci Technol 19:8769–8782. https://doi.org/10.1007/s13762-021-03741-0
14. Zdeb M, Papciak D, Zamorska J (2018) An assessment of the quality and use of rainwater as the basis for sustainable water management in suburban areas. E3S web of conferences vol 45. INFRAEKO, p 00111. https://doi.org/10.1051/e3sconf/20184500111
15. Wang YH, Ger TH, Lou JR (2022) Water-saving strategies in the face of water shortage crisis: a case study of science museum in Taiwan. IOP Conf Ser: Earth Environ Sci 987:012014. https://doi.org/10.1088/1755-1315/987/1/012014
16. Pacheco M (2020) Harvesting rainwater: an adaptation strategy for peace and the climate crisis. Chapter 6. In: International rainwater catchment system experiences. IWA Publishing. https://doi.org/10.2166/9781789060584_0083
17. Grzeciak W, Guzdek P, Kulawik J, Maćków P, Piekarski J, Guzik P, Kulawik P (2021) Semiconductor UV-C LED radiation sources for selected procedures used in food technology. Electrotechnical Rev 97:2/2021. https://doi.org/10.15199/48.2021.02.07 (in Polish)
18. Latif S, Alim MA, Rahman A (2022) Disinfection methods for domestic rainwater harvesting systems: a scoping review. J Water Process Eng 46:102542. https://doi.org/10.1016/j.jwpe.2021.102542
19. Zamorska J, Domoń A, Zdeb M (2018) Disinfection of rainwater as a way to their microbiological stability and safe use, 44:00199, 1–7. https://www.e3s-conferences.org/articles/e3sconf/abs/2018/19/e3sconf_eko-dok2018_00199/e3sconf_eko-dok2018_00199.html
20. Wolska M (2015) Removal of nutrients in water purification technology for human consumption: Wrocław, ISBN 978-83-7493-895-2 (in Polish)

Check for updates

Research of Modified Cementitious System for Non-Autoclave Aerated Concretes

Oksana Pozniak, Andriy Melnyk, Uliana Marushchak(✉), and Igor Margal

Department of Building Production, Lviv Polytechnic National University, Lviv, Ukraine
uliana.d.marushchak@lpnu.ua

Abstract. The modern building construction technologies are based on the use of effective structural and thermal insulation materials with high technological, physical, mechanical and operational properties in accordance with requirements for energy effectiveness and thermal protection of buildings. The aerated concrete is characterized by such structural properties as low density, thermal conductivity, reduced self-load. It has been carried out investigation of properties and structure of cementitious system for non-autoclaved aerated concrete. It was established that the use of metakaolin and polycarboxylate superplasticizer accelerates the development of plastic strength of cementitious system to the level of 250 kPa from 109 min to 97 min. The adding of metakaolin into cementitious system allows to increase it strength after 28 days by 15.5% and adding of metakaolin and polycarboxylate superplasticizer—by 1.9 time. The possibility of directional regulation of the microstructure of partitions between pores due to directional phase formation of the modified cementitious system was revealed. The structurally active stability hexagonal hydrate products are formed as a result of the interaction between metakaolin, Portland cement hydration products and aluminium powder.

Keywords: Aerated concrete · Portland cement · Metakaolin · Polycarboxylate superplasticizer · Hydrate phase · Compressive strength

1 Introduction

Aerated concrete is a known building technology which aids the design and construction of lightweight concrete structures. Aerated concrete is characterized by moderate strength, low density, and high thermal properties since it combines the positive characteristics of construction and insulation materials. It is categorized as lightweight concrete, a kind of concrete created by sealing air gaps in the mortar through the aid of an appropriate foaming agent. Density of foam concrete varies from 300 to 1000 kg/m^3. Its distinctive qualities include low aggregate consumption, high porosity and flowability, effective heat insulation, low self-weight, sound insulation, fire resistance and desired compressive strength [1, 2].

Different applications of aerated concrete include block elements, wall elements, insulation and filling materials. It is extensively used for wall and roof insulation, floor leveling filling, etc. as foamed concrete exhibits a low specific density, it contributes

Z. Blikharskyy et al. (Eds.): CEE 2023, LNCE 438, pp. 308–315, 2024.
https://doi.org/10.1007/978-3-031-44955-0_31

to saving the input of mineral resources, reducing the environmental impact of the building sector [3, 4]. Such concretes can be useful for construction in regions covered by problematic soil, where building foundation load is often minimized [2]. Aerated concrete is indispensable in the reconstruction of buildings when there is no possibility to strengthen the foundations [5].

When the density of concrete decreases, their strength decreases too. Typically the mechanical properties are enhanced by autoclaved treatment and thus result in high energy consumption [1]. Given the urgency of reducing energy for the production of autoclaved aerated concrete, there is a constant search for alternative raw materials and additives to reduce the environmental consequences of its production, in particular CO_2 emissions and energy consumption [6, 7]. It is possible to achieve reasonably acceptable strength performance achieve by modifying of the cement matrix, in particular by incorporating of supplementary cementitious materials such as silica fume, fly ash, metakaoline in the mix design [8–11], use of full cement potential due to mechanical activation [12] or use of polypropylene fibers [13–15].

In terms of cutting technology of aerated concrete, it is necessary to solve the problem of accelerated development of the plastic strength of the aerated concrete massif for its edging and cutting. At the same time, an important role is given precisely to mineral additives of multifunctional action, which will not only contribute to the increase in the strength of the final product and the reduction of cement costs. Thus, many studies devoted to the use of gypsum stone, active mineral additives of natural and artificial origin in aerated concrete [16–20].

It is known that the strength of aerated concrete is mainly determined by the strength of interpore partitions [2]. Therefore, it is necessary to take into account the negative impact on their strength of an excess amount of mixing water. The optimum water-to-powder ratio of aerated concrete 500 kg/m^3 is usually between 0.5 and 0.75 [21]. A large amount of water is required in the casting process so that the suspension has the proper fluidity, which provides the necessary conditions for gas expansion. The free water eventually evaporates to form larger sized voids, which reduce the strength of the aerated concrete.

One of the main ways to increase the strength of interpore partitions is a reducing of the water-solid (water-cement) ratio, which leads to a decrease in the capillary porosity of the material and an increase in its strength. In concrete products, polycarboxylic acid superplasticizer [22–25] can be used to improve the workability of the concrete product. The effect of polycarboxylate superplasticizer on the density and strength of standard curing palm oil aerated concrete was studied, moreover the conclusion that polycarboxylate superplasticizer can improve early strength [24]. The incorporation of polycarboxylate superplasticizers has a positive effect on the rheological properties of the slurry, the gas generation, and the physical properties of the aerated concrete with using of circulating fluidized bed combustion ash with high water absorption [25].

2 Experimental Program

2.1 Materials

An ordinary Portland cement CEM-I 42.5R was used in the experimental program which is in conformity with EN 197-1. The aluminium powder PAP-1 (content of active aluminium is 91.1–93.9%, specific surface is 500–600 m^2/kg) was used to obtain an aerated structure. The metakaolin (MK) as supplementary cementitious materials and high-range water reducing admixture based on polycarboxylate polymer (PCE) were used to prepare cementitious systems. The specific surface of Portland cement CEM-I 42.5R is 320 m^2/kg and metakaolin is 1860 m^2/kg. Content of MK was 10% by mass of cement and PCE content was 1.0%. The amount of aluminium powder was 0.1% of the dry mixture.

2.2 Experimental Process and Methods

The fluidity measurement of modified Portland cement system was measured using Suttard viscometer. Plastic strength and compressive strength tests were carried out to evaluate the influence of admixtures on the Portland cement properties. The gas foaming rate was recorded by the changed volume value of paste until the volume no longer changed after the 100 ml of the paste was poured into a 250 ml measuring cylinder. A cone penetrometer was used to determine the plastic strength. Cube-shaped samples (20 × 20 × 20 mm) were prepared to research of strength development kinetics of cementitious systems. The specimens were demolded 24 h after casting and were subjected to normal curing condition (relative humidity 95 ± 5%) in the room temperature. Samples are tested after 1, 2, 7 and 28 days. The peculiarities of the hydration processes of cement matrix of aerated concrete were investigated by X-ray powder diffraction (XRD).

3 Results and discussion

Casting technology of aerated concrete requires the use of good workability slurry (spread-flow diameter 180–340 mm). In the case of Portland cement without additives, the required fluidity (spread-flow is 218 mm) was achieved at W/C = 0.5. The adding of metakaolin with an increased specific surface causes some decreasing of spread-flow to 190 mm at W/C = 0.5. The using of an effective polycarboxylate superplasticizer ensures a 22% reduction in water consumption while maintaining the necessary fluidity. The water consumption of the Portland cement system is reduced by 26% when the MK + PCE additive complex is used.

The value of plastic strength is important in the technology of non-autoclaved aerated concrete. Both the swelling kinetics of the aerated concrete massif and the rate of hardening and accordingly, the reversibility of forms depend on plastic strength. The plastic strength of the aerated concrete must be sufficient for cutting and edging the massif and completely exclude the possibility of its deformation and the appearance of cracks. The optimal plastic strength of the matrix depends on the composition of the aerated concrete mixture and will be different for different compositions of the mixtures. The aerated concrete mass is sent for cutting when the plastic strength is set to 200–250 kPa [18].

An experimental study was conducted to determine the effect of metakaolin and PCE superplasticizer on the plastic strength of cementitious systems (Fig. 1). The use of metakaolin leads to an acceleration of plastic strength gain. Thus, the plastic strength of Portland cement 250 kPa is reached after 109 min. The time of achieving the same plastic strength is reduced to 105 min, when metakaolin is introduced. The time to acquire the plastic strength of the cementitious system modified by MK + PCE is reduced to 97 min.

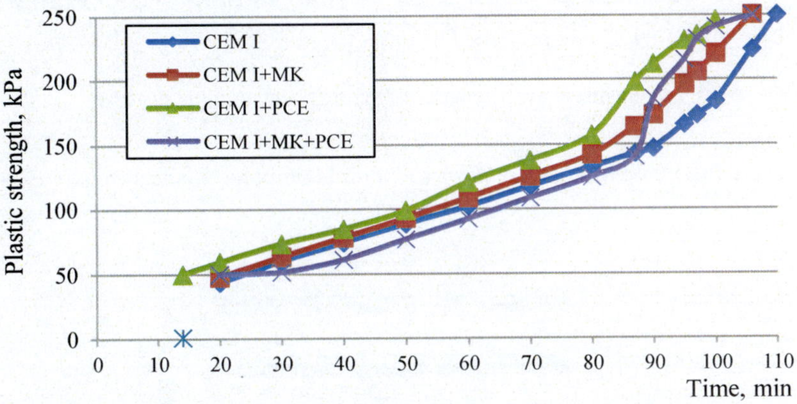

Fig. 1. Plastic strength development of modified cementitious systems.

The results of compressive strength testing of Portland cement paste indicate that its increased water consumption to ensure the necessary fluidity causes a decreasing of the strength of the cement paste during the completely hardening period (Fig. 2). The addition of metakaolin, particles of which fill the voids of the system, causes an increasing of strength. The compressive strength of the paste based on CEM I 42.5 + MK increases by 25.0% after 2 days, and by 15.5% after 28 days. The high water-reducing effect of the PCE superplasticizer causes convergence of the particles in the system and increasing of its strength by 1.8–2.9 times compared to paste based on Portland cement. The cementitious system modified by MK + PCE is characterized by the highest strength, both in the early stages of hardening and after 28 days.

In the technology of non-autoclaved aerated concrete, it is necessary to achieve the optimal ratio of the end of gas generation and the beginning of hardening of the aerated concrete massive. It is one of the factors that determines the average density of non-autoclaved aerated concrete. The effect of additives on the processes of gas generation in aerated concrete mixtures was studied. The early gas-foaming rate is the lowest for the CEM I-42.5 based system (Fig. 3). It may be because the slurry thickening rate is too slow to fix the gas in the paste and the final volume is low.

The use of metakaolin leads to an increasing of the slurry viscosity, which leads to an increasing of the early gas-foaming rate and the final volume of the paste. The rate of foaming of the slurry based on the cementitious system CEM I 42.5 + MK + PCE

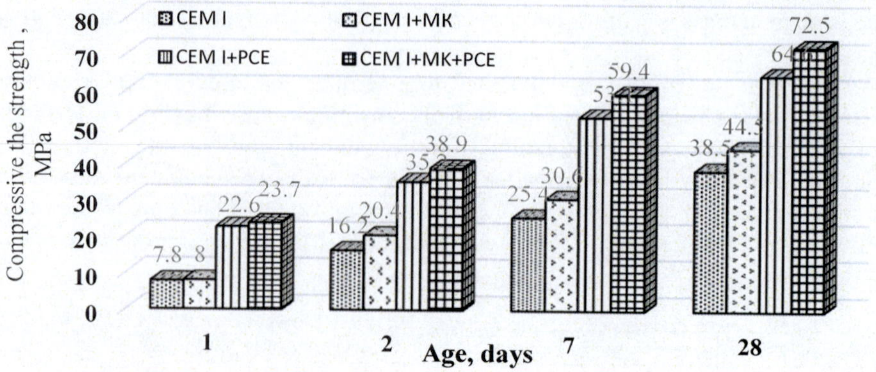

Fig. 2. Compressive strength of modified cementitious systems.

increases, but the final volume decreases to 105 ml compared to the system CEM I 42.5 + MK (112 ml).

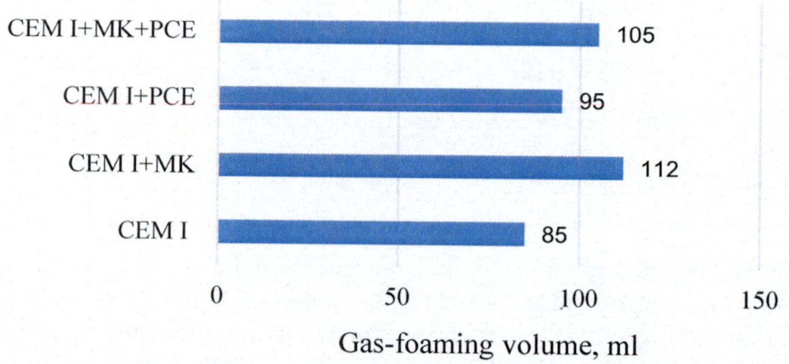

Fig. 3. Gas-foaming volume of cementitious system.

Supplementary cementitious materials play an important role in formation of the properties and structure of binders during their hardening. The model systems "$Ca(OH)_2$—aluminium powder" and "$Ca(OH)_2$—metakaolin—aluminium powder" were used to study the peculiarities of the hydration interaction between the components of the cementitious matrix. Such systems reproduce the hydration processes in real aerated concrete mixtures.

According to the data of the XRD analysis of the model system containing $Ca(OH)_2$ and aluminium powder, after 4 h of hardening under normal conditions (Fig. 4), portlandite lines (d/n = 0.493; 0.263; 0.193 nm) and metastable hexagonal calcium aluminate hydrate $4CaO{\cdot}Al_2O_3{\cdot}19H_2O$ (d/n = 1.07; 0.532; 0.288 nm) are recorded. The intensity of hexagonal calcium aluminate hydrate lines increase over time. After 7 days the lines of cubic calcium hydroaluminate $3CaO{\cdot}Al_2O_3{\cdot}6H_2O$ (d/n = 0.442; 0.520 nm) appear, which was formed as a result of the conversion of metastable hexagonal calcium aluminate hydrate $4CaO{\cdot}Al_2O_3{\cdot}19H_2O$.

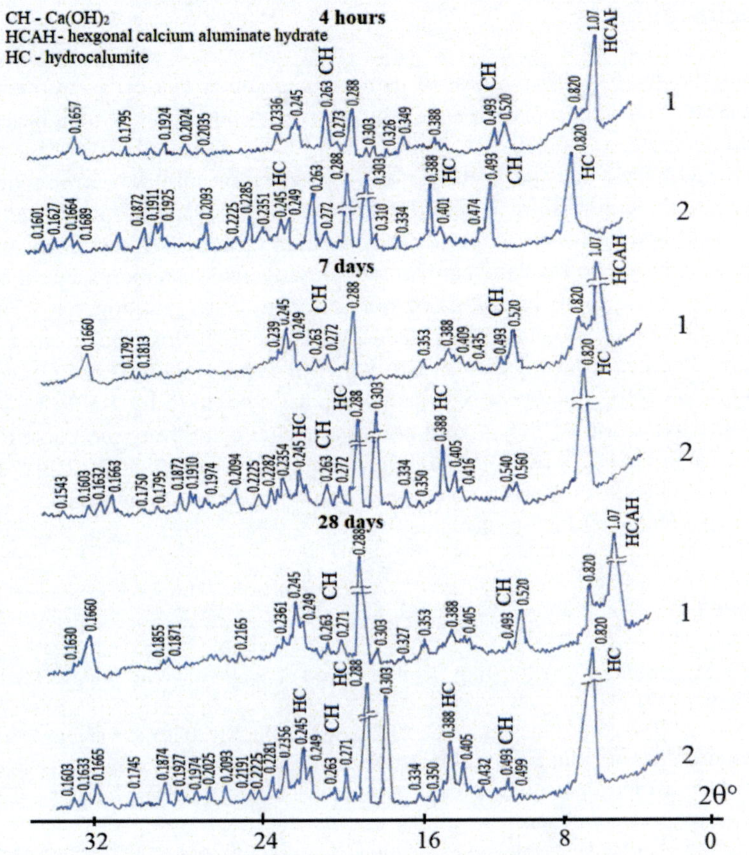

Fig. 4. XRD pattern of model system samples: 1—"$Ca(OH)_2$—aluminium powder"; 2—"$Ca(OH)_2$—metakaolin – aluminium powder".

The XRD patterns of the model system "$Ca(OH)_2$—metakaolin—aluminium powder" show that lines of aluminate hydrate C_4AH_{19} are absent, instead intense lines corresponding to hydrocalumite $Ca_4Al_2(OH)_{14}\cdot 6H_2O$ (d/n = 0.820; 0.388; 0.288; 0.245 nm) appear. Hydrocalumite belongs to the family of layered calcium aluminium double hydroxides (Ca–Al–LDH), also known as AF_m phases, can be represented as $[Ca_2Al(OH)_6](X)(x)\cdot yH_2O$, in which X can be replaced by OH^-, Cl^-, NO_3^-, CO_3^{2-}, or SO_4^{2-} [26]. In addition, the lines of calcium carbonate (d/n = 0.303; 0.227; 0.208 nm) and portlandite (d/n = 0.493; 0.263; 0.193 nm) are observed. It should be noted that after 28 days of hardening on the XRD patterns of model system with metakaolin, the intensity of portlandite lines is significantly lower, and the intensity of the lines of hydrocalumite increases, which indicates an intense interaction between the components of the system.

4 Conclusions

The studies showed the effectiveness of using of supplementary cementitious materials and polycarboxylate superplasticizer to regulate the properties of cement based system for aerated concrete.

It was established that the introduction of metakaolin and polycarboxylate superplasticizer into the composition of modified cementitious system provides reduction of its water consumption by 26% and increasing of strength by 1.9–3.0 times, compared to cement paste based on Portland cement, which will ensure an increase in the strength characteristics of interpore partitions of non-autoclaved aerated concrete. The use of metakaolin and polycarboxylate superplasticizer is the factor of regulation of the plastic strength of cementitious system due to which it is possible to shorten the duration needed by an aerated concrete mix to acquire plastic strength by 10–20%. The regularities of the hydration of the system containing metakaolin are associated with the formation of stable AF_m phases—hexagonal calcium hydroaluminates. The presence of these phases contributes to the compaction of the microstructure of interpore partitions of non-autoclaved aerated concrete.

References

1. Kalpana M, Mohith S (2020) Study on autoclaved aerated concrete: review. Mater Today: Proc 22:894–896
2. Priyatham BPRVS, Lakshmayya MTS, Chaitanya DVSRK (2020) Review on performance and sustainability of foam concrete. Mater Today: Proc 67(3):284–329
3. Yüksek I (2015) The evaluation of building materials in terms of energy efficiency. Periodica Polytech Civil Eng 59:48–58
4. Liu Y-L, Li C-F, Zhai H-X, Ahmad MR, Guo D, Dai J-G (2023) Production and performance of CO_2 modified foam concrete. Constr Build Mater 389:131671
5. Marushchak U, Pozniak O, Mazurak O (2023) Assessment of wall structures for reconstruction of buildings. Lect Notes Civil Eng 290:270–276
6. Scrivener KL, John VM, Gartner EM (2018) Eco-efficient cements: potential economically viable solutions for a low-CO_2 cement-based materials industry. Cem Concr Res 114:2–26
7. Sanytsky M, Kropyvnytska T, Fic S, Ivashchyshyn H (2020) Sustainable low-carbon binders and concretes. E3S Web Conf 166:06007
8. Zhang S, Qi X, Guo S, Zhang L, Ren J (2022) A systematic research on foamed concrete: the effects of foam content, fly ash, slag, silica fume and water-to-binder ratio. Constr Build Mater 339:127683
9. Lashari AR, Kumar A, Kumar R, Rizvi SH (2023) Combined effect of silica fume and fly ash as cementitious material on strength characteristics, embodied carbon, and cost of autoclave aerated concrete. Environ Sci Pollut Res 30:27875–27883
10. Ha Y, Yang G, Liang K (2022) Development of fly ash and slag based high-strength alkali-activated foam concrete. Cement Concr Compos 128:104447
11. Pozniak O, Melnyk V, Margal I, Novosad P (2021) Production of fly ash aerated concrete and efficiency of its application. Lect Notes Civil Eng 100:347–352
12. Sanytsky M, Usherov-Marshak A, Marushchak U, Kabus A (2021) The effect of mechanical activation on the properties of hardened Portland cement. Lect Notes Civil Eng 100:378–384

13. Poznyak O, Sanytsky M, Zavadsky I, Braichenko S, Melnyk A (2018) Research into structure formation and properties of the fiber-reinforced aerated concrete obtained by the non-autoclaved hardening. Eastern-Eur J Enterp Technol 3(6–93):39–46
14. Marushchak U, Sanytsky M, Sydor N, Braichenko S (2018) Research of nanomodified engineered cementitious composites. Proceedings of the 2018 IEEE 8th international conference on nanomaterials: applications and properties, NAP 8914835
15. Ardhira PJ, Ardra R, Harika M, Sathyan D (2023) Study on fibre reinforced foam concrete-a review. Mater Today: Proc 83
16. Demirboğa R (2007) Thermal conductivity and compressive strength of concrete incorporation with mineral admixtures. Build Environ 42(7):2467–2471
17. Kearsley EP, Wainwright PJ (2002) Ash content for optimum strength of foamed concrete. Cem Concr Res 32(2):241–246
18. Serdyuk V, Rudchenko D, Dyuzhilova N (2020) The use of low clinker binders in the production of autoclaved aerated concrete by cutting technology. Eastern-Eur J Enterp Technol 6(1–108):63–71
19. Esmaily H, Nuranian H (2012) Non-autoclaved high strength cellular concrete from alkali activated slag. Constr Build Mater 26(1):200–206
20. Yang L, Yan Y, Hu Z (2013) Utilization of phosphogypsum for the preparation of non-autoclaved aerated concrete. Constr Build Mater 44:600–606
21. Hussin MW, Muthusamy K, Zakaria F (2010) Effect of mixing constituent toward engineering properties of POFA cement-based aerated concrete. J Mater Civ Eng 22(4):287–295
22. Novosad P, Pozniak O, Melnyk V, Braichenko S (2020) Porous thermal insulation materials on organic and mineral fillers. Lect Notes Civil Eng 47:354–360
23. Marushchak U, Sanytsky M, Pozniak O, Mazurak O (2019) Peculiarities of nanomodified portland systems structure formation. Chem Chem Technol 13(4):510–517
24. Pasupathy K, Ramakrishnan S, Sanjayan J (2022) Enhancing the chemical foaming process using superplasticizer in aerated geopolymer concrete. Constr Build Mater 324:126535
25. Wu R et al (2020) Utilization of the circulating fluidized bed combustion ash in autoclaved aerated concrete: effect of superplasticizer. Constr Build Mater 237:117644
26. Matschei T, Lothenbach B, Glasser FP (2007) The AFm phase in Portland cement. Cem Concr Res 37(2):118–130

Assessment of Water Volume Allocation in Network Water Supply Tanks Using Hulbert Method

Janusz R. Rak, Krzysztof Boryczko, and Magdalena Stręk(✉)

The Faculty of Civil and Environmental Engineering and Architecture, Department of Water Supply and Sewage Systems, Rzeszow University of Technology, Rzeszow, Poland
magdas@prz.edu.pl

Abstract. The collective water supply system (CWSS) belongs to the critical infrastructure. This requires the appropriate level of safety of the water supply system functioning. Diversification of the number of water intakes and allocation of water volume related to the number of network tanks in the distribution system is important to achieve this. Water in water network tanks is stored when the amount of water supplied by the high pressure pumping station exceeds the demand for water, and it is used when there is an increased demand for water in the supplied area. This paper presents a method to evaluate water allocation in network water tanks according to the Stuart H. Hulbert index. The calculation of allocation ratios was presented for 11 selected cities in south-eastern Poland. A scale to assess the degree of water volume allocation in network water tanks were developed. On this basis, an assessment of the degree of allocation of water volume in network water tanks was made in the examined cities.

Keywords: Water supply · Water tanks · Allocation degree

1 Introduction

One of the greatest challenges faced by modern agglomerations is to ensure water supply for the constantly growing number of inhabitants. In case of a failure, the ability to obtain alternative sources of water is crucial in maintaining the continuity of water supply. Water supply from several sources is a way to achieve diversification of supplies [1, 2, 12]. However, when the water supply is cut off, the volume of water stored in the water network tanks plays a key role. The number and diameter of collective pressure pipelines associated with high pressure pumping stations which affect the water supply, are also important [10, 11, 20, 21]. On the other hand, reduced water supply or total lack of water can threaten the lives or health of water consumers, with possible financial losses for recipients and water companies. In this kind of situation, the allocation of water in the network water tanks becomes necessary to ensure the safety and reliability of water supply to consumers. The allocation strategy is considered to be one of the methods of maintenance of reliable and safe operation of collective water supply systems (CWSS) [13, 15, 18].

Z. Blikharskyy et al. (Eds.): CEE 2023, LNCE 438, pp. 316–324, 2024.
https://doi.org/10.1007/978-3-031-44955-0_32

The most effective allocation strategies concern the number of water intakes, the number of network water tanks, mains and the number of pumping stations [1, 16, 17]. The implementation of this task requires the allocation (placement, distribution) of reserve volumes of water in the network water tanks. The main purpose of allocating water resources is to ensure a supply of water of appropriate quality and in the necessary volume [1, 14]. The network water tanks play an important role in the functioning of each CWSS [7–9]. Their existence, volume and location affect the nature of operation of most water supply facilities and equipment, increasing the reliability and safety of the CWSS. The main aim of the network water tanks is to equalize the water supply in a daily cycle, stabilize the pressure in the supply area and store water in case of fire and failure. The total volume of the water network tanks consists of the amount of the usable water volume and volume of water for crisis supply (safety water volume). The tanks collect water during the hours of reduced demand and return it to the system during increased consumption. In the case of critical failures, if the water volumes accumulated in the water network tanks are large enough, they can be treated as an additional source of water [8, 9]. Emergency situations, which may be, for example, water contamination in the source, failures at intakes, failures of equipment or pumping stations, may be compensated by the supply of water from the water network tanks in a limited period of time. It depends on the amount of reserve volume of water stored in the tanks. According to various sources the total water volume in the water network tanks should be 30–50% of maximum daily demand Q_{dmax}, with the safety volume of 25–33% of Q_{dmax} [8, 9].

In order to minimize the effects and range of potential failures, apart from increasing the volume of reserved water in the system, the appropriate location of the water network tanks is equally important. Diversification in relation to collective water supply systems according to K. Knapik and R. Plaskonka consists in different arrangement of network water supply tanks, taking into account area water intakes [8]. J. Rak and K. Boryczko presented methods for assessing diversification in relation to sources of water intake, water mains and network water tanks [1]. Currently, allocation strategies are the subject of research all over the world [3, 5, 6, 19, 22].

The aim of the paper is to present a method for assessing the degree of water vol-ume allocation in network water tanks, which has a significant impact on the functioning of the entire CWSS, especially in crisis situations.

2 Research Methodology

The assessment of water allocation in the network water tanks was performed using the Hulbert index. Stuart H. Hulbert graduated in biology from Amherst College in 1961 and received his PhD in zoology from Cornell University in New York state seven years later. Since 1978 he has worked as a professor at San Diego State University and the director of the Center for Inland Waters. His scientific interests include invertebrates, fish and water birds, lake plankton, biostatistics and mathematical ecology.

The index developed by him to determine the diversification of various animal species was adapted to indicate the allocation of water in the water network tanks [1, 4]:

$$d_H = \frac{m}{m+1}\left(1 - \sum_{i=1}^{m} u_i^2\right) \tag{1}$$

where:

m—number of water tanks,

u_i—share of the i-th tank in the total volume of network water tanks.

The share of the volume of i-th tank is defined by the formula:

$$u_i = \frac{V_i}{V} \tag{2}$$

where:

V_i—i-th water network tank capacity [m^3],

V—total capacity of water in water network tanks [m^3],

$$\sum_{i=1}^{m} u_i = 1 \tag{3}$$

Tables 1, 2, 3 and 4 present the values of the Hulbert index for a different number of tanks and different water volume shares calculated according to Eq. (1).

Table 1. Hulbert's index values for two network tanks of different volumes.

m = 2	$u_1 = 0.5$ $u2 = 0.5$	$u_1 = 0.6$ $u_2 = 0.4$	$u_1 = 0.7$ $u_2 = 0.3$	$u_1 = 0.8$ $u_2 = 0.2$	$u_1 = 0.9$ $u_2 = 0.1$	$u_1 = 0.95$ $u_2 = 0.05$	$u_1 = 0.99$ $u_2 = 0.01$
dH	0.333	0.320	0.280	0.213	0.120	0.0633	0.0132

Table 2. Hulbert's index values for three network tanks of different volumes.

m = 3	$u_1 = 0.33$ $u_2 = 0.33$ $u_3 = 0.33$	$u_1 = 0.4$ $u_2 = 0.3$ $u_3 = 0.3$	$u_1 = 0.5$ $u_2 = 0.3$ $u_3 = 0.2$	$u_1 = 0.6$ $u_2 = 0.3$ $u_3 = 0.1$	$u_1 = 0.6$ $u_2 = 0.2$ $u_3 = 0.2$	$u_1 = 0.7$ $u_2 = 0.2$ $u_3 = 0.1$	$u_1 = 0.8$ $u_2 = 0.1$ $u_3 = 0.1$
d_H	0.5	0.495	0.465	0.405	0.42	0.345	0.255

Table 3. Hulbert's index values for four network tanks of different volumes.

m = 4	$u_1 = 0.25$ $u_2 = 0.25$ $u_3 = 0.25$ $u_4 = 0.25$	$u_1 = 0.3$ $u_2 = 0.3$ $u_3 = 0.2$ $u_4 = 0.2$	$u_1 = 0.4$ $u_2 = 0.3$ $u_3 = 0.15$ $u_4 = 0.15$	$u_1 = 0.5$ $u_2 = 0.3$ $u_3 = 0.1$ $u_4 = 0.1$	$u_1 = 0.6$ $u_2 = 0.2$ $u_3 = 0.1$ $u_4 = 0.1$	$u_1 = 0.7$ $u_2 = 0.1$ $u_3 = 0.1$ $u_4 = 0.1$
d_H	0.6	0.592	0.564	0.512	0.464	0.384

An analysis of Tables 1, 2, 3 and 4 shows that the growing number of network tanks does not mean that the d_H indicator will also increase. The indicator reaches its maximum values for no or small inequality of shares. Any combination of the transition from two tanks to three or more gives an increase in the degree of allocation, provided

Table 4. Hulbert's index values for five network tanks of different volumes.

m = 5	$u_1 = 0.2$ $u_2 = 0.2$ $u_3 = 0.2$ $u_4 = 0.2$ $u_5 = 0.2$	$u_1 = 0.3$ $u_2 = 0.3$ $u_3 = 0.2$ $u_4 = 0.1$ $u_5 = 0.1$	$u_1 = 0.4$ $u_2 = 0.3$ $u_3 = 0.1$ $u_4 = 0.1$ $u_5 = 0.1$	$u_1 = 0.5$ $u_2 = 0.2$ $u_3 = 0.1$ $u_4 = 0.1$ $u_5 = 0.1$	$u_1 = 0.6$ $u_2 = 0.1$ $u_3 = 0.1$ $u_4 = 0.1$ $u_5 = 0.1$
d_H	0.667	0.633	0.6	0.567	0.5

that their shares are relatively balanced. In cases of high schedule of shares, an increase in the number of tanks does not always result in an increase in the degree of allocation. Table 5 presents the numerical values of the Hurlbert indexex for equal shares of the tanks in the total volume of network water tanks.

Table 5. Hulbert's index values for network tanks of the same volume.

m	2	3	4	5	6	8	10	20
u_i	0.5	0.33	0.25	0.20	0.167	0.125	0.1	0.05
d_H	0.333	0.5	0.6	0.667	0.714	0.778	0.818	0.905

The categorisation and the scale of the assessment of the degree of allocation of water is presented below:

- lack of allocation $d_H = 0$ (for $m = 1$),
- low allocation $0 < d_H \leq 0.200$,
- average allocation $0.200 < d_H \leq 0.400$,
- good allocation $0.400 < d_H \leq 0.500$,
- excellent allocation $d_H \geq 0.500$.

3 Results and Discussion

The allocation index d_H was calculated for 11 cities of the Podkarpackie Voivodship. For the city of Rzeszow, the calculations were carried out in two variants. The city has 18 network water tanks, of which 14 are in operation, and 4 remain in hot reserve (they will operate the Rzeszow-Dworzysko Science and Technology Park in the future).

Calculation for Rzeszow (variant I): number of water tanks—m = 18

Individual shares of water supply network tanks are:

$u_1 = 0.056$	$u_7 = 0.056$	$u_{13} = 0.006$
$u_2 = 0.056$	$u_8 = 0.056$	$u_{14} = 0.006$
$u_3 = 0.286$	$u_9 = 0.333$	$u_{15} = 0.014$

(continued)

(*continued*)

$u_4 = 0.011$	$u_{10} = 0.003$	$u_{16} = 0.014$
$u_5 = 0.034$	$u_{11} = 0.003$	$u_{17} = 0.014$
$u_6 = 0.034$	$u_{12} = 0.004$	$u_{18} = 0.014$

The d_H index is calculated according to Eq. (1):
$d_H = 0.75$.
Allocation category according to the Hulbert index: excellent.

Calculations for Rzeszow (variant II): number of water tanks—m = 14

Individual shares of water supply network tanks are:

$u_1 = 0.059$	$u_7 = 0.060$	$u_{13} = 0.006$
$u_2 = 0.059$	$u_8 = 0.060$	$u_{14} = 0.006$
$u_3 = 0.303$	$u_9 = 0.353$	
$u_4 = 0.012$	$u_{10} = 0.003$	
$u_5 = 0.036$	$u_{11} = 0.003$	
$u_6 = 0.036$	$u_{12} = 0.004$	

The d_H index is calculated according to Eq. (1):
$d_H = 0.715$.
Allocation category according to the Hulbert index: excellent.

Calculations for other cities

Table 6 presents calculations of d_H index according to Eq. (1) for the other analyzed cities.

Figure 1 presents the values of d_H index and number of water network tanks in all examined CWSS.

The results shows that:

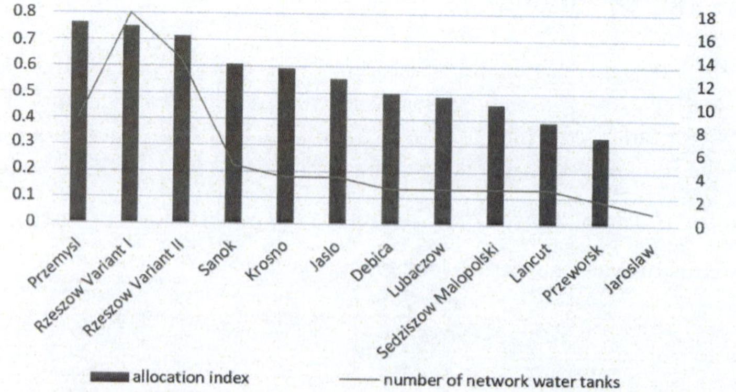

Fig. 1. Hulbert's allocation index values for selected cities.

- the highest value of the allocation index was obtained for the city of Przemysl, which has 9 water network tanks,
- an excellent category of allocation was also established for the cities of Jaslo, Krosno and Sanok,
- for city of Rzeszow, the two-variant analysis did not show a significant impact on the value of the allocation index in the case of a smaller number of water tanks (variant II), an excellent category of allocation was obtained in each variant,
- for three cities Debica, Lubaczow and Sedziszow Malopolski a good degree of allocation was found,
- the average level of allocation was set for the SZZW of the cities of Przeworsk and Lancut,
- in the Jaroslaw city there was no allocation of water in the network water tanks.

Table 6. Calculations of the water allocation index for selected cities.

City	m	u_i	Allocation index d_H	Allocation category
Jaslo	4	$u_1 = 0.132$ $u_2 = 0.132$ $u_3 = 0.368$ $u_4 = 0.368$	0.555	Excellent
Krosno	4	$u_1 = 0.292$ $u_2 = 0.292$ $u_3 = 0.21$ $u_4 = 0.21$	0.595	Excellent
Debica	3	$u_1 = 0.333$ $u_2 = 0.333$ $u_3 = 0.333$	0.5	Good
Lubaczow	3	$u_1 = 0.385$ $u_2 = 0.385$ $u_3 = 0.230$	0.488	Good
Przemysl	9	$u_1 = 0.055$ $u_2 = 0.055$ $u_3 = 0.093$ $u_4 = 0.093$ $u_5 = 0.231$ $u_6 = 0.231$ $u_7 = 0.093$ $u_8 = 0.093$ $u_9 = 0.056$	0.765	Excellent

(continued)

Table 6. (*continued*)

City	m	u_i	Allocation index d_H	Allocation category
Sedziszow Malopolski	3	$u_1 = 0.144$ $u_2 = 0.428$ $u_3 = 0.428$	0.459	Good
Lancut	3	$u_1 = 0.179$ $u_2 = 0.179$ $u_3 = 0.642$	0.393	Average
Jaroslaw	1	$u_1 = 1$	0	Lack
Przeworsk	2	$u_1 = 0.5$ $u_2 = 0.5$	0.333	Average
Sanok	5	$u_1 = 0.170$ $u_2 = 0.339$ $u_3 = 0.339$ $u_4 = 0.051$ $u_5 = 0.101$	0.607	Excellent

4 Conclusions

The allocation of the water volume in the network water tanks is particularly important in crisis situations, allowing to ensure the continuity of water supply to consumers.

The value of the degree of allocation is influenced by three factors: the number of tanks, the share of the volume of individual tanks in relation to the total volume of water accumulated in network water tanks, and the equability of the volume distribution of individual tanks.

A single tank, or its absence, means no water allocation. Transition from one tank to two tanks increases the degree of allocation. Any combination of the transition from two to more tanks contributes to an increase in the degree of allocation, provided that the volumes of individual tanks are balanced as much as possible. In the case of a large diversification of water volume shares, an increase in the number of water supply tanks does not always lead to an increase in the degree of allocation.

References

1. Boryczko K, Rak J (2017) Bezpieczeństwo Systemów Wodociągowych. Dywersyfikacja zasobów wody (Safety of water supply systems. Diversification of water resources). Oficyna Wydawnicza Politechniki Rzeszowskiej, Rzeszow
2. Boryczko K, Rak J (2020) Method for assessment of water supply diversification. Resources 9(7):87
3. He H, Yin M, Chen A et al (2018) Optimal allocation of water resources from the "wide-mild water shortage" perspective. Water 10(10):1289

4. Hurlbert SH (2011) Immigration control and biodiversity in North America. Soc Contract 20:21–22
5. Humphrey JE, Benson KL, Low RKY, Lee W-L (2015) Is diversification always optimal? Pac Basin Fin J Part B 35:521–532
6. Ke W et al (2016) Dynamic simulation of water resource management focused on water allocation and water reclamation in Chinese mining cities. Water Policy 18(4):844–861
7. Knapik K, Bajer J (2011) Wodociągi (Waterworks). Oficyna Wydawnicza Politechniki Krakowskiej, Kraków
8. Knapik K, Ploskonka R (2008) Metoda oceny funkcjonowania systemu zaopatrzenia w wodę przy uwzględnieniu dywersyfikacji zbiornikowych rezerw wody (A method of assessing the functioning of the water supply system, taking into account the diversification of water tanks reserves). In: Elektrowicz M, Sozański MM (eds) Zaopatrzenie w wodę, jakość i ochrona wód, tom ii. viii międzynarodowa konferencja naukowo-techniczna 2008. PZiTS O/Wielkopolski, Poznań-Gniezno, pp 439–512
9. Kulbik M (2012) Rola i znaczenie zbiornika sieciowego w kształtowaniu hydrauliki i jakości wody w systemie wodociągowym (The role and importance of the water network tank in shaping the hydraulics and water quality in the water supply system). In: Elektrowicz M, Sozański MM (eds) Zaopatrzenie w wodę, jakość i ochrona wód, tom ii. międzynarodowa konferencja naukowo-techniczna 2012. PZiTS O/Wielkopolski, Poznań, pp 93–106
10. Piegdon I, Tchorzewska-Cieslak B, Szpak D (2017) The use of geographical information system in the analysis of risk of failure of water supply network. In: Pawlowski L, Pawlowska M (eds) Environmental engineering v 2017 proceedings of the 5th national congress of environmental engineering. CRC Press, London, pp. 7–14
11. Pietrucha-Urbanik K, Studzinski A (2019) Qualitative analysis of the failure risk of water pipes in terms of water supply safety. Eng Fail Anal 95:371–378
12. Rak JR (2017) Metoda oceny stopnia dywersyfikacji dostawy wody dla wybranych miast w Polsce (A method of assessing the degree of water supply diversification for selected cities in Poland). Instal 5:68–70
13. Rak J (2014) Problematyka dywersyfikacji dostaw wody (Problems of water supply diversification). Technol Wody 1:14–16
14. Rak J (2015) Propozycja oceny dywersyfikacji objętości wody w sieciowych zbiornikach wodociągowych (A proposal for the assessment of water volume diversification in the network water tanks). Czasopismo Inżynierii Lądowej, Środowiska i Architektury 62:339–349
15. Rak J, Boryczko K (2017) Assessment of water supply diversification using the Pielou index. In: Pawlowska M, Pawlowski L (eds) Environmental engineering v proceedings of the 5th national congress of environmental engineering. CRC Press, London, pp 53–58
16. Rak J, Boryczko K (2018) Diversification of water supply. E3S Web Conf 59:00006. https://doi.org/10.1051/e3sconf/20185900006
17. Rak JR, Pietrucha – Urbanik K (2015) Diversification of water network tanks volume. Appl Mech Mater 811:395-401
18. Rak J, Wloch A (2015) Models of level diversification assessment of Water Supply Subsystems. In: Kolonko A, Madryas C, Nienartowicz B, Szot A (eds) Underground infrastructure of urban areas 3 2015. Taylor & Francis Group, London, pp 237–244
19. Roozbahani R, Schreider S, Abbasi B (2015) Optimal water allocation through a multi-objective compromise between environmental, social, and economic preferences. Environ Model Softw 64:18–30
20. Studzinski A (2014) Amount of labour of water conduit repair. In: Van Gelder PHAJM, Steenbergen RDJM, Miraglia S, Vrouwenvelder ACWM (eds) Safety, reliability and risk analysis: beyond the horizon. Taylor & Francis Group, London, pp 2081–2084

21. Szpak D, Tchorzewska-Cieslak B (2015) Water producers risk analysis connected with collective water supply system functioning. In: Mazurkiewicz J, Zamojski M, Sugier J, Walkowiak T, Kacprzyk J (eds) Dependability engineering and complex. systems. advances in intelligent systems and computing. Springer: Cham, Switzerland, pp 479–489
22. Wang J, Hou B, Jiang D et al (2016) Optimal allocation of water resources based on water supply security. Water 8(6):237

Climatic Loads on Triple-Glazed Insulating Glass Units Used as an External Horizontal Building Envelope

Zbigniew Respondek(✉)

Faculty of Civil Engineering, Czestochowa University of Technology, Częstochowa, Poland
zbigniew.respondek@pcz.pl

Abstract. Insulating glass units (IGUs) have tight cavities filled with gas in their structure, which results in a specific manner of transferring climatic loads. Under the influence of these loads, the gas in the cavities changes its parameters, which affects the resultant load on each of the glass panes in the unit. In the computational models described in the literature concerning this phenomenon, vertical location of IGUs is usually assumed. However, nowadays glass roof-ceilings and rooflights with triple-glazed IGUs located diagonally or horizontally are more and more often implemented. In this case, an additional factor influencing the static quantities in IGUs is the own weight of glass. The aim of the article is to analyze the influence of the horizontal location of triple-glazed IGUs on their deflection and stress under climatic load. The analysis was carried out using own analytical model. Based on the presented examples, conclusions were drawn about the resultant loading, deflection and stress in IGUs of various structure and dimensions with climatic factors. It was shown that extreme annual temperature changes and a large increase in atmospheric pressure are the most dangerous in the context of the possibility of excessive deflections. The horizontal location, on the other hand, mitigates the effects of atmospheric pressure drop and wind suction.

Keywords: Glass in construction · Insulating glass units · Climatic loads · Horizontal building envelopes

1 Introduction

Insulating glass units (IGUs) are designed to reduce heat loss through windows and other transparent building envelopes. The tight cavity between the glass panes limits the convective heat exchange and besides, it is commonly filled with a better thermal insulation gas in comparison to air. The most often argon gas is used [1, 2].

The use of such a structure caused a side effect in the form of a specific way of transferring climatic loads. Any change in external atmospheric pressure p_a [kPa] or temperature T [kPa] of the gas in the cavity generates a load acting on the component glass panes. If the atmospheric pressure increases or the gas temperature decreases, the component panes start to "bulge", whereas the reverse action leads to a concave form of IGU deflection (Fig. 1) [3, 4]. Since the action of these loads is related to the deflection of

Z. Blikharskyy et al. (Eds.): CEE 2023, LNCE 438, pp. 325–336, 2024.
https://doi.org/10.1007/978-3-031-44955-0_33

the component panes, the gas enclosed in the cavities changes its volume and pressure, partially compensating climatic loads (gas interaction). Nevertheless, this compensation is not complete, so the existence of these loads should be considered as an adverse effect of cavity tightness [5–7].

The situation becomes different in the case of wind load. Pressure changes in the cavities caused by the deflection of a directly loaded pane cause partial unloading of that pane and partial external load transfer to the other IGU component panes (gas coupling) [5–7].

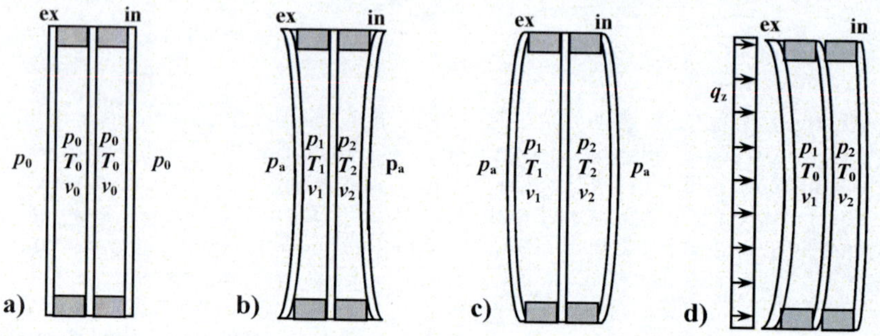

Fig. 1. Deformations in IGUs: **a** unloaded unit, **b** increase of atmospheric pressure or decrease of gas temperature in cavities, **c** decrease of atmospheric pressure or increase of gas temperature in cavities, **d** wind pressure. Subscript "0" denotes initial conditions, whereas subscripts "1" and "2"—operating conditions.

There are several analytical models covering described phenomenon. The Feldmaier's works are the best known e.g. [5, 6]. The basic assumption was made that the gas in the cavities satisfies the ideal gas law. It was also assumed that the component panes deflect in accordance with the theory of a linear-elastic Kirchhoff's plate which is simply supported at the edges. Similar assumptions were made in the models described in [8–11]. Galuppi and Royer-Carfagni [12, 13] proposed models using Betti's Analytical Method and Green's functions. In technical terms, the method of determining the design climatic loads is described in the standard [7]. However, there is an opinion [3, 10] that the assumption of simply supported unit is the most unfavorable borderline case, but confirmation of this requires a separate analysis.

The problem was the subject of experimental research described across the literature. Hart et al. [14] measured deflections of window panes in operating conditions using laser devices, Penkova et al. [15] and Buddenberg et al. [3] simulated temperature changes in a climatic chamber. McMahon [16] simulated the loading by inducing a pressure difference on both sides of the IGUs. Stratiy [4] and Kozłowski et al. [17] simulated climatic loads by injecting/extracting gas into/out of the cavity.

The issue of climatic loads was underestimated for many years, because double-glazed IGUs were commonly used. Currently, with the tightening of legal regulations regarding the reduction of heat loss in buildings, triple-glazed units have become a standard, however the possibility of further increasing the number of glass panes is being considered [18, 19]. Model calculations [20] showed that deflections and stresses in IGUs increase approximately proportionally to the sum of the cavity thicknesses in IGUs, which may be the reason of increasing the risk of damage to the operating glazes [21, 22].

One of the factors affecting the deflection and stress in the operated IGUs is their location in the structure. At present, horizontal and diagonal roof windows, rooflights, glass roof-ceilings become more and more popular. In the literature, this issue is generally ignored—vertical location of the IGUs is assumed, i.e. the own weight of the glass panels is omitted. Own analysis of horizontal location influence on static quantities in double-glazed IGUs is presented in [23].

The aim of the article is to analyze the influence of the horizontal location of triple-glazed IGUs on their deflection and stress. Simulation studies were carried out using authorship analytical model [9, 10]. For typical climatic loads (changes in atmospheric pressure, temperature, wind) the static quantities in IGUs located vertically and horizontally were compared.

2 Methodology of Research

As already mentioned, the static quantities (maximum deflection w [mm] and maximum stress σ [MPa] in the center of the IGU component pane) were determined using an authorship analytical model described in earlier publications, e.g. [9, 10]. The problem of multi-glazed IGUs requires solving a system of quadratic equations describing the temporary equilibrium between external loads and operating gas parameters in the cavities (pressure, temperature, volume). The appropriate system of equations, taking into account the own weight of glass panes, is presented in [10]. The solution to the system are operational pressure values in each of the cavities. Then, on the basis of the load balance, the resultant load of each of the component panes is calculated, and then the resulting values of deflection and stress are obtained. According to the handbook [24], the linear model is accurate enough for deflections not exceeding the pane thickness. With larger deflections, membrane stresses start to expose, thus the pane susceptibility to deflections is lower than it results from the linear model.

In the considered case (triple-glazed IGU), the "ex" index denotes the results for the panel located on the external air side, whereas the "in" index panel on the room side, respectively. Resulting values are not given for the middle pane as they are not extreme ones in regard to the entire IGU.

The following initial conditions were assumed: pressure $p_0 = 100$ kPa, temperature $T_0 = 20$ °C. These parameters reflect the conditions prevailing in the production hall during the assembly of an IGU. In other words, it is assumed that this is a state without stresses and deflections of the component panes, in the case of a vertically located IGU (Fig. 1a).

Two cases of location were considered: vertical and horizontal. Of course, even with horizontal plane glass roofs, some slight inclination is designed, however, this is a borderline case where, under initial conditions, there are some deflections and stresses in the IGU due to the constant action of own weight. In the case of IGUs located diagonally, the vertical component of the own weight should be considered. Thus, the static quantities will assume intermediate values between the borderline cases.

Two IGU designs were considered (Fig. 2). The first is a symmetrical system of three monolithic glass panes (standard or toughened), each 4 mm thick with 16 mm gas-filled cavities (the symbol "SYM" is used). Based on the standard [25], the following glass parameters were adopted: Young's modulus 70 GPa, Poisson's ratio 0.23, glass weight 25 kN/m^3. Although in glass flat roof-ceilings, for safety reasons, the "in" panel should be made of laminated glass, nevertheless, this case was considered for comparative purposes. The second structure (the symbol "LAM") is a very often used system both in glass roofs and in some vertical glazing: a monolithic pane "in" has been replaced with a laminated glass (two 4 mm panes connected by a 0.76 mm foil). Since the laminated board is a sandwich panel and has viscoelastic properties, an equivalent thickness was assumed in the calculations. The standard [7] provides for the pane described above, following equivalent thickness values for: load and deflection equal to 6.63 mm and stress equal to 7.32 mm simultaneously taking into account the coefficient for the shear transfer of an interlayer in laminated glass equal to 0.3. For both structures, the calculations were carried out for two sizes (length × width): square IGUs 60 × 60 cm and 120 × 120 cm.

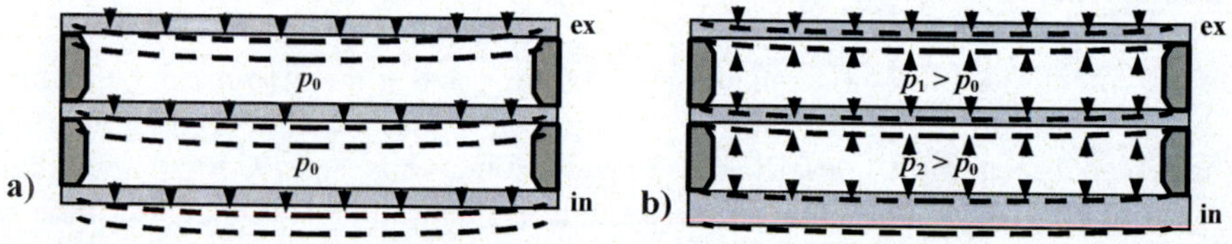

Fig. 2. Deflections in a horizontal IGU: **a** IGU with component panes of equal thickness (SYM), **b** IGU with laminated pane "in" (LAM).

Three typical climatic loads were considered: changes in atmospheric pressure Δp_a [kPa], changes in gas temperature in the cavity ΔT [°C] and loads per area (snow, wind) acting on the glass pane "ex" q_z [kN/m^2]. The load values considered are described in Sect. 3. Deflection is assumed to be positive if the pane ("ex" or "in") takes the form of convex deformation and absolute values are given for stresses. The external load is positive for pressure and negative for suction.

3 Results and Discussion

3.1 IGUs Loaded with a Change in Atmospheric Pressure

The atmospheric pressure change $\Delta p = p_a - p_0$ was considered in the range of $\Delta p = \pm 4$ kPa. Such load may result from natural weather changes. The resulting maximum deflections for the cases described in Sect. 2 are shown in Fig. 3 (IGUs 60 × 60 cm) and Fig. 4 (IGUs 120 × 120 cm).

First of all, one can notice that in vertical SYM units, the component panes deflect symmetrically. When increasing the size of the IGU, the deflection of the component panes does not increase significantly and the stress decreases. This is a result of large panes are more susceptible to deflection, so that the load Δp is compensated to a

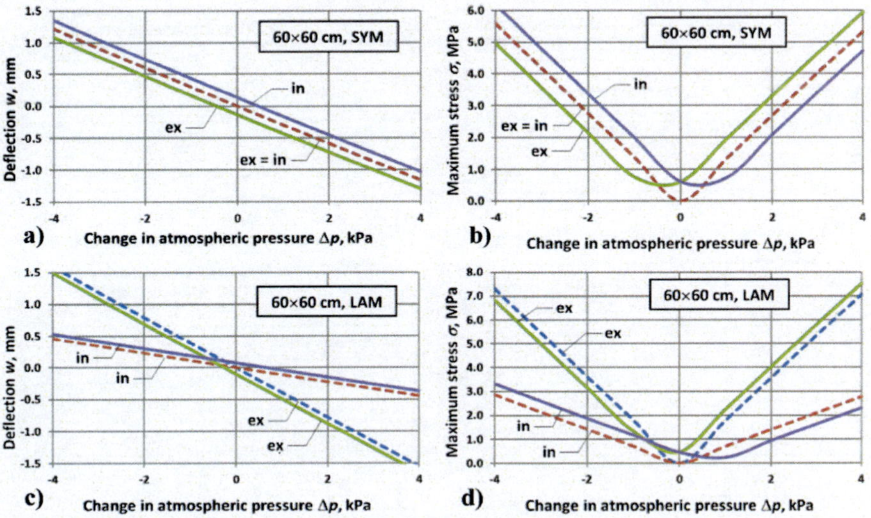

Fig. 3. Effect of atmospheric pressure change on: **a** deflection in the SYM unit, **b** stress in the SYM unit, **c** deflection in the LAM unit, **d** stress in the LAM unit. IGU dimensions: 60 × 60 cm. The dashed line indicates the vertical position, the solid line indicates the horizontal position.

much greater extent by the pressure change in the cavity: for IGUs 60 × 60 cm this compensation is 78.3%, while for 120 × 120 cm it is 98.3%.

Horizontal location of SYM units is unfavorable. In the case of $\Delta p = 0$, all panes are subjected to the own weight load $q_\gamma = 0.1$ kPa. Hence, the panes deflect, however the initial pressure p_0 does not change—the pane "ex" becomes concave, the pane "in" is convex (Fig. 2a). As a consequence, under the load Δp, the deflections in the component panes are differentiated, which is particularly visible at larger IGUs sizes. As can be seen from Fig. 4a, b, at the size of 120 × 120 cm extreme deflections (absolute maximum values for component panes) approach the limit of large deflections at $\Delta p = 4$ kPa. The stresses also become different and are greater than in vertical IGUs. With smaller IGU sizes (60 × 60 cm), the differences between the horizontal and vertical location less significant (Fig. 3a, b).

In the case of the LAM unit, a different phenomenon occurs. When located vertically, the thickened pane "in" is less susceptible to deflection, which adversely affects the compensation of loads in the entire IGU. For the pane "ex", size of 120 × 120 cm and the load $\Delta p = 4$ kPa one obtains 97.4% however for the pane "in" it is 96.5%, respectively. This seemingly small difference primarily affects the increase in static quantities in the pane "ex" compared to the SYM unit. Deflection increases from 1.6 to 2.53 mm, whereas stress from 1.66 to 2.53 MPa. This is a certain paradox and a certain general property of IGUs loaded with a change in pressure. Thickness differences (stiffness) of the component panes (by making one of them thicker) is unfavorable.

The horizontal location of the LAM unit is in most cases unfavorable compared to the vertical IGUs. As can be seen from Fig. 4c, d, this applies in particular to the increase in atmospheric pressure—then the own weight increases the concavity of the glass pane.

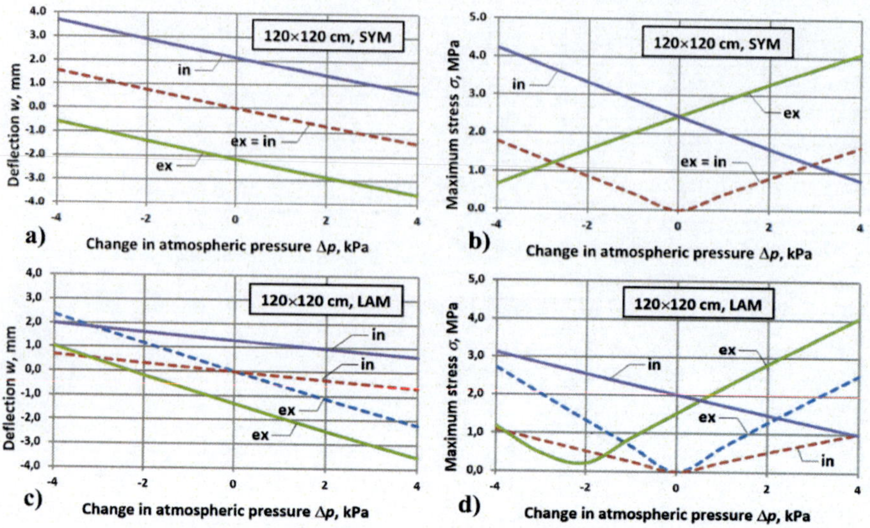

Fig. 4. Effect of atmospheric pressure change on: **a** deflection in the SYM unit, **b** stress in the SYM unit, **c** deflection in the LAM unit, **d** stress in the LAM unit. IGU dimensions: 120 × 120 cm. The dashed line indicates the vertical position, the solid line indicates the horizontal position.

However, the horizontally located LAM unit has some advantages over the horizontally located SYM unit. The reduced susceptibility of the laminated pane "in" to deflections causes that at $\Delta p = 0$, some overpressure is created in the cavities, which additionally loads the pane "in", but relieves the other panes (Fig. 2b). The additional load of the pane "in" does not cause a dangerous increase in static quantities due to its thickening. As a consequence, extreme deflections in the LAM unit in the range $\Delta p < 0$ are smaller than in the SYM unit. For example at $\Delta p = -4$ kPa an extreme deflection $w = 3.70$ mm for SYM unit and $w = 1.99$ mm for LAM unit is obtained (taking into account size of 120 × 120 cm).

As beforehand, it can be seen that the described phenomena are significant for large IGUs. For the size of 60 × 60 cm, the differences between the horizontal and vertical location are not that large (Fig. 3c,d).

3.2 IGUs Loaded with Change in Gas Temperature in Cavities

The problem of gas temperature change in the cavities was analyzed for two borderline cases: gas heating in insolation and gas cooling in winter conditions. Results are given only for LAM units.

The temperature distribution in IGUs under the influence of insolation was analyzed by Wüest and Luible in [26]. The authors found that in triple-glazed IGUs, all the component panes get hotter, but the middle pane heats up the most, as its energy release is hindered by gas in the cavities and coated on the panes "ex' and "in". Exemplary calculations of IGU, using the standard [27] showed that the average gas temperature (the average temperature of the surface of the glass panes limiting the cavity) in sunlight conditions is similar for both cavities which is approx. 40 °C.

Table 1. Static quantities in IGUs loaded with change of gas temperature in cavities.

Load	Size of IGU [cm]	Location	Deflection [mm]		Stress [MPa]	
			"ex"	"in"	"ex"	"in"
$T_1 = 40$ °C, $T_2 = 40$ °C	60 × 60	Vertical	2,63	0,76	12,08	4,73
		Horizontal	2,53	0,83	11,61	5,19
	120 × 120	Vertical	3,88	1,15	4,45	1,79
		Horizontal	2,54	2,44	2,91	3,81
Winter conditions	60 × 60	Vertical	-3,12	-0,70	14,31	4,37
		Horizontal	-3,62	-0,72	16,62	4,48
Reduction T_0 from 20 °C to 15 °C						
$T_1 = 40$ °C, $T_2 = 40$ °C	60 × 60	Horizontal	3,23	1,03	14,83	6,45
	120 × 120	Horizontal	3,60	2,75	4,12	4,29
Winter conditions	60 × 60	Horizontal	-2,98	-0,53	13,68	3,31

For winter conditions, own calculations were made, also using the standard [27]. Assuming temperature on both sides of the IGU −20 °C and 20 °C, the resulting gas temperatures are $T_1 = -9.7$ °C and $T_2 = 7.2$ °C. These temperatures were considered valid for vertical location. For the horizontal location an additional factor was taken into account—temperature drop on the horizontal surfaces of building envelopes due to the so-called radiative cooling. This phenomenon consists in increased heat exchange by radiation between the surface of the partition and the atmosphere in the conditions of a cloudless sky. This is especially intensified in windless weather. Since the horizontal surface exchanges heat with the so-called "cold sky", the temperature on that surface can drop below the temperature of the external air [28–31]. Calculations made using the model described in [28] showed that the atmosphere temperature in the analyzed case is −45.6 °C, while on the surface of the IGU −21.0 °C. After substituting these data, the resulting gas temperatures are $T_1 = -13.4$ °C, $T_2 = 5.5$ °C. These temperatures were considered valid for horizontal location. The results of the deflection and stress calculations in the component panes are shown in Table 1.

The analysis of obtained results showed that in the conditions of sunlight, the increase in gas temperature in the cavities generates overpressure, which loads the component panes. The lower susceptibility to deflection of the laminated pane "in" causes the overpressure in the cavity adjacent to this pane to be greater than in the cavity adjacent to the pane "ex". For example, in vertical IGU 120 × 120 cm, the overpressure acting on "in" is 0.245 kPa, while on "ex" it is 0.182 kPa. Nevertheless, the static quantities are greater in the pane "ex", because it is much less rigid. When located horizontally, the greater own weight of the pane "in" further increases the overpressure acting on it, but at the same time the increased gas interaction relieves the pane "ex". In the horizontal IGU 120 × 120 cm, the overpressure acting on "in" is 0.520 kPa, while on "ex" it is 0.119 kPa. As a consequence, when the size of the pane is increased, the static quantities in both panes are more uniform and their extreme values are smaller than in the case of vertical location.

Through the calculations it has been also showed that the temperature load in winter conditions is mostly dangerous for IGUs with laminated glass pane "in". In this case, in addition to the structural asymmetry, there is an unfavorable load asymmetry—the cavity adjacent to the pane "ex" is colder than the one adjacent to the pane "in", which results in additional underpressure acting on the pane "ex". The effect can be additionally enhanced by radiative cooling. This results in a high value of deflection in the pane "ex". Table 1 presents the results only for the IGU 60 × 60 cm, because with the size of 120 × 120 cm there are evidently larger deflections (both in the vertical and horizontal position), which goes beyond the model's applicability range.

A certain action partially compensating large loads of IGUs in winter conditions could be their production at a lower temperature (or at higher atmospheric pressure). Table 1 analyzes the case of reducing the initial temperature from 20 to 15 °C. This leads to a decrease in static quantities in winter conditions—for example, the deflection in IGU 60 × 60 cm decreases from 3.62 to 2.98 mm. However, deflection and stress under sunlight starts to increase.

3.3 IGUs Loaded Per Area of the Glass Pane "Ex"

The effects of the external load per area were analyzed in the range of $q_z = \pm 0.3$ kN/m^2. Positive value of q_z means pressure, negative—suction. In the case of vertical IGUs, there may be wind pressure or wind suction. In the case of horizontal ones, there is basically only wind suction, the pressure can be caused, for example by the snow weight.

The results were analyzed only for IGUs 120 × 120 cm, because with the increase in IGU size, the static quantities increase exponentially. Gas coupling causes that, with large dimensions, the external load is distributed almost equally on all panes in the SYM unit, while in the LAM unit it is proportional to the stiffness of the panes (the stiffer pane takes the greater part of the load). However, these favorable phenomena is not eliminating the scale effect. Calculation results are shown in Fig. 5.

In the case of the SYM unit, the deflections and stresses (absolute values) in both panes are nearly equal (taking into account absolute value). In the horizontal location, in the case of pressure, the external load is added to the own weight of glass—at the value of $q_z \approx 0.25$ kN/m^2, large deflections occur in the IGU. It should be remembered that the SYM units are currently rarely used for glazing flat roofs. Moreover, as already mentioned, in the case of wind load on horizontal envelopes, we are dealing with suction. In this case, the suction minimizes the own weight, as shown in Fig. 5a, b.

It has also been shown that thickening the pane "in" reduces deflection and stress (compare Fig. 5a–d). As stated before, one can notice that the pressure load is more dangerous than the suction load, despite it has insignificant influence on deflection and stress.

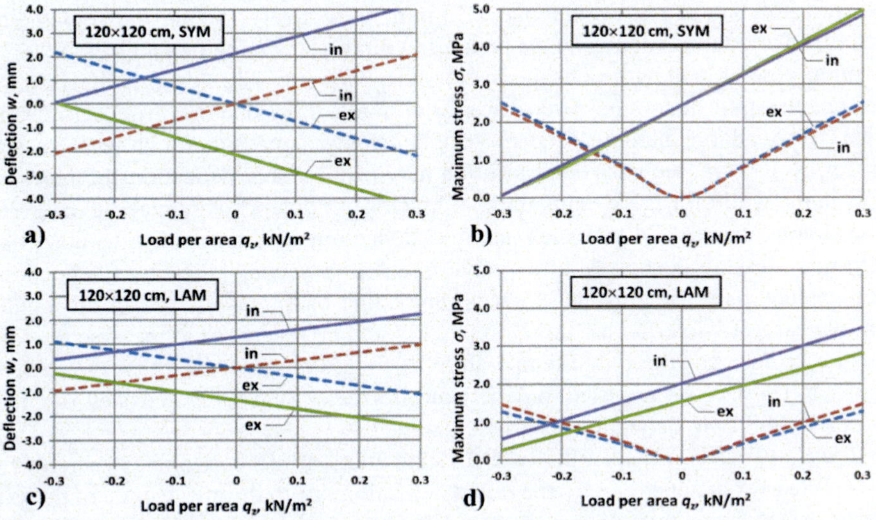

Fig. 5. Effect of load per area of the glass pane "ex" on: **a** deflection in the SYM unit, **b** stress in the SYM unit, **c** deflection in the LAM unit, **d** stress in the LAM unit. IGU dimensions: 120 × 120 cm. The dashed line indicates the vertical position, the solid line indicates the horizontal position.

4 Conclusions

The influence of climatic loads on deflection and stress in IGUs is a complex problem. The gas interaction in a loaded IGU depends on many factors: the structure of IGU, its dimensions, the initial conditions (which are not always known, so some assumptions must be made), type of load, etc. In principle, each factor must be considered individually.

In this work, the influence of typical loads on the static work of two horizontally located triple-glazed IGUs structures was considered. Nowadays, the IGU with a laminated bottom pane is currently the most frequently used in glass roof-ceilings due to the people safety.

It was found that the greatest threat from the structural safety point of view is the load caused by drop in gas temperature in the cavities under severe frost conditions. Some unfavorable phenomena accumulate: asymmetry of the IGU structure, asymmetry of the load and the possibility of the radiative cooling. In this case, already in IGUs 60 × 60 cm, the estimated deflection of the pane "ex" is 3.62 mm. With larger sizes of IGUs, the occurrence of large deflections is therefore more expected. A way to minimize that load may be, for example, the production of IGUs with lower initial temperature conditions (or higher initial pressure), lowering the temperature inside the building during severe frosts, or using valves (membranes) to equalize the pressure between the cavities and the environment. Despite that discussed latter method requires a separate analysis.

An unsafe load is also increase in gas temperature between the cavities during sunlight. In this case, however, the horizontal position of the glass has a positive effect on the static values. For example, in a vertical IGU 120 × 120 cm, the estimated deflection is

3.88 mm, however in a horizontal one, 2.53 mm, respectively. Despite that, in large-area glazing, solar control coatings must be used to limit gas heating in the recesses.

In the case of load with a change in of atmospheric pressure, for small IGUs, e.g. 60 × 60 cm, the static quantities of vertical and horizontal IGUs does not differ significantly. In the case of a large increase or decrease of atmospheric pressure for larger IGUs sizes, e.g. 120 × 120 cm, the horizontal location may lead to large deflections in the case of SYM units, while for LAM units only an increase in this pressure may be dangerous. For example, for $\Delta p = 4$ kPa, the pane "ex" deflection is 3.70 mm. On the other hand, when this pressure decreases, a favorable interaction occurs—increased deflection of the thickened pane "in" results in partial unloading of the pane "ex". In this context, as in the case of winter temperature drops, it would be beneficial to increase the initial pressure or decrease the initial temperature. It should also be noted that the stress for smaller IGUs sizes (as opposed to deflection) is greater than for larger ones, due to the low susceptibility to deflection for small panes.

It was also shown that in the case of wind suction, the horizontal location of the IGUs is clearly advantageous, as the initial deflection due to the own weight of the panes and the external load minimize each other. The unfavorable effect of horizontal location is revealed under the case of pressure load. However, for flat roofs there is virtually no wind pressure. Therefore a heavy snow load can be dangerous. Hence, it is important to prevent snow from accumulating on glass roof-ceilings.

References

1. Van Den Bergh S, Hart R, Jelle BP, Gustavsen A (2013) Window spacers and edge seals in insulating glass units: A state-of-the-art review and future perspectives. Energy and Buildings 58:263–280. https://doi.org/10.1016/j.enbuild.2012.10.006
2. Arıcı M, Karabay H, Kan M (2015) Flow and heat transfer in double, triple and quadruple pane windows. Energy and Buildings 86:394–402. https://doi.org/10.1016/j.enbuild.2014.10.043
3. Buddenberg S, Hof P, Oechsner M (2016) Climate loads in insulating glass units: comparison of theory and experimental results. Glass Struct. Eng. 1:301–3013. https://doi.org/10.1007/s40940-016-0028-z
4. Stratiy, P.: Numerical-and-Analytical Method of Estimation Insulated Glass Unit Deformations Caused by Climate Loads. In: Murgul, V., Popovic, Z. (eds.) International Scientific Conference Energy Management of Municipal Transportation Facilities and Transport EMMFT 2017, Advances in Intelligent Systems and Computing, vol. 692, pp. 970–979. Springer, Cham, Switzerland (2017). https://doi.org/10.1007/978-3-319-70987-1
5. Feldmeier F (2006) Klimabelastung und Lastverteilung bei Mehrscheiben-Isolierglas. Stahlbau 75(6):467–478. https://doi.org/10.1002/stab.200610050
6. Feldmeier, F.: Bemessung von Dreifach-Isolierglas. Stahlbau Spezial 2011 – Glasbau/Glass in Building, 75–80 (2011). DOI: https://doi.org/10.1002/stab.201120012
7. EN 16612:2020 Glass in building. Determination of the lateral load resistance of glass panes by calculation. CEN, Brussels, Belgium (2020)
8. Curcija, C., Vidanovic, S.: Predicting Thermal Transmittance of IGU Subject to Deflection. Lawrence Berkeley National Laboratory, Environmental Energy Technologies Division, Berkeley, CA, USA (2012)
9. Respondek Z (2020) Heat Transfer Through Insulating Glass Units Subjected to Climatic Loads. Materials 13(2):286. https://doi.org/10.3390/ma13020286

10. Respondek Z, Kozłowski M, Wiśniowski M (2022) Deflections and Stresses in Rectangular, Circular and Elliptical Insulating Glass Units. Materials 15(7):2427. https://doi.org/10.3390/ma15072427
11. Heiskari J, Romanoff J, Laakso A, Ringsberg JW (2022) On the thickness determination of rectangular glass panes in insulating glass units considering the load sharing and geometrically nonlinear bending. Thin-Walled Structures 171:108774. https://doi.org/10.1016/j.tws.2021.108774
12. Galuppi L, Royer-Carfagni G (2020) Betti's Analytical Method for the load sharing in double glazed units. Compos Struct 235:111765. https://doi.org/10.1016/j.compstruct.2019.111765
13. Galuppi L, Royer-Carfagi G (2020) Green's functions for the load sharing in multiple insulating glazing units. Int J Solids Struct 206:412–425. https://doi.org/10.1016/j.ijsolstr.2020.09.030
14. Hart R, Goudey H, Arasteh D, Curcija DC (2012) Thermal performance impacts of center-of-glass deflections in installed in-sulating glazing units. Energy and Buildings 54:453–460. https://doi.org/10.1016/j.enbuild.2012.06.026
15. Penkova N, Krumov K, Zashkova L, Kassabov I (2017) Heat transfer and climatic loads at insulating glass units in window systems. International Journal of Advances in Science Engineering and Technology 5(2):22–28
16. McMahon S, Norville HS, Morse SM (2018) Experimental investigation of load sharing in insulating glass units. J Archit Eng 24(1):04017038. https://doi.org/10.1061/(ASCE)AE.1943-5568.0000297
17. Kozłowski M, Respondek Z, Wiśniowski M, Cornik D (2023) Zemła K: Experimental and Numerical Simulations of Climatic Loads in Insulating Glass Units by Controlled Change of Pressure in the Gap. Applied Sciences-Basel 13(3):1269. https://doi.org/10.3390/app13031269
18. Hardtke R (2016) Die Zukunft von Qualitäts-Isolierglas ist multifunktional. BAU flash 1(2):14–16
19. Kralj A, Drev M, Žnidaršič M, Černe B, Hafner J, Jelle BP (2019) Investigations of 6-pane glazing: Properties and possibilities. Energy and Buildings 190:61–68. https://doi.org/10.1016/j.enbuild.2019.02.033
20. Respondek Z (2018) Influence of Insulated Glass Units Thickness and Weight Reduction on their Functional Properties. Open Engineering 8(1):455–462. https://doi.org/10.1515/eng-2018-0056
21. Sack, N., Rose A.: Untersuchungen zur Umsetzbarkeit von druckentspanntem Mehrscheiben-Isolierglas. Institute for Window Technology Rosenheim (2014). Available online: https://www.ift-rosenheim.de/documents/10180/671018/FA_GR1406.pdf/2d0f2a7b-4c25-4133-900f-48953b6bfdad, last accessed 2023/03/17
22. Rose, A.: Zmiana ciśnienia wewnątrz szyb zespolonych (IGU). Wyniki badań, praktyczna realizacja i perspektywy. Świat Szkła, 1, 12–17 (2017)
23. Respondek Z (2019) Rozkład obciążeń środowiskowych w jednokomorowych szybach zespolonych usytuowanych niepionowo. Construction of Optimized Energy Potential. 8(1):147–154. https://doi.org/10.17512/bozpe.2019.1.16
24. Klindt LB, Klein W (1997) Glas als Baustoff: Eigenschaften, Anwendung. Bemessung. Verlagsgesellschaft R. Müller, Köln-Braunsfeld, Germany
25. EN 572–2:2004 Glass in buildings – basic soda lime silicate glass products. CEN, Brussels, Belgium (2004)
26. Wüest T, Luible A (2016) Increased thermal induced climatic load in insulated glass units. Journal of Facade Design and Engineering 4:91–113. https://doi.org/10.3233/FDE-161146
27. EN 673:2011. Glass in Building – Determination of Thermal Transmittance (U Value) – Calculation Method. CEN, Brussels, Belgium (2011)

28. A Building and Its Physical Environment. Śliwowski, L. (ed.). Mon. 28. Wydawnictwo Politechniki Wrocławskiej, Wrocław, Poland (1992)
29. Adelard L, Pignolet-Tardan F, Mara T, Lauret P, Garde F, Boyer H (1998) Sky temperature modelisation and applications in building simulation. Renew Energy 15(1–4):418–430. https://doi.org/10.1016/S0960-1481(98)00198-0
30. Cucumo M, De Rosa A, Marinelli V (2006) Experimental testing of correlations to calculate the atmospheric ""transparency window"" emissivity coefficient. Sol Energy 80:1031–1038. https://doi.org/10.1016/j.solener.2005.06.012
31. Respondek Z (2021) Condensation of water vapor on the external surfaces of building envelopes. Constr Optim Energy Potential 10(1):119–126. https://doi.org/10.17512/bozpe.2021.1.12

Improving the Efficiency of Using Water Resources for the Production of Electricity in Market

Oleksandr Riabenko[1], Anna Panasenko[2], Dmytro Olefir[2], and Volodymyr Tymoshchuk[1](✉)

[1] National University of Water and Environmental Engineering, Rivne, Ukraine
v.s.tymoshchuk@nuwm.edu.ua
[2] PJSC Ukrhydroenergo, Vyshgorod, Ukraine

Abstract. According to the methods of calculating the flow of floods, which is based on the characteristics of the flood wave and the mode of filling the reservoir, calculations and recommendations for early operation of water through hydroelectric turbines when passing floods of different security. The method is designed to effectively pass flood waters and reduce the maximum discharge costs through the hydro. The introduction of the method when calculating the flow of floods will avoid idle water discharges through the spillways of the hydroelectric power plant and the most efficient use of flood water resources, which is especially relevant in today's market conditions. Modern technologies of electricity production using water resources have been studied and the possibilities of improving the efficiency of these technologies in market conditions have been determined. The problems of restoration of water resources are considered and the ways of their solution are proposed. The results of the study can be useful for companies engaged in the production of electricity using water resources, as well as for government bodies that regulate this branch of industry.

Keywords: Flood · Flood wave · Reservoir · Hydroelectric power station · Water consumption · Efficient use of hydro resources · Electricity generation

1 Introduction

The recurrence of abnormal rainfall, which leads to floods or droughts, is a normal component of natural climate variability. However, the devastating effects of floods often have significant socio-economic and environmental consequences, and can lead to flooding of riverside areas, damage to land and the environment, and even death and loss of property.

The frequency of floods is subject to certain patterns, which are manifested in the alternation of periods of high and low water content. It is during the period of high water levels that floods become threatening and sometimes catastrophic. Floods also have a significant impact on the efficiency of hydropower companies that generate electricity from hydroelectric power plants. The urgency of the task of efficient use of water resources

Z. Blikharskyy et al. (Eds.): CEE 2023, LNCE 438, pp. 337–350, 2024.
https://doi.org/10.1007/978-3-031-44955-0_34

during the period of floods is increasing in the conditions of functioning of the liberalized model of the electricity market (the new electricity market in Ukraine started working on July 1, 2019). In the new market conditions, market participants generating electricity at HPPs have departed from the traditional model of electricity tariffs by the Regulator, and must effectively sell electricity in different market segments, making maximum use of all available hydro resources and avoiding idle water discharges through hydropower plants.

Due to the lack of reliable advance forecasting of precipitation in the upper reaches of rivers, insufficient number of meteorological posts and lack of automation of hydrometeorological information collection, the degree of vulnerability to such dangerous natural phenomena is high in Ukraine.

Therefore, the development of effective measures to protect the territory from the harmful effects of flood waters has become a major task of national importance [1, 2].

Analysis of European experience [3–9] shows that in almost all countries of the world there is a gradual transition from passive flood protection to the active phase, which is based on the use of floods and reservoirs in hydropower plants. In addition, work is underway to introduce a sufficient number of hydrological meteorological posts, automate the collection and transmission of data from these posts to settlement centers, which use powerful software and hardware systems to manage and optimize the use of water resources [10–13]. These measures make it possible to reduce the levels of high floods to the level of floods of 10% security, and to pass them within the riverbed without unregulated flooding of the surrounding areas and with minimal economic damage.

In Ukraine, the level of flood regulation with the help of hydropower facilities is quite high. For example, the creation of the Dniester Reservoir in the Dniester hydroelectric power station reduced the damage from flooding of the areas adjacent to the Dniester River during floods. The Dniester Reservoir has a special anti-flood volume of 570 million m^3 to regulate floods. Forcing in the Dniester reservoir can be carried out within the limits of the normal retaining level (NRL)—121.00 m and forced retaining level (FRL)—125.00 m. During high floods, such a flood complex reduces the impact of flood waters by redistributing runoff by accumulating excess water in flood volumes, and water levels in the Dniester due to flood control are reduced to two meters.

This article presents a comparative analysis of the actual flood pass through the Dniester hydroelectric power station, which took place in June 2020 and the results of calculations of different flood pass scenarios built using the method [1, 2], which allows to prepare the reservoir for receiving and regulating large waters. And efficient use of water resources.

2 Statement of the Problem

Insufficient level of automation of collection and transmission of actual data on meteorological situation in the upper reaches of the Dniester and other rivers of Western Ukraine, receipt of actual data on water levels and flows in rivers with insufficient timeliness, lack of effective tools for water management lead to the rapid rise of water levels in rivers comes with a significant delay, and anti-flood actions are aimed more at reducing the impact of natural phenomena than at preventing their occurrence [14, 15].

In the period from June 23 to June 29, 2020, the rivers of the Carpathian region were flooded with 5% security, maximum water flow in the Dniester River:

- June 24, 12:00 at the Galich hydrological post 4200 m^3/s;
- June 26 00:00 at the hydrological post Zalishchyky 3920 m^3/s;
- June 26, 21:00 at the Dniester hydrological post 3300 m^3/s.

The task was to prepare different scenarios and make appropriate calculations for selected scenarios to reduce the impact of floods and increase the efficiency of flood water resources in the production of electricity at HPPs for further sale in different segments of the electricity market of Ukraine.

3 Output Data

The main morphometric characteristics and design levels of the reservoir are given in Table 1.

Water levels and flows in the Dniester River during the flood in June 2020 are given in Table 2.

Initial conditions:

- the reservoir before the flood was filled to the mark of 119.55 m;
- the volume of water in the reservoir was 2408.5 million m^3;
- 5 hydraulic units with a maximum capacity of 585 MW were in operation at the Dniester HPP.

In order to prevent a sharp rise in the water level in the Dniester Reservoir in the period from 24 to 26 June 2020, idle water discharges were carried out through the Dniester hydroelectric power station with an average volume of 450 m^3/s.

4 Results of Calculation of Flood Passage in June 2020 Under Different Scenarios

Three scenarios were developed to assess the reduction of flood impact on the hydrological regime of the Dniester hydroelectric power station and increase the efficiency of flood water resources in the production of electricity at HPPs for further sale in different segments of the Ukrainian electricity market.

Each of the scenarios envisages the operation of five hydro units of the Dniester HPP for different periods of time during the day and with different loads.

Flood transformation calculations were performed on static tanks. The change in the volume of water in the reservoir bowl for the estimated time interval was determined using static morphometric curves in full compliance with the requirements of the "Rules of operation of reservoirs of the Dniester cascade of HPPs and PSPs" [16].

Throughputs through the Dniester hydroelectric power station are assigned based on the forecast of water consumption near the hydrological post Zalishchyky daily advance. The calculation took into account the fact that the forecast comes during the flood at least 3 times a day every 8 h.

Table 1. Characteristic design levels and morphometric characteristics of the Dnistrian reservoir

Indicator	Indicator value
1.Waterworks structure	Drainage type hydroelectric power plant, stone-earth dam
2. Run-off control by a reservoir	Seasonal, with the transition to perennial
3.Reservoir purpose	Hydropower,. water supply, irrigation, flood protection
4.Characteristic design levels in the reservoir: – normal retaining level (NRL) – dead volume level (DVL) – forced retaining level (FRL) when discharging flood with a probability of 0.01%	121.00 102.50 125.00
5. Reservoir morphometric characteristics: 5.1.Mirror area, km^2 – at NRL – at DVL	136 67.9
5.2. Static volume, mln.m^3: – at FRL – at NRL – at DVL – anti-flood volume (between FRL and NRL) – usable volume (between NRL and DVL)	3227 2657 750 570 1907
6. Length, km	194 (along fairway at NRL = 121.0 m)
7. Mean width, m	701
8. Depth, m: – maximal – mean	54 19,5
9. Designed flow rate 9.1. Discharge regulated flow rate of rain floods, m^3/s, with a probability of: – 0.01% – 0.1% – 1% – 10%	13260 8320 2600 2600
9.2. Hydropower plant maximal flow rate, m^3/s	1970
10. Maximum power, MW	702
11. Number of units, pcs	6

Scenario 1 of the flood pass envisages the operation of the Dniester HPP with maximum capacity (operation of 5 hydropower units with a load of 585 MW) for 24 h throughout the study period. Cost through turbines in this mode does not exceed 1440 m^3/s (based on actual data).

Table 2. Main indicators of actual levels and water consumption during the flood period of June 23–29, 2020

Date	Hour	Zalishchyky station inflow, m^3/s	Dniester Reservoir inflow, m^3/s	Discharges the turbines of the Dniester HPP, m^3/s	Idle discharges from the Dniester Reservoir, m^3/s	Water level near the Dniester HPP dam site, m
23.06	8	1270	750	851		119.63
	12	1280	900	339		119.61
	16	1270	1000	1448		119.58
	20	1280	1050	1456		119.56
24.06	0	1390	1100	1483		119.51
	4	1560	1100	1470		119.47
	8	1820	1100	1455	360	119.43
	12	1980	1120	1500	550	119.34
	16	2190	1150	1500	550	119.25
	20	2340	1180	1500	550	119.17
25.06	0	2540	1250	1500	545	119.14
	4	2820	1300	1505	545	119.10
	8	3130	1300	1500	550	119.07
	12	3420	2050	1500	550	119.07
	16	3680	2100	1480	370	119.08
	20	3870	2300	1480	370	119.14
26.06	0	3920	2300	1480	370	119.22
	4	3900	2500	1480	370	119.31
	8	3740	3100	1480	370	119.42
	12	3630	3200	1440		119.55
	16	3410	3200	1430		119.69
	20	3280	3200	1430		119.84
27.06	0	3040	3300	1440		120.02
	4	2830	3000	1430		120.14
	8	2550	2900	1430		120.26
	12	2220	2600	1425		120.38
	16	1860	2300	1420		120.46

(*continued*)

Table 2. *(continued)*

Date	Hour	Zalishchyky station inflow, m^3/s	Dniester Reservoir inflow, m^3/s	Discharges the turbines of the Dniester HPP, m^3/s	Idle discharges from the Dniester Reservoir, m^3/s	Water level near the Dniester HPP dam site, m
	20	1540	2200	1415		120.52
28.00	0	1390	2300	1432		120.56
	4	1260	1900	1253		120.60
	8	1240	1800	894		120.60
	12	1180	1300	1167		120.64
	16	1120	1200	1187		120.65
	20	1070	1200	1114		120.64
29.06	0	1000	1050	1412		120.63
	4	938	1050	898		102.64
	8	913	1050	860		120.65
	12	866	950	1135		120.64
	16	850	930	1078		120.62
	20	800	900	1022		120.60

The results of calculations, in particular the water levels in the upper reservoir near the Dniester HPP, are given in Table 3.

Scenario 2 envisages the operation of the Dniester HPP with maximum capacity in the period from 7:00 to 23:00 during the entire study period. Expenses through turbines in this mode do not exceed 1440 m3/s.

The results of calculations in scenario 2 are shown in Table 4.

Scenario 3 envisages the operation of the Dniester HPP with a maximum capacity (585 MW) during peak load hours (from 8:00 to 23:00) and a capacity of 234 MW at night (from 24:00 to 8:00) throughout the study period.

The results of calculations in scenario 3 are shown in Table 5.

The analysis of the actual results of the flood pass shows that the costs through the Dniester hydroelectric power station were reduced from 3300 m^3/s to 2050 m^3/s, ie the flood was reduced from 5% security to 20% security. At the same time, the maximum water level in the Dniester Reservoir rose to 120.65 m. However, in the period from 8:00 on June 24 to 11:00 on June 26 there were idle discharges of water in the amount of 86.706 million m^3, which in terms of electrical indicators corresponded to about 9232 MWh. In fact, this electricity was not sold on the electricity market of Ukraine.

Analysis of flood calculation calculations on the Dniester River according to different scenarios shows that the most effective scenario is 1, according to which the water level near the Dniester HPP would not exceed the maximum value of 120,727 m. The worst

Table 3. Results of calculations of levels and water consumption during flood passage on June 23–29, 2020 under scenario 1

Date	Hour	Dniester reservoir inflow, m^3/s	Discharge a the turbines of the Dniester HPP, m^3/s	Load, MWh	Water level near the Dniester HPP dam site, m
23.06	8	750	1440	585	119.466
	12	900	1440	585	119.432
	16	1000	1440	585	119.403
	20	1050	1440	585	119.380
24.06	0	1100	1440	585	119.361
	4	1100	1440	585	119.343
	8	1100	1440	585	119.324
	12	1120	1440	585	119.305
	16	1150	1440	585	119.288
	20	1180	1440	585	119.272
25.06	0	1250	1440	585	119.260
	4	1300	1440	585	119.249
	8	1300	1440	585	119.241
	12	2050	1440	585	119.261
	16	2100	1440	585	119.299
	20	2300	1440	585	119.344
26.06	0	2300	1440	585	119.392
	4	2500	1440	585	119.445
	8	3100	1440	585	119.530
	12	3200	1440	585	119.627
	16	3200	1440	585	119.725
	20	3200	1440	585	119.826
27.06	0	3300	1440	585	119.930
	4	3000	1440	585	120.049
	8	2900	1440	585	120.206
	12	2600	1440	585	120.345
	16	2300	1440	585	120.444
	20	2200	1440	585	120.530
28.00	0	2300	1440	585	120.615

(continued)

scenario is Scenario 2, where the water level near the Dniester HPP would reach a

Table 3. (*continued*)

Date	Hour	Dniester reservoir inflow, m^3/s	Discharge a the turbines of the Dniester HPP, m^3/s	Load, MWh	Water level near the Dniester HPP dam site, m
	4	1900	1440	585	120.676
	8	1800	1440	585	120.714
	12	1300	1440	585	120.727
	16	1200	1440	585	120.705
	20	1200	1440	585	120.680
29.06	0	1050	1440	585	120.641
	4	1050	1440	585	102.601
	8	1050	1440	585	120.558
	12	950	1440	585	120.506
	16	930	1440	585	120.451
	20	900	1440	585	120.335

maximum mark of 122,387 m, but is acceptable. At the same time, calculations show that skipping the flood under scenario 2 in the period 23.06.2020–29.06.2020 would increase electricity generation by 7% to 94,613.4 MWh, and marketable products by 16%.

An important factor is that in all three scenarios idle water discharges through the Dniester hydroelectric power station were avoided.

In all calculation scenarios, the maximum costs through the Dniester hydroelectric power station are 1440 m^3/s, which corresponds to costs close to 50% of the flood.

Scenarios 2 and 3 managed to reduce the night hours of the Dniester HPP, and to accumulate flood volumes in the Dniester Reservoir to operate during the day in order to ensure the sale of electricity at better prices in the electricity market.

From a technical point of view, the most effective option is to skip the flood in scenario 1, however, from an economic point of view, it is inferior to the option in scenario 2.

Figure 1 shows the actual course of water levels in the Dniester Reservoir under different scenarios of flooding through the Dniester hydroelectric power station (on the example of the flood that took place in June 2020). Volumes of electricity production at the Dniester HPP for each scenario are shown in Fig. 2.

In the calculations of the transformation of floods, it is necessary to ensure the dynamics of the movement of the flood wave in the reservoir. The exclusion of stationary motion allows to obtain more accurate results when analyzing the behavior of water during flood events. To achieve this, it is necessary to use mathematical models that take into account the non-stationary nature of the flood wave movement and the interaction with the terrain and other factors of the reservoir. This approach allows for more accurate

Table 4. Results of calculations of levels and water consumption in the occasion of flood 23–29 June 2020 under scenario 2

Date	Hour	Dniester reservoir inflow, m^3/s	Discharge a the turbines of the Dniester HPP, m^3/s	Load, MWh	Water level near the Dniester HPP dam site, m
23.06	8	750	1440	585	119.590
	12	900	1440	585	119.556
	16	1000	1440	585	119.528
	20	1050	1440	585	119.504
24.06	0	1100	0	0	119.505
	4	1100	0	0	119.568
	8	1100	1440	585	119.590
	12	1120	1440	585	119.571
	16	1150	1440	585	119.554
	20	1180	1440	585	119.538
25.06	0	1250	0	0	119.546
	4	1300	0	0	119.618
	8	1300	1440	585	119.650
	12	2050	1440	585	119.669
	16	2100	1440	585	119.706
	20	2300	1440	585	119.753
26.06	0	2300	0	0	119.822
	4	2500	0	0	119.956
	8	3100	1440	585	120.153
	12	3200	1440	585	120.336
	16	3200	1440	585	120.526
	20	3200	1440	585	120.713
27.06	0	3300	0	0	120.951
	4	3000	0	0	121.279
	8	2900	1440	585	121.507
	12	2600	1440	585	121.640
	16	2300	1440	585	121.738
	20	2200	1440	585	121.822
28.00	0	2300	0	0	121.940

(*continued*)

Table 4. (*continued*)

Date	Hour	Dniester reservoir inflow, m^3/s	Discharge a the turbines of the Dniester HPP, m^3/s	Load, MWh	Water level near the Dniester HPP dam site, m
	4	1900	0	0	122.148
	8	1800	1440	585	122.259
	12	1300	1440	585	122.268
	16	1200	1440	585	122.251
	20	1200	1440	585	122.227
29.06	0	1050	0	0	122.225
	4	1050	0	0	122.333
	8	1050	1440	585	122.387
	12	950	1440	585	122.319
	16	930	1440	585	122.268
	20	900	1440	585	122.192

results and a better understanding of the dynamics of flood processes, which helps in the further implementation of effective measures for flood risk management.

5 Conclusion

In order to effectively pass floods through hydropower units, when making appropriate calculations and choosing the option of its passage, it is necessary to pay attention to both technical and economic factors.

From a technical point of view, it is necessary to investigate the probability of existing weather forecasts. In addition, it is necessary to take organizational and technical measures for the timely receipt of factual data and management decisions to effectively cut the maximum cost through the hydro.

From an economic point of view, the choice of the flood scenario should be influenced by maximizing the profit from increasing electricity production and its further sale on the electricity market by avoiding idle water discharges through hydropower plants.

In the future, calculations of the transformation of floods should be performed using the method of taking into account the unsteady movement of the flood wave in the reservoir.

Improving the accuracy of weather forecasting and automating the collection and transmission of actual data on meteorological conditions around Ukraine's rivers, timely receipt of actual data on water levels and flows in rivers, introduction of effective tools for water management will help reduce the impact of natural phenomena and prevent them.

Table 5. Results of calculations of levels and water costs in the occasion of flood 23–29 June 2020 in scenario 3

Date	Hour	Dniester reservoir inflow, m^3/s	Discharge a the turbines of the Dniester HPP, m^3/s	Load, MWh	Water level near the Dniester HPP dam site, m
23.06	8	750	1440	585	119.553
	12	900	1440	585	119.519
	16	1000	1440	585	119.490
	20	1050	1440	585	119.467
24.06	0	1100	576	234	119.459
	4	1100	576	234	119.490
	8	1100	1440	585	119.507
	12	1120	1440	585	119.488
	16	1150	1440	585	119.471
	20	1180	1440	585	119.455
25.06	0	1250	576	234	119.455
	4	1300	576	234	119.494
	8	1300	1440	585	119.524
	12	2050	1440	585	119.544
	16	2100	1440	585	119.582
	20	2300	1440	585	119.627
26.06	0	2300	576	234	119.687
	4	2500	576	234	119.790
	8	3100	1440	585	119.907
	12	3200	1440	585	120.011
	16	3200	1440	585	120.198
	20	3200	1440	585	120.387
27.06	0	3300	576	234	120.614
	4	3000	576	234	120.881
	8	2900	1440	585	121.108
	12	2600	1440	585	121.243
	16	2300	1440	585	121.343
	20	2200	1440	585	121.423
28.00	0	2300	576	234	121.526

(*continued*)

Table 5. (*continued*)

Date	Hour	Dniester reservoir inflow, m^3/s	Discharge a the turbines of the Dniester HPP, m^3/s	Load, MWh	Water level near the Dniester HPP dam site, m
	4	1900	576	234	121.675
	8	1800	1440	585	121.784
	12	1300	1440	585	121.794
	16	1200	1440	585	121.775
	20	1200	1440	585	121.749
29.06	0	1050	576	234	121.731
	4	1050	576	234	121.784
	8	1050	1440	585	121.821
	12	950	1440	585	121.761
	16	930	1440	585	121.706
	20	900	1440	585	121.650

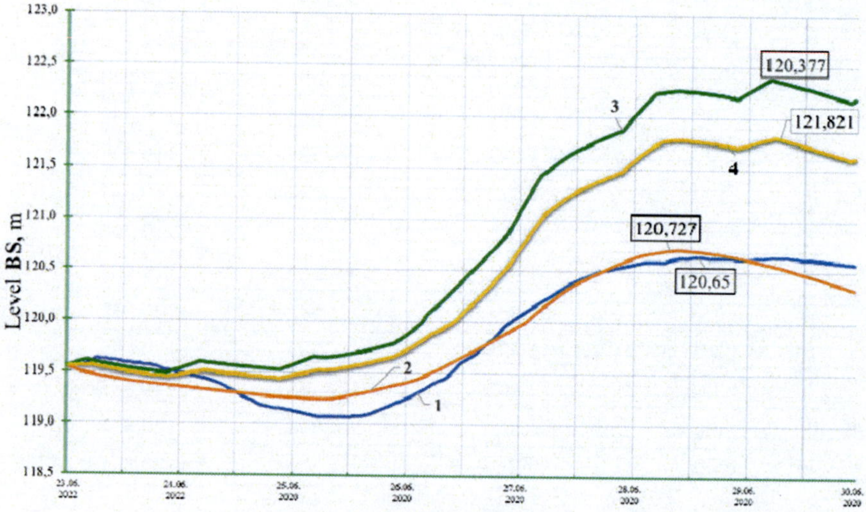

Fig. 1. The course of water levels in the Dniester Reservoir in the period June 23–30, 2020 during the flood: 1—Actual. 2—Option 1. At maximum power (585 MW) 24 h. 3—Option 2. When working with maximum capacity (585 MW) from 7:00 to 23:00. 4—Option 3. When operating with a maximum capacity (585 MW) during peak load hours (8:00–23:00) and a capacity of 234 MW at night (24:00–8:00).

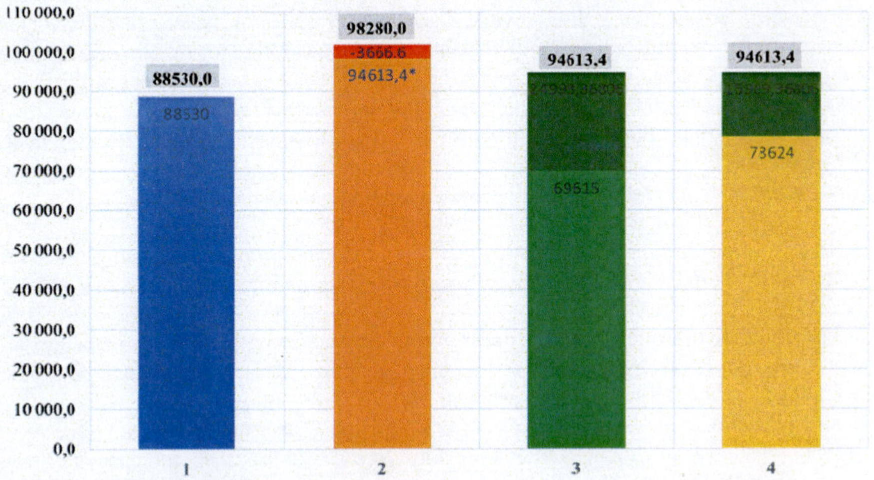

Fig. 2. Electricity generation by the Dniester HPP in the period 23–30.06.2020 during the flood according to different scenarios: 1—Actual. 2—Option 1. At maximum power (585 MW) 24 h. 3—Option 2. When working with maximum capacity (585 MW) from 7:00 to 23:00. 4—Option 3. When operating with a maximum capacity (585 MW) during peak load hours (8:00–23:00) and a capacity of 234 MW at night (24:00–8:00).

References

1. Olefir D, Panasenko A (2021) Aprocedure to forecast and manage water resources and to redistribute runoff water flow when passing floods. Eastern-Euro J Enterpr Technol 2(110):6–17
2. Ryabenko A, Tatarintsev A, Semenukha A (2016) Passing floods through the Dniester hydroelectric complex based on inflow forecasts, taking into account the unsteady movement of the flood wave through the reservoir. Hydropower Ukraine 1–2:62–66
3. Dottori F, Baldassarre GD, Todini E (2013) Detailed data is welcome, but with a pinch of salt: accuracy, precision, and uncertainty in flood inundation modelling. Water Resour 49(9):6079–6085
4. Song X, Zhang J, Zhan C, Xuan Y, Ye M, Xu C (2015) Global sensitivity analysis in hydrological modeling: review of concepts, methods, theoretical framework, and applications. J Hydrol 523:739–757
5. Ward PJ, Jongman B, Salamon P, Simpson A, Bates P, De Groeve T, Muis S, De Perez EC, Rudari R, Trigg MA, Winsemius HC (2015) Usefulness and limitations of global flood risk models. Nat Clim Change 5(8):712–715
6. Hall J, Solomatine D (2008) A framework for uncertainty analysis in flood risk management decisions. Int J River Basin Manag. 6(2):85–98
7. Saltelli A, Tarantola S, Campolongo F, Ratto M (2014) Sensitivity analysis in practice: a guide to assessing scientific models. Wiley
8. Mason DC, Bates PD, Dall'Amico JT (2009) Calibration of uncertain flood inundation models using remotely sensed water levels. J Hydrol 368(1):224–236
9. Hunter NM, Bates PD, Horritt MS, Wilson MD (2007) Simple spatially-distributed models for predicting flood inundation: a review. Geomorphology 90:208–225

10. Loboda N, Shestopalova V, Loginova V, Osadchego V (2011) Influence of climate change on water resources of Ukraine (modeling and forecasts according to climate scenarios). Glob Regional Clim Change, pp 340–352
11. Klapoushchak OI (2011) Modern devices, materials and technologies for non-destructive testing and technical diagnostics of machine-building and oil and gas industry equipment, pp 209–214
12. Skakun SV, Neural network method of flood mapping based on satellite images
13. Horseman SV (2001) Scientific works of Don-NTU. 10(153):52–58
14. Stefanishin DV (2005) The use of the method of extrapolations in forecasting water levels in the river, where the transformation of the channel, taking into account the risk, vol 30, pp 107–116
15. Transcarpathian information and business portal "Mukachevo.net". Transcarpathians will be protected from floods by the Beregivska polder system: Transcarpathians will be protected from floods by the Beregiv polder system, 5 May 2023
16. Basin management of water resources of the river Tisza. [Electronic resource]: State Agency of Water Resources of Ukraine. Official site. http://www.vodhosp.uzhgorod.ua/aivs_tisa.html. Last accessed 5 May 2023
17. Rules of operation of reservoirs of the Dniester hydro power plant complex. 589-39-T36. (1987)

Investigation of Reinforced Concrete Pipe Deformability by Reinforcement Frame Under Static Loads

Roman Rybak[1], Vitalii Kovalchuk[1], Bohdan Parneta[1](✉), and Ihor Karnakov[2]

[1] Lviv Polytechnic National University, S. 12 Bandery Str, Lviv 79013, Ukraine
bz_parneta@ukr.net

[2] National Transport University, Omelianovych-Pavlenko Str, Kyiv 01010, Ukraine

Abstract. It was established that significant service life of pipes, exposure to aggressive environment, and overload from vehicles lead to deteriorating the technical condition of pipes. As a result, defects and damages, which require immediate elimination to ensure safe operation, develop in the pipes. The technology of strengthening a defective concrete pipe using a reinforcement frame and a metal clip with filling the space with concrete mortar is presented. Statistical experimental studies of the deformability of reinforced concrete pipe by a spatial reinforcement frame and a metal clip are carried out. It is established that reinforcement of a defective pipe with a reinforcement frame leads to increasing the pipe bearing capacity. The maximum amount of deformation of the pipe reinforced by a metal clip, at which the pipe was destroyed, was 6.61 mm. When reinforcing the defective pipe by a reinforcement frame and a metal clip, it was found that the maximum amount of deformation at which the pipe was destroyed was 9.44 mm. It is found that reinforcement of a defective pipe with a reinforcement frame can be effectively used in practice in order to increase (restore) the bearing capacity of defective pipes operated on roads and railways.

Keywords: Concrete pipe · Reinforcement frame · Reinforcement · Deformation · Crack

1 Introduction

The technical condition of transport facilities plays an important role in ensuring high throughput and safe traffic on roads and railways. The passage of transport units at the set speeds and ensuring traffic safety is the main criteria for the efficiency of the infrastructure of both the railway track [1–3] and the road infrastructure [4]. Pipes are an important element of the road infrastructure [5]. The use of reinforced concrete pipes has a number of advantages over pipes made of other materials. They play a leading role and are endowed with the greatest technical and economic advantages in comparison with cast iron or metal pipes [6].

Z. Blikharskyy et al. (Eds.): CEE 2023, LNCE 438, pp. 351–361, 2024.
https://doi.org/10.1007/978-3-031-44955-0_35

The most common construction of concrete and reinforced concrete pipes is cylindrical. The use of such a design is the most rational and convenient for manufacturing and installation [5]. However, the biggest disadvantage of the cylindrical structure is its low crack resistance. In addition, the significant service life of pipes, exposure to aggressive environments, and overloading from vehicles lead to deteriorating their technical condition. As a result, defects and damages, which require immediate elimination to ensure safe operation, develop in the pipes [7, 8]. For this purpose, various technologies of pipe repair and reinforcement are used. A common technology for repairing pipes under operating conditions is the use of the "lining" method. The method involves restoring a defective pipe by installing a new metal pipe inside, followed by filling the inter-pipe space with concrete mortar [8]. In addition, in work [9], a pseudo-elastic damper SMA was developed, it can be effectively used in structures that are subjected to cyclic loading in terms of yield strength, energy dissipation, and the ability to re-load.

A schematic representation of the "lining" method is shown in Fig. 1.

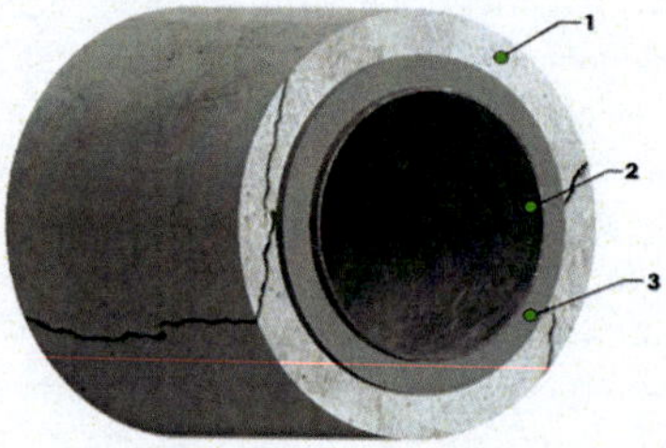

Fig. 1. The "lining" method for repairing defective pipes: 1—adefective pipe; 2—a metal pipe; 3–cement mortar

The studies presented in work [8] show that the "lining" method can increase the defective pipes' bearing capacity to 30%.

It should be noted that reinforcing the concrete pipe in operational conditions is more advantageous compared to replacing it with a new pipe because the work can be carried out without interrupting the movement of vehicles.

However, there are no other reasonable methods for strengthening defective pipes under operating conditions. There is also no experimental evaluation of the influence of pipe reinforcement on the deformability of concrete pipe cracks. Therefore, an important and urgent task of scientific research is the implementation of methods for determining the influence of the reinforcement frame on the deformability of reinforced concrete pipes.

2 Analysis of Literature Data and Problem Statement

To date, a number of methods and means of restoring the bearing capacity and researching the stressed-deformed state of concrete and reinforced concrete pipes and structures are known. In work [10], the destruction of concrete and reinforced concrete beams with different reinforcement coefficients is analyzed. The author cites the causes of cracks and their dependence on the reinforcement coefficient of the structure. The effect of reinforcement on the deformability of reinforced concrete elements is not considered. It is noted, however, a deeper knowledge of crack development will help improve the design of structural reinforcement.

In work [11], the real bearing capacity of combined metal structures united in joint work with a reinforced concrete slab was determined.

In work [12], a mathematical model and an algorithm for the spatial calculation of combined steel-reinforced concrete sprengel systems, which allow determining the parameters of the stress-strain state in structural elements, were developed. However, the reinforcement influence on the deformation of structures was not carried out in the work.

In work [13] the model for determining the formation of cracks in reinforced concrete elements and structures, which is based on the laws of reinforcement adhesion to concrete, is proposed. However, the role of the influence of reinforcement on the deformability of structures is not considered.

In work [14], the results of experimental and theoretical studies of the mechanisms of crack formation in reinforced concrete beams are presented. The method for vibration diagnostics of reinforced concrete structures, which provides a lot of information about the level of cracking and the residual service life of a concrete structure, is developed. The developed method allows assessing the degree of effectiveness of measures to restore and strengthen the structure. The obtained data provide a general algorithm for creating and performing vibration diagnostics procedures for cracking reinforced concrete structures. However, no studies have been conducted on the effect of reinforcement on the concrete structures' bearing capacity after the bearing capacity was exhausted by concrete.

In work [15], the reliability of reinforced concrete pipes in the limit state of crack opening was evaluated, and an experimental study was conducted to obtain reliability indicators of reinforced concrete pipes. However, the values of displacement under critical load and the effect of reinforcement on these values have not been established.

In work [16], a quantitative analysis of the influence of the length, depth and location of cracks on the bearing capacity and mechanical properties of concrete pipes was performed. Conventional concrete pipes and pipes with existing cracks were tested under an external compressive load by computer modeling. However, the concrete pipes without reinforcement are studied in the work, and full-scale experimental tests are not carried out.

In work [17], the behavior of reinforced concrete beams with cracks exposed to constant and repeated load is compared. The surface deformations are found to be significantly higher both in the tensile zones and in the compression zones of beams exposed to repeated types of loads. It is shown that a progressive long-term increase in deflection is the result of the deformation development mainly in the compression zone. The authors conclude that the effect of reinforcement of reinforced concrete structures, which occurs due to the stretching of reinforcement, is overestimated.

In work [18], the effect of cracks on deformations and damage is analyzed. A comparison between experimental and simulated deflections confirms that the impact of cracks should be taken into account to assess the stiffness of reinforced concrete under static load.

In work [19], an experimental study of the cracking process in light-reinforced concrete beams is carried out. It is noted that the presence of reinforcement prevents premature destruction and leads to the branching of the crack, where one crack bifurcates in the compression zone. In addition, in work [20], the relationship between the crack opening width and the level of operational load is established. Based on a parametric study of the crack width, it was found that the development of cracks and the distance between them is affected by the size of the rod and the effective area of concrete surrounding the reinforcement.

In work [21], an experimental study of the bearing capacity of two types of round reinforced concrete pipes was carried out. It is noted that the hypothesis that reinforced concrete pipes behave like a round ring under load is valid.

In works [22, 23], a method for assessing the deformed state of wooden structures is presented.

From the analysis of scientific research works [10–23], it was established that experimental studies of the deformed state of reinforced concrete pipes were not carried out. In addition, the influence of the reinforcement frame on the deformability of reinforced pipes has not been established. Therefore, conducting experimental studies of the deformed state of reinforced pipes by reinforcement frame in order to determine the effect of reinforcement on the deformability of the pipe when opening cracks is an urgent task. Such studies will allow justifying the effectiveness of strengthening concrete pipes under the influence of static loads by a reinforcement frame.

3 Purpose and Objectives

The aim of the work is to assess the deformed state of a restored concrete pipe with a spatial reinforcement frame and determine its effect on deformability during crack formation. This will provide an opportunity to obtain reasonable data on the deformed state of the restored concrete pipes. To achieve this objective, the following tasks were set up:

- to provide a method for strengthening a defective concrete pipe with a spatial reinforcement frame;
- to conduct experimental studies of deformations that occur in the restored concrete pipe by a reinforcement frame.

4 Materials and Methods of the Research

4.1 Geometric Parameters of Reinforced Concrete Pipe

Experimental studies of the effect of the reinforcement frame on deformability during cracks opening were carried out on a reinforced concrete ring by the "lining" method with adding a reinforcement frame. The design scheme of pipe reinforcement is shown in Fig. 2.

Geometric parameters of the defective concrete ring: outer diameter—725 mm; inner diameter—585 mm; wall thickness—70 mm; ring length—400 mm; weight—130 kg; concrete class—C25/30.

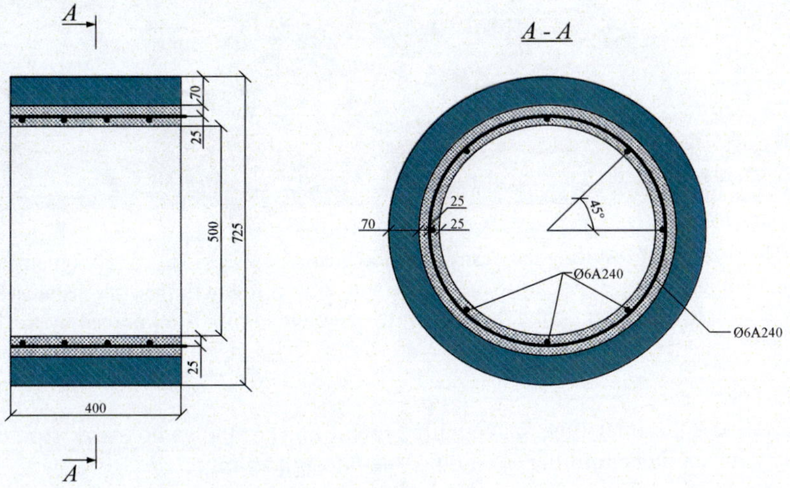

Fig. 2. A structural diagram of a reinforced concrete ring with adding a reinforcement frame

Parameters of the metal liner: material—steel St3; diameter—500 mm; thickness—2 mm, length—450 mm.

The material for filling the structural gap is a self-sealing solution C32/40. Reinforcement class of the reinforcement frame-Ø6A240.

4.2 Method for Strengthening a Defective Concrete Pipe

Reinforcing the defective concrete pipe is carried out by the "lining" method by adding a reinforcement frame. The frame consisted of four reinforcement rings Ø6A240 in 100 mm increments, which were welded to eight longitudinal reinforcement rods Ø6A240. To produce a reinforcement frame sample, 10.5 m of reinforcement of class Ø6A240 was used.

The manufactured reinforcement frame for reinforcing the defective concrete ring is shown in Fig. 3A.

For lining, a welded metal pipe of Ø500 mm and 450 mm long is used, which is made of steel St3 with a thickness of 2 mm (Fig. 3b).

Self-compacting cement mortar C32/40 was used to fill the space between the defective concrete ring and the metal pipe (Fig. 3c).

The inner surface of the defective concrete ring was treated with a deep-penetrating primer to improve adhesion, after which a metal pipe and the reinforcement frame were installed and centered. After installing and centering the reinforcement frame and metal pipe, the mixture for filling the structural gap was poured into a container, water was added and mixed until a uniform consistency was obtained. After 5 min, the mixture was thoroughly mixed again. The resulting cement mortar was applied with a spatula to the structural gap until it was completely filled. Compaction of the mixture was performed by manual docking.

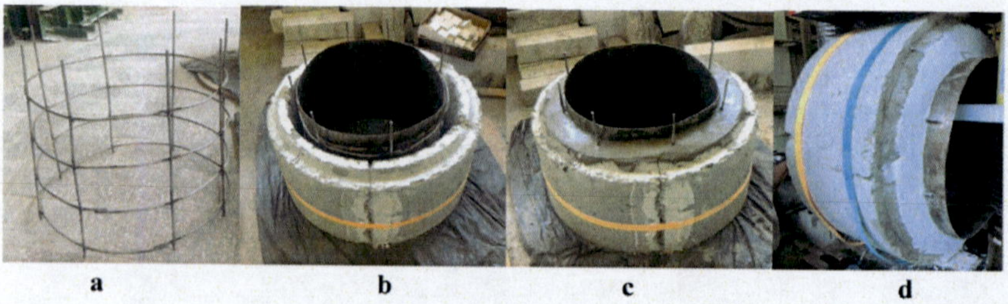

Fig. 3. The process of restoring the destroyed pipe bearing capacity: **a**—a reinforcement frame for strengthening concrete pipes; **b**—a metal pipe and reinforcement frame after installation; **c**—a defective concrete pipe after laying the mortar; **d**—a concrete ring is reinforced by the "lining" method with adding a reinforcement frame.

After filling the inter-tube space with cement mortar, the sample was stored in the laboratory for 28 days until the solution gained design strength.

4.3 Procedure for Experimental Tests of Reinforced Pipe

In order to assess the influence of the reinforcement frame on the deformability of the pipe, experimental studies were conducted in accordance with the goals and objectives set in the work.

The load on the reinforced concrete pipe was set using a hydraulic test press P-250 (Fig. 4).

Fig. 4. General view of pipe testing with hydraulic press P-250

To improve the accuracy of measuring the actual load, a multi-turn dynamometer with a clock-type indicator was used.

Using a crane beam with an electric winch, the restored concrete ring was mounted on a test press. After mounting, electronic digital sensors were fixed, which are necessary for measuring movements that occur during the loading process. For the convenience of fixing on the metal elements of the press and the stand, magnetic bases were used, which provided a reliable position of the sensors in the horizontal and vertical directions. The sensor resolution is 0.01 and 0.001 mm. The layout of digital sensors on the pipe is shown in Fig. 5.

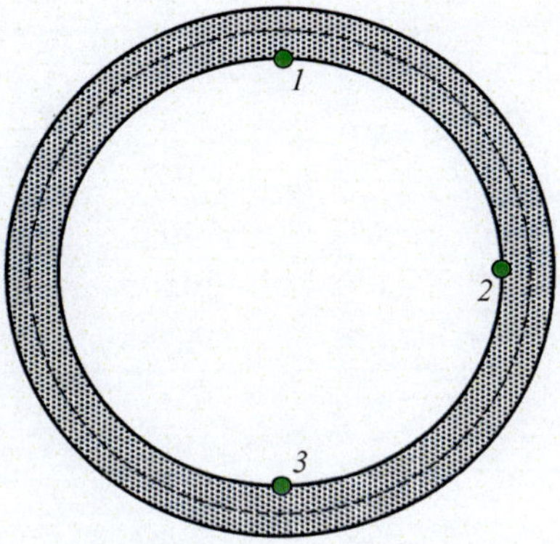

Fig. 5. Layout of digital sensors

The load was set with press P-250, under the action of which, pipe deformations were formed. They were recorded continuously by electronic digital sensors, transmitted to an ADC (analog-to-digital converter), and stored on a computer.

The formation of new cracks was recorded visually and noted graphically. The crack opening width was measured at the maximum opening point with microscope MPB-2 with a division rate of 0.1 mm. In order to improve visual observation of the appearance and development of cracks, lime mortar was applied to the edge of the pipe.

To ensure safe conditions for the experiment and preserve the integrity of the equipment loose ratchet tie belts were applied.

5 The Results of Deformability Studies of Reinforced Concrete Pipe

The results of the maximum deformations obtained of a concrete pipe reinforced with a metal liner and a reinforcement frame, at which cracks occur in the concrete pipe, were made at point 1—6.611 mm, at point 2—–6.83 mm, and at point 3—–5.77 mm. At the same time, the crack opening width was 1.3 mm, and the load was 16.67 kN.

After that, the load on the pipe from the hydraulic press fell to 6.37 kN. Further studies of pipe deformations were carried out from a load of 6.37 kN with a pitch of 0.49 kN until the complete destruction of the pipe. The results are shown in Table. 1.

To justify the reinforcement of the defective pipe with a metal clip and reinforcing frame, research was conducted on three samples. The strain distribution diagrams at certain load levels are shown in Fig. 6a–c.

From the conducted experimental studies, it was established that the beginning of crack formation in a concrete pipe reinforced by a reinforcement frame and a metal clip occurred at a load on the pipe equal to 16.67 kN.

Table 1. The results of deformations that occur in a restored concrete pipe with reinforcement frame under the static loads' influence

No.	Load (kN)	Deformations (mm)			Crack opening width (mm)
		1	2	3	
1	6.37	7.188	−2.94	−8.04	2.1
2	6.86	8.689	−2.63	−8.71	3.2
3	7.07	9.447	−2.23	−9.41	4.2
4	7.26	9.446	−2.05	−9.87	4.2
5	7.35	9.447	−1.91	−10.67	5.1
6	7.50	9.445	−1.63	−11.73	5.5
7	7.55	9.443	−1.14	−13.1	6.1

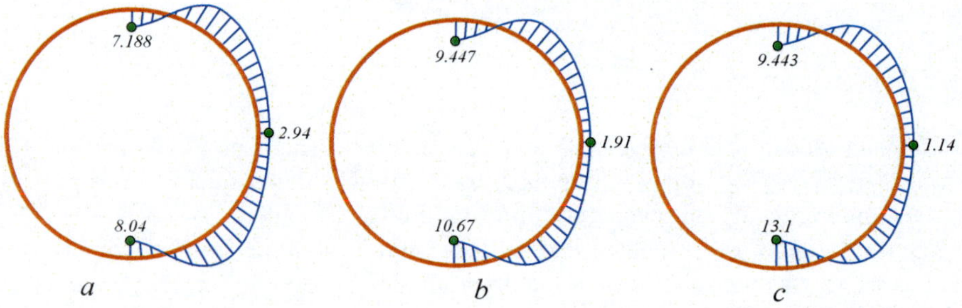

Fig. 6. Strain distribution diagrams of the reinforced concrete pipe as a function of the applied load magnitude: **a**—6.37 kN, **b**—7.35 kN, **c**—7.55 kN

After that, the pipe was unloaded and the pressure force on the pipe was 6.37 kN. The concrete of the pipe practically ceases to provide the bearing capacity of the pipe. Then a metal clip and a reinforcement frame are included in the work. The vertical displacements of the pipe at point 1 are 7.188 mm, on the horizontal side at point 2—−2.94 mm. Further, the load from the press was gradually increased from 6.37 kN to the complete destruction of the concrete ring of 7.55 kN. At a load of 7.55 kN, the value of movements of the reinforced pipe in the vertical direction was at point 1—9.44 mm, at point 3—−13.1 mm and on the horizontal side of the pipe at point 2—−1.14 mm.

From the provided diagrams (Fig. 6), it can be observed that as the load increases, there is an increase in vertical deformations of the reinforced concrete pipe. However, horizontal deformations decrease which is associated with the widening of the crack opening.

Therefore, by carrying out static experimental studies of the deformability of a reinforced concrete pipe with a spatial reinforcing frame and a metal cage, it was established that the reinforcement of a defective pipe leads to an increase in deformability by more than 40%, while the use of only a metal liner up to 30% [8]. The maximum value of deformations of the pipe reinforced by a metal clip, at which the reinforced concrete pipe was destroyed, was 6.61 mm, and when the defective pipe was reinforced by a reinforcement frame, the maximum value of deformation at which the pipe was destroyed was 9.44 mm. Therefore, reinforcement of defective pipes by a reinforcement frame is effective when applied in practice, in order to increase (restore) the defective pipes' bearing capacity operated on the transport highways.

6 Conclusions

1. A practical method of reinforcing a defective concrete pipe by the "lining" method with reinforcement by a spatial reinforcement frame has been developed.

The use of metal pipes coated with zinc coating is recommended to protect the pipe from corrosion. PVC pipes can also be used. It should also be noted that reinforcing defective pipes using the "lining" method leads to a reduction in the cross-sectional area of the pipe. Therefore, this should be taken into account when designing such reinforcement and calculating the flow capacity of the repaired pipe in each specific case.

2. The results of experimental studies showed that the reinforcement of a defective concrete pipe leads to an increase in the deformability of the pipe by more than 40%. The maximum amount of deformations of the pipe reinforced by a metal clip, at which the reinforced concrete pipe was destroyed, was 6.61 mm, and when the defective pipe was reinforced by a reinforcement frame, the deformations were 9.44 mm. Therefore, when using a spatial reinforcement frame in combination with reinforcement by the "lining" method, it allows increasing defective pipes' bearing capacity under operating conditions almost twice. This will lead to an extension of their service life and ensure an increase in the state of vehicle safety.

References

1. Sysyn M, Gerber U, Nabochenko O, Kovalchuk V (2019) Common crossing fault prediction with track based inertial measurements: statistical vs mechanical approach. Pollack Periodica Int J Eng Inf Sci 14(2):15–26. https://doi.org/10.1556/606.2019.14
2. Sysyn M, Nabochenko O, Kluge F, Kovalchuk V, Pentsak A (2019) Common crossing structural health analysis with track-side monitoring. Commun Sci Lett Univ Zilina 21(3):77–84
3. Kovalchuk V, Sysyn M, Hnativ Y, Bal O, Parneta B, Pentsak A (2018) Development of a promising system for diagnosing the frogs of railroad switches using the transverse profile measurement method. Eastern-Eur J Enterpr Technol Kharkov 2/1(92):33–42. https://doi.org/10.15587/1729-4061.2018.125699
4. Kovalchuk V, Sysyn M, Hnativ Y, Onyshchenko A, Koval M, Tiutkin O, Parneta M (2021) Restoration of the bearing capacity of damaged transport constructions made of corrugated metal structures. Baltic J Road Bridge Eng 16(2):90–109. RTU Press. ISSN 1822-427X/eISSN 1822-4288
5. Kovalchuk V, Sobolevska Y, Onyshchenko A, Bal O, Kravets I, Pentsak A, Parneta B, Kuzyshyn A, Boiarko V, Voznyak O (2022) Investigating the influence of the diameter of a fiberglass pipe on the deformed state of railroad transportation structure «embankment-pipe». Eastern-European J Enterpr Technol Kharkov 116(2/7):35–43

6. Andriychuk OV (2011) Pidvyshchennya mitsnosti ta trishchynostiykosti zalizobetonnykh beznapirnykh vodoprovidnykh trub. Mistobuduvannya ta terytorial′ne planuvannya 40(1):52–57. http://www.irbisnbuv.gov.ua/cgibin/irbis_nbuv/cgiirbis_64.exe?I21DBN=LINK&P21DBN=UJRN&Z21ID=&S21REF=10&S21CNR=20&S21STN=1&S21FMT=ASP_meta&C21COM=S&2_S21P03=FILA=&2_S21STR=rmkbs_2013_27_5
7. Koval PM, Babyak IP, Sitdykova TM (2010) Normuvannya pry proektuvanni I budivnytstvi sporud z metalevykh hofrovanykh konstruktsiy. Visnyk Dnipropetr. nats. un-tuzal. transp. im. ak. V. Lazaryana 33:114–117. https://doi.org/10.15802/stp2010/13184
8. Kovalchuk V, Rybak R, Parneta B, Onyshchenko A, Kvasnytsya R (2022) Determining patterns of the deformed state of the transport concrete pipe reinforced with a metal clamp under the action of static load. Eastern-European J Enterpr Technol Kharkov 119(5/7):54–60
9. Iasnii V, Yasniy O, Homon S, Budz V, Yasniy P (2023) Capabilities of self-centering damping device based on pseudoelastic NiTi wires. Eng Struct 278, article number 115556. https://doi.org/10.1016/j.engstruct.2022.115556
10. Słowik M (2018) The analysis of failure in concrete and reinforced concrete beams with different reinforcement ratio. Arch Appl Mech 89(5):885–895. https://doi.org/10.1007/s00419-018-1476-5
11. Vybranets Y, Vikhot S (2023) Spatial calculation of metal truss structure in joint work with reinforced concrete slab. In: Lecture notes in civil engineering, 290 LNCE, pp 450–458
12. Ivanyk I, Vikhot S, Vybranets Y, Ivanyk Y (2018) Theoretical research in tospatial work of a steel-reinforced concrete statically in determinate combined structure. Eastern-European J Enterpr Technol 5(7–95):13–22
13. Mashimo H, Isago N, Kitani T, Endou T (2006) Effect of fiber reinforced concrete on shrinkage crack of tunnel lining. Tunn Undergr Space Technol 21(3):382–383. https://doi.org/10.1088/1757-899X/708/1/012069
14. Shardakov IN, Shestakov AP, Glot IO, Bykov AA (2016) Process of cracking in reinforced concrete beams (simulation and experiment) Fratturaed Integrità Strutturale 38:339–350. https://doi.org/10.3221/IGF-ESIS.38.44
15. Silva JD, El Debs MK, Beck AT (2008) Reliability evaluation of reinforced concrete pipes in crack opening limit state. Revista IBRACON de Estruturas e Materiais 1:314–330. https://doi.org/10.1590/S1983-41952008000400001
16. Zhang Z, Fang H, Li B, Wang F (2020) Mechanical properties of concrete pipes with pre-existing cracks. Appl Sci 10(4):1545. https://doi.org/10.3390/app10041545
17. Higgins L, Forth JP, Neville A, Jones R, Hodgson T (2013) Behaviour of cracked reinforced concrete beams under repeated and sustained load types. Eng Struct 56:457–465. https://doi.org/10.1016/j.engstruct.2013.05.034
18. Xu T, Castel A (2016) Modeling the dynamic stiffness of cracked reinforced concrete beams under low-amplitude vibration loads. J Sound Vib 368:135–147. https://doi.org/10.1016/j.jsv.2016.01.007
19. Fayyad TM, Lees JM (2017) Experimental investigation of crack propagation and crack branching in lightly reinforced concrete beams using digital image correlation. Eng Fract Mech 182:487–505. https://doi.org/10.1016/j.engfracmech.2017.04.051
20. Soltani A, Harries KA, Shahrooz BM (2013) Crack opening behavior of concrete reinforced with high strength reinforcing steel. Int J Concr Struct Mater 7(4):253–264. https://doi.org/10.1007/s40069-013-0054-z
21. da Silva JL, El Debs MK, Kataoka MN (2018) A comparative experimental investigation of reinforced-concrete pipes under three-edge-bearing test: Spigot and Pocket and Ogee Joint pipes. Acta Scientiarum. Technol 40(1):e30860. https://doi.org/10.4025/actascitechnol.v40i1.30860

22. Gomon S, Gomon P, Korniychuck O, Homon S, Dovbenko T, Kulakovskyi L, Boyarska I (2022) Fundamentals of calculation of elements from solid and glued timber with repeated oblique transverse bending, taking into account the criterion of deformation. Acta Facultatis Xylologiae Zvolen 64(2):37–47. https://doi.org/10.17423/afx.2022.64.2.04
23. Homon S, Litnitsky S, Gomon P, Kulakovskyi L, Kutsyna I (2023) Methods for determining the critical deformations of wood with various moisture content. Sci Horiz 26(1):73–86. https://doi.org/10.48077/scihor.26(1).2023.73-86

Nanomodified Ultra High-Performance Fiber Reinforced Cementitious Composites with Enhanced Operational Characteristics

Myroslav Sanytsky(✉), Tetiana Kropyvnytska, Orest Vakhula, and Yuriy Bobetsky

Department of Building Production, Lviv Polytechnic National University, S. Bandera Str. 12, Lviv 79013, Ukraine

myroslav.a.sanytskyi@lpnu.ua

Abstract. This study presents nanomodified ultra-high-performance fiber-reinforced cement composites, which are characterized by increased technical characteristics in extreme operating conditions. It is shown that the synergistic combination of mixtures based on Portland cement with high early strength CEM I, ultra fine zeolite, nano-SiO_2, abrasive aggregate, basalt fibers using a superplasticizer of the polycarboxylate type allows obtaining rapid-hardening, high-intensity compounds. Such cementitios composites, after aging at temperatures up to T = 400 °C due to the phenomenon of self-autoclaving, are characterized by extremely high compressive strength (over 150 MPa), in this case low wear resistance is achieved—0.06 g/cm^2, and that compounds acquire the characteristics of a chemically bonded ceramics In the cementitious composite, the amount of residual portlandite is 2.75% or 66 kg/m^3, which meets the requirements for the strength index of UHPC. The high efficiency of cementitious composites is ensured by the creation of a monolithic dense structure at the meso-, micro- and nanostructural levels by improving the interphase interaction. It was established that as a result of the interaction of nano-SiO_2 with calcium hydroxide in the cement paste, calcium hydrosilicates of the CSH(I) type are formed, which settle on the surface of fine-grained abrasive aggregate, that contributes to the strengthening of the contact zone. Nanomodified rapid-hardening ultra-high-strength fiber-reinforced cementitious composites can be used to create protective structures in industries where increased abrasion resistance at high temperatures is required, particularly in the coal, cement, metallurgical or mining industries.

Keywords: Ultra high-performance fiber reinforced cementitious composites · Nanomodifier · Portland cement · Super fine zeolite · Operational characteristics · Engineering applications

1 Introduction

Ultra high-strength fiber reinforced rapid-hardening cementitious composites are materials used in construction and have properties that make them particularly strong and stress-resistant [1, 2]. Compared to ordinary concrete, ultra high-performance concrete

Z. Blikharskyy et al. (Eds.): CEE 2023, LNCE 438, pp. 362–371, 2024.
https://doi.org/10.1007/978-3-031-44955-0_36

(UHPC) is characterized by high compressive strength (≥120…150 MPa), increased tensile strength (≥8 MPa) and durability [3–5]. Typical UHPC materials consist of a high cement content (800…1200 kg/m^3), additional cementing materials (SCMs) (fly ash, microsilica), fine aggregate, and high-grade superplasticizer [6–8]. The addition of fibers to ultra high-performance concrete (UHFRPC) significantly improves its mechanical properties and increases resistance to deformation [9–13].

One of the most widely used additives is microsilica [14–16]. At the same time, highly dispersed microsilica due to agglomeration processes leads to the problem of its uniform distribution in the mass of concrete. In addition, the introduction of polycarboxylate ethers with microsilica slows down hydration processes. The use of highly effective superplasticizers ensures an ultra low water/binder ratio (0.15…0.25), which allows obtaining a dense, low-porosity UHPC matrix [17]. At the same time, the high-strength matrix of cementitious composites is capable of cracking, brittleness and reduced wear resistance, especially after heat treatment [18–20].

It should be noted the importance of using natural active mineral additives that belong to superpozzolans [21, 22]. Such an additive is a highly dispersed zeolite tuff—super fine zeolite, which in terms of pozzolanic activity approaches microsilica [23]. On the other hand, the addition of super fine zeolite largely determines the operational properties of the cementitious matrix of ultra high-strength composites, especially when exposed to high temperatures. This is due to the structural features of zeolite, the main mineral of which is clinoptilolite. As a result of the specific structure of the three-dimensional aluminosilicate lattice of clinoptilolite, a developed system of micropores and channels filled with water molecules and exchangeable cations (Na, Ca) is formed. As a result, after the temperature treatment of such cement stone, the effect of "self-autoclaving" is created, which to a large extent ensures an increase in strength without the formation of cracks [24].

Increasing the packing density of cement paste due to mixing with ultrafine active mineral additives plays a major role in improving the properties of cementitious composites [25, 26]. At the same time, nanomaterials are widely used in cementitious composites to improve mechanical properties and give concrete multi-functionality [27–29]. As nanoadditives, nanosilica has become the most widespread [30–33]. Due to its high reactivity, nanosilica contributes to the formation of nuclei and the filling effect, providing strong contacts at different structural levels, high mechanical properties and wear resistance of nanoengineered composites [34–37].

2 The Aim and Objectives of the Study

Aims of this study is creating multifunctional ultra high-strength cementitious composites of a new generation that provide long-term high performance characteristics, including in conditions of high temperatures and strong abrasive wear due to increasing the effectiveness of cementitious systems with using such mineral additives of super fine zeolite and nanomosifiers and polycarboxylate ethers.

3 Experimental Program

3.1 Raw Materials

Portland Cement CEM I 42.5 R (Manufacturer—JSC "Ivano-Frankivsk Cement", Ukraine) was used in experiments. The contents of the main clinker phases were as follows, mass. %: C_3S—60.35; C_2S—13.79; C_3A—6.96; C_4AF—12.35. The Blaine specific surface area of CEM I is 3800 cm^2/g.

Zeolite tuff from Sokyrnytsky quarry (Ukraine) were utilized as SCMs. Highly dispersed zeolite (SSA = 11500 cm^2/g) was used as an active pozzolanic additive, which was obtained in ball mills with the latest generation separators with simultaneous drying of zeolite tuff at JSC Ivano-Frankivskcement. Technologically obtained highly dispersed zeolite belongs to the super fine zeolite (SPZ) class. It can be seen from the microphotograph (Fig. 1a) that SPZ is characterized by the content of fractions less than 1.0 μm, 1.0…20.0 μm. X-ray spectral analysis data (Fig. 1b) showed that the zeolite contains 70.0 mass.% SiO_2.

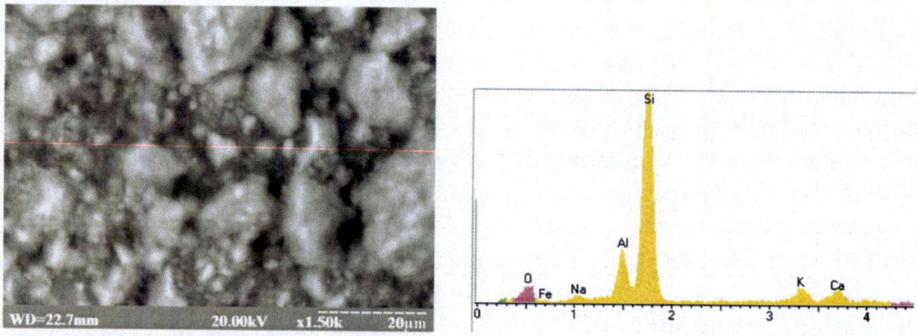

a b

Fig. 1. Microstructure (**a**) and spectra of characteristic X-ray (**b**) super fine zeolite

Basalt fiber RBR-18-t10 (BF) (l = 5 mm and l = 12 mm) of TechnobasaltInvest LLC was used as a reinforcing component. Natural fine quartz sand (modul fineness 1.3) and abrasive material based on N/FeNi/G components (AP) with a size of 0.315…3.0 mm (Ukraine) was used as an aggregate. As a nano additive, a dispersion of nano-SiO_2 Levasil CB8 colloidal solution (dry matter content—50% by mass) with spherical particles of 10–150 nm in size was used. Polycarboxylate superplasticizer (PCE) MasterGlenium ACE 430 based on polymers of a comb structure with nanoengineered chains (BASF, Germany) was used.

3.2 Experimental Process

The compressive and flexural strengths of the tested nanomodified fiber reinforced cementitious composites under study were determined on prisms (40 × 40 × 160 mm). Compressive and flexural strengths of the cementitious composites were measured at the following early ages: 1, 2 and 4 days. Workability of fresh mortars (consistency measured on a flow table) was measured in accordance with EN 1015–3.

Determination of the effect of temperatures (up to T = 400 °C) on the strength of cementitious composites was carried out using a SNOL 8.2/1100 type muffle furnace (Ukraine). The compressive strength test was carried out using the Controls AUTOMAX 5 automatic machine. Abrasion of cementitious composites was determined using the LKY-3 abrasion wheel in accordance with DSTU B B.2.7–212:2009 and the device in accordance with DSTU B EN 1338:2016.

The shape and size of superfine zeolite particles, the microstructure nanomodified fiber reinforced cementitious composites were studied using a SEM 106Y scanning electron microscope (SELMI, Ukraine). The microscope is equipped with a chemical composition microanalyzer. The macrostructure of the samples was studied using a Sigeta Forward 10-500x digital microscope (matrix type C-MOS-5 MPa, focal length from 10 to 300 mm).

Sample mass loss was investigated by thermogravimetry on a Q-1500 derivatograph in the temperature range of 20–600 °C with free air access to the furnace (m = 500 mg, average heating rate—8–8.5 C/min). Determination of the quantitative phase composition of the cementitious matrix of nanomodified ultra high-performance cementitious composites was carried out using a PANalytical AERIS Macrolab diffractometer.

4 Results and discussion

Nanomodified ultra high-performance fiber reinforced cementitious composites (NUH-PFRCC) based on Portland cement CEM I 42.5 R were studied. Material consumption per 1 m^3 of concrete CEM—786 kg, abrasive material AP—1180 kg, sand—118 kg, SFZ—78 kg, basalt fiber $BF_l = 5$ mm—2.36 kg, $BF_l = 12$ mm—2.36 kg, PCE—15.7 kg. Additionally, nanosilica was introduced in the amount of 0.8% of the cement mass. As can be seen from Fig. 2, for the cementitious composite with PCE (W/C = 0.25, Flow = 200 mm), the compressive strength after 1, 2, and 3 days of hardening under normal conditions reaches 87.2, 97.6, and 99.2 MPa, respectively; the flexural strength is 8.2, 9.3, 11.9 MPa.

The introduction of nano-SiO_2 ensures an intensive increase in the early compressive strength of the nanoengineered composite (W/C = 0.25, Flow = 200 mm). Thus, after 1 day, the strength increases by 1.3 times, after 2 and 3 days by 1.2 times, and reaches—114.0, 120.4 and 122.0 MPa. At the same time, the flexural strength is increases by 5–9% and is 8.7, 10.2, 12.9 MPa. It was established that after the thermal exposure (heating to T = 400 °C), the compressive strength of the cementitious composite increases to $f_{cmT} = 124.4$ MPa, while for the nanomodified cementitious composite it increases to $f_{cmT} = 155.6$ MPa. Thus, the use of nano-SiO_2 provided an increase in compressive strength $\Delta f_{cmT} = 20\%$. That is, in conditions of high temperatures (up to T = 400 °C) a strength of more than 150 MPa is achieved, that it can be classified as ultra high-strength cementitious composite.

Using the method of thermogravimetric analysis, it was established that when a sample of a nanomodified cementitious composite was heated to T = 600 °C, the total mass loss was $\Delta m = 6.51\%$. As can be seen from Fig. 3, in the temperature range of

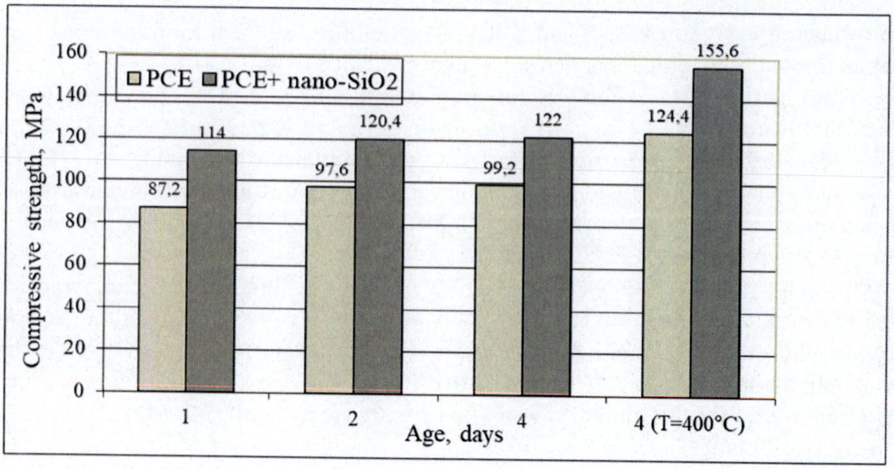

Fig. 2. Compressive strengths of the ultra high-performance fiber reinforced cementitious composites

20–155 °C, samples lose physically adsorbed moisture; in the same temperature range, partial thermal decomposition of ettringite occurs (mass loss is $\Delta m = 3.83\%$). Stepwise dehydration of hydroaluminates and hydrosilicates takes place in the temperature range of 155–438 °C. This process corresponds to a gradual loss of sample mass ($\Delta m = 2.01\%$).

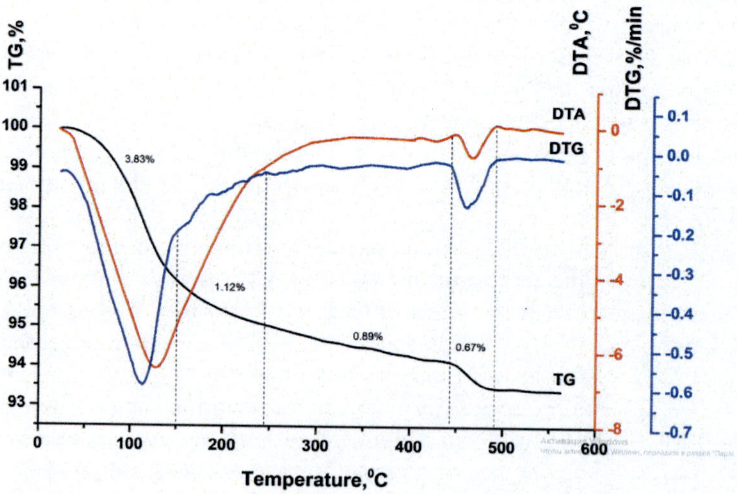

Fig. 3. Thermogram of a sample of nanomodified ultra high-performance fiber reinforced cementitious composite

As evidenced by the thermogravimetric analysis of the cementitious composite, the decomposition of portlandite ($\Delta m = 0.67\%$) occurs in the temperature range $T = 460$–490 °C. In this case, the amount of $Ca(OH)_2$ in the cementitious composite is 2.75% or 66 kg/m^3. Such content of residual portlandite in the cementitious composite meets the requirements for the UHPC durability indicator [7].

According to data from electron microscopy, the ultra high-strength composite after heat treatment ($T = 400$ °C) is characterized by a dense microstructure and strong contacts between the grain of the abrasive aggregate and the cementing matrix (Fig. 4a, b, c). At the same time, the strengthening of the contact zone and improvement of interphase interaction is ensured by the introduction of nano-SiO_2. As a result, pozzolanic reactions of super fine zeolite and nano-SiO_2 synthesize zeolite-like alkaline hydroaluminosilicates with a framework structure. At the same time, calcium hydroxide interacts with active SiO_2 and binds to such nano-dispersed phases as low-base calcium hydrosilicates C-S-H(I). These data are confirmed by the method of microprobe X-ray spectral analysis (Fig. 4d).

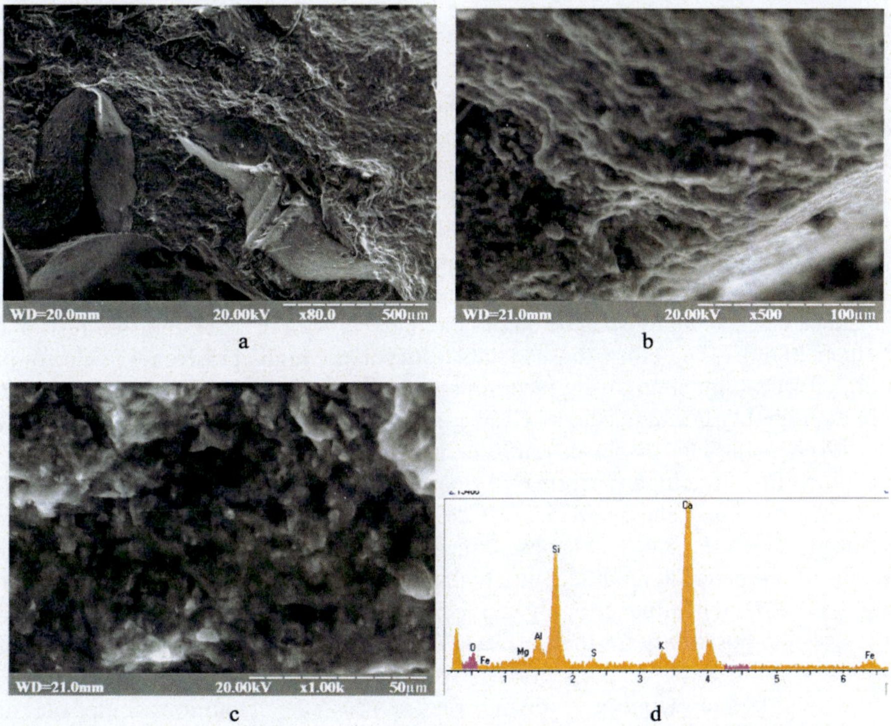

Fig. 4. Microstructure (**a–c**) and spectra of characteristic X-ray (**d**) nanomodified ultra high-performance fiber-reinforced cementitious composites

As can be seen from the photographs (Fig. 5a, b), the contact zone between the aggregate grains and the cementing matrix at the macro- and mesolevels is dense and monolithic. The sample of nanomodified cementitious composite is characterized by a homogeneous structure without voids and capillary pores. In the early period of structure formation, due to the deposition on the grain surface of the abrasive aggregate of low-basic calcium hydrosilicates, the contact zone is strengthened.

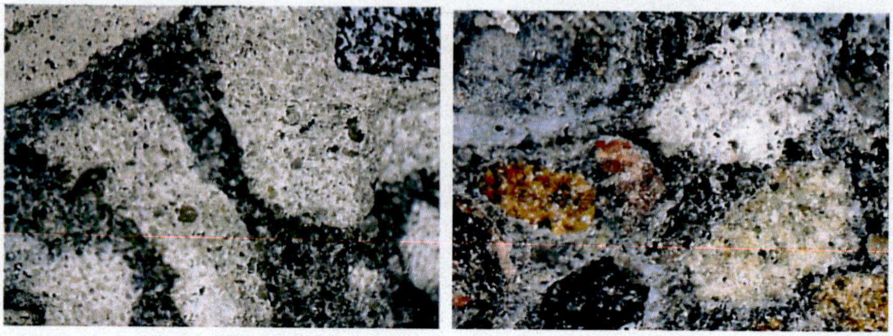

Fig. 5. Photographs of macro- and mesostructures (**a**, **b**) nanomodified ultra high-performance fiber reinforced cementitious composites (300×)

Abrasion testing of cementitious composites was carried out by the BEME method. It was established that for cementitious composite without nano-additive, the abrasion resistance is 0.10 g/cm^2. At the same time, for the nanomodified composite, the value decreases to 0.06 g/cm^2.

Tests in accordance with DSTU B EN 1338:2016 established that the minimum value of the groove width is 15.6 mm for the ultra-high-strength cementitious composite, and the value is reduced to 13.5 mm for the nano-modified composite. According to their indicators, the composites belong to the highest—4th class of abrasion resistance (less than 20 mm). Therefore, these results testify to the high abrasive resistance of the developed ultra-high-strength engineering composites. The above data indicate that the use of high-quality Portland cement CEM I, mineral additive super fine zeolite in a sub-base with nanosilica and dispersed reinforcement with basalt fiber ensures the production of nanomodified ultra high-performance fiber reinforced cementitious composites with enhanced operational characteristics. At the same time, the stability in conditions of extreme abrasive wear and heat exposure increases. This determines the effectiveness of using the developed nanomodified ultra high-performance fiber reinforced cementitious composites under the influence of high temperatures in the short-term or long-term mode and creates the possibility for their use as a lining abrasive material at temperatures up to 400 °C.

Nanomodified ultra high-performance fiber reinforced cementitious composites are characterized by high heat resistance, minimal deformations when heated, which allows them to work under high loads; high fire resistance allows maintaining structural integrity in case of fire. Such cementitious compounds can be used to create protective structural elements in industries that require increased abrasive resistance at high temperatures, in particular, the coal industry, cement plants, metallurgical or mining enterprises.

5 Conclusions

The synergistic combination of components in mixtures based on Portland cement CEM I 42.5 R, ultra fine zeolite, nano-SiO_2 nanomodifier, abrasive aggregate and basalt fiber with admixtures of polycarboxylate superplasticizer allows to obtain high-intensity compounds with increased operational characteristics. For nanomodified ultra high-performance fiber reinforced cementitious composites after exposure at temperatures up to T = 400 °C due to the phenomenon of self-autoclaving, the compressive strength exceeds 150 MPa, while a low wear resistance of 0.06 g/cm^2 is achieved. In the cementitious composite, the amount of residual portlandite is 2.75% or 66 kg/m^3, which meets the requirements for the UHPC durability indicator. The high efficiency of nanomodified cement composites is ensured by a monolithic dense structure at the meso-, micro- and nanostructural levels due to the improvement of interphase interaction and strengthening of the contact zone.

References

1. Mouloud B, Batoz J-F (2009) UHPFRC development on the last two decades: an overview Lafarge cement division. In: Designing and building with UHPFRC: state of the art and development UHPFRC 2009—Nov 17th & 18th—Marseille, France
2. Sadiq M, Soroushian P, Bakker M (2021) Ultra-high-performance cementitious composites with enhanced mechanical and durability characteristics. SN Appl Sci 3:676. https://doi.org/10.1007/s42452-021-04628-y
3. Wang JJ, Zhang SS, Nie XF (2023) Compressive behavior of FRP-confined ultra-high performance concrete (UHPC) and ultra-high performance fiber reinforced concrete (UHPFRC). Compos Struct 312:116879. https://doi.org/10.1016/j.compstruct.2023.116879
4. Fan D, Zhu J, Fan M, Lu J-X (2023) Intelligent design and manufacturing of ultra-high performance concrete (UHPC)—A review. Constr Build Mater 385:131495. https://doi.org/10.1016/j.cemconres.2017.10.013
5. Lampropoulos A, Tsioulou O, Mina A (2023) Punching shear and flexural performance of ultra-high performance fibre reinforced concrete (UHPFRC) slabs. Eng Struct 281:115808. https://doi.org/10.1016/j.engstruct.2023.115808
6. Chu SH, Kurumisawa K, Kong YK (2023) Physically explicable mathematical model for strength prediction of UHPFRC. Eng Struct 275(Part B):115191. https://doi.org/10.1016/j.engstruct.2022.115191
7. Jasiczak J, Wdowska A Rudnicki T (2008) Betony ultrawysokowartosciowe. SPC, Kraków, 157 s
8. Fan DQ, Yu R, Shui ZH (2021) A new design approach of steel fibre reinforced ultra-high performance concrete composites: experiments and modeling. Cement Concr Compos 110:103597. https://doi.org/10.1016/j.cemconcomp.2020.103597
9. Zhang M, Deng G, Du Y (2023) Experimental investigation of the anti-penetration mechanism of UHPC reinforced with high-strength steel bars subjected to projectile impact. Constr Build Mater 372:130820. https://doi.org/10.1016/j.conbuildmat.2023.130820
10. Chen Z, Wang X, Ding L (2023) Mechanical properties of a novel UHPC reinforced with macro basalt fibers. Constr Build Mater 377:131107. https://doi.org/10.1016/j.conbuildmat.2023.131107
11. Kang S-T, Kim J-K (2011) The relation between fiber orientation and tensile behavior in an ultra high performance fiber reinforced cementitious composites (UHPFRCC). Cem Concr Res 41(10):1001–1014. https://doi.org/10.1016/j.cemconres.2011.05.009

12. Kroviakov S, Volchuk V, Zavoloka M (2019) Search for ranking approaches of expanded clay concrete quality criteria. Mater Sci Forum 968:20–25. https://doi.org/10.4028/www.scientific.net/MSF.968.20
13. Blikharskyy Y, Selejdak J, Kopiika N (2021). Study of concrete under combined action of aggressive environment and long-term loading. Materials 14(21):6612. https://doi.org/10.3390/ma14216612
14. Mohamed S, Shanour AS, Abdelaziz G (2023) Influence of blended powders on properties of ultra-high strength fibre reinforced self compacting concrete subjected to elevated temperatures. Case Stud Constr Mater 18:e01793. https://doi.org/10.1016/j.cscm.2022.e01793
15. Sharma R, Jang JG, Bansal PP (2022) A comprehensive review on effects of mineral admixtures and fibers on engineering properties of ultra-high-performance concrete. J Build Eng 45:103314. https://doi.org/10.1016/j.jobe.2021.103314
16. Yuan Y, Zhang W, Sun G (2023) Study on the effect of mineral admixture on the water stability of UHPC under long-term immersion. Constr Build Mater 380:131276. https://doi.org/10.1016/j.conbuildmat.2023.131276
17. Shaikh UA, Salmabanu L, Hasan S (2020) Performance evaluation of ultrahigh performance fibre reinforced concrete—A review. Constr Build Mater 232:117152. https://doi.org/10.1016/j.conbuildmat.2019.117152
18. Krivenko P, Petropavlovskyi O, Kovalchuk O, Rudenko I, Konstantynovskyi O (2020) Enhancement of alkali–activated slag cement concretes crack resistance for mitigation of steel reinforcement corrosion. E3S Web of Conferences this link is disabled 166:06001. https://doi.org/10.1051/e3sconf/202016606001
19. Horszczaruk E, Sikora P, Cendrowski K (2017) The effect of elevated temperature on the properties of cement mortars containing nanosilica and heavyweight aggregates. Constr Build Mater 137:420–431. https://doi.org/10.1016/j.conbuildmat.2017.02.003
20. Pluhin O, Plugin A, Plugin D (2017) The effect of structural characteristics on electrical and physical properties of electrically conductive compositions based on mineral binders. MATEC Web Conf 116:01013. https://doi.org/10.1051/matecconf/201711601013
21. Jaskulski R, Jóźwiak-Niedźwiedzka D, Yakymechko Y (2020) Calcined clay as supplementary cementitious material. Materials 13:4734. https://doi.org/10.3390/ma13214734
22. Krivenko P, Petropavlovskiy O, Kovalchuk O (2018) Comparative study on the influence of metakaolin and kaolin additives on properties and structure of the alkali activated slag cement and concrete. Eastern-European J Enterpr Technol 1(6–91):33–39. https://doi.org/10.15587/1729-4061.2018.119624
23. Chen JJ, Li LG, Ng PL (2017) Effects of superfine zeolite on strength, flowability and cohesiveness of cementitious paste. Cement Concr Compos 83:101–110. https://doi.org/10.1016/j.cemconcomp.2017.06.010
24. Sanytsky M, Kropyvnytska T, Heviuk, Sikora P, Braichenko S (2021). Development of ultra-high strength cementitious composites using superzeolite and N-C-S-H-PCE alkaline nanomodifier. Eastern European J Enterpr Technol 113(5/6):62–72. https://doi.org/10.15587/1729-4061.2021.242813
25. Kropyvnytska T, Sanytsky M, Rucinska T, Rychlitska O (2019) Development of nanomodified rapid hardening clicker–efficient concretes based on Portland–composite cements. Eastern–European J Enterpr Technol Technol Organic Inorganic Substances 100(4/6):38–48. https://doi.org/10.15587/1729-4061.2019.185111
26. Maruchchak U, Sanytsky M, Olevych Y (2019) Effects of elevated temperatures on the properties of nanomodified rapid hardening concretes. MATEC Web Conf 116. https://doi.org/10.1051/matecconf/201711601008

27. Sumariuk O, Romankevych V, Halunka O (2020) Influence of polyfunctional nanomodificators on the microstructure of concrete composites of high strength and density. Phys Chem Solid State 21(1):19–26. https://doi.org/10.15330/pcss.21.1.19-26
28. Su Y, Li J, Wu C (2016) Influences of nano-particles on dynamic strength of ultra-high performance concrete. Compos B Eng 91:595–609. https://doi.org/10.1016/j.compositesb.2016.01.044
29. Sikora P, Elrahman MA, Stephan D (2018) The influence of nanomaterials on the thermal resistance of cement–based composites: a review. Nanomaterials 8(465):1–33. https://doi.org/10.3390/nano8070465
30. Li M, Sun J, Li L (2023) Effect of nanosilica on fiber pullout behavior and mechanical properties of strain hardening ultra-high performance concrete. Constr Build Mater 367:130255. https://doi.org/10.1016/j.conbuildmat.2022.130255
31. Heikal M, Al-Duaij OK, Ibrahim NS (2015) Microstructure of composite cements containing blast-furnace slag and silica nano-particles subjected to elevated thermally treatment temperature. Constr Build Mater 93:1067–1077. https://doi.org/10.1016/j.conbuildmat.2015.05.042
32. Bastami M, Baghbadrani M, Aslani F (2014) Performance of nano-Silica modified high strength concrete at elevated temperatures. Constr Build Mater 68:402–408. https://doi.org/10.1016/j.conbuildmat.2014.06.026
33. Guler S, Türkmenoğlu ZF, Ashour A (2020) Performance of single and hybrid nanoparticles added concrete at ambient and elevated temperatures. Constr Build Mater 250:118847. https://doi.org/10.1016/j.conbuildmat.2020.118847
34. Pischanska V, Voytiuk H, Pitak Y (2014) Influence of modifier on hardening cement stone and refractory concrete properties. Eastern-European J Enterpr Technol 3(6(69)):51. https://doi.org/10.15587/1729-4061.2014.24811
35. Kropyvnytska T, Sanytsky M, Kaminskyy A, Vakhula O (2022) Effect of layered double hydroxides Ca-Al LDHs and polycarboxylate ethers on the hardening of portland limestone cement. Eastern-European J Enterpr Technol this link is disabled 5(6–119):30–40. https://doi.org/10.15587/1729-4061.2022.266269
36. Szczesniak A, Siwinski J, Stolarski A (2022) Effect of aggregate type on properties of ultra-high strength concrete. Materials 15:5072. https://doi.org/10.3390/ma15145072
37. Arora A, Almujaddidi A, Kianmofrad F, Barzin Mobashera B, Neithalatha N (2019) . Material design of economical ultra-high performance concrete (UHPC) and evaluation of their properties. Cem Concr Compos 104:103346. https://www.sciencedirect.com/journal/cement-and-concrete-composites

Architectural Self-Compacting Concrete Based on Nano-Modified Cementitious Systems

Myroslav Sanytsky(✉), Bohdan Rusyn, Iryna Kirakevych, and Andriy Kaminskyy

Lviv Polytechnic National University, Stepana Bandery St, 12, Lviv, Ukraine
myroslav.a.sanytskyi@lpnu.ua

Abstract. The article presents the results of research architectural self-compacting concrete based on cementitious systems "Portland cements with mineral additives—microfillers—superplasticizers—nanomodifiers—alkaline hardening accelerator", which provide such properties as ease of laying, durability and surface cleanliness. The expediency of using such cementitious systems for the production of architectural concrete, which simultaneously performs constructive and artistic-aesthetic and design functions, is shown. Physical and chemical features of structure formation of superplasticized cementitious systems have been established. It is shown that the formation of an increased amount of finely dispersed ettringite as a result of the interaction of sodium sulfate with calcium hydroxide and the formation of structurally active AFm-phases contributes to increasing the strength of the cement system in the early period of hardening. Modification of the cement matrix with nanomodifier type X-SEED contributes to the strengthening of the transition zone, which makes it possible to obtain rapid-hardening self-compacting concrete. As a result, it is possible to introduce innovative technologies in architecture and design to create architectural concepts taking into account urban and functional planning. The introduction of rapid-hardening architectural self-compacting concrete allows to reduce the cost of construction, increase the quality of construction works, increase the durability of building structures, and implement industrial design projects.

Keywords: Architectural self-compacting concrete · Cementitious systems · Nano-modifiers · Alkaline activation · Design functions

1 Introduction

Architectural self-compacting concrete is an innovative material used in modern construction. The main advantage of architectural self-compacting concrete is its ability to create complex geometric forms and surfaces without the need for traditional forming methods such as vibration or rebar placement [1–3]. This allows architects and designers to implement the most creative ideas, such as unique architectural elements, complex facades and unusual forms. In addition, architectural self-compacting concrete has improved aesthetics, thanks to its ability to fill complex forms, it allows you to

Z. Blikharskyy et al. (Eds.): CEE 2023, LNCE 438, pp. 372–380, 2024.
https://doi.org/10.1007/978-3-031-44955-0_37

create smooth surfaces without visible welds or joints [4, 5]. At the same time, high strength and stability of structures is ensured, this allows it to be used in various types of buildings. Due to its self-compacting ability, architectural self-compacting concrete reduces the time and effort normally required to place traditional concrete. This makes it possible to reduce labor costs and increase labor productivity. It should be noted that architectural self-compacting concrete is made from environmentally friendly materials and can be completely recycled, and this allows to reduce its impact on the environment [6–8].

Architectural self-compacting concrete is produced by combining superplasticized cementitious systems with fine-grained aggregate. According to [9–12], in the composition of self-compacting concrete, an important role is played by highly dispersed mineral additives (fly ash, granulated blast furnace slag, etc.), as well as complex chemical admixtures of multifunctional action. This allows such concretes to evolve from ordinary to architectural self-compacting concretes, which are characterized by increased technological indicators, such as ease of application, viscosity and resistance to segregation. At the same time, the use of an increased amount of superplasticizers based on polycarboxylate ethers can provide a slowdown in the early structure formation of cementitious systems, which leads to decrease in the early strength of concrete [13, 14].

The analysis of data in the field of rapid-hardening concrete shows that the directed control of the early structure formation processes of the cement matrix is achieved with the help of multi-level modification of the energy state through the use of superplasticizers based on nano-engineered polycarboxylate ethers and ultra-dispersed mineral components that contain energetically active nano-sized elements (synthetic calcium hydrosilicates. Such components perform the function of crystallization centers and do not require energy for nucleation), which allows significantly accelerate the kinetics of strength growth and obtain a stone with regulated properties [15–17]. According to this, the use of X-SEED type hardening nanoaccelerator in concrete technologies is of great interest. Nanomodifier X-SEED is a nanoscale hardening accelerator that contributes to rapid hydration of cement and early strength of concrete [18]. The rapid development of strength and shorter hardening time obtained with the help of the X-SEED nanomodifier can lead to an increase in structural efficiency. This makes it possible to remove the formwork faster, apply the load earlier, and ensure faster and better workability of the concrete mixture [19–21].

Accelerated development of strength and provision of fast work are achieved due to the use of cementitious systems of ultra-rapid hardening [22–24]. Despite the properties of quick setting, such cementitious systems can demonstrate increased durability of self-compacting concrete. They usually form dense microstructures that are resistant to chemical influence, abrasion, and other environmental factors [25–27]. As a result, such construction works as casting, removal of formwork and decoration are accelerated, which leads to increased efficiency and reduced labor costs [28–31].

A significant acceleration of the processes of early structure formation of cement systems is also achieved due to alkaline activation [32, 33]. At the same time, sodium sulfate can be used as a hardening accelerator for Portland cement systems. Sodium salts can significantly accelerate the hydration reaction, which leads to faster strength development compared to traditional portland cement systems [34, 35].

2 The Aim and Objectives of the Study

The aim of the research was to develop rapid-hardening architectural self-compacting concrete based on cementitious systems "Portland limestone cement—mineral filler—superplasticizers—nanomodifier—alkaline hardening accelerators", which provide such properties as ease of laying, durability and surface cleanliness, etc.

3 Materials and Methods

Portland limestone cement with high early strength CEM II/A-LL 42.5 R produced by PJSC "Ivano-Frankivsk Cement" based on the clinker of next mineralogical composition (% by mass: C_3S—60.82; C_2S—14.62; C_3A—6.76; C_4AF—12.32, Na_2O_e—0.8) was used as binder. Finely ground limestone bright white powder was used as microfiller. The specific surface of Portland cement was determined on Blaine's device. The particle size distribution of the CEM II/A-LL and finely ground limestone powder was measured using Malvern Mastersizer.

The innovative acceleration nanomodifier Master X-SEED 100 (BASF) (2.0 mass.%) according to Crystal Speed Hardening concept, which contains synthetically produced C-S-H nanoparticles suspended in a liquid, was used.

Superplasticizers based on polycarboxylate esters (PCE) type Master Glenium SKY 695 (2.0 mass.%) and innovative superplasticizer GLENIUM ACE 430 (2.0 mass.%) with nanodesigned chains were used for obtaining self-compacting concretes (SCC). The admixture Master X-SEED 100 (2.0 mass.%) and the superplasticizers were added together with mixing water. Sodium sulphate (2.5 mass.%) was used as alkaline hardening accelerator.

Quartz sand (fineness modulus—1.32), granite crushed stone of 2–5 and 5–10 mm fractions were used as aggregates for concrete. Technological and mechanical properties of self-compacting architectural concrete based on super plasticized cementitious systems were determined according to special quality assessment methods, current standards and generally accepted methods [36, 37].

4 Results and Discussion

According to laser granulometry data, Portland limestone cement is characterized by a high content (9.01%) of finely dispersed fraction less than 1.0 μm; for fractions less than 5.0 μm, the content of particles is 31.92%. It should be noted that the specific surface area of Portland cement CEM II/A-LL 42.5 R (SSA = 4300 cm^2/g) is mainly determined by the fine fraction of limestone (Fig. 1a), which contributes to its higher chemical reaction activity. As can be seen from Fig. 1b, limestone powder contains a fine fraction in the range of up to 10 μ and a coarser fraction in the range of 10…500 μ in a ratio of 1:2. In this case, the fine fraction of limestone helps to increase the flowability of the cement paste, and the remaining fractions perform the function of a microfiller, increasing the density of the cementitious matrix.

The procedure for determining the slump-flow and T_{500} time for self-compacting concrete was carried out according EN 12350–8:2010. The slump flow of designed

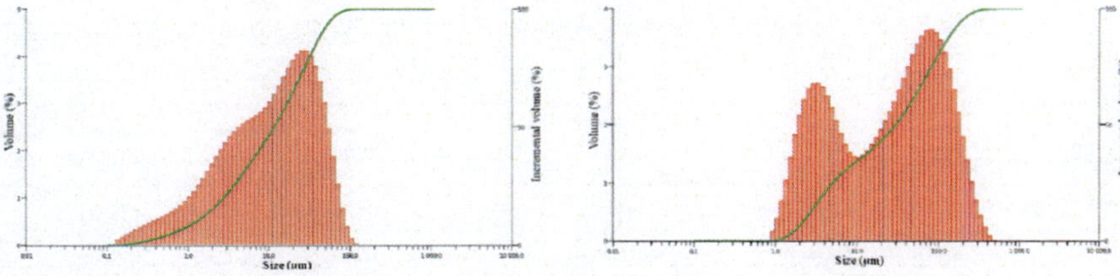

a b

Fig. 1. Particle size distribution of Portland cement CEM II/A-LL 42.5 R (**a**) and limestone powder (**b**)

concrete was 710 and 660 mm what refers to SF2 and class. The t_{500} parameter is 4.0 and 4.5 s correspondently. The J-Ring test, in combination with the slump flow test, is used together to determine the passing ability of SCC (Fig. 2).

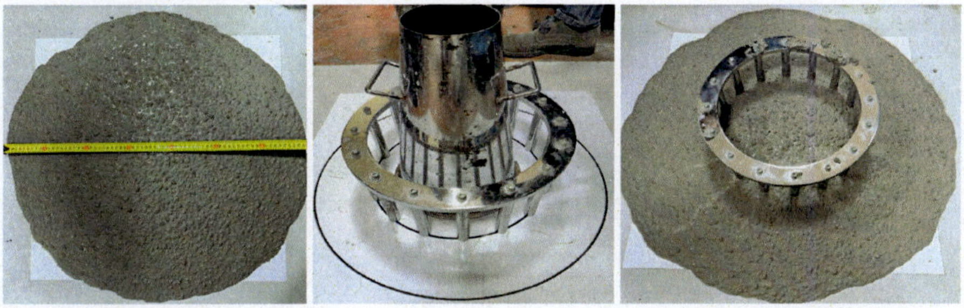

a b c

Fig. 2. The slump flow (**a**) and J-ring test (**b**, **c**) of concrete mixtures based on superplasticized cementitious systems

Self-compacting concrete mixtures (C:S:CA = 1:1.85:2.71; C = 350 kg/m^3) with the optimal consumption of mineral components are characterized by the following technological indicators: SF-2 spreading class (cone spreading diameter 660–730 mm), viscosity class VS2 (T_{500} = 5–8 s), penetration ability class PJ1 (blocking degree PJ = 8–9 mm), segregation resistance class SR1 (segregation index SR = 10–15%), the volume of entrained air is 2.1%.

The conducted studies established that after testing the concretes based on cementitious systems for architectural self-compacting concrete modified with nano-modifier X-SEED and hardening accelerator in accordance with EN 206, is characterized by high early strength of concrete, 27.5 MPa after 1 day of hardening. Compressive strength of concretes after 2 days of hardening are respectively 3.6 MPa for the reference and 33.6 MPa for the concrete modified by X-SEED, PCE and hardening accelerator. The strength of concrete after 7 days of hardening is 44.7 MPa for SCC-1 and 46.8 MPa for SCC-2, respectively. At the age of 28 days, the strength of SCC-1 concrete is 69.5 MPa, and the strength of alkali-activated concrete SCC-2 with nanomodifiers is 75.3 MPa (Table 1).

Table 1. Composition and properties of designed self-compacting architectural concretes

Composition	Unit	SCC-1	SCC-2
CEM II/A-LL 42,5 R	kg/m^3	350	350
Quartz sand		650	650
Granite crushed stone 2–5 mm		285	285
Granite crushed stone 5–10 mm		665	665
Limestone powder		200	200
Water	kg/m^3	175	161
MasterGlenium SKY 695	kg/m^3	7	–
GLENIUM ACE 430		–	7
X-SEED 100		–	7
Sodium sulpate salt (HA)		–	8.75
W/B	–	0.32	0.30
W/C	–	0.50	0.46
Air entraining	%	2.7	2.1
Average density	kg/m^3	2333	2334
SF	mm	710	660
SF_J	mm	705	650
T_{500}	mm	4.0	4.5
T_{500j}	mm	4.5	4.0
Compressive strength, MPa, at the age, days	1d	–	27.5
	2d	3.6	33.6
	7d	44.7	48.7
	28d	69.5	75.3

Due to the combination of physical and chemical approaches through the use of alkaline activator, polycarboxylate superplasticizer and X-SEED nanomodifier, it is possible to obtain fast-hardening, high-strength cementitious systems based on CEM II/A-LL 42.5 R. When sodium sulfate interacts with calcium hydroxide, a reaction occurs with the formation of alkaline hydroxide and chemically active gypsum, which contributes to the additional formation of early ettringite. As a result, the microstructure of the nanomodified alkali-activated cement stone is denser, and it is also filled with finely dispersed (0.2–1.8 μm) ettringite crystals (Fig. 3). The presence of a finely dispersed fraction of limestone leads to the formation of structurally active hexagonal hydrocarboaluminates $C_4A{\cdot}CO_2{\cdot}12H_2O$, which, as a result of epitaxial joints, provide good internal cohesion between the components of cement stone. At the same time, the larger fraction of $CaCO_3$ particles, as a result of the "fine powder" effect, acts as a microfiller, which contributes to the acceleration of hydration processes, the compaction of cement stone and the growth of its strength.

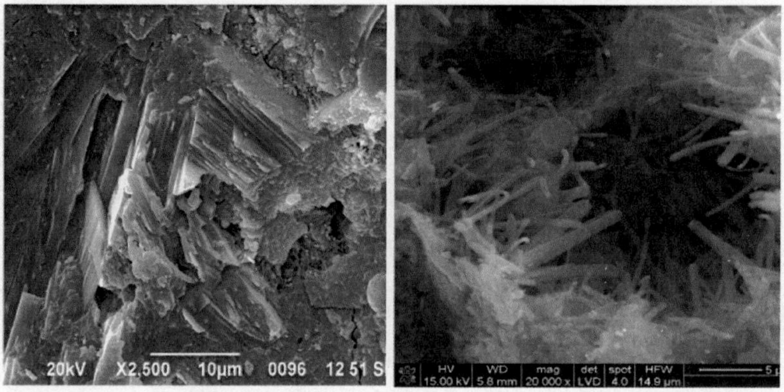

a b

Fig. 3. Crystals of portlandite in Portland cement stone with the addition of PCE (**a**) and ettringite in the interpore space of the stone based on the superplasticized cementitious system (**b**) after 28 days of hardening

The SCC based on superplasticized cementitious systems with limestone microfiller, nanomodifier and alkaline activator are characterized by dense structure and high technical properties. Thus, volumetric water absorption, which characterizes the open capillary porosity, is $W_V = 2.7–3.1\%$ and the pore distribution index $\alpha = 0.9–1$, this indicates a high degree of their distribution in the volume of concrete. High density causes the increasing of SCC waterproof to the class W10. Frost resistance of concrete was in range of 300 freezing–thawing cycles according to DSTU B V. 2.7-49-96.

Nanotechnological modification of cementitious systems opens up the prospects of creating a new class of architectural self-compacting concretes with a given functionality—(Defined Performance architectural self-compacting concrete—DF ASCC).

5 Conclusions

The possibility of obtaining fast-hardening architectural self-compacting concrete is achieved due to the directed control of the structure formation processes of superplasticized cementitious systems "Portland cement CEM II /A LL 42.5 R—limestone filler—polycarboxylate ethers—nanomodifier—alkaline activator". The combined use of sodium sulphater, nanomodifier type X-SEED and polycarboxylate ethers with nanodesigned chains in cementitious systems provides creation of high workability, viscosity, homogeneity and prevents segregation of mixtures. During the hydration process in the superplasticized cementitious systems there is a reaction in the unclinker part with formation of an additional amount of ettringite and structurally active AFm-phases, which provides the compaction of cementitious matrix and increasing early strength of self-compacting concrete.

References

1. Kuniczuk K (2011) Beton architektoniczny. Krakow, Stowarzyszenie producentow cementu
2. Pereira C, Hamadyk E, Silva A (2020) Probabilistic analysis of the durability of architectural concrete surfaces. Appl Math Model 77(1):199–215. https://doi.org/10.1016/j.apm.2019.07.031
3. López A, Guzmán GA, Sarli ARD (2016) Color stability in mortars and concretes. Constr Build Mater 123:248–253. https://doi.org/10.1016/j.conbuildmat.2016.06.147
4. Szwabowski J, Golaszewski J (2010) Technologia betonu samozageszczalnego. Krakow, Stowarzyszenie Producentow Cementu
5. Ghalehnovi M, Shamsabadi EA, Khodabakhshian A, Sourmeh F, Brito J (2019) Self-compacting architectural concrete production using red mud. Constr Build Mater 226:418–427. https://doi.org/10.1016/j.conbuildmat.2019.07.248
6. Ali RA, Kharofa OH (2021) The impact of nanomaterials on sustainable architectural applications smart concrete as a model. Mater Today Proc 42(Part 5):3010–3017. https://doi.org/10.1016/j.matpr.2020.12.814
7. Tayeh BA, Hakamy AA, Fattouh MS (2023) The effect of using nano agriculture wastes on microstructure and electrochemical performance of ultra-high-performance fiber reinforced self-compacting concrete under normal and acceleration conditions. Case Stud Constr Mater 18:e01721. https://doi.org/10.1016/j.cscm.2022.e01721
8. Diniz HAA, Anjos MAS, Rocha KA, Ferreira RLS (2022) Effects of the use of agricultural ashes, metakaolin and hydrated-lime on the behavior of self-compacting concretes. Constr Build Mater 319:126087. https://doi.org/10.1016/j.conbuildmat.2021.126087
9. Borziak OS, Plugin AA, Chepurna SM, Zavalniy OV, Dudin OA (2019) The effect of added finely dispersed calcite on the corrosion resistance of cement compositions. IOP Conf Ser Mater Sci Eng 708:012080. https://doi.org/10.1088/1757-899X/708/1/012080
10. Kropyvnytska T, Semeniv R, Ivanyshchyshyn H (2017) Increase of brick masonry durability for external walls of buildings and structures. MATEC Web Conf 116:01007. https://doi.org/10.1051/matecconf/201711601007
11. Ivashchyshyn H, Sanytsky M, Kropyvnytska T, Rusyn B (2019) Study of low-emission multi-component cements with a high content of supplementary cementitious materials. Eastern-European J Enterpr Technol 4(6(100)):39–47. https://doi.org/10.15587/1729-4061.2019.175472
12. Borziak O, Vandolovskyi S, Chajka V, Perestiuk V, Romanenko O (2017) Effect of microfillers on the concrete structure formation. MATEC Web Conf 116:01001. https://doi.org/10.1051/matecconf/2011711601
13. Sanytsky M, Kropyvnytska T, Kotiv R (2014) Modified plasters for restoration and finishing works. Adv Mater Res 923:42–47. https://doi.org/10.4028/www.scientific.net/AMR.923.42
14. Kropyvnytska T, Semeniv R, Kotiv R, Kaminskyy A, Hots V (2018) Studying the efect of nanoliquids on the operational properties of brick building structures. Eastern-European J Enterpr Technol 95(5/6):27–32. https://doi.org/10.15587/1729-4061.2018.145246
15. Chakravarthy PK, Namratha K (2022) Strength and durability properties of high strength self compacting concrete. Mater Today. Proc 69(Part 3):896–900. https://doi.org/10.1016/j.matpr.2022.07.365
16. Winnefeld F, Becker S, Pakusch J, Gotz T (2007) Effects of the molecular architecture of comb-shaped superplasticizers on their performance in cementitious systems. Cement Concr Compos 29:251–262. https://doi.org/10.1016/j.cemconcomp.2006.12.006
17. Sabet FA, Libre NA, Shekarchi M (2013) Mechanical and durability properties of self consolidating high performance concrete incorporating natural zeolite, silica fume and fly ash. Constr Build Mater 44:175–184. https://doi.org/10.1016/j.conbuildmat.2013.02.069

18. Sanytsky M, Marushchak U, Olevych Y, Novytskyi Y (2019) Nano-modified ultra-rapid hardening portland cement compositions for high strength concretes. In: International conference current issues of civil and environmental engineering Lviv—Košice—Rzeszów. CEE 2019: Proceedings of CEE 2019, pp 392–399. https://doi.org/10.1007/978-3-030-27011-7_50
19. Kroviakov S, Volchuk V, Zavoloka M (2019) Search for ranking approaches of expanded clay concrete quality criteria. Mater Sci Forum 968:20–25. https://doi.org/10.4028/www.scientific.net/MSF.968.20
20. Shilar A, Ganachari SV, Patil VB (2022) Advancement of nano-based construction materials. Constr Build Mater 359:129535. https://doi.org/10.1016/j.conbuildmat.2022.129535
21. Hemalatha T, Gunavadhi M, Bhuvaneshwari B, Sasmal S, Iyer NR (2015) Characterization of micro- and nano- modified cementitious system using micro analytical techniques. Cement Concr Compos 58:114–128. https://doi.org/10.1016/j.cemconcomp.2015.01.004
22. Stel'makh S, Shcherban E, Beskopylny A, Mailya L (2022) Development of high-tech self-compacting concrete mixtures based on nano-modifiers of various types. Materials 15:2739. https://doi.org/10.3390/ma15082739
23. Hemalatha T, Gunavadhi M, Bhuvaneshwari B, Sasmal S, Nagesh R (2015) Characterization of micro- and nano- modified cementitious system using micro analytical techniques. Cement Concr Compos 58:114–128. https://doi.org/10.1016/j.cemconcomp.2015.01.004
24. Sikora P, Chung S-Y, Liard M, Lootens D, Dorn T, Kamm PH, Stephan D, Elrahman MA (2021) The effects of nanosilica on the fresh and hardened properties of 3D printable mortars. Constr Build Mater 281:122574. https://doi.org/10.1016/j.conbuildmat.2021.122574
25. Shilar FA, Ganachari VS, Patil VB (2022) Advancement of nano-based construction materials-A review. Constr Build Mater 359:129535. https://doi.org/10.1016/j.conbuildmat.2022.129535
26. Kostrzanowska-Siedlarz A, Gołaszewski J (2015) Rheological properties and the air content in fresh concrete for self-compacting high performance concrete. Constr Build Mater 94:555–564. https://doi.org/10.1016/j.conbuildmat.2015.07.051
27. Kostrzanowska-Siedlarz A, Gołaszewski J (2016) Rheological properties of high performance self-compacting concrete: effects of composition and time. Constr Build Mater 115:705–715. https://doi.org/10.1016/j.conbuildmat.2016.04.027
28. Niewiadomski P, Hoła J, Ćwirzeń A (2018) Study on properties of self-compacting concrete modified with nanoparticles. Arch Civ Mecha Eng 1(8):877–886. https://doi.org/10.1016/j.acme.2018.01.006
29. Blikharskyy Y, Selejdak J, Kopiika N (2021) Study of concrete under combined action of aggressive environment and long-term loading. Materials 14(21):6612. https://doi.org/10.3390/ma14216612
30. Gołaszewski J, Kostrzanowska-Siedlarz A, Cygan G, Drewniok M (2016) Mortar as a model to predict self-compacting concrete rheological properties as a function of time and temperature. Constr Build Mater 124:1100–1108. https://doi.org/10.1016/j.conbuildmat.2016.08.136
31. Krivenko P, Runova R, Rudenko I et al (2017) Analysis of plasticizer effectiveness during alkaline cement structure formation. Eastern–European J Enterpr Technol 4/6(88):35–41. https://doi.org/10.15587/1729-4061.2017.106803
32. Krivenko P, Petropavlovskyi O, Kovalchuk O (2018) A comparative study on the influence of metakaolin and kaolin additives on properties and structure of the alkali activated slag cement and concrete. Eastern-European J Enterpr Technol 1(6 (91)):33–39. https://doi.org/10.15587/1729-4061.2018.119624
33. Kropyvnytska TP, Kaminskyy AT, Semeniv RM, Chekaylo MV (2019) The effect of sodium aluminate on the properties of the composite cements. IOP Conf Ser Mater Sci Eng 708(1):012091. https://doi.org/10.1088/1757-899x/708/1/012091

34. Kovalchuk O, Grabovchak V, Govdun Y (2018) Alkali activated cements mix design for concretes application in high corrosive conditions. MATEC Web Conf 230:03007. https://doi.org/10.1051/matecconf/201823003007
35. Sanytsky M, Kropyvnytska T, Shyiko O (2023) Effect of potassium sulfate on the portland cement pastes setting behavior. Chem Chem Technol 17(1):170–178. https://doi.org/10.23939/chcht17.01.170
36. DIN EN 12350-8 (2010) Testing fresh concrete—Part 8: self-compacting concrete—Slump-flow test
37. DIN EN 12350-12 (2010) Testing fresh concrete—Part 12: Self-compacting concrete—J-ring test

Analysis of Current Use of Renewable and Alternative Energy Sources by European Countries

Stepan Shapoval[1], Stepan Mysak[2], Pavlo Shapoval[3], and Halyna Matiko[2](✉)

[1] Institute of Building and Environmental Engineering, Lviv Polytechnic National University, Lviv, Ukraine

[2] Institute of Power Engineering and Control Systems, Lviv Polytechnic National University, Lviv, Ukraine
halyna.f.matiko@lpnu.ua

[3] Institute of Chemistry and Chemical Technologies, Lviv Polytechnic National University, Lviv, Ukraine

Abstract. The paper considers the contribution of renewable and alternative energy sources in gross final energy consumption for some European countries. In particular, the share of biomass, biofuel and waste, wind energy, solar energy, energy from heat pumps, and geothermal energy in gross final energy consumption was analyzed for the Netherlands, Germany, France, Poland, and Ukraine. Data on various renewable and alternative sources for these countries are also compared. In addition, the prospects for increasing the use of these energy sources in gross final energy consumption of the specified European countries are analyzed and compared. The technique for analyzing the data of climate and energy programs of studied countries has been developed, which can be used for further similar research. The obtained graph-analytical dependences are of scientific and practical importance because they helped to analyze the achievements of the leading European countries in using renewable and alternative energy sources and further prospects for their development.

Keywords: Emission reduction · Greenhouse gases · Energy and climate strategy · Energy saving · Energy efficiency

1 Introduction

The consequences of recent climate changes regarding the impact of high rates of global warming on safe modern life, forced the world to look for ways of solving these problems. In 2015, the Paris Agreement was adopted as a part of the UN Framework Convention on Climate Change (UNFCCC) to increase the ability to adapt to the negative effects of climate change, to prevent an increase in the average global temperature of more than 2 °C (if possible, no more than 1.5 °C) relative to the indicators of the pre-industrial era [1]. The countries and associations of countries that signed this agreement have committed to develop their national plans following the policy of reduction in greenhouse gas (GHG) emissions into the atmosphere, to implement them in their economy, and update them every 5 years to ensure sustainable development [2].

Z. Blikharskyy et al. (Eds.): CEE 2023, LNCE 438, pp. 381–391, 2024.
https://doi.org/10.1007/978-3-031-44955-0_38

According to the goals of the European Green Deal [3], in 2019, the Clean Energy for All Europeans Package was adopted and implemented into the legislation of the EU member states to fulfill the commitments made within the framework of the Paris Agreement. In addition to the main goal of reducing GHG emissions by at least 40% from the volume of emissions in 1990 defined in the 2030 Climate and Energy Framework of the EU [4], this set of energy rules contains the following most important goals by 2030:

- according to the EU Directive 2018/2002 on energy efficiency, increasing the energy efficiency over the current level by at least 32.5%;
- according to EU Directive 2018/2001 on encouraging the use of energy from renewable energy sources (RES), ensuring the share of energy from RES in gross final energy consumption at the level of 32%;

In August 2021, a set of proposals "Fit for 55" was presented to review and update EU legislation, as well as the implementation of new initiatives to ensure compliance of the Union's policy with climate goals [5]. This package requires by 2030:

- reduction of GHG emissions by at least 55% compared to the level of emissions in 1990;
- increasing the share of produced energy from RES from 32 to 40%;
- reduction of the share of primary energy consumption to 39–41%, and final energy consumption—to 36–37%;
- to reform the EU Emissions Trading Scheme (EU ETS), etc.

Because of the aggression of the Northeastern neighbor to Ukraine on February 24, 2022, and the destruction of the international energy market, the European Commission approved the plan for energy saving, clean energy production, and diversification of energy resources—REPowerEU [6]. This plan aims to increase the energy freedom of EU member states from unreliable suppliers and unstable supplies of imported fossil fuels, speed up the transition to clean energy, and help to solve the climate crisis as proposed in the Paris Agreement. RePowerEU by 2030 offered:

- to increase the share of RES at the level of 40–45% in gross final energy consumption due to more active development of solar and wind power plants;
- the recommendations for the reduction of energy consumption and increasing the all-European efficiency target from 9 to 13%;
- to oblige the installation of photovoltaic panels and solar collectors on the roofs of newly built buildings;
- to increase the production of biomethane and intensify the development of low-carbon hydrogen energy due to the construction of new electrolyzers, the capacity of which will allow the production of 10 million tons of renewable hydrogen; to solve the problems of its transportation and consumption.

Considering these changes, EU member states, according to [7], must submit their updated integrated national energy and climate plans, which will contain developed policies for the implementation of the above goals, for consideration by the European Commission by the end of June 2023.

2 Formulation of the Goal of the Research

Compared to Ukraine, the EU is introducing strict rules for the reduction of GHG emissions. Namely, it is setting high carbon taxes and reforming its Emissions Trading Scheme (EU ETS) to give the industry an incentive to modernize. A flexible credit system (grants for covering part of the funds spent on the purchase of energy-efficient equipment or optimization of technological processes), and support for the European Green Transition initiative allow countries to attract financing on favorable terms.

Therefore, the goal of the paper is to investigate the contribution of renewable and alternative energy sources in gross final energy consumption for some European countries, particularly, to analyze the use of the share of biomass, biofuel and waste, wind energy, solar energy, energy from heat pumps, geothermal energy in gross final energy consumption by such European countries as the Netherlands, Germany, France, Poland, and Ukraine.

3 Analysis of Recent Research and Publications

According to [8, 9], bioenergy has one of the greatest development potentials among all RES due to the good climate, the advantages of the agricultural sector, and the presence of the necessary workforce. Bioenergy is for about 70% of energy consumption from renewable sources. The most popular are agricultural residues formed during harvest gathering and processing, animal manure, as well as energy plants, from which solid fuel and biogas are obtained [10, 11]. According to data for 2020, the dynamics of electricity production from biomass is inferior to the electricity generation by other sources of renewable energy, so the total electricity production from this source of renewable energy was 18% of the planned (actual 755 GW h versus planned 4220 GW h [12]).

Today, new technologies of solar energy are being actively implemented [13–15], and new highly energy-efficient materials for covering solar panels are being developed [16]. Wind energy, energy from heat pumps, and geothermal energy are also promising [8, 17, 18]. According to scientists' calculations, with the maximum use of wind power, it would be possible to obtain electricity in the necessary amount to provide 12% of the total world energy consumption [8, 12].

The authors analyzed numerous documents containing data on shares of biomass, biofuel and waste, wind energy, solar energy, energy from heat pumps, and geothermal energy in gross final energy consumption by such European countries as the Netherlands, Germany, France, Poland, and Ukraine [8–21].

For the Netherlands, biomass energy data were obtained from [19], for Germany—from [20], for France—from [21], for Poland—from [22], and for Ukraine the actual values were obtained from [12], and the forecast values—from [8].

For the Netherlands, wind energy data were obtained from [19], for Germany—from [20], for France, actual values—from [23] and forecasted values—from [21], for Poland—from [22], for Ukraine actual data were obtained from [24] and [12] and forecast—from [8].

For the Netherlands, data on energy from heat pumps were obtained from [19], for Germany—from [20], for France—from [21], for Poland—from [22], and for Ukraine actual values were obtained from [12] and forecast values—from [8].

For the Netherlands, geothermal energy data were obtained from [19], for Germany the actual values—from [24], and the forecast values—from [25], for France—from [21], for Poland—from [22], for Ukraine, the actual data were obtained from [12], and the forecast values—from [8].

For the Netherlands, data on energy from hydropower sources were obtained from [19], for Germany—from [20], for France the actual data were obtained from [23], and the forecast ones—from [21], for Poland—from [22], for Ukraine the actual values were obtained from [24], and the forecast values—from [8].

The Integrated National Energy and Climate Plan of France [21] provides data on the annual final energy consumption for all sectors of the economy. Therefore, the data of the annual gross final energy consumption, necessary for the calculations of the share of energy consumption of each renewable energy source (RES) for the corresponding year, were determined by multiplying the annual final energy consumption in this year by 1.08 (the arithmetic mean of the ratio of the annual gross final energy consumption to the annual final energy consumption of all sectors of the economy in 2023 and 2028, respectively [21]).

4 Presentation and Discussion of the Research Results

To achieve these goals, for the first time, a methodology was developed to analyze the energy and climate plans of European countries, which allows determining the share of renewable energy consumption in a selected sector of the economy.

The share of the source (or technology) of renewable energy in gross final energy consumption for each year was determined as the ratio of the annual final energy consumption from this renewable energy source to gross final energy consumption of all sectors of economy:

$$k_{tRES} = \frac{E_{tRES}}{E} \cdot 100, \%, \tag{1}$$

where E_{tRES} is the final energy consumption by the renewable energy source, PJ, TWh h or ktoe; E is the gross final energy consumption of the economy, PJ, TWh h or ktoe.

Since Table B10, and Table B28 in [20] contain the shares of renewable energy sources (technologies) in the energy consumption of the relevant sectors of economy—electric power, heating and cooling, and transport sector, then we propose to determine the total shares of RES energy consumption in the gross final energy consumption k_{tRES} as described below.

The share of final energy consumption from the renewable energy source in the sector of the economy equals to:

$$k^s_{tRES} = \frac{E^s_{tRES}}{E^s} \cdot 100, \%, \tag{2}$$

where E^s_{tRES} is the final energy consumption from the renewable energy source in the sector of economy, PJ; E^s is the final energy consumption of the sector of the economy, PJ.

Since $E_{tRES} = E^s_{tRES}$ for a renewable energy source, from which all produced energy is consumed only by one sector of the economy, and $E_{tRES} = \sum E^s_{tRES}$—by several economy sectors, then the share of the renewable energy source in gross final energy consumption k'_{tRES}, related to the consumption of all energy produced by one economy sector can be written as:

$$k'_{tRES} = k^s_{tRES} \cdot \frac{E^s}{E} \tag{3}$$

The share of RES final energy consumption in the corresponding total sectoral energy consumption:

$$k^s_{tRES} = \frac{E^s_{RES}}{E^s} \cdot 100, \%, \tag{4}$$

where E^s_{RES} is the final energy consumption from RES in the relevant sector of the economy [20, Table C2.7.1], PJ.

So, we obtain:

$$k'_{tRES} = k^s_{tRES} \cdot \frac{E^s_{RES}}{k^s_{tRES} \cdot E} = \frac{k^s_{tRES}}{k^s_{RES}} \cdot \frac{E^s_{tRES}}{E}. \tag{5}$$

Considering (1), the share of the RES in gross final energy consumption, related to the consumption of all energy produced by one economy sector equals

$$k'_{tRES} = \frac{k^s_{tRES} \cdot k_{RES}}{k^s_{RES}} \cdot \frac{E^s_{RES}}{E_{RES}}. \tag{6}$$

The share of the renewable energy source in gross final energy consumption, related to the consumption of all energy produced by several sectors of the economy is determined as:

$$k_{tRES} = \frac{\sum E^s_{RES}}{E^s} \cdot 100 = \sum k'_{tRES} \tag{7}$$

Using the proposed technique for the analysis of energy and climate plans, the actual and forecast shares of energy from biomass, biofuel, and waste (%) in gross final energy consumption were analyzed in Fig. 1.

The actual values of the share are the values of the share recorded by the state monitoring services in the past years, which are represented by solid lines in the graphs, and by straight, normal font saturation—in the tables.

The predicted values of the share are the present and future values of the share based on the adopted model of the development of the state energy policy. They are shown by dotted lines in the graphs, and by semi-rough font saturation—in the tables.

We can see from Fig. 1, that Ukraine plans to triple energy consumption from biomass by 2030 compared to 2020 and double it compared to 2023, reaching a share of 19.5%

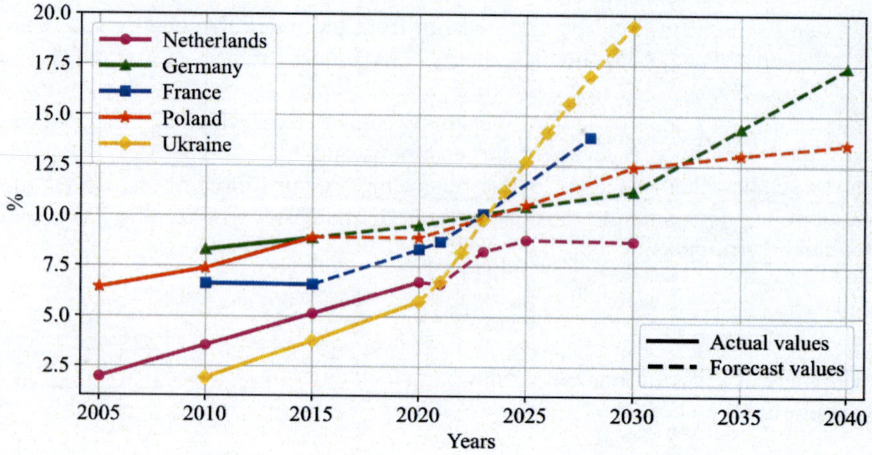

Fig. 1. Shares of energy from biomass, biofuel and waste in gross final energy consumption in Ukraine and the EU countries

in the country's gross final energy consumption. Germany is going to reach a 10.5% share of energy consumption in 2025, 11.2% in 2030, 14.4% in 2035, and 17.4% in 2040, almost doubling it from 2015. France plans 10.1% of biomass energy in gross final energy consumption in 2023 and 13.9% in 2028, i.e., to increase the share by 2.6 times and 3.7 times, respectively, compared to 2015. Poland plans to consume 10.6% in 2025, 12.4% in 2030, 13.0% in 2035, and 13.6% in 2040. The Netherlands is not going to increase energy consumption from biomass significantly, planning to increase the share only by 2% from 2020 to 2030.

Then the actual and forecast shares of wind energy (%) in the gross final energy consumption were analyzed (see Fig. 2).

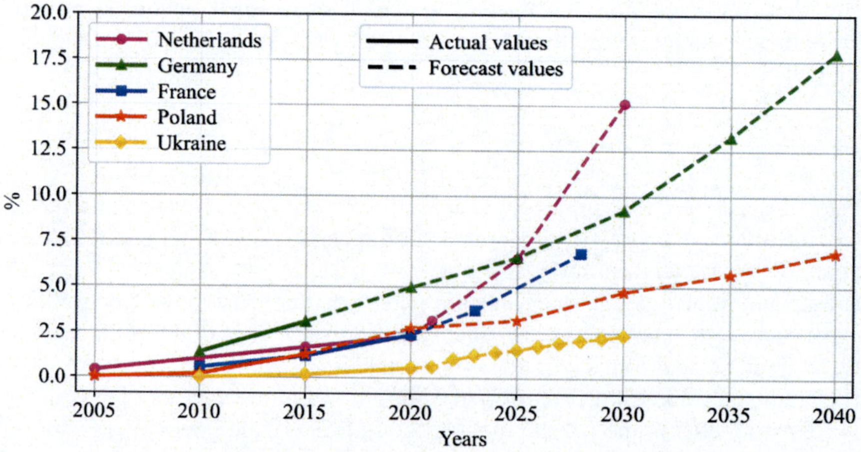

Fig. 2. Shares of wind energy in gross final energy consumption in Ukraine and the EU countries

Shares of renewable sources of solar energy in gross final energy consumption in Ukraine and EU countries are presented in Table 1, where the actual values are shown by straight, normal font saturation, predicted values—by semi-rough font saturation, and missing values—with a dash. Based on the data in Table 1, an analysis of the actual and forecast shares of solar energy in gross final energy consumption was carried out.

Table 1. Shares of solar energy (%) in gross final energy consumption of Ukraine and the EU countries

Country	2005	2010	2015	2020	2021	2022	2023	2025	2028	2030	2035	2040
Netherlands	0	0.5	1.0	1.5	1.9	2.3	**2.6**	3.5	**3.9**	4.2	–	–
Germany	–	0.6	1.7	2.4	2.7	2.9	**3.2**	**3.7**	**4.5**	**5.1**	**6.8**	**8.7**
France	–	0	0.4	0.9	1.1	1.3	**1.6**	**2.4**	**3.7**	–	–	–
Poland	0	0	0.1	0.4	0.5	0.6	**0.7**	**0.9**	**1.3**	**1.5**	**2.2**	**2.7**
Ukraine	–	0	0.1	1.0	1.3	1.5	**1.7**	**2.1**	**2.6**	**2.9**	–	–

Then an analysis of the actual and forecast shares of energy produced from heat pumps in gross final energy consumption was carried out (see Fig. 3).

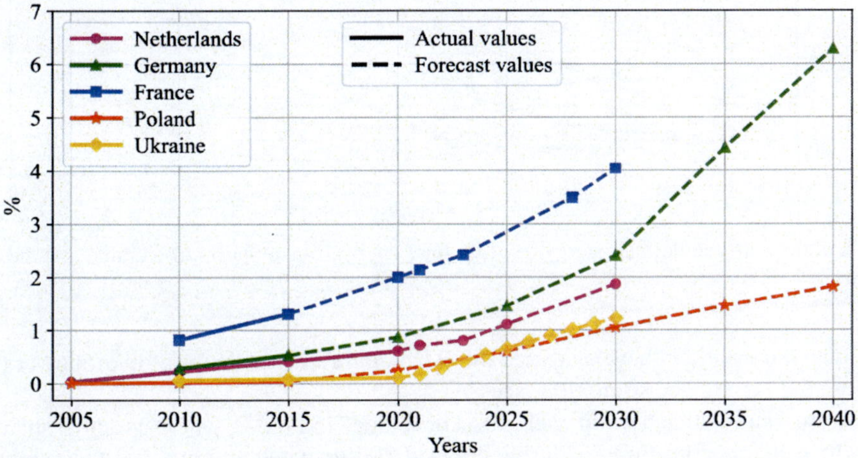

Fig. 3. Shares of energy from heat pumps in gross final energy consumption in Ukraine and the EU countries

Table B10, and Table B28 in [20] contain the total shares of solar thermal energy, geothermal energy, and heat pumps in the total energy consumption of Germany without dividing them into separate technologies. Therefore, the data for heat pumps were determined by subtracting the shares of solar thermal energy [20, Table. C2.1.1] and geothermal energy [24, 25] from the total shares.

According to Fig. 3, the low-temperature heat from heat pumps is most used in France, where its share in gross final energy consumption in 2010 and 2015 was more than 2.5 times higher than in Germany and was 0.8% and 1.3%, respectively. In the future, a 2.3% share of consumption is forecast in 2023, 3.5%—in 2028, and 4%—in 2030 [21].

The next step was the analysis of the actual and forecast shares of geothermal energy (%) in gross final energy consumption carried out in Fig. 4.

Figure 4 shows that geothermal energy does not take a significant share in gross final energy consumption of all investigated countries, but it is best developed in the Netherlands, where it provided 0.3% of all energy in 2020. In the future [19], a gradual increase in obtaining heat from this technology is predicted and it will reach a share in energy consumption of 0.8% in 2028. This type of thermal energy is somewhat worse developed in France, which plans to double heat production in 2023, reaching a share of 0.2% and triple in 2028, providing 0.3% of gross final energy consumption.

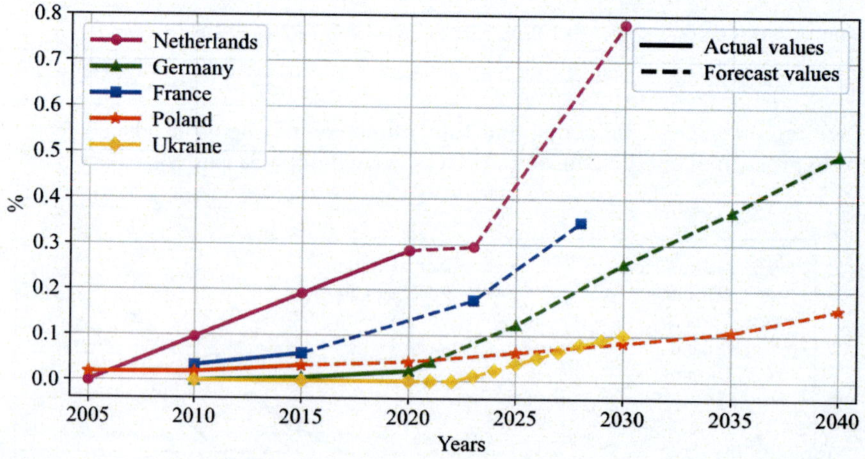

Fig. 4. Shares of geothermal energy in gross final energy consumption in Ukraine and the EU countries

Finally, the authors have analyzed the actual and forecast shares of hydropower (%) in gross final energy consumption (see Fig. 5).

Figure 5 shows that hydropower is best developed in France, where its actual share in gross final energy consumption during 2010–2020 fluctuated within 3.5%. In the future, the French government plans to refuse fossil fuels in the production of electrical and thermal energy, and to continue developing hydropower, reaching 4.2% of gross final energy consumption in 2028.

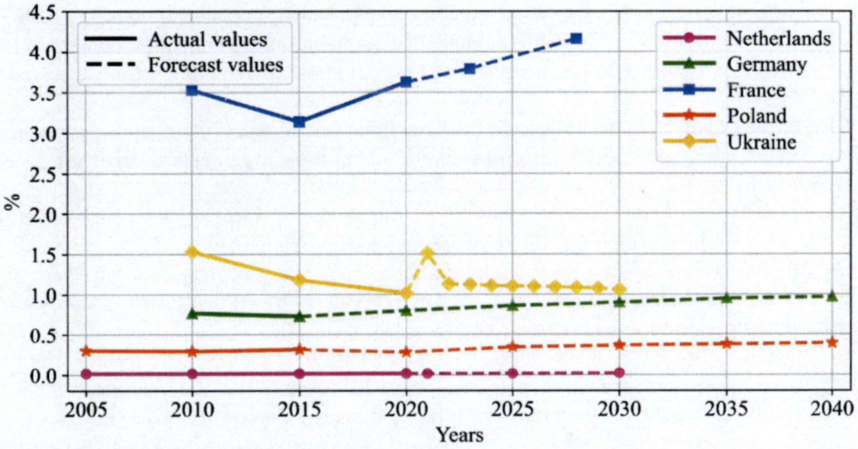

Fig. 5. Shares of hydropower in gross final energy consumption in Ukraine and the EU countries

5 Conclusion

In the paper, a new methodology was proposed for analyzing the shares of final energy consumption from the renewable energy source in the respective total sectoral energy consumption. Based on the developed methodology the authors have analyzed the current and forecast use of different renewable energy sources for some European countries. The research results show that the Netherlands and Germany are the leaders in the implementation of various renewable energy sources. The Netherlands prefers using wind energy (5% now and 15% till 2030) and solar energy (2.6% now and 4.2% till 2030). Germany uses 6% of wind energy now and approximately 18% till 2040, 3.2% of solar energy now and 8.7% till 2040 as well as 10% energy from biomass now and 17.5% till 2040. Poland leads in the use of biomass—10% now and 13% till 2040, however, according to forecasts, Ukraine is going to occupy a leading position in this sector (11% now and approximately 20% till 2030). France also prefers the use of biomass—10% now and 14% till 2028, heat pumps (2.5% now and 4% till 2030) and hydropower (3.5% now and 4.2% till 2028). The shares of geothermal energy in gross final energy consumption are the least for all investigated EU countries, for example, 0.3% now and 0.8% till 2030 for the Netherlands.

Now the European Union countries are working on the improvement of their national energy and climate plans to achieve the Union's main goals. Implementation of REPowerEU for energy saving will help the EU countries to increase the share of energy from renewable sources to 40–45% in gross final energy consumption by 2030 due to more active development of wind and solar power plants, use of biomethane and low-carbon hydrogen energy.

References

1. Paris Agreement. United Nations. https://treaties.un.org/doc/Treaties/2016/02/20160215%2006-03%20PM/Ch_XXVII-7-d.pdf. Last accessed 25 Mar 2023

2. Transforming our world: the 2030 Agenda for Sustainable Development. A/RES/70/1. United Nations, https://sustainabledevelopment.un.org/content/documents/21252030%20Agenda%20for%20Sustainable%20Development%20web.pdf. Last accessed 25 Mar 2023
3. A European Green Deal. Striving to be the first climate-neutral continent, https://commission.europa.eu/strategy-and-policy/priorities-2019-2024/european-green-deal_en. Last accessed 01 Apr 2023
4. The 2030 climate and energy framework. https://www.consilium.europa.eu/en/policies/climate-change/2030-climate-and-energy-framework/. Last accessed 29 Mar 2023
5. Fit for 55. https://www.consilium.europa.eu/en/policies/green-deal/fit-for-55-the-eu-plan-for-a-green-transition. Last accessed 14 Mar 2023
6. REPowerEU: affordable, secure and sustainable energy for Europe. European Commision, https://commission.europa.eu/strategy-and-policy/priorities-2019-2024/european-green-deal/repowereu-affordable-secure-and-sustainable-energy-europe_en. Last accessed 2 Apr 2023
7. Establishing the framework for achieving climate neutrality and amending Regulations (EC) No 401/2009 and (EU) 2018/1999 ('European Climate Law'). 30 June 2021, https://eur-lex.europa.eu/legal-content/EN/TXT/?uri=CELEX:32021R1119. Last accessed 25 Mar 2023
8. About the National Renewable Energy Action Plan for the period up to 2030. Edition of 21.09.2022, https://saee.gov.ua/sites/default/files/DraftNPDVE_2030_SAEE_21_09_2022pdf. Last accessed 1 Apr 2023
9. Wiśniewska M, Pusz A, Rogalski D (2020) Development of renewable energy sources (RES) in the European Union and Poland. Pol J Sustain, pp 101–112
10. Sobczak A, Chomac-Pierzecka E, Kokiel A, Rózycka M, Stasiak J, Sobón D (2022) Economic conditions of using biodegradable waste for biogas production, using the example of Poland and Germany. Energies 15:1–18
11. Slusarz G, Gołebiewska B, Cierpiał-Wolan M, Gołebiewski J, Twaróg D, Wójcik S (2021) Regional diversification of potential, production and efficiency of use of biogas and biomass in Poland. Energies 14:1–20
12. National Renewable Energy Action Plan (NREAP) for the period up to 2020, Draft, https://saee.gov.ua/documents/NpdVE.pdf. Last accessed 25 Mar 2023
13. Mysak Y, Pona O, Shapoval S, Kuznetsova M, Kovalenko T (2017) Evaluation of energy efficiency of solar roofing using mathematical and experimental research. Eastern-European J Enterpr Technol 3:26–32
14. Venhryn I, Shapoval S, Voznyak O, Datsko O, Gulai B (2021) Modelling of optical characteristics of the thermal photovoltaic hybrid solar collector. In: International scientific and technical conference on computer sciences and information technologies, vol 1, pp 255–258
15. Stec M, Grzebyk M (2022) Statistical analysis of the level of development of renewable energy sources in the countries of the European Union. Energies 15:1–18
16. Guminilovych R, Shapoval P, Yatchyshyn I, Shapoval S (2015) Modeling of chemical surface deposition (CSD) of CdS and CdSe semiconductor thin films. Ch&ChT 9(3):287–292
17. Miłek D, Nowak P, Latosińska J (2022) The development of renewable energy sources in the European Union in the Light of the European Green Deal. Energies 15:1–17
18. Angheluta SP, Burlacu S, Diaconu A, Curea CS (2019) The energy from renewable sources in the European Union: achieving the goals. Eur J Sustain Dev 8:57
19. Klimaat- en Energieverkenning 2022. PBL Planbureau voor de Leefomgeving. Den Haag, 2022. PBL-publicatienummer: 4838, https://www.pbl.nl/sites/default/files/downloads/pbl-2022-klimaat-en-energieverkenning-4838.pdf. Last accessed 1 Apr 2023
20. Integrated National Energy and Climate Plan. Germany, https://energy.ec.europa.eu/system/files/2022-08/de_final_necp_main_en.pdf. Last accessed 30 Mar 2023

21. Integrated National Energy and Climate Plan for France (2020) https://energy.eceuropa.eu/system/files/2020-04/fr_final_necp_main_fr_0.pdf. Last accessed 01 Apr 2023
22. Energy Policy of Poland until 2040. Appendix 2. Conclusions from forecast analyses for the energy sector. Ministry of Climate and Environment, https://www.gov.pl/web/climate/energy-policy-of-poland-until-2040-epp2040. Last accessed 5 Apr 2023
23. Our World in Data. Electricity production by source, France, https://ourworldindata.org/energy/country/france. Last accessed 27 Mar 2023
24. Eurostat. Production of electricity and derived heat by type of fuel, https://ec.europa.eu/eurostat/databrowser/view/NRG_BAL_PEH__custom_4992657/default/table?lang=en. Last accessed 13 Apr 2023
25. Germany's current climate action status. Bundesministerium fur Wirtschaft und Klimaschutz, https://www.bmwk.de/Redaktion/EN/Downloads/E/germany-s-current-climate-action-status.html. Last accessed 5 Apr 2023

Integrated Solar Collectors in External Protection for Energy-Efficient Buildings

Stepan Shapoval[1](✉), Vasyl Zhelykh[2], Oleksandr Pryimak[3], and Bogdan Gulai[1]

[1] National University "Lviv Polytechnic", Lviv, Ukraine
stepan.p.shapoval@lpnu.uamailto
[2] Czestochowa University of Technology, Częstochowa, Poland
[3] Kyiv National University of Construction and Architecture, Kyiv, Ukraine

Abstract. Heating energy-efficient buildings by using renewable energy sources, particularly solar energy, is an important task. Solar collectors cannot always be accommodated on buildings due to lack of space, so it is promising to use solar heating systems, the elements of which are integrated into the structures of external enclosures of buildings and structures. This article presents a study of the thermal-physical characteristics of the proposed integrated solar collectors in the external protection of energy-efficient buildings. Additionally, the study of these integrated solar window constructions was conducted in the presence or absence of blinds. Specifically, the study presents the changes in the useful efficiency coefficient of the integrated solar window depending on the angle of blind opening, which was 0, 45, and 90 degrees, respectively. The maximum value of the useful efficiency coefficient was 45%. Furthermore, the instantaneous thermal power studies of the developed solar window in the heating system were conducted and analyzed. The studies were conducted for different designs of solar windows, including a heat sink in the form of a coil and a heat exchanger in the form of a ribbed heat exchanger. Additionally, it was found that the presence of blinds increased the useful efficiency coefficient of the integrated solar window with a ribbed heat exchanger by 7–11%.

Keywords: Solar collector · Solar panel · Energy-efficient building · Renewable energy sources

1 Introduction

In recent years, the European Union has been paying more attention to energy efficiency issues. The use of energy-efficient technologies is not only envisaged by energy and climate plans of countries but also by building standards. Increasing energy efficiency is closely related to the use of renewable energy sources and fuel and energy resource savings. Therefore, it is an important task today because all these measures will not only reduce greenhouse gas emissions, reduce the use of traditional energy sources but also improve the future of our planet. Today, many technologies are being developed for the

Z. Blikharskyy et al. (Eds.): CEE 2023, LNCE 438, pp. 392–398, 2024.
https://doi.org/10.1007/978-3-031-44955-0_39

use of non-traditional and renewable energy sources, including solar energy systems. Solar thermal and power generation systems are becoming increasingly popular due to their ease of use. However, they require additional space for installation. Therefore, the development of hybrid and integrated systems is important today. Solar energy systems can be an effective alternative to traditional systems in spring-summer periods. And in winter periods, they can complement existing building and structure energy supply systems.

2 Objectives the Formulation of the Problem

In these days, along with the need to intensify the use of renewable energy sources and increase the energy efficiency of buildings, due to high building density, there is a need for integrated solar heating systems. In such systems, their elements can be part of external building protections and structures, including windows. Therefore, it is necessary to develop such designs and conduct their thermophysical research.

3 The Analysis of Recent Research and Publications

Today, it is important not only to achieve energy efficiency of the building but also effective management of all energy processes associated with its functioning [1]. After all, in addition to the use of non-traditional and renewable energy sources, it is necessary to apply smart technologies, which will help increase the energy efficiency of buildings and effective management of energy processes in them [2, 3]. Therefore, today it is important to introduce the most modern technologies in construction, in particular such as Smart Energy Networks, to achieve smart energy systems [4, 5], namely cogeneration systems in combination with renewable energy sources [6]. Therefore, it is important to develop energy-efficient measures and technologies, in particular using the practical experience of European countries [7]. It is necessary to take into account the amount of solar energy that comes on surfaces of different orientation taking into account their reflected component [8]. It is important to take into account the amount of solar energy that reaches the roofs of buildings [9], as well as other elements of buildings, this energy is significant and can be used for solar energy supply systems of energy-efficient buildings and structures [10, 11].

Therefore, it is necessary to develop and implement energy-efficient solar heating systems whose elements are integrated into external protection of buildings and structures [12], as well as to combine thermal and photovoltaic solar collectors [13]. This can increase the efficiency of such systems and contribute to the growth of energy efficiency of buildings and structures [14].

4 The Main Material

A design of a window that can simultaneously act as a thermal solar collector was proposed for research purposes. Research was conducted on the thermal efficiency of a heliowindow with a coiled heat exchanger. Experimental studies of the thermal efficiency of the heliowindow with a heat absorber in the form of a polyvinyl chloride coil contained inside the window were carried out to establish its thermal characteristics. The experimental setup consisted of a transparent cover (1), a casing with a thermal insulation layer (2), and tubes for the heat carrier (3) located between them (Fig. 1).

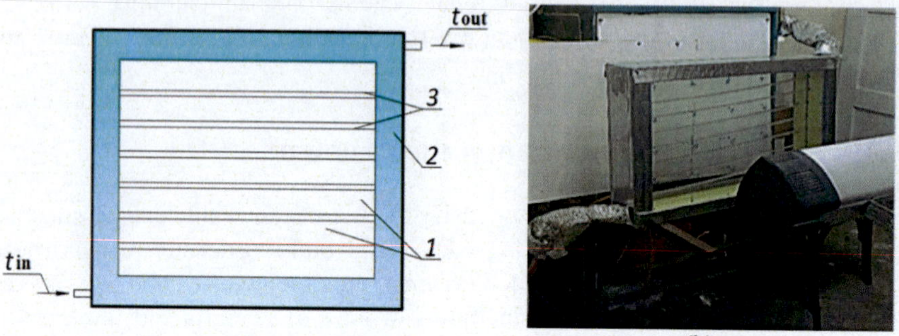

(a) (b)

Fig. 1. Scheme (**a**) and photo (**b**) of a heliowindow with a coil: 1—transparent covering; 2—casing with a thermal insulation layer; 3—tubes for heat transfer medium

To conduct effective research, an experimental design matrix was created that took into account the interaction effect between factors. The factors were radiation flux density I_B, coolant flow rate G, and mass in the thermal accumulator m_{TA}. Figure 2 shows the thermal characteristics of the heliowindow during the experiment.

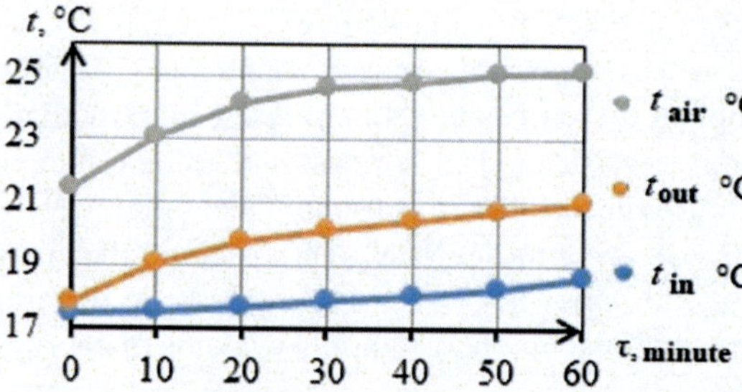

Fig. 2. Temperature of the heat carrier in the inlet and outlet pipes of the heliowindow and the temperature of the surrounding air environment for $I_B = 300$ W/m^2

Figure 3 shows the changes in the temperature of the heat transfer fluid at the inlet and outlet of the heliowindow, as well as the surrounding air temperature during circulation. At the beginning of the experiment, the temperature at the inlet of the heliowindow was 17.5 °C and by the end, it increased to 18.6 °C. Correspondingly, at the outlet of the heliowindow, the temperature was 17.8 °C and 21.0 °C, respectively. The temperature of the heat transfer fluid at the outlet of the heliowindow increased by 12.7% relative to the inlet temperature at the end of the study. The ambient temperature changed during the experiment from 21.4 °C to 25.1 °C, and increased by 17.2%.

The efficiency coefficient of the heliowindow can be calculated according to the Eq. (1):

$$\eta_{hw} = \frac{Q_{hw}}{I_{hw}} \cdot 100\%, \tag{1}$$

where, Q_{hw}—the value of the specific instantaneous thermal power of the heliowindow, W/m^2; I_{hw}—the value of the specific intensity of solar radiation in the plane of the heat receiver, W/m^2.

Figure 3 shows the summarized data on the change in the efficiency factor of the solar window η_{hw} during the experiment as a function of the mode of fluid flow, either direct flow (DF), forced circulation (FC), or natural circulation (NC).

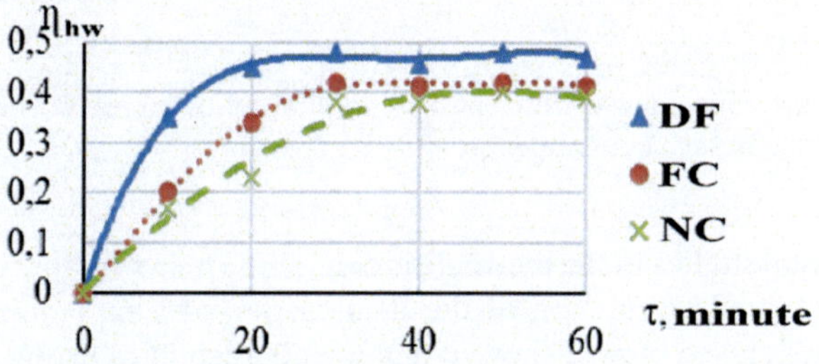

Fig. 3. Change of solar window efficiency η_{hw} during the experiment

On Fig. 3, the change in the efficiency coefficient of the heliowindow is shown throughout the experiment without blinds for the radiation flux density of 300 W/m^2. Within 20 min, there was a rapid increase in the efficiency coefficient up to 34%, and by the end of the experiment, the efficiency coefficient reached a maximum of 0.48, with its average value being 47%.

The next stage was the study of the thermal efficiency of the heliowindow with a coil and blinds. The heliowindow scheme differs from the scheme used in previous studies in that white blinds (4) have been added to the rear part of the heliowindow (Fig. 4).

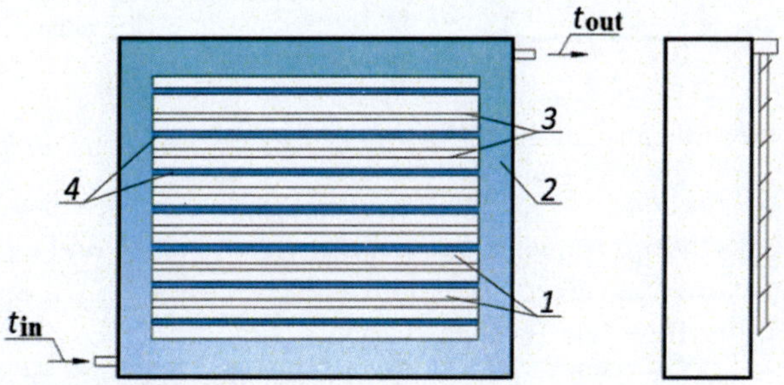

Fig. 4. The scheme of the heliowindow with a serpentine and shutters: 1—transparent covering; 2—housing with a heat-insulating layer; 3—pipes of the heat carrier; 4—shutters

The research was conducted for different angles of the blinds, which were 0° (closed); 45°, 90° (open) degrees (see Fig. 5).

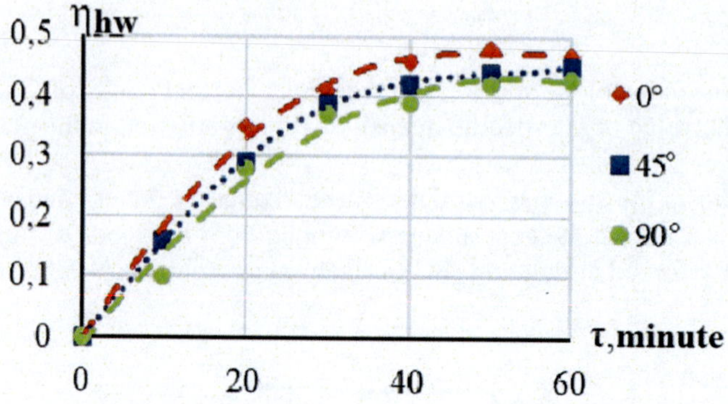

Fig. 5. Change in the efficiency coefficient η_{hw} of the heliowindow during the experiment depending on the angle of the shutter opening

The experimental studies of the thermal efficiency of a heliowindow with a serpentine and blinds under forced circulation of the heat carrier, at a radiation flux density of 300 W/m^2 and with closed blinds, showed that the coefficient of useful effect increased on average to 15% within 10 min and gradually reached 45% during the rest of the experiment. The change in the instantaneous power of the heliowindow Qhw, W/m^2, during the experiment is presented in Fig. 6.

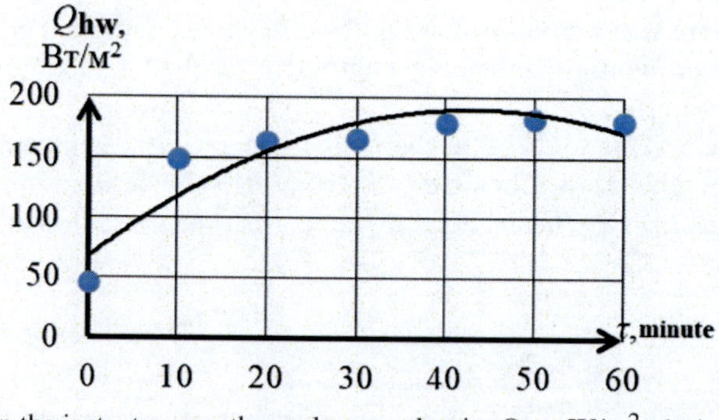

Fig. 6. Change in the instantaneous thermal power density Q_{hw}, W/m^2, during the experiment.

The instantaneous thermal power averaged 170 W/m^2. After conducting experiments on the heliowindow with open and closed blinds, it was reasonable to compare the values of the coefficient of useful effect for these variants (Fig. 7).

The conclusion that can be drawn from Fig. 7 is that the efficiency of the heliowindow with closed blinds is 10–15% higher.

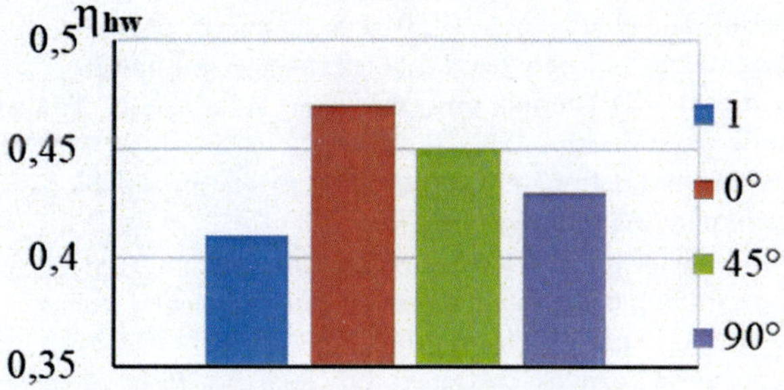

Fig. 7. Average value of the efficiency coefficient of the heliowindow (1—window without blinds) depending on the angle of blind opening

5 Conclusion

Research conducted on the proposed designs of the heliowindow showed that its coefficient of performance reached 50%. In addition, it was found that the presence of blinds increased the coefficient of performance of the heliowindow with a ribbed heat exchanger by 7–11%. This indicates the promising nature of the proposed solutions and the need for further research to improve them and develop calculation methods for such systems. Therefore, today it is important to study and improve integrated energy supply systems for more efficient use of alternative energy sources. As well as the development of combined energy supply systems, which will include the possibility of attracting various non-traditional energy sources, is necessary.

References

1. Andreas K (2012) NSERC smart net-zero energy buildings strategic research network. SNEBRN Newsl 1. (in English)
2. Korendiy V, Kachur O (2023) Dynamic behavior of a vibratory plate compactor working on a horizontal elastic-viscous-plastic surface. In: Lecture notes in mechanical engineering, pp 434–443. https://doi.org/10.1007/978-3-031-16651-8_41
3. Zhelykh V, Shapoval P, Shapoval S, Kasynets M (2020) Influence of orientation of buildings facades on the level of solar energy supply to them. In: Lecture notes in civil engineering, vol 100: Proceedings of 2nd international scientific conference on ecocomfort and current issues of civil engineering ecocomfort, Lviv, Ukraine, 16–18 Sept 2020, pp 499–504
4. Lund H, Andersen A, Ostergaard P, Mathiesen B, Connolly D (2012) From electricity smart grids to 63—a market operation based approach and understanding. Energy 42:96–102 (in English)
5. Sig Chai D, Wena J, Nathwani J (2013) Simulation of cogeneration within the concept of smart energy networks. Energy Convers Manage 75:453–465 (in English)
6. Mathiesen B, Lund H, Connolly D, Wenzel H (2015) Smart energy systems for coherent 100% renewable energy and transport solutions. Energy 145:139–154 (in English)
7. Synytsia S (2010) Energy efficiency in Germany-opportunities for Ukraine. Friedrich Ebert Stiftung 21 (in Ukraine)
8. Shapoval S (2015) The potential of solar energy in Ukraine. In: Litteris et Artibus: proceedings of the 5th international youth science forum. Lviv Polytechnic National University, pp. 122–123 (in English)

9. Horvath M, Kassai-Szoó D, Csoknyai T (2016) Solar energy potential of roofs on urban level based on building typology. Energy Build 111:278–289 (in English)
10. Daffi D, Bekman U (1977) Thermal processes using solar energy. Ttranslated from English by the editor Malevskyi Yu. Myr 420
11. Martynov V (2013) Dissertation of doctor of science «Optimization of geometric parameters of energy-efficient buildings (in Ukraine)
12. Mysak Y, Pona O, Shapoval S, Kuznetsova M, Kovalenko T (2017) Evaluation of energy efficiency of solar roofing using mathematical and experimental research. Eastern-European J Enterpr Technol 3(8–87):26–32. https://doi.org/10.15587/1729-4061.2017.103853
13. Venhryn I, Shapoval S, Voznyak O, Datsko O, Gulai B (2021) Modelling of optical characteristics of the thermal photovoltaic hybrid solar collector. In: International scientific and technical conference on computer sciences and information technologies, vol 1, pp 255–258
14. Kasynets M, Kozak K, Piznak B, Venhryn I (2022) Enhancing of efficiency of solar panels combined with buildings coating. In: Lecture notes in civil engineering, vol 290: Proceedings of the 3rd international scientific conference ecocomfort and current issues of civil engineering. ecocomfort 2022, Lviv, 14–16 Sept 2022, pp 136–149
15. Guminilovych R, Shapoval P, Yatchyshyn I, Shapoval S (2015) Modeling of chemical surface deposition (CSD) of CdS and CdSe semiconductor thin films. Ch&ChT 9(3):287–292

Experimental Research on the Effectiveness of Using Cellular Polycarbonate Translucent Shields for Cultivation Facilities

Vasyl Shtets, Vasyl Zhelykh, Yurii Furdas(✉), and Khrystyna Myroniuk

Lviv Polytechnic National University, Lviv 79013, Ukraine
yurii.v.furdas@lpnu.ua

Abstract. The article deals with the study of the effectiveness of using translucent cellular polycarbonate shields for cultivation facilities. A greenhouse model is presented, which consists of two plates of double-layer cellular polycarbonate serving as an external coating and a solar energy absorber. This structure is connected to distributing, intermediate and accumulating collectors for supply and discharge of the heat carrier. The sequence of conducting the experimental research is presented, that was taking place with the help of an infrared heater, which warmed up the surface of the translucent coating of the model, with the radiation intensity measurements performed using an albedometer, temperature sensors used to measure the external air temperature, the air temperature inside the model, and the temperature in the supply and return pipelines. The findings of the experimental research are presented, namely diagrams demonstrating the dependence of the temperature of the heat carrier on its flow rate; the value of the appropriate flow rate of the heat carrier is obtained, which makes up 11.5 l/min, at which the temperature of the heat carrier will remain constant and equal to 28.1 °C; the graphical dependence of the average temperature of the heat carrier on the temperature inside the greenhouse and the flow rate is presented as well.

Keywords: Translucent structure · Cellular polycarbonate · Cultivation facility · Greenhouse · Infrared heater

1 Introduction

In today's world, with environmental issues becoming even more urgent, the search for energy- and resource-saving technologies in agriculture is becoming particularly important. One of the most promising innovative solutions is the use of translucent shields made of cellular polycarbonate [1–3].

Translucent shields have been widely used in greenhouses to provide plants with the necessary amount of light, and to protect them from negative environmental factors. One of these popular materials is cellular polycarbonate.

From the perspective of energy saving and the use of cellular polycarbonate as a solar collector, it is quite an efficient solution. The structure of cellular polycarbonate with its air chambers allows it to be used as a solar collector. The polycarbonate surface is able to

Z. Blikharskyy et al. (Eds.): CEE 2023, LNCE 438, pp. 399–408, 2024.
https://doi.org/10.1007/978-3-031-44955-0_40

collect solar energy and supply it inside the greenhouse. So, cellular polycarbonate can function as a reliable source of free heating for greenhouses, reducing dependence on traditional energy sources. Also, due to its high thermal insulation and ability to collect solar energy, cellular polycarbonate can significantly increase the energy efficiency of the greenhouse [4–8].

Using solar heat collected by polycarbonate helps reduce heating costs and ensure constant temperature inside the structure. This is especially relevent in the cold season when there is a need in additional heating [9–11].

As a solar collector, cellular polycarbonate allows to use the renewable energy of the sun. Solar energy is a free and safe source of energy, which allows to reduce the use of traditional fuels and contributes to the conservation of natural resources. Using cellular polycarbonate as a solar collector helps reduce harmful gas emissions associated with the use of traditional energy sources. This contributes to the creation of environmentally friendly and energy-efficient systems for cultivation facilities [12–14].

Scientific research on the use of polycarbonate as translucent shields in greenhouses demonstrates significant amount of experiments aimed at determining its effectiveness and feasibility. Referring to the published books, we can find the following information about research conducted in this area:

1. Polycarbonate performance compared to other materials: studies compare the performance of polycarbonate with other materials such as glass, polyethylene film, and acrylic coating. They take into account parameters such as light transmission, thermal insulation, durability and service life. Most studies demonstrate that cellular polycarbonate has advantages over other materials, especially in terms of its thermal insulation properties and impact resistance.
2. Effects of polycarbonate on plant growth and development: many studies have focused on the effects of translucent polycarbonate shields on plant growth, development, and yields. They analyze parameters such as illumination, light distribution, temperature conditions, and humidity. Most studies prove that polycarbonate provides optimal conditions for plant growth, promotes photosynthesis, and increases yields.
3. Adaptation to different climates: some studies examine the effectiveness of polycarbonate in different climates. They investigate how polycarbonate affects the heat balance inside the greenhouse at different times of the year, how temperature is regulated, and how ventilation is provided. These studies help determine the best parameters for using polycarbonate in different climate zones.
4. Durability and environmental resistance: another important aspect of the research is the study of the service life and durability of polycarbonate in the demanding conditions of cultivation facilities. Studies evaluate the material's resistance to UV radiation, chemicals, atmospheric conditions, and mechanical stress. They help determine the service life and options for improving the material.

In general, scientific research confirms the feasibility of using cellular polycarbonate as translucent shields for cultivation facilities. It helps understand its advantages over other materials, establish the best parameters for its using and prove its effectiveness in increasing yields and providing comfortable conditions for plant growth.

2 Theoretical Bases

The relevance of the paper: one of the most pressing issues of today is a significant reduction in energy costs in various sectors of industrial and agricultural production.

Greenhouse cultivation structures are one of the most energy-intensive agricultural facilities. Current level of energy costs for growing greenhouse products, which has developed in the agricultural complex of Ukraine, significantly exceeds the corresponding indicators of similar industries in developed countries of the world. Due to the high cost of energy resources, the development of new integrated solutions that will increase the energy efficiency of greenhouse cultivation facilities is becoming important and relevant.

The purpose of the paper is a theoretical and experimental justification and development of effective methods for using solar energy by cultivation facilities with cellular polycarbonate enclosing structures.

Using translucent shields made of cellular polycarbonate in cultivation facilities is becoming even more popular, due to their advantages, such as high level of transparency for light to pass through, thermal insulation properties, UV resistance, as well as durability and easy installation.

However, there is a need for more detailed research aimed at determining the effectiveness of this type of shields. To date, there are insufficient research data and numerical results that would allow to objectively assess the impact of translucent cellular polycarbonate shields on plant growth, temperature conditions and energy efficiency in cultivation facilities.

So, the problem statement lies in the need to conduct experimental research in order to assess the effectiveness of using translucent cellular polycarbonate shields for cultivation facilities. These studies will allow to find out the real impact of this material on plant growth conditions, optimize temperature parameters and improve energy efficiency, which in turn will contribute to improving cultivation technologies and ensuring stable yields.

One of the key aspects is to determine the impact of the internal air temperature in the greenhouse (t-internal) and the external air temperature (t-external) on the average temperature of the heat carrier (t-average) at a certain flow rate of the heat carrier (L).

This impact can be determined using the following Eq. (1):

$$t_{average} = f\ (t_{internal},\ t_{external},\ L), \tag{1}$$

where:

$t_{average}$ the average temperature of the heat carrier, °C.
$t_{internal}$ the temperature inside the greenhouse, °C.
$t_{external}$ the outdoor air temperature, °C.
L the heat carrier's flow rate, l/min.

This formula allows to estimate how changes in the temperature of indoor and outdoor air, as well as the flow rate of the heat carrier, affect the average temperature of the heat carrier in cultivation facilities with translucent cellular polycarbonate shields.

3 Experimental Research of the House Model

Experimental setup for studying the efficiency of using a translucent greenhouse coating made of cellular polycarbonate (Fig. 1).

An arched greenhouse structure was used for this study, the principle of operation of which is that water circulating through cellular polycarbonate is heated by solar energy.

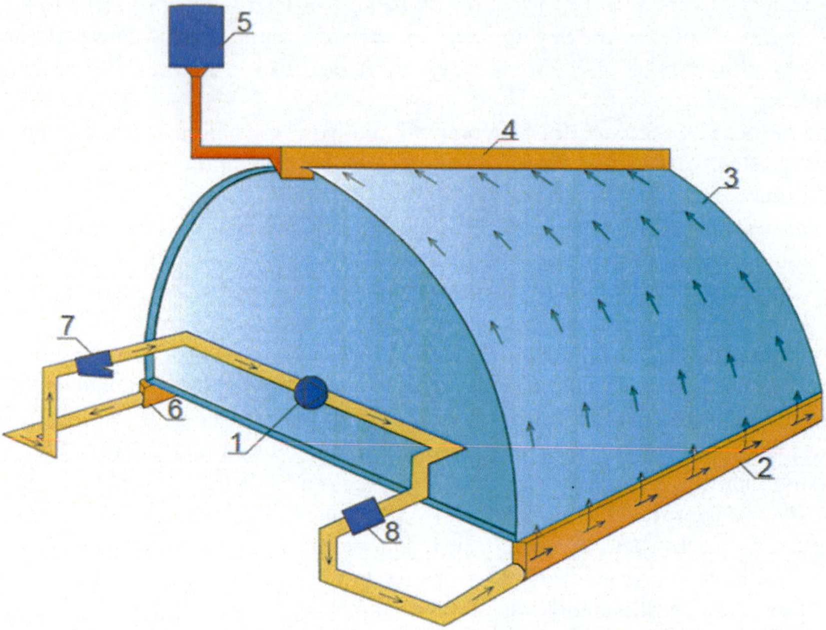

Fig. 1. Structure of the model under study1—circulation pump; 2—distributing collector; 3—protective coating; 4—intermediate collector;5—expansion tank; 6—accumulating collector; 7—mesh filter; 8—meter.

The greenhouse model consists of two plates of double-layer cellular polycarbonate 3 serving as an external coating and a solar energy absorber, which are connected to the distributing 2, intermediate 4, accumulating 6 collectors for the supply and discharge of heat carrier (water), circulation pump 1, mesh filter 7, which protects the system from contamination and various deposits, and a meter 8 to determine the flow rate.

The system operates as follows:

The heat carrier starts to run in the system with the help of a circulation pump 1, after which the heat carrier enters the distributing collector 2, where it is distributed along the entire length of the collector and gets evenly pumped into the cells of the first polycarbonate plate; passing through the cells of translucent protective coating 3, the rays are repeatedly reflected in the faces of the cell polymer thus warming up the heat carrier, after which it enters the intermediate collector 4 with the expansion tank 5, then the heat carrier runs along the cells of the second plate to the accumulating collector 6.

After that, the heat carrier, which has already been warmed up, runs through the pipeline and a mesh filter 7 to the circulation pump 1.

3.1 Experiment Planning and Processing of Experimental Data

To implement the experiment, we carried out its planning, since the main advantage of mathematical methods of experiment planning lied in reducing the share of experimental costs (by 2–10 times) due to the proper organization of research, minimizing the number of tests and the share of analytical work.

Reducing the number of tests in laboratory and semi-industrial experiments is achieved mainly due to the rejection of the traditional principle of studying multi-factor processes by reviewing all factors according to a certain most efficiently designed program.

The temperature of the heat carrier ($t_{average}$, °C) is affected by the temperature inside the greenhouse ($t_{internal}$, °C) and the heat carrier's flow rate (L, l/min). These values which will vary depending on the temperature of the heat carrier are input parameters for planning the experiment [15].

The number of tests is determined using the dependence (Eq. (2)):

$$N = P^K \tag{2}$$

P the number of factor levels, ($P = 2$).
K the number of factors. ($K = 2$).

Creating a FFE type 2^2 planning matrix (Table 1).

Table 1. FFE type 2^2 planning matrix

No	x_0	x_1	x_2	$x_1 . x_2$	y
1	+	+	+	+	27
2	+	–	+	–	28
3	+	+	–	–	29
4	+	–	–	+	30

$N = 2^2 = 4$ experiments
$x_1 = t_{internal}$–air temperature in the greenhouse, °C;
x_2 = L–the heat carrier's flow rate, l/min;
$y = t_{average}$–average temperature of the heat carrier, °C.

The regression equation is as follows (Eq. 3):

$$y = b_0 + b_1x_1 + b_2x_2 + b_{12}x_1x_2, \tag{3}$$

where b_0, b_1, b_2, b_{12}—coefficients are determined by the least squares method.

Regression equation (Eq. 4):

$$y = 28.5 - 2x_1 - 4x_2 \tag{4}$$

3.2 Conducting the Experimental Research

The experimental research was carried out using an infrared heater, which warmed up the surface of the translucent coating of the model; radiation intensity measurements were conducted using an albedometer, and temperature sensors to measure the outdoor temperature, the air temperature in the middle of the model, and the temperature at the supply and return of the system.

Figure 2 shows the structure of a greenhouse model for conducting experimental research to establish the dependence of the temperature of the heat carrier in polycarbonate channels on its flow rate. In the process, the thermal irradiation power of the greenhouse surface was 800 W/m^2.

Fig. 2. Experimental setup.

The experiment was conducted as follows:

1. At the beginning of the research, the temperature indicators of the air and heat carrier were taken.
2. Then we switched on the infrared heater, which warmed up the surface of the translucent coating of the model.
3. We had the pump running, set it to a certain capacity, and recorded the time; upon obtaining stable temperature indicators, we had them recorded, and changed the pump performance.
4. Measurements were taken several times at different pump flow rates.
5. We put down data to Table 2.

Based on experimental data, we obtained a diagram of the temperature dependence on the flow rate of the heat carrier (Fig. 3).

To increase the efficiency of the system, we determine the minimum flow rate at which the temperature of the heat carrier is constant on the supply and return system collectors. The value of the appropriate heat carrier's flow rate at which the supply and return temperatures will remain constant has been determined, that makes up 11.5 l/min.

Figure 4 shows the dependence of the internal temperature of the greenhouse on the average temperature of the heat carrier. Based on experimental research, the following

Table 2. Experimental data

Time	Flow rate, l/min	$t_{external}$, °C	$t_{internal}$, °C	t_{supply}, °C	t_{return}, °C	$t_{average}$, °C
Initial parameters		11.1	11.3	13.3	13.4	13.3
19:50	33.0	13.2	16.8	28	27.7	27.8
20:00	32.5	13.4	16.9	28	27.7	27.8
20:25	24.5	13.4	17.3	28.3	28.1	28.2
21:00	11.5	13.6	17.4	28.3	28.1	28.2
21:45	3.5	13.9	18.4	29.5	28.1	28.8
22:45	1.4	14.0	18.6	30.9	27.1	29
23:13	1.2	14.0	19.2	32.9	25.8	29.4

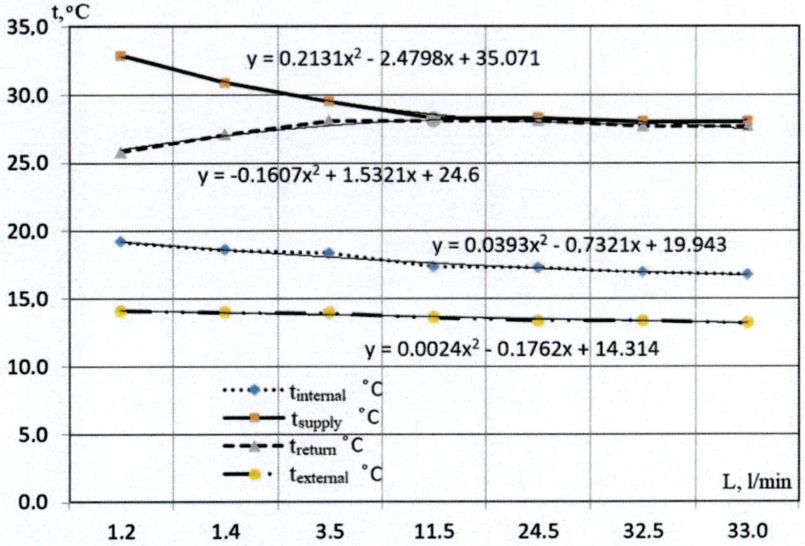

Fig. 3. Diagram of the dependence of the internal air temperature in the middle of the model and the temperature at the supply and return of the system on the flow rate of the hear carrier.

analytical dependence was obtained (Eq. 5):

$$t_{internal} = 1.497 \cdot t_{average} - 24.82, \quad (5)$$

where $t_{internal}$ is the temperature inside the greenhouse, °C; $t_{average}$ is the average temperature of the heat carrier, °C.

Figure 4 shows that as the average temperature of the heat carrier increases, the temperature in the greenhouse increases as well.

Figure 5 demonstrates the dependence of the average temperature of the heat carrier on its flow rate. On the basis of experimental studies, numerical results have been

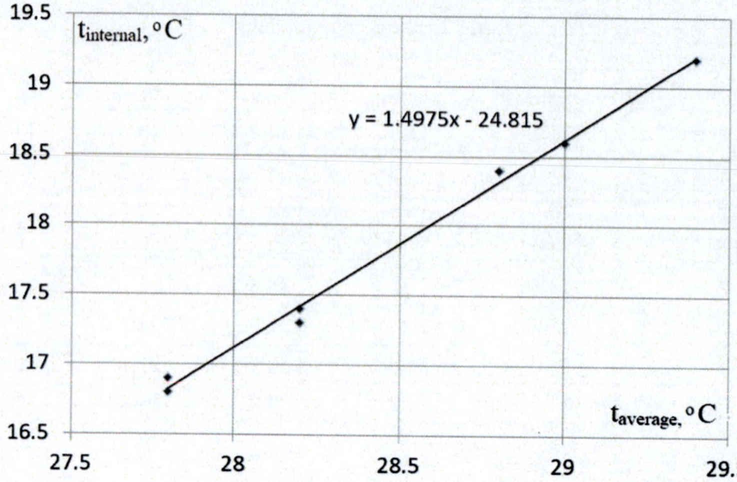

Fig. 4. Dependence of the air temperature inside the greenhouse on the average temperature of the heat carrier.

processed, with the following analytical dependence obtained (Eq. (6)):

$$t_{average} = -0.039 \cdot L + 29.07, \tag{6}$$

where L is the flow rate of the heat carrier, l/min.

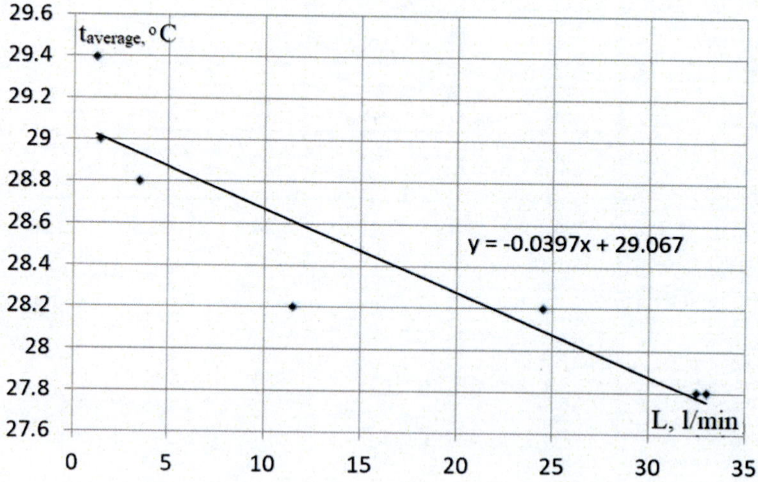

Fig. 5. Dependence of the average temperature of the heat carrier on its flow rate

Based on the data obtained in Fig. 5, we can see that with an increase in flow, the temperature of the heat carrier decreases, thereby regulating the temperature in the greenhouse.

4 Conclusions

1. Experimental research of the effectiveness of using translucent cellular polycarbonate shields for cultivation facilities have been conducted. In particular, the impact of the air temperature inside the greenhouse (t-internal), the outdoor air temperature (t-external) and the flow rate of the heat carrier (L) on the average temperature of the heat carrier (t-average) in cultivation facilities has been established.
2. A diagram of the dependence of the heat carrier's temperature on its flow rate has been obtained, which demonstrates that with an increase in the flow rate of the heat carrier, the temperature of the heat carrier decreases.
3. The value of the appropriate flow rate of the heat carrier has been determined, which makes up 11.5 l/min. At this flow rate of the heat carrier, its temperature remains constant and equal to 28.1 °C.
4. A graphical dependence of the average temperature of the heat carrier on the air temperature inside the greenhouse and the flow rate has been built. The relationship between input factors has been demonstrated, and the best values for maintaining the required temperature of the heat carrier have been established.
5. The effectiveness of using translucent cellular polycarbonate shields for cultivation facilities has been proved, with practical recommendations on the best temperature parameters of the heat carrier provided to achieve effective results in the cultivation process.

References

1. Venhryn I, Shapoval S, Voznyak O, Datsko O, Gulai B (2021) Modelling of optical characteristics of the thermal photovoltaic hybrid solar collector. International Scient Tech Conf Comp Sci Info Tech 1:255–258. https://doi.org/10.1109/CSIT52700.2021.9648738
2. Mysak Y, Pona O, Shapoval S, Kuznetsova M, Kovalenko T (2017) Evaluation of energy efficiency of solar roofing using mathematical and experimental research. Eastern-European J Enterprise Tech 3(8–87):26–32. https://doi.org/10.15587/1729-4061.2017.103853
3. Guminilovych R, Shapoval P, Yatchyshyn I, Shapoval S (2015) Modeling of chemical surface deposition (CSD) of CdS and CdSe semiconductor thin films. Ch&ChT 9(3):287–292. https://doi.org/10.23939/chcht09.03.287
4. Ulewicz M, Zhelykh V, Kozak K, Furdas Y (2020) Application of thermosiphon solar collectors for ventilation of premises. Lecture Notes in Civil Engineering, 47. https://doi.org/10.1007/978-3-030-27011-7_23
5. Ulewicz M, Zhelykh V, Furdas Y, Kozak K (2021) Assessment of the economic feasibility of using alternative energy sources in Ukraine. Lecture Notes in Civil Engineering this link is disabled, 2021, 100 LNCE, pp 482–489. https://doi.org/10.1007/978-3-030-57340-9_59
6. Freris L (2008) Renewable energy in power systems (Freris L, Infield D). Wiley, 284 p. ISBN-10 1118649931
7. Pluta Z (2007) Solar energy installations (Pluta Z). Publishing House of the Warsaw University of Technology, Warsyava, 246 p. ISBN: 978-83-7207-672-4
8. Lorenz P, Pinner D, Seitz T (2008) The economics of solar power. The McKinsey Quart, 19
9. Maczulak AE (2010) Renewable energy: sources and methods (Maczulak A). Infobase Publishing, 206 p. ISBN-978-1-4381-2748-4

10. Kots IV, Hrytsun AV, Bernyk IM, Yu MY (2011) Mathematical modeling of heat and mass exchange processes of a greenhouse using energy-saving technologies. Collect Scient Works of VNAU 8:54–59
11. Dieleman A, Marcelis LFM, Elings A et al (2006) Energy saving in greenhouses: optimal use of climate conditions and crop management. ISHS Acta Horticulturae 718: In: III International symposium on models for plant growth, Environmental control and farm management in protected cultivation, 203–209. https://doi.org/10.17660/ActaHortic.2006.718.22
12. Vox G (2010) Sustainable greenhouse systems. (Vox G, Teitel M, Pardossi A). Nova Science Publishers. Inc., Italy, 79 p. ISBN: 978-1-60876-269-9
13. Voznyak O (2015) Method of determining the efficiency of heliocollector in system with forced circulation of coolant. In: Voznyak O, Pona O, Eltman A, Shapoval S, Spodyniuk N (eds), Scientific and technical collection "Energy efficiency inconstruction and architecture", 7-K: KNUBA, pp 10–11
14. Pahlavan R (2011) Modeling and sensitivity analysis of energy inputs for greenhouse cucumber production (Pahlavan R, Omid M, Akram A). J Agri Tech 7(6):1509–1521. SSN 1686-9141
15. Voznyak O, Zhelykh V (2003) Basics of scientific research in construction: education. Publication of Lviv Polytechnic University, Lviv, 176 p. ISBN: 966-553-362-2

Bitumen Emulsion with Hydrochloric and Ortophosphoric Acid in the Technology of Cold Recycling of Road Pavement—Comparison

Iurii Sidun(✉), Khrystyna Sobol, Yurii Novytskyi, and Andriy Helesh

Lviv Polytechnic National University, Lviv 79013, Ukraine
yurii.v.sidun@lpnu.ua

Abstract. Investigated is the use of orthophosphoric acid (which is unconventional for road cationic bitumen emulsions) with a special emulsifier for the cold recycling technology of road pavement. Road cationic bitumen emulsions with hydrochloric and orthophosphoric acids were selected and compared with the help of establishing their physical and mechanical parameters. The study was performed for the special interaction of road cationic bitumen emulsions with orthophosphoric acid and Portland cement, which manifests itself during the determination of parameters of bitumen emulsion disintegration. The difference was determined between the indicators of storage-stability for cationic bitumen emulsions using hydrochloric and orthophosphoric acid. Mix for cold recycling technology were designed on bitumen emulsions with hydrochloric and orthophosphoric acids and with stone materials (virgin aggregate) of acidic granite and alkaline carbonate rocks, reclaimed asphalt, Portland cement and water. The physical-mechanical properties of the prepared mixtures were investigated for water-saturation and strength parameters and the corresponding dependences were obtained. It was established that cationic bitumen emulsions on orthophosphoric acid are advisable for use with aggregates of alkaline rock, while bitumen emulsions made on hydrochloric acid should be used with aggregates of acidic rock.

Keywords: Bitumen emulsion · Orthophosphoric acid · Reclaimed asphalt · Cement · Cold recycling

1 Introduction

The usage of hydrochloric acid for the production of various-purpose cationic bitumen emulsions is widely reported in scientific literature and Andriy possesses the significant practical spread on the territory of Ukraine and abroad. In particular, bitumen emulsions based on this acid are used for the technology of cold recycling for road pavement [1–5].

As regards the usage of ortophosphoric acid for cationic bitumen emulsions, it is mainly limited to its application with special emulsifiers for Slurry Surfacing [6–12]. Not often in scientific literature one can meet the usage of bitumen emulsions with

Z. Blikharskyy et al. (Eds.): CEE 2023, LNCE 438, pp. 409–418, 2024.
https://doi.org/10.1007/978-3-031-44955-0_41

orthophosphoric acid for other road technologies, particularly also for the technology of cold recycling for road pavement [13–15].

This technology is one of the promising directions in solving the problem of the road network status improvement. Both the theoretical research and the practical experience in application of the said technology facilitate reaching the substantial technical, economic and ecologic benefits. The limited mineral resources & lack of land areas on the one hand and the pressing need to reduce adverse environmental impacts on the other hand—have stimulated in recent decades the implementation of road pavement recycling technology into the practice of roads reconstruction and repair, including those ones performed in Ukraine. The usage of orthophosphoric acid in bitumen emulsion formulation intended for this technology may give both economic and technical benefit.

2 Aim and Objectives of the Research

Therefore, the aim of investigation is studying the possibility of using the cationic bitumen emulsions with orthophosphoric acid and local aggregates for the technology of cold recycling. The objectives of research were as follows: to select the raw materials for bitumen emulsions production and perform the cold mix design for the technology of recycling, while thereafter—to study their physical-mechanical parameters.

3 Methods and Materials

The main parameters of grade 70/100 bitumen intended for the emulsion production were determined according to [16–21] by the indices presented in Table 1. Cationic bitumen emulsions were tested by physical-technical indices according to State Standard of Ukraine DSTU B V.27-129:2013 [22]. Samples for the technology of cold recycling were tested according to DSTU 8976:2020 [23] and DSTU 8977:2020 [24].

As a raw material for the bitumen emulsions production there was used the pavement-grade crude-oil oxidized bitumen (grade 70/100), a sample being taken at PJSC UkrTatNaphta (city of Kremenchuk, Ukraine).

Presented in Table 1 are the main parameters of bitumen for getting bitumen emulsion.

Besides of bitumen for bitumen emulsion production there were used the following ingredients: drinking water, emulsifiers in a form of amine surfactant under the trade names REDICOTE E-11 and REDICOTE C-320E, produced by Nouryon (Sweden), orthophosphoric acid (85% solution) and hydrochloric acid (13% solution).

The emulsifiers choice was not a random one, as far as REDICOTE E-11 is compatible with hydrochloric acid, while REDICOTE C-320E—with ortophosphoric acid (for the production of cationic bitumen emulsions).

For the cold mix design there was taken reclaimed asphalt (RA), namely dense graded hot-mix asphalt concrete, while binder content in it (above the 100% of mineral components) constituted 6.55% w/w, and virgin aggregate, namely dense graded (stone and sand mix) with grading from 0 to 20 mm originating from the following quarries [25]:

Table 1. Physical-mechanical parameters of bitumen 70/100

Index	Value	Determination method
Penetration at 25 °C, (0,1 mm)	79	EN 1426 [17]
Softening point, °C	47.8	EN 1427 [18]
Fraas breaking point, °C	−13	EN 12593 [19]
Penetration index	−0.70	EN 12591 [20]
Resistance to hardening at 163 °C		EN 12607–1 [21]
Retained penetration	45.0	
Softening temperature	54.0	
Increase in softening point,	5.6	
Change of mass	0.12	
Residual penetration, %	56	

- Burdiakivskii special quarry (carbonate alkaline aggregate that reacts with hydrochloric acid);
- Novograd-Volynskii quarry (granite acidic aggregate that does not react with hydrochloric acid).

It is proposed in this article to use virgin aggregate of acidic rocks for the investigation of interaction between the said rocks with cationic bitumen emulsions on ortophosphoric acid.

As a binder (for the cold recycling technology) there was used so-called combined or complex binder, consisting of cationic bitumen emulsions (organic binder) and Portland cement (mineral binder). As a mineral binder there was used Portland cement (CEM II/A-S) by requirements of EN 197-1 [26]. The bitumen emulsions formulations are presented in Table 2, while their physical-technical indices are presented in Table 3.

Table 2. Bitumen emulsions formulations

No.	Bitumen		Emulsifier		Acid	Water phase pH
	Producer	Content %	Type	Content %		
1	PJSC	61	REDICOTE E-11	1.3	HCI	2.5
2	UkrTatNafta		REDICOTE C-320E	1.3	H_3PO_4	

In general, bitumen emulsions (by their content and physical-technical indices) are suitable for the usage in technology of cold recycling. The unique feature of ortophosphoric-acid-based emulsions is higher breaking index comparatively to that

Table 3. Physical-technical indices of bitumen emulsions

Name of index	Bitumen emulsions formulations, no.	
	1	2
Particle charge test, pH	2.73	3.15
Residue on sieving (sieve No 014),%;	0.10	0.30
Binder content, %	60.94	60.97
Viscosity–Efflux time, 4 mm –25 °C	6.8	7.2
Stability during storage: residue on sieving (sieve No 014), %, not more than: – after 7 days – after 30 days	0.19 0.25	0.29 0.30
Breaking behavior (index), %	190	242

of hydrochloric-acid-based emulsions, along with the worse values for sieve-residue and stability during storage for 7 and 30 days.

4 Results and Discussions

By means of virgin aggregate (alkaline or acidic) and RA the cold mix design (Table 4) was done in accordance with the requirements towards grading prescribed in DSTU 8976:2020 [23].

To avoid the influence of virgin aggregate grading upon the design and properties of cold mix—for each of mix-design versions there was chosen the identical particle-size distribution, providing for the aggregate content in mineral part of the mix at a rate of 40% w/w, taking into account the RA grading accordingly.

Due to similar particle-size distribution of the mineral part of cold mix with virgin aggregate (alkaline or acidic) and RA, as well as due to one and the same bitumen content in cationic bitumen emulsions used —the basic mix-design was done for the cold mix (Table 5).

Formed according to DSTU 8977:2020 [24] cold-mix cylinder-samples presented a simulation of reclaimed asphalt pavement (RAP). Their testing was performed with certain curing time intervals, namely after 3, 7 and 28 days after their production (Table 6).

Table 4. Aggregate grading for mineral part of cold mix

% w/w	Sieve orifice size, mm										
	20	10	5	2.5	1.25	0.63	0.315	0.14	0.071	0.071 >	%
Partial sieve residues, %											
Virgin aggregate	0.0	53.2	39.3	3.3	1.8	0.8	0.4	0.7	0.3	0.2	100.0
RA	0.0	4.2	24.3	16.4	15.1	9.3	7.2	8.9	3.3	11.3	100.0
Partial fractions content according to adopted %											
Virgin aggregate	–	21.3	15.7	1.3	0.7	0.3	0.2	0.3	0.1	0.1	40
RA	–	2.5	14.6	9.8	9.1	5.6	4.3	5.3	2.0	6.8	60
Virgin aggregate + RA	0.0	23.8	30.3	11.1	9.8	5.9	4.5	5.6	2.1	6.8	100.0
Complete residues, %											
Virgin aggregate + RA	0.0	23.8	54.1	65.3	75.1	81.0	85.4	91.1	93.2	100.0	
Requirements of DSTU 8976:2020	10.0	40.0	65.0	80.0	85.0	90.0	95.0	97.0	100.0	100.0	max
	0.0	20.0	35.0	50.0	65.0	70.0	75.0	80.0	85.0	100.0	min

When analyzing the results presented in Table 6, one may conclude that RAP samples tested for water-saturation, after all the adopted for testing time-intervals (3, 7 and 28 days), do not show any significant difference—both when the samples-used are produced on different virgin aggregates (alkaline or acidic), and when the emulsions-used are obtained on different acids (ortophosphoric or hydrochloric) (Fig. 1).

As per the experiments results, the pattern is established, which can be expressed as follows: compression strength at 20 °C and 50 °C on 3, 7 and 28 day (Fig. 2) on RAP samples with ortophosphoric-acid-based emulsion is higher when using alkaline aggregate, but not the acidic one. In its turn, compression strength at 20 °C and 50 °C on 3, 7 and 28 day on RAP samples with hydrochloric-acid-based emulsion is higher with acidic aggregate used. As a confirmation of this dependence there are the results with water sensitivity.

Table 5. Basic mix-design for the cold mix

Indices	Component content in cold mix, % w/w					
	Virgin aggregate (alkaline or acidic)	RA	Bitumen emulsion (No 1 or 2)	CEM II/A-S-	Drinking water	Total
Laboratory mix-design	40.0	60.0	3.0	3.0	3.0	109.0
Actual mix-design	36.6	55.0	2.8	2.8	2.8	100
Residual organic binder content	–	3.72	1.71	–	–	5.43
Designed content	Mineral part	Residual organic binder		Mineral binder	Water from bitumen emulsion and drinking water	
	87.88	5.43		2.8	3.89	100

5 Conclusions

There was established that (by characteristics of compression strength and water sensitivity) it is expedient to use bitumen emulsion on ortophosphoric acid in technology of cold recycling of road pavement with virgin aggregate used – both with alkaline and acidic rocks. Still, the usage of alkaline aggregate is more efficient, as far as the compression strength at 20 °C and 50 °C on the 28th day is higher by 0.1 MPa in two cases, while water sensitivity is higher by 0.07 (comparatively to RAP with emulsion on hydrochloric acid).

In general, cationic bitumen emulsions on ortophosphoric acid shall be used in RAP when alkaline virgin aggregate is used, while cationic bitumen emulsions on hydrochloric acid shall be used for RAP containing acidic virgin aggregate.

Table 6. Physical-mechanical indices of RAP

Name of RAP indices	Indices				
	Average density (g/cm^3)	Water-saturation (vol %)	Compression strength at (MPa):		Water sensitivity
			20 °C	50 °C	
On the 3rd day					
Alkaline aggregate + bitumen emulsion no. 1 (HCI)	2.27	3.4	1.3	0.4	–
Alkaline aggregate + bitumen emulsion no. 2 (H_3PO_4)	2.27	2.8	1.4	0.4	–
Acidic aggregate + bitumen emulsion no. 1 (HCI)	2.27	3.8	1.5	0.6	–
Acidic aggregate + bitu-men emulsion no. 2 (H_3PO_4)	2.27	4.0	1.4	0.5	–
On the 7th day					
Alkaline aggregate + bitumen emulsion no. 1 (HCI)	2.27	2.8	1.6	0.7	–
Alkaline aggregate + bitumen emulsion no. 2 (H_3PO_4)	2.27	2.8	1.6	0.6	–

(continued)

Table 6. *(continued)*

Name of RAP indices	Indices				
	Average density (g/cm^3)	Water-saturation (vol %)	Compression strength at (MPa):		Water sensitivity
			20 °C	50 °C	
Acidic aggregate + bitumen emulsion no. 1 (HCl)	2.27	3.4	1.7	0.8	–
Acidic aggregate + bitu-men emulsion no. 2 (H_3PO_4)	2.27	3.9	1.6	0.7	–
On the 28th day					
Alkaline aggregate + bitumen emulsion no. 1 (HCl)	2.27	3.4	2.3	1.1	0.78
Alkaline aggregate + bitumen emulsion no. 2 (H_3PO_4)	2.27	3.6	2.2	1.0	0.75
Acidic aggregate + bitumen emulsion no. 1 (HCl)	2.27	4.7	2.4	1.1	0.82
Acidic aggregate + bitu-men emulsion no. 2 (H_3PO_4)	2.28	5.8	2.0	0.9	0.70
Requirements of DSTU 8976:2020 [23] for the grade M20 with complex binder used	–	Not more than 9.0	From 2.0 to 3.9 inclusive	Not less than 0.9	Not less than 0.65

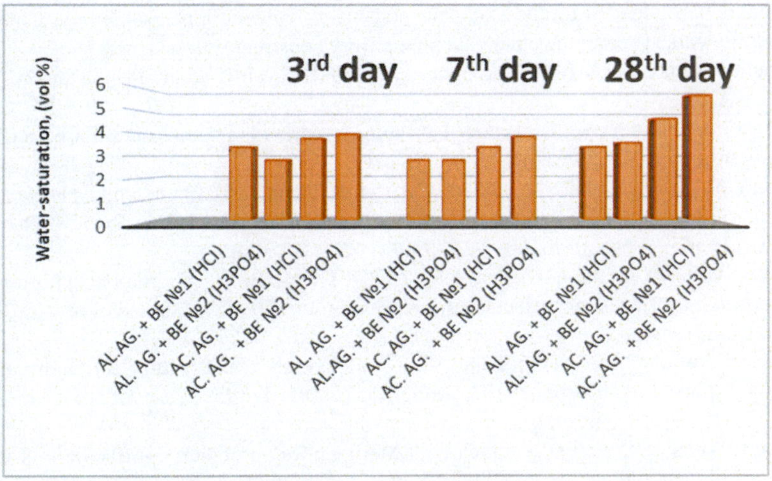

Fig. 1. Water-saturation indices of RAP on 3, 7 and 28 day.

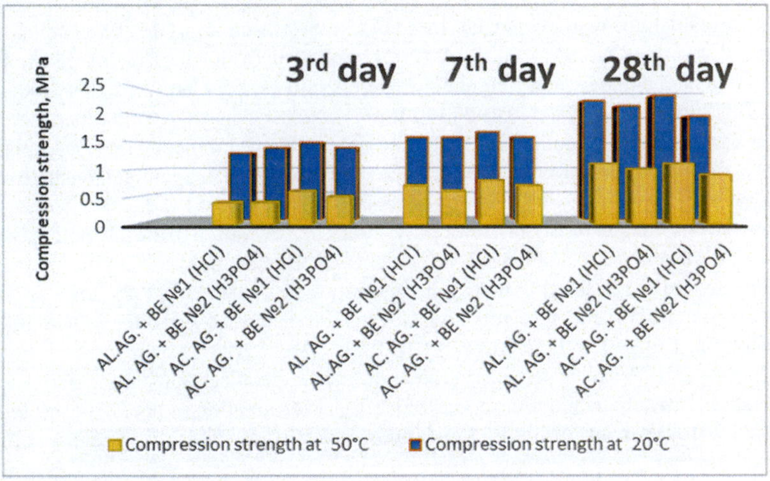

Fig. 2. Compression strength indices of RAP on 3, 7 and 28 day.

References

1. Xiao F, Yao S, Wang J, Li X, Amirkhanian S (2018) A literature review on cold recycling technology of asphalt pavement. Constr Building Mat 180:579–604. ISSN 0950-0618. https://doi.org/10.1016/j.conbuildmat.2018.06.006
2. Wang Y, Leng Z, Li X, Hu C (2017) Cold recycling of reclaimed asphalt pavement towards improved engineering performance. J Clean Produc 171:1031–1038. ISSN 0959-6526. https://doi.org/10.1016/j.jclepro.2017.10.132
3. Dołżycki B, Jaskuła P (2019) Review and evaluation of cold recycling with bitumen emulsion and cement for rehabilitation of old pavements. J Traffic Transportation Eng 6(4):311–323. ISSN 2095-7564. https://doi.org/10.1016/j.jtte.2019.02.002

4. Ilchenko V (2006) Restoration of pavement using cold milling technology, Collection of scientific works (branch mechanical engineering, construction). Issue 18
5. Stepura V, Osadchuk V (2011) Application of cold recycling technology (Doctoral dissertation). NAU
6. Sidun I, Solodkyy S, Vollis O (2019) Acids in bitumen emulsions. JCEEA 653(18):83–90. https://doi.org/10.7862/rb.2018.45
7. Sidun I, Solodkyy S, Vollis O, Gunka V, Pyryk R, Shits I (2020) Orto-phosphoric acid as an alternative to hydrochloric acid—for cationic bitumen road emulsions. Review. Theory Build Pract 2(1):88–93. https://doi.org/10.23939/jtbp2020.01.088
8. Sidun I, Vollis O, Gunka V, Ivasenko V (2020) Hydrochloric and orthophosphoric acids use in the quick-traffic slurry surfacing mix. Ch&ChT 14:380–385. https://doi.org/10.23939/chcht14.03.380
9. Sidun I, Vollis O, Hidei V, Bidos V (2021) Quick-traffic slurry surfacing mix with orthophosphoric acid. Product Eng Archives 27(3):191–195. https://doi.org/10.30657/pea.2021.27.25
10. Sidun I, Vollis O, Solodkyy S, Gunka V (2020) Cohesion of slurry surfacing mix with slow setting bitumen emulsions. In: Blikharskyy Z, Koszelnik P, Mesaros P (eds) Proceedings of CEE 2019. CEE 2019. Lecture Notes in Civil Engineering. Springer, Cham, p 47. https://doi.org/10.1007/978-3-030-27011-7_53
11. Demchuk Y et al (2020) Slurry surfacing mixes on the basis of bitumen modified with phenol-cresol-formaldehyde resin. Ch&ChT 14:251–256. https://doi.org/10.23939/chcht14.02.251
12. Sidun I, Solodkyy S, Vollis O, Gunka V (2021) Cohesion of slurry surfacing mix on bitumens of different acid numbers at different curing temperatures. EcoComfort LNCE 100:429–435. https://doi.org/10.1007/978-3-030-57340-9_52
13. Garilli E, Autelitano F, Giuliani F (2019) Use of bending beam rheometer test for rheological analysis of asphalt emulsion-cement mastics in cold in-place recycling. Constr Build Mater 222:484–492. https://doi.org/10.1016/j.conbuildmat.2019.06.141
14. Godenzoni C, Graziani A, Bocci E, Bocci M (2018) The evolution of the mechanical behaviour of cold recycled mixtures stabilised with cement and bitumen: field and laboratory study. Road Mat Pavement Design 19(4):856–877. https://doi.org/10.1080/14680629.2017.1279073
15. Teja K (2011) Improvement of silty soil as subgrade material by stabilizing with bituminous emulsion. Int J Engg Res Indu Appls 8(III). ISSN 0974-1518
16. DSTU 4044:2019 Viscous petroleum road bitumens. Specifications
17. EN 1426:2015Bitumen and bituminous binders—determination of needle penetration
18. EN 1427:2015 Bitumen and bituminous binders—determination of the softening point—Ring and Ball method
19. EN 12593:2015 Bitumen and bituminous binders—determination of the Fraass breaking point
20. EN 12591:2009 Bitumen and bituminous binders—specifications for paving grade bitumens
21. EN 12607-1:2014 Bitumen and bituminous binders—determination of the resistance to hardening under influence of heat and air—Part 1: RTFOT method
22. DSTU B V.2.7-129:2013 Bituminous road emulsions. Specifications
23. DSTU 8976:2020 Road materials manufactured using cold recycling technology. Specifications
24. DSTU 8977:2020 Road materials manufactured using cold recycling technology. Test methods
25. DSTU 9177-2:2022 Crushed and gravel materials for road construction. Specifications. Part 2. Materials are unreinforced
26. EN 197-1:2011 Cement—part 1: Composition, specifications and conformity criteria for common cements

Modeling of the Cryogenic Tank to Warehouse Liquefied Natural Gas (LNG) in the Event of the Earthquake

Przemysław Smela(✉), Rafał Szozda, and Leonard Ziemiański

Department of Structural Mechanics, Rzeszow University of Technology, Rzeszow, Poland
p.smela@prz.edu.pl

Abstract. The main objective of this article is description of modeling of a cryogenic tank which is fitted for warehousing Liquefied Natural Gas in the event of earthquake. The tank consists of two other tanks: the inner steel tank and the external reinforced prestressed concrete tank. The model of the tank is made in the FEM program—ADINA. The designed structure is a shell structure where walls of the tank are designed as single layer elements and the bottom of the inner tank is designed as multi-layered element. The emphasis in the article is put on the work of the inner tank and the interaction of the tank with the warehoused liquid. Earthquake is an accidental action that is taken into account during the analysis. In the analysis a New Zealand earthquake signal was taken as an excitation signal. The analysis is executed in the time domain. Stress graphs, pressure graphs, displacements graphs maps of reduced stresses and yield stress graphs are shown in the article.

Keywords: Cryogenic tank · Liquefied natural gas · Modeling · Earthquake · ADINA · Shell

1 Introduction. Characteristics of Cryogenic Tanks

Tanks are used in water distribution systems and in industrial plants for the storage and/or processing of a variety of liquids and liquid-like materials, such as oil, liquefied natural gas, chemical fluids and wastes of different forms [1]. Tanks with circular section are more economic than tanks with rectangular section because walls, roof and bottom of tanks with circular section carry mainly tension and compression stresses. The values of bending moments are small [2]. In the subsequent part of this chapter, the particular attention on the cryogenic tanks will be paid.

Cryogenic tanks are engineering structures whose geometry comes from technological functions. Their character is based on the tightness issue. Cryogenic tanks are built with full containment technology. It means that they consist of two tanks: the inner tank which is made of cryogenic steel and the external tank which is made of reinforced concrete. In addition, the external tank is prestressed in two perpendicular directions: longitudinally and latitudinally [3].

Z. Blikharskyy et al. (Eds.): CEE 2023, LNCE 438, pp. 419–426, 2024.
https://doi.org/10.1007/978-3-031-44955-0_42

The inner tank is responsible for the transfer of hydrostatic loads. Then the external tank protects the inner tank and the insulation against external factors. In addition, the external tank takes over the inner tank features in case of an accident in the interior tank [3].

There is always insulation between the inner and external tanks made of pearlite or glasswool. The main function of insulation is keeping the appropriate low temperature [3].

Design of cryogenic tanks inconveniences designers due to the effects of large temperature loads and the change in the strength parameters of the materials under cryogenic conditions. Due to the serious negative consequences for people and the natural environment of the potential accident, particular attention must be given to consulting the exceptional events (for example floods, earthquakes, hurricanes, fires of gas, solid strikes). Other dangers are frost heaves under the foundation plate. Therefore, the bottom of the steel inner tank is insulated and the bottom of the external tank is heated [3].

2 The Purpose of the Article and the Most Important Parameters of the Cryogenic Tank

The cryogenic tank designed in [4] is the above-ground tank, which consists of two integrated tanks: the inner steel tank and the external reinforced concrete tank with prestressing. The bottom of the inner tank is designed as a multi-layered component (exactly 4-layered component). The bottom of the external tank is made from reinforced concrete. There is a layer of insulation between these bottoms which was modelled as a three-dimensional elastic body. This cryogenic tank is designed according to the Eurocode norm [5]. The main purpose of this article is to show the results of the analysis of the cryogenic tank model as exemplified by the chosen earthquake. Dynamic analysis is very useful because if the earthquake damaged the cryogenic tank, it would be really hazardous to the natural environment and people's lives. Furthermore, dynamic analyses are very crucial because ground vibrations during an earthquake can severely damage structures and equipment housed in them. The ground acceleration, velocity, and displacement (referred to as ground motion) when transmitted through a structure, are in most cases amplified. This amplified motion can produce forces and displacements, which may exceed those the structure can sustain [6]. Moreover, the effortfulness, the time-consuming and the cost of the restoration of the tank would be very high. The model of the cryogenic tank was made in the FEM program—ADINA according to [7] and [8]. Earthquakes were the most important loads which were analysed in this research. The dynamic analysis was conducted in the time domain.

The most important parameters of the designed cryogenic tank are shown in Table 1.

3 Dynamic Model of the Cryogenic Tank and Exceptional Actions

3.1 Description of the Dynamic Model

The designed cryogenic tank consists of an inner steel tank, an external reinforced concrete tank with prestressing and fluid, LNG. The cooperation of both tanks holds according to proper boundary conditions and technological solutions. Therefore, the

Table 1. Parameters of the analysed cryogenic tank

Parameter	Value of the parameter
Radius of the external tank	40.00 m
Radius of the inner tank	36.00 m
Height of the external tank	42.50 m
Height of the inner tank	41.00 m
Overall height of the tank	52.60 m
Capacity	160.0 m^3

conditions of cooperation were determined. In the numerical model, two tanks were designed as independent tanks as batch files. The documentation which allowed to create these files, is included in [7]. The analysis during the earthquakes primarily is based on the work of the inner tank.

Afterwards, an impact of the dynamic actions was analysed. These dynamic actions were described as a function of time f(t). Vibrations were analysed as a result of the reaction of the numerical model, for which forced displacements were placed in the directions x, y and z.

The rigid restraint of the cryogenic tank in the ground was applied to simplified nonlinear dynamic analysis. The walls of the external tank bring the loads from the roof and the tank equipment to the foundation plate. All the above elements are fully welded due to the continuity of the reinforcement and prestressing.

The vertical wall of the external tank was divided into segments where different prestressing loads are located. The main feature of prestressing is to minimise the negative results of the accident, when the inner tank would be damaged and the LNG would load the external tank.

In addition, change of the parameters of materials and the thickness of the inner tank was assumed.

The view of the designed cryogenic tank is shown in the Fig. 1 on the next page.

3.2 The Description of the Exceptional Actions—The Earthquake

Earthquake is the result of a rapid relative deformation, called the seismic deformation. The cause of the movement can be found in a thermo-mechanical equilibrium of the Earth's materials. In fact, the upper part of the mantle is in contact with the cooler portion of the crust while the lower portion is located near to the hot outer core. The temperature differences inside the mantle produce a variation of material's density that leads to an unstable situation in which the denser material sinks under the action of gravity, while the less dense material tends to rise up. This movement continuously occurs since the cooler material gradually becomes less dense and vice versa for the warmer materials. Thus, convection currents occur inside the mantle and they lead to the movement of the plate tectonics [9].

In the analysis a New Zealand earthquake signal was taken as an excitation signal [10]. The graphs of the displacements of the ground in the foundation plate mirror the earthquake for the location above.

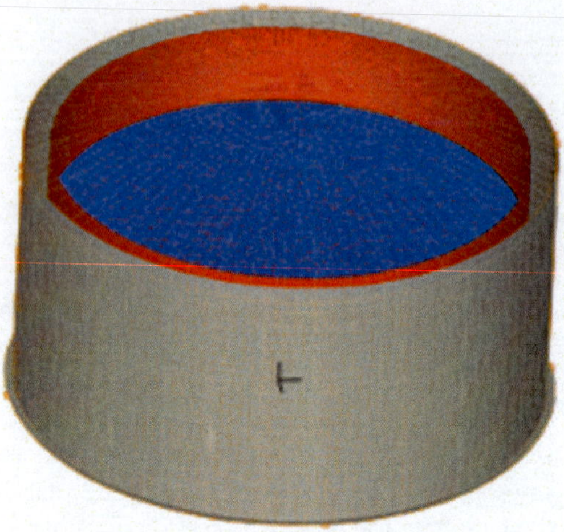

Fig. 1. The view of the designed cryogenic tank. The external tank is marked in gray, the inner tank is marked in orange and the fluid is marked in blue. *Source* Own elaboration

4 Results of the Analysis

In the following chapter, results of the analysis were set, including stress graphs, pressure graphs, displacements graphs, maps of reduced stresses and conclusions. Reduced stress were calculated according to the Huber-Mises-Hencky hypothesis (Figs. 2, 3, 4, 5, 6, and 7).

During the analysis of the graphs, the greatest values of displacements in the x-direction which are equal 25 cm (the positive value) and 30 cm (the negative value) were observed. The greatest positive value was indicated in the 12th second of the analysis and the greatest negative value was indicated in the 15th second of the analysis. In all researched directions (x, y and z) both positive and negative values were observed. Since the 9th second of the analysis, the changes of values of displacements were abrupt.

The values of reduced stresses were in the range from 50 to 225 MPa. In the stresses graph we can observe the significant increase of values of shear stresses about the 13th second of the analysis. This phenomenon shows appearance of yield stresses. Minor changes of reduced stresses agrees with minor changes of shear stresses and major changes of reduced stresses agrees with major changes of shear stresses.

During the analysis of the graph of relative displacements, the values in the z-direction are roughly constant (maximum displacement in this direction is equal about 1 cm). The greatest values of relative displacements were included in the x-direction (about 9 cm), a shade lower values were observed in the y-direction (about 6 cm).

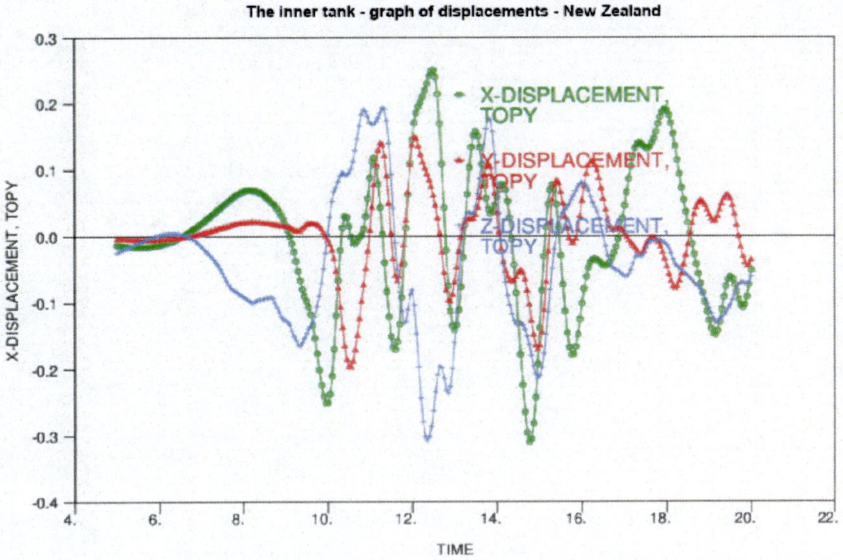

Fig. 2. Displacement graph of the inner tank for three directions: x, y and z. *Source* Own elaboration

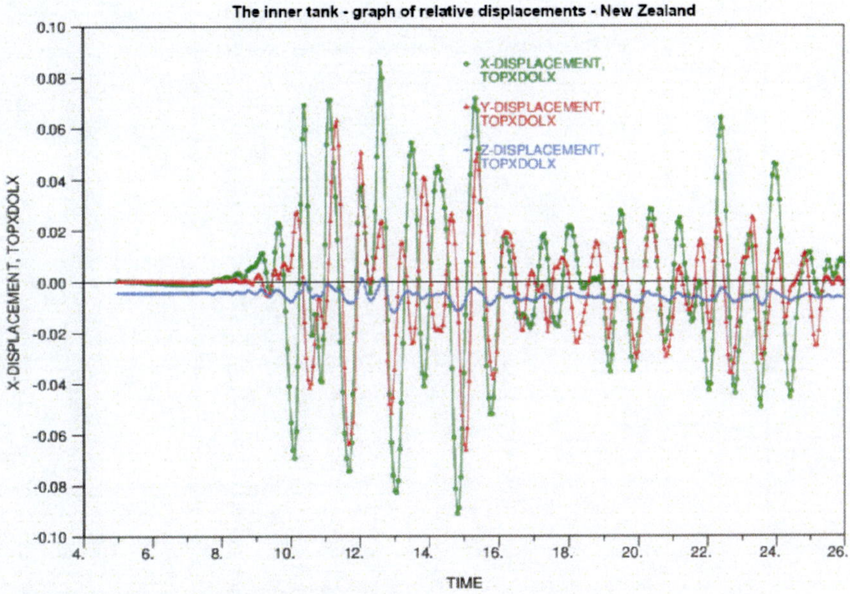

Fig. 3. Relative displacements graph of the inner tank for three directions: x (green line), y (pink line) and z (blue line). *Source* Own elaboration

In the pressure graph, we can observe that values of pressure in the bottom part of the inner tank are higher than in the top part of the inner tank, which is connected with the

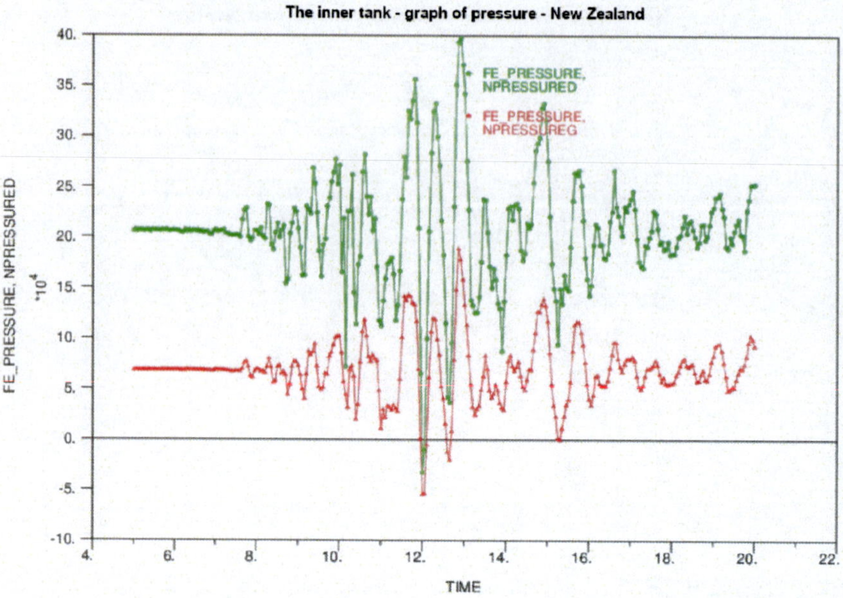

Fig. 4. Pressure graph in the inner tank (in the top part of the tank—pink line, in the bottom part of the tank—green line). *Source* Own elaboration

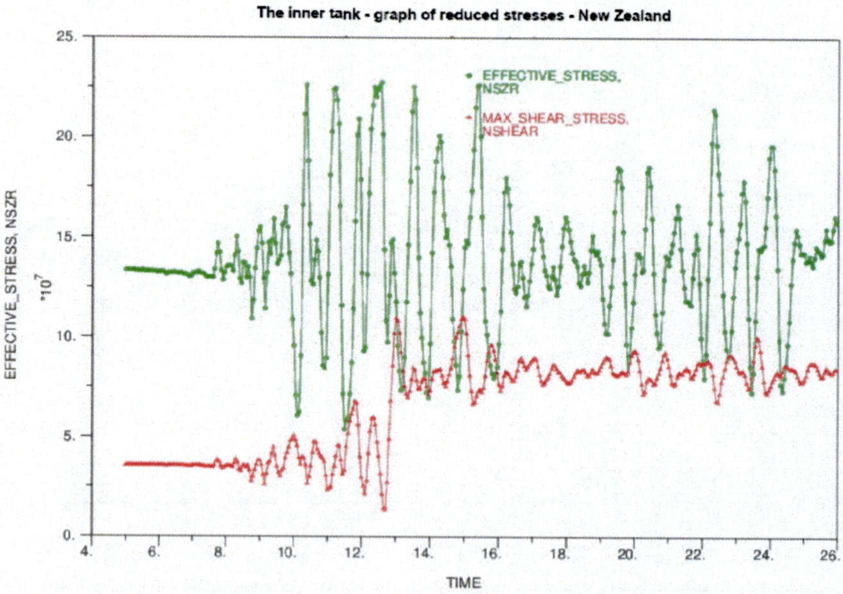

Fig. 5. Stress graph in the inner tank (green line—reduced stresses, pink line—maximum shear stresses). *Source* Own elaboration

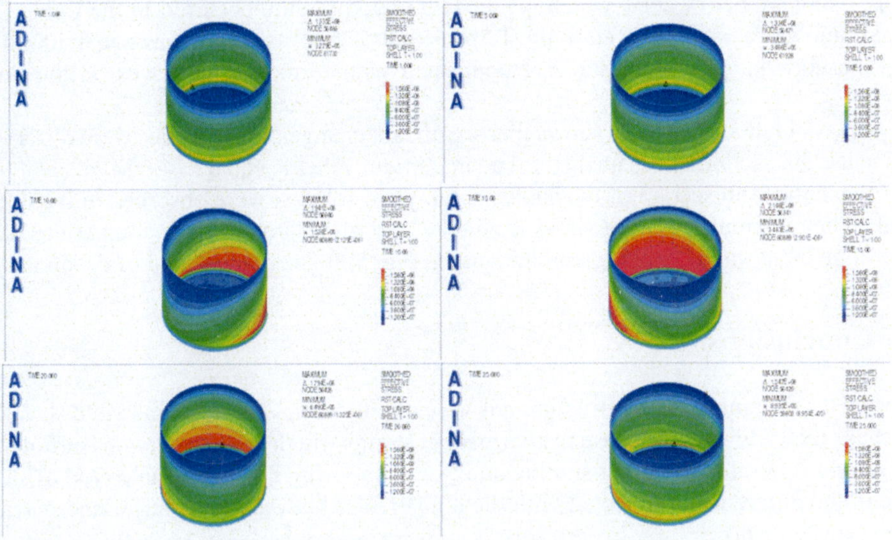

Fig. 6. Maps of reduced stresses of the inner tank in the determined times. *Source* Own elaboration

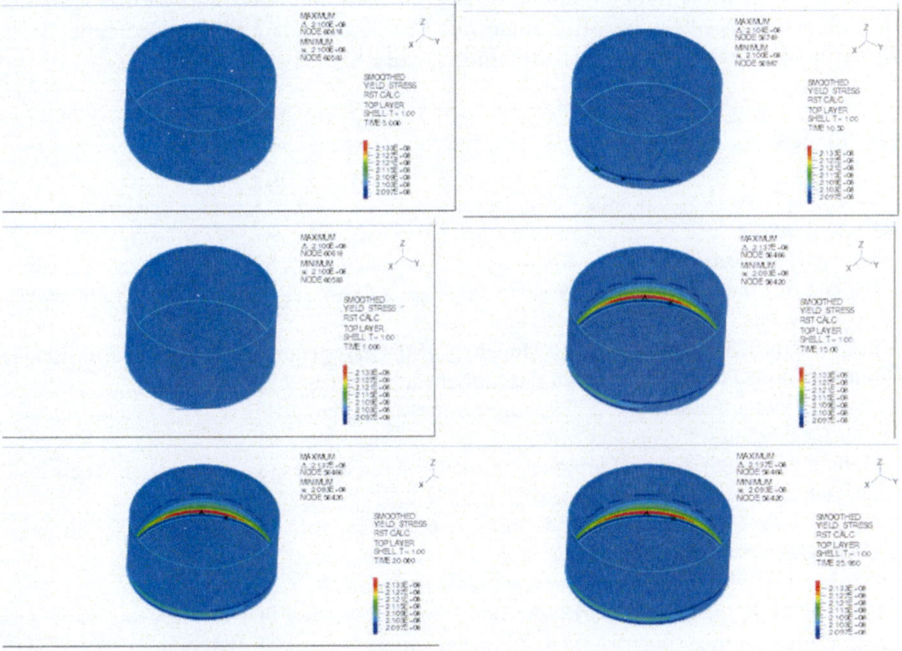

Fig. 7. Maps of yield stresses of the inner tank in the determined times. *Source* Own elaboration

hydrostatic pressure. For the first eight seconds of the analysis, the values of pressure are roughly constant. About the 12th and the 13th second of the analysis, the values of pressure are below zero, which means that underpressure occurred in the fluid. The maximum value was observed in the 13th second of analysis and it was equal 400 kPa. The abrupt changes of pressure are connected with formation of waves on the fluid surface.

Maps of reduced stresses shows that significant changes of the values were observed since the 9th second of the analysis. The maximum value is equal 214.6 MPa.

Since the 10th second of the analysis, the yield stresses were observed in the inner tank. The maximum value of stress is equal 214.6 MPa. This value is compatible with the value in the stress graph (green line on the Fig. 5) in the 10th second of the analysis.

5 Conclusions

During the designing of any structure it is very important not to exceed the yield strength because the yield stresses cause the permanent strains which could be very hazardous to users and structure. The other significant issue is choosing the appropriate type of steel due to the differences connected with the temperature between the interior and outside the cryogenic tank.

Displacements in the z-direction are not equal 0. The nonzero displacements are caused by an isolation layer which was modelled as a three-dimensional calling between the bottom of the inner tank and the bottom of the external tank. This calling is marked by a high elasticity of the isolation material. The other reason of displacements on the z-direction is the weight of the liquid which is very high.

References

1. Jia J (2017) Modern earthquake engineering. Springer-Verlag. https://doi.org/10.1007/978-3-642-31854-2
2. Kobiak J, Stachurski W (1991) Konstrukcje żelbetowe (vol 4, 5th edn). Arkady, Warsaw
3. Halicka A, Franczak D (2013) Projektowanie zbiorników żelbetowych (Vol 2) Zbiorniki na ciecze, 1st edn. PWN, Warsaw
4. Szozda R (2023) Graduate work "Modeling of the cryogenic tank to warehouse Liquefied Natural Gas (LNG) in the event of the earthquake", Rzeszow
5. PN-EN 1992-1-1 Eurocode 2 "Design of concrete structures" Part 1-1: General rules and rules for buildings, European Committee for Standardization
6. Mohraz B, Sadek F (2003) Earthquake ground motion and response spectra. In: The seismic design handbook. Springer Science, Chapter 2
7. ADINA User Interface, Commands Reference Manual, vol I. ADINA Solids & Structures Model Definition, Watertown
8. Bathe K-J (1996) Finite element procedures. Prentice Hall, New Jersey
9. Cimellaro GP, Marasco S (2018) Introduction to dynamics of structures and earthquake engineering. Springer International Publishing. https://doi.org/10.1007/978-3-319-72541-3
10. Database Strong-motion virtual data center (VDC). https://strongmotioncenter.org/vdc/scripts/earthquakes.plx, last access: 2023/10/05

Effect of Chemical Modification of the Coal Fly Ash onto Adsorption of Chromium(III) Ions in the Presence of Cobalt(II) Ions in a Single and Binary System

Eleonora Sočo[1(✉)] and Andżelika Domoń[2]

[1] Department of Inorganic and Analytical Chemistry, Faculty of Chemistry, Rzeszow University of Technology, 35-959 Rzeszow, Poland
eleonora@prz.edu.pl

[2] Department of Water Purification and Protection, Faculty of Civil, Environmental Engineering and Architecture, Rzeszow University of Technology, 35-959 Rzeszow, Poland

Abstract. The purpose of the research was to determine the effect of chemical modification of waste coal fly ash (FA) with solutions of $HClO_4$, KOH, sodium diethyldithiocarbamate (NaDDTC), and ammonium formate ($HCOONH_4$) on the adsorption of chromium(III) ions in the presence of cobalt(II) in a single and binary system. In the study, the morphology of ash grains before and after modification was determined using scanning electron microscopy (SEM-EDS), mineral composition using X-ray diffraction (XRD), and chemical properties of sorbents using Fourier transform infrared spectroscopy (FT-IR). Evaluation of the sorption properties of the tested materials was determined using 8 models of adsorption isotherms. The adsorption of Cr(III) and Co(II) ions in a one-component system with FA, FA-NaDDTC, and FA-KOH is well described by the Langmuir and Redlich-Peterson models. The highest value of the maximum sorption capacity was obtained for FA-KOH (235 $mg{\cdot}g^{-1}$ and 65 $mg{\cdot}g^{-1}$ for Cr(III) and Co(II) ions, respectively). Studies have shown that FA-KOH has a higher selectivity for Cr(III) ions, which is related to the size of the radius of the hydrated metal ion and the value of the first equilibrium constant of the hydrolysis reaction. The obtained adsorption kinetic data were well expressed by the pseudo-second-order model ($R^2 = 0.998$) but showed a very poor fit to the pseudo-first-order model ($R^2 < 0.680$). The Elovich model and the intraparticle sorption model were also analyzed, which showed that the sorption process of Cr(III) and Co(II) ions is controlled by the diffusive boundary layer and diffusion inside the pores.

Keywords: Coal fly ash · Heavy metals · Adsorption · Isotherms · Kinetic

1 Introduction

Production of electricity from solid minerals is associated with the production of slag and fly ash in combined heat and power plants, power plants, and boiler houses. The generated waste poses a serious threat to the natural environment and human health,

Z. Blikharskyy et al. (Eds.): CEE 2023, LNCE 438, pp. 427–439, 2024.
https://doi.org/10.1007/978-3-031-44955-0_43

especially due to its dustiness and emissions of soluble components to groundwater [1, 2]. The storage of combustion by-products on heaps or in wet settling tanks is particularly troublesome and expensive, mainly due to the necessary safeguards to prevent dust from being blown away and soil insulation [3, 4]. To limit the surface storage of ashes, new possibilities for their economic and industrial use are still being sought. However, this requires a good knowledge of the physical and chemical properties of the ashes and their susceptibility to further conversion [3]. Increasing environmental pollution associated with landfills and the migration of mobile forms of toxic substances results in the contamination of waters and land, thus causing irreversible changes in the ecosystem [5]. Therefore, the development of methods that will ensure a balance between further economic development and environmental protection becomes a priority. One of the ways of purifying wastewater from harmful compounds is the use of various sorbents. Waste coal ash seems to be an excellent material for this purpose [6, 7]. Fly ash characterized by a zeolite structure (containing significant content of silicon and aluminum oxides) has good sorption properties towards heavy metal ions.

The increasing demand for thermal energy led to an increase in the use of coal, which in turn contributed to the production of large amounts of fly ash (a waste product) [1]. Therefore, the increasing amount of waste ashes creates the need to continuously investigate and solve various environmental problems that arise due to the unused surplus of this waste product. Despite the use of a significant proportion of fly ash in building construction or agro-technical industry, there is still a large proportion of unmanaged waste in landfills that directly impacts the environment. This type of waste management is uneconomical, moreover, the accumulated heaps after drying show strong dusting and emissions of soluble components to groundwater, causing damage to the biosphere. They have an adverse effect on plants, i.e. they reduce the intensity of assimilation and thus slow down plant growth [9, 10].

Therefore, this work is devoted to obtaining information on the adsorption efficiency of heavy metal ions by chemically modified coal fly ash. In the study, laboratory adsorption experiments were performed to study adsorption equilibrium and adsorption kinetics in a single and binary system.

2 Method

2.1 Subject of Study

The fly ash used in the research came from the Rzeszów-Załęże substation in the south-eastern part of Poland. The tested ash was a powdered residue obtained as a result of fluidized bed combustion of hard coal in a furnace with an exhaust gas temperature of 1200 °C. The chemical conversion of fly ash was carried out in an apparatus with temperature programming in the range of 25 to 450 °C (model DK6) with a set of coolers (model TMD6 by VELP). Oxidation was carried out by treating the ash sample with a solution of 2 M $HClO_4$, 4 M KOH, 2 M NaDDTC, and 1 M $HCOONH_4$ in a weight ratio of fly ash to the solution of 1:10. The ash suspension was heated at 100 °C for 12 h. After the modification process, the solution was decanted and the remaining sorbent was washed with deionized water until the pH of the water was 6.7. Then, the ash sample was dried in an electric dryer at 110 °C until a constant weight was obtained.

2.2 Analysis of Physicochemical Properties of Fly Ash

FT-IR analysis. FT-IR spectra of fly ash samples before and after chemical modification were obtained using a Fourier transform infrared spectrometer (Alpha model by Bruker). The tests were carried out using the transmission method using the technique of pressing samples under pressure with potassium bromide. The compressed sorbent samples were mixed with KBr at a constant ratio of sorbent mass to KBr mass equal to 0.25%.

Chemical composition of fly ash. The content of SiO_2, Al_2O_3, Fe_2O_3, K_2O, MgO, CaO, and Na_2O was determined by the Polish Standard PN-G-04528. Silica (SiO2) content was determined by weight using hydrofluoric acid. To determine Al_2O_3, Fe_2O_3, K_2O, MgO, CaO, and Na_2O, the ash sample was dissolved in a mixture of concentrated sulfuric(VI), nitric(V), and hydrofluoric acid. Determination of individual components was performed by flame atomic absorption spectrometry (FAAS) using a PERKIN-ELMER model 3100 spectrometer.

Wide-angle X-ray diffraction (XRD) analysis. The mineral composition of the ash before and after the oxidation process was assessed using an X-ray diffractometer (BRUKER, type D2-phaser) with CuKα radiation ($\lambda = 0.15406$ nm) in the scanning range from 5 to 100°2Θ, measuring step 0.02°2Θ and pulse counting time 1 s/step. The apparatus was equipped with a graphite reflective monochromator with a tube voltage and current of 30 kV and 10 mA, respectively.

SEM-EDS analysis. The morphology and chemical composition of the tested sorbents was performed using a scanning electron microscope (SEM) equipped with a scintillation photoelectron multiplier EDS (Energy Dispersive Spectroscopy) (VEGA 3, Tescan).

2.3 Analysis of the Effectiveness of the Sorption Process

50 cm^3 of a solution of metal ions of various concentrations (10–500 mg/dm^3) was introduced into 100 cm^3 of conical glass flasks. Then 0.5 g of FA/FA-modified was added to each flask and stirred until equilibrium was reached. The resulting suspension was centrifuged and filtered, and the supernatant was assayed for Cr(III) and Co(II) content. The concentration of Cr(III) and Co(II) after the sorption process was determined using a flame atomic absorption spectrometer (FAAS) PERKIN-ELMER model 3100.

2.4 Adsorption Isotherms

The following empirical isotherms were used to assess the fly ash sorption properties in the single system: Freundlich, Langmuir, Redlicha-Peterson, Jovanovic, Tóth, Elovich, Halsey, and Brunauer-Emmett-Teller (Table 1). Sorption properties in a binary system were determined based on extended Langmuir, extended Langmuir-Freundlich, and Jain-Snoeyink isotherms (Table 2).

The estimations of the tested isotherms to the experimental points were carried out based on the reduced chi-square test (χ^2/DoF) and the determined coefficient of determination (R2). The reduced chi-square test determines the compliance of the examined variable distribution with its theoretical distribution. Deviations between experimental data and model data were determined based on the following equation:

$$\chi^2/\mathrm{DoF} = \frac{1}{d}\sum_{i=1}^{N}\frac{\left(q_{ei}\ q_{ei,m}\right)^2}{q_{ei,m}} \tag{1}$$

where:

q_e—adsorbate content in the solid phase determined experimentally [mg/g],
$q_{e,m}$—adsorbate content in the solid phase obtained from model calculations [mg/g],
d—number of degrees of freedom.

2.5 Adsorption Kinetics

To investigate the mechanism and determine the rate of the process controlling the sorption stage of Cr(III) and Co(II) ions, 4 kinetic models were used. Process rate constants were calculated using pseudo-first-order, pseudo-second-order, Elovich, and Weber-Morris kinetic models (Table 3).

3 Results

3.1 Chemical Composition of the Initial Fly Ash

Determination of the chemical composition of fly ash showed that it was dominated by SiO_2, Al_2O_3, and Fe_2O_3, the other components were: K_2O, MgO, CaO, and Na_2O. The SiO_2:Al_2O_3 ratio is 1.54, and the sum of the main oxides indicates that the tested ash is classed as F 1 according to ASTM C618–89. However, according to the chemical classification, fly ash belongs to aluminate-silicate ($SiO_2:Al_2O_3 < 2.0$, $CaO \leq 15.0\%$) and is the result of conventional combustion.

3.2 FT-IR Analysis

The conducted FT-IR spectroscopic studies showed that in all spectra a broad absorption band was observed in the region of approximately 3440 cm^{-1}, 1630 cm^{-1}, and 1050 cm^{-1} (Fig. 1). The band at 3440 cm^{-1} is attributed to the stretching vibrations of the –OH group. Because the spectra were recorded in the air atmosphere, it is impossible to make an unambiguous interpretation, i.e. assigning the band to a specific type of vibrations occurring in the structure. It may come from both the –OH group, which was chemically bound to the ash surface, as well as from the H–O–H stretching vibrations of water molecules adsorbed on the ash surface. The very weak band at 2900 cm^{-1} is attributed to the stretching vibrations of the –OH group bound to the methyl radical.

In the case of the band in the region of about 1630 cm^{-1}, it can be stated that this band is characteristic of the O–H stretching vibrations of water adsorbed on the ash surface [11]. The next band around 1450 cm^{-1} can be attributed to the CH structure in aromatic systems and the formation of carboxylic-carbonate structures. The most intense band in the analyzed spectra was found in the region of about 1050 cm^{-1}, which can

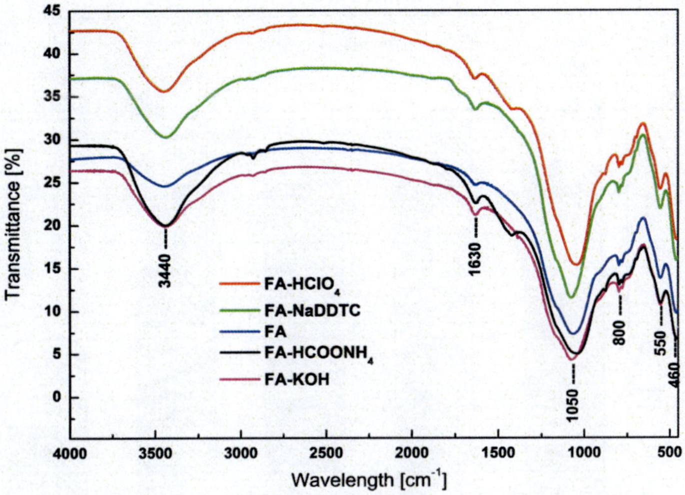

Fig. 1. FT-IR spectra of fly ash samples before and after modification.

be attributed to asymmetric stretching vibrations of Si-O-Si(Al) bridges, the presence of which is associated with the aluminosilicate skeleton [11].

In the spectra of samples after the process of chemical modification with $HClO_4$, an increase in band intensity in the region of 3440 cm^{-1} is additionally observed. This proves the increase in the hydrophilicity of the oxidized fly ash samples about the initial ash sample. The band at 800 cm-1 is associated with the presence of $SiO_4{}^{4-}$ and $AlO_4{}^{5-}$ in tetrahedral aluminosilicate systems. The intensity of the bands at 550 cm^{-1} and 460 cm^{-1} is related to the Si-O-Al and Si-O-Si deformation vibrations, respectively [12].

Modification of the fly ash with KOH resulted in changes in the IR spectra compared to the spectrum of the initial ash sample. After the modification, a significant increase in permeability was observed in the entire recorded spectral range and an increase in the intensity of the absorption bands in comparison with the unmodified sample. The increase in sorption after modification can be attributed to the formation of basic surface groups. The resulting hydroxyl ions on the surface of the adsorbent give rise to the sorption of both cations (Cr(III) and Co(II)).

3.3 Effect of Fly Ash Modification on Adsorption Process Efficiency

The effectiveness of the adsorption process of Cr(III) and Co(II) ions on modified fly ashes is shown in Fig. 2. Based on the tests carried out, it was found that the ash modified with NaDDTC and $HClO_4$ achieved worse results than the initial fly ash. FA-KOH has the best sorption properties (Cr(III) and Co(II) ions were removed at the level of 85% and 75%). For this reason, an ash sample was taken into account for further research.

SEM microscopic observations allowed to characterize the morphology of FA and FA-KOH. Based on the obtained images, the presence of microparticles in the shape of

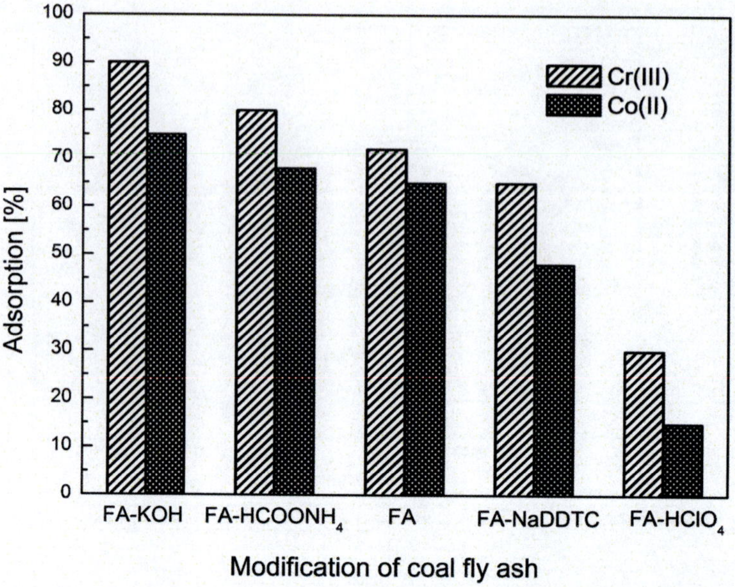

Fig. 2. Effect of flu ash modification on Cr(III) and Co(II) ion adsorption efficiency.

smooth spheres, microspheres, and plenospheres of various sizes was found. Modification of the fly ash with KOH solution caused changes in the structure of the ash (the process of crystallization was observed (Fig. 3). The elemental composition of the fly ash before and after KOH conversion was analyzed using the EDS detector (Fig. 4).

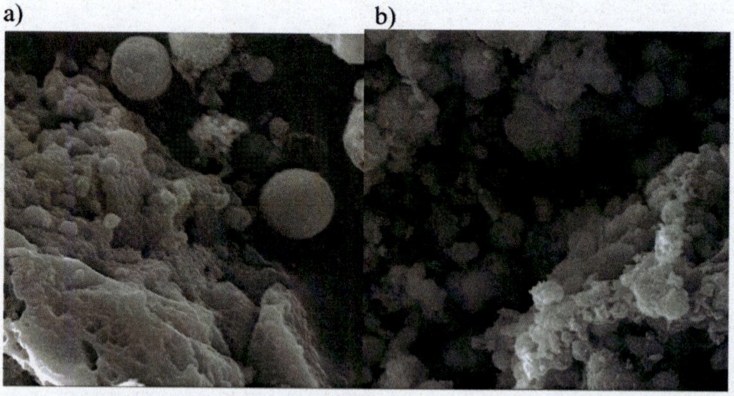

Fig. 3. SEM analysis of **a** initial fly ash, **b** fly ash after the KOH modification.

To determine the mineral composition of the sample of initial ash and modified ash (Fig. 5), the locations of reflections characteristic of the crystalline phases present in the material were determined and then compared with the values of standards from the files. The obtained results of the analyzed sorbents showed the presence of quartz, magnetite,

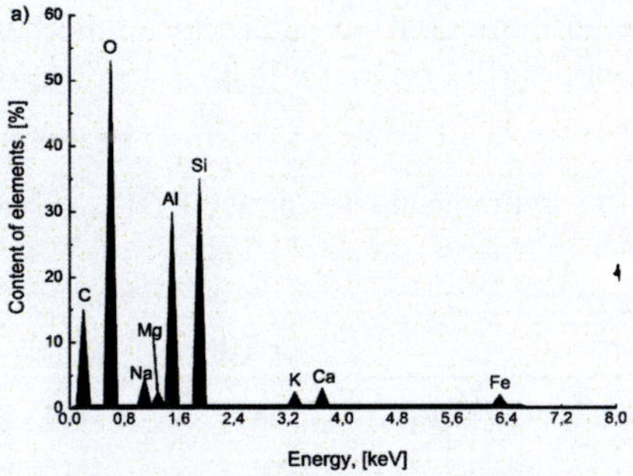

Fig. 4. EDS spectrum of **a** initial fly ash, **b** fly ash after the KOH modification.

and kaolinite, while after KOH modification, the presence of crystalline phases of sodalite and X-type zeolite was also found.

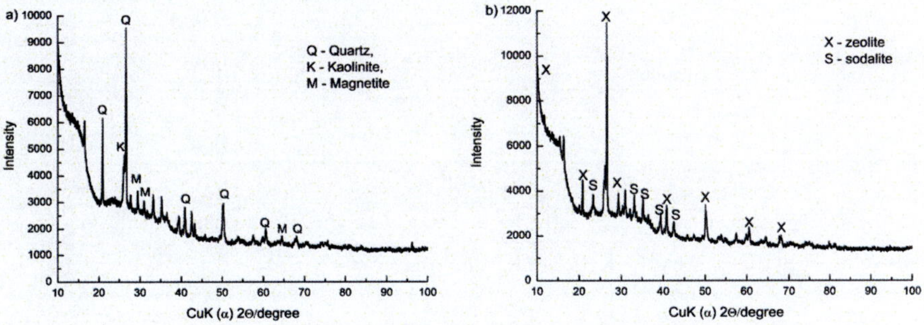

Fig. 5. Diffractogram of **a** initial ash, **b** ash after the KOH modification process.

3.4 Models of Sorption Isotherms in a Single System

Isotherms define the balance between the adsorbate concentration in the solid phase and its concentration in the liquid phase. Based on the course of isotherms, we can obtain information on the maximum sorption capacity. In addition, they provide information on the power of the sorbent and the amount of sorbent required to remove a unit mass of pollutants in the tested conditions [13, 14]. Experimental data were processed using various models of sorption isotherms in the solid-liquid system. The results are summarized in Table 1.

Based on the results summarized in Table 1, it was found that the sorption of Cr(III) and Co(II) in the single system on FA and FA-KOH is well described by the Langmuir and Redlich-Peterson models. In the case of Cr(III) ions, this is indicated by the lowest value of the error determined using the chi-square test ($\chi 2/\text{DoF} < 0.2$) and a high value of the coefficient of determination ($R^2 = 0.998$). The values of the maximum saturation of the sorbent surface determined based on the Langmuir isotherm for Cr(III) and Co(II) ions are 235 and 65 mg/g, respectively.

The dimensionless parameter n in the Freundlich equation determines whether the sorption process is efficient. Its values calculated based on static tests are in the range

Table 1. Parameter values of Cr(III) and Co(II) ion sorption isotherms on FA-KOH sample in a single system.

Isotherm	Parameter	Cr(III)	Co(II)
Freundlich $q_e = K_F C_e^{1/n}$	K_F $(mg^{1-1/n} \cdot dm^{3/n})/g$	2.4	18.4
	n	2.2	1.3
	χ^2/DoF	2.0	15.5
	R^2	0.987	0.963
Langmuir $q_e = q_{max} \frac{K_L C_e}{1+K_L C_e}$	K_L dm^3/mg	0.52	0.06
	q_{max} mg/g	235	65
	χ^2/DoF	0.17	13.7
	R^2	0.998	0.974
Redlich-Peterson $q_e = \frac{K_{RP} C_e}{1+a_{RP} C_e^B}$	K_{RP} dm^3/g	0.83	0.56
	a_{RP} $(dm^3/mg)^B$	0.13	0.05
	B	0.56	0.01
	χ^2/DoF	0.02	18.2
	R^2	0.998	0.855
Jovanović $q_e = q_{max}(1 - \exp(-K_J C_e))$	K_J dm^3/mg	0.12	0.19
	q_{max} mg/g	23.1	16.8
	χ^2/DoF	2.4	19.2
	R^2	0.996	0.955
Tóth $q_e = \frac{q_{max} K_{Th} C_e}{\left[1+(K_{Th} C_e)^{Th}\right]^{\frac{1}{Th}}}$	K_{Th} $(dm^3/mg)^{Th}$	0.05	0.01
	q_{max} mg/g	9.8	5.3
	Th	0.78	0.45
	χ^2/DoF	0.09	21.5
	R^2	0.997	0.936
Elovich $q_e = q_{max} K_E C_e \exp\left(-\frac{q_e}{q_{max}}\right)$	K_E dm^3/mg	0.25	0.95
	q_{max} mg/g	6.2	3.3
	χ^2/DoF	36.5	33.2
	R^2	0.589	0.689
Halsey $q_e = (K_H/C_e)^{1/n}$	K_H $mg^{n+1}/(g \cdot dm^3)$	0.09	0.2
	n	2.1	0.8
	χ^2/DoF	79.3	96.5
	R^2	< 0.2	< 0.2

(*continued*)

Table 1. (*continued*)

Isotherm	Parameter	Cr(III)	Co(II)
Brunauer-Emmett-Teller	K_{BET}	2.36	3.56
$q_e = q_{max}\frac{K_{BET}\frac{C_e}{C_0}}{\left(1-\frac{C_e}{C_0}\right)\left[1+(K_{BET}-1)\frac{C_e}{C_0}\right]}$	q_{max} mg/g	8.5	6.3
	χ^2/DoF	98	101
	R^2	0.856	0.756

where, C_e—equilibrium concentration [mg/dm^3], C_0—initial concentration [mg/dm^3], q_e—sorbate content in the solid phase [mg/g], q_{max}—maximum occupation of sorbate active centers in the solid phase [mg/g], Ki, a_{RP}, n, Th, B—constants of the tested sorption isotherms

of $1 < n < 10$. The inverse of the n parameter characterizes the sorbent surface, i.e. it informs about the degree of diversity of sorption sites on the sorbent surface (the parameter assumes the values $0 < 1/n < 1$). The closer the 1/n value is to zero, the more heterogeneous the sorbent surface is, while for values close to 1 it is defined as homogeneous. In the experiment, the value of n determined for Cr(III) and Co(II) ions is 2.2 and 1.3, respectively, and 1/n 0.45 and 0.77. Based on the obtained results, it can be concluded that the surface of FA-KOH has more available active sites for Cr(III) ions than for the larger Co(II) cation. The FA-KOH sorbent has a higher selectivity for Cr(III) ions than for Co(II), which is related to the size of the radius of the hydrated Cr(III) and Co(II) ions. An analogous relationship is presented by the parameter B in the Redlich-Peterson equation, which is related to the parameter 1/n in the Freundlich equation: $1/n = 1-B$. The sorption of Cr(III) ions using FA-KOH is very well described by the Redlich-Peterson model (χ^2/DoF = 0.05; $R^2 = 0.998$), for which the parameter B is 0.56, which also indicates a large number of active sites on the surface of the sorbent for ions Cr(III). The other isotherm models tested showed a larger error in the calculated values in relation to the experimental data. A weak fit of the models to the experimental values was shown by the Elovich, Halsey, and Brunaumer-Emmett-Teller isotherms, for which the multilayer sorption was not met in the case of the description of the surface phenomenon.

3.5 Models of Sorption Isotherms in a Binary System

For two-component mixtures, isotherm models allow to direct calculation of the concentrations of adsorbed components based on the knowledge of isotherms in a single system [13, 15]. The total adsorbed amount can be determined based on the sum of the concentrations of the adsorbed components (Table 2).

The analysis of sorption isotherm parameters using FA-KOH in the Cr(III)-Co(II) binary system showed that the extended Langmuir model has the best fit of the calculated values to the experimental data for both sorbates (Table 2). The maximum occupation of the sorbent surface determined based on the extended Langmuir model for Cr(III) and Co(II) ions is 56.2 and 23.5 mg/g, respectively.

Table 2. Parameter values of Cr(III) and Co(II) ion sorption isotherms on FA-KOH sample in a binary system.

Isotherm	Parameter	Cr(III)	Co(II)
Extended Langmuir $q_{ei} = q_{max} \frac{K_{Li}C_{ei}}{1+\sum_{j=1}^{N} K_{Lj}C_{ej}}$	K_{Li} dm^3/mg	0.36	0.26
	q_{maxi} mg/g	56.2	23.5
	χ^2/DoF	1.1	3.6
	R^2	0.989	0.986
Extended Langmuir-Freundlich $q_{ei} = q_{max} \frac{(K_{Li}C_{ei})^{1/n_i}}{1+\sum_{j=1}^{N} (K_{Lj}C_{ej})^{1/n_j}}$	K_{Li} dm^3/mg	0.12	0.25
	q_{maxi} mg/g	28.5	9.3
	n_i	0.12	0.09
	χ^2/DoF	44.6	54.2
	R^2	0.598	0.469
Jain-Snoeyink $q_{el} = (q_{max\,1} - q_{max\,2})\frac{K_{L1}C_{el}}{1+K_{L1}C_{el}} + q_{max\,2}\frac{K_{L1}C_{el}}{1+K_{L1}C_{el}+K_{L2}C_{e2}}$	K_{Li} dm^3/mg	0.56	0.23
	q_{maxi} mg/g	45.4	10.2
	χ^2/DoF	45.9	69.5
Where: $q_{max1} > q_{max2}$ $q_{e2} = q_{max\,2}\frac{K_{L2}C_{e2}}{1+K_{L1}C_{el}+K_{L2}C_{e2}}$	R^2	0.478	0.596

Based on the obtained results of the studied isotherm models in one- and two-component systems, a high competitiveness of the sorption process of smaller Cr(III) ions was found in comparison to the larger Co(II) cation.

3.6 Study of the Kinetics of the Sorption Process

The initial sorption rates for Cr(III) and Co(II) ions were calculated based on pseudo-second order process rate constants of 35 and 17 [mg/(g•min)], respectively, indicating a higher rate of Cu(III) ion sorption compared to the Co(II) ion on FA-KOH (Table 3).

The Weber and Morris model of intragrain diffusion shows that there are two separate steps in the sorption process such as external diffusion and intergranular diffusion. The non-linear course of the sorption process indicates the multi-stage sorption of Cr(III) and Co(II) ions using FA-KOH. The sorption step for heavy metals on porous sorbents is generally very fast compared to the external or internal diffusion step. It is known, however, that sorption equilibrium can be reached within a few minutes in the absence of internal diffusion [16]. The total kinetics of Cr(III) and Co(II) ion sorption was described by a pseudo-secondary model [16]. However, it was not possible to identify the rate-limiting stage of the sorption process. The rate-limiting factor (the slowest stage of the process) may be the boundary layer (film) or diffusion inside the particles (pores) of the solid surface, substances dissolved in the solution. In the case of the obtained results, the long time of reaching sorption equilibrium (approx. 1.5 h) indicates that internal diffusion may dominate over the general sorption kinetics.

Table 3. Constants of kinetic models and linear correlation coefficients for the studied sorption system.

Kinetic model	Parameter	Cr(III)	Co(II)
Pseudo-first order $dq_t/dt = k_1(q_e - q_t)$	R^2 k_1 [1/min]	0.678 $2.1 \cdot 10^{-3}$	0.659 $6.3 \cdot 10^{-3}$
Pseudo-second order $dq_t/dt = k_2(q_e - q_t)^2$	R^2 k_2 [g/(mg·min)]	0.998 0.35	0.998 0.17
Elovich $dq_t/dt = \alpha \exp(-\beta q_t)$	R^2 α [mg/(g·min)] β [g/mg]	0.898 21.3 0.75	0.796 17.5 0.42
Weber-Morris $q_t = k'_i t^{1/2} + b_i$	R^2 k'_1 [mg/(g·min$^{1/2}$)] b_1 [mg/g]	0.997 0.45 0.96	0.999 0.78 0.51
	R^2 k'_2 [mg/(g·min$^{1/2}$)] b_1 [mg/g]	0.998 6.5·10–5 5.1	0.995 4.1·10–5 5.5

where, q_t—the amount of adsorbed substance at time t [mg/g], q_e—the content of adsorbate in the solid phase at equilibrium [mg/g], k_1—rate constant of the pseudo-first-order kinetic equation, k_2—rate constant of pseudo kinetic equation second-order; k'_i—internal diffusion rate constant; b_i—layer (film) thickness; α, β—constants of the Elovich equations

4 Scientific Novelty and Practical Significance

Adsorption is probably one of the most attractive and effective separation processes, and due to its simplicity and efficiency, it is widely used in industry [13, 17–20]. Even though commercial activated carbon has a large specific surface area, porous form, and high adsorption capacity for heavy metals from industrial wastewater, it is currently not recommended as an adsorbent due to high operating costs. On this basis, there is a growing need to find a cheap and efficient adsorbent to remove heavy metal ions from aqueous solutions [5, 8]. Due to its high efficiency, sorption is one of the most popular methods of eliminating pollutants from industrial wastewater [13, 21]. The development of methods for the industrial use of fly ash could solve the problem of storing waste from the power industry, and help in getting rid of pollutants from water reservoirs, i.e. using one waste to get rid of another [1].

5 Conclusions

The chemical conversion of fly ash showed a significant effect on the sorption capacity of the sorbent towards Cr(III) and Co(II) ions. The best results were obtained for fly ash modified with KOH. The sorption equilibrium was well described in the single system by the Langmuir and Redlich-Peterson isotherm models, and in the binary system by the extended Langmuir model. The maximum sorption capacity for FA-KOH in the single-component system was 235 and 65 mg/g, respectively for Cr(III) and Co(II) ions, while in the binary system, it was 56.2 and 23.5 mg/g, respectively. The sorption kinetic data obtained were well expressed by a pseudo-second-order model. The Elovich model and the intragranular sorption model were analyzed, which showed that the sorption process of Cr(III) and Co(II) ions is controlled by the diffusion boundary layer and diffusion inside the pores. Studies have shown that FA-KOH has a higher selectivity for Cr(III) ions than for Co(II) ions, which is related to the size of the radius of the hydrated metal ion and the value of the first equilibrium constant of the hydrolysis reaction. It was found that the chemical conversion of fly ash with FA-KOH effectively increases its sorption capacity and becomes an attractive material for wastewater treatment containing Cr(III) and Co(II) ions.

References

1. Ruhl L, Vengosh A, Dwyer GS, Hsu-Kim H, Deonarine A (2010) Environmental impacts of the coal ash spill in Kingston, Tennessee: an 18-month survey. Environ Sci Technol 44:9272–9278
2. Dai S, Ren D, Chou C-L, Finkelman RB, Seredin VV, Zhou Y (2012) Geochemistry of trace elements in Chinese coals: a review of abundances, genetic types, impacts on human health, and industrial utilization. Int J Coal Geol 94:3–21
3. Hycnar JJ (2009) EUROCOALASH—weryfikacja popiołów ze spalania węgla. Energetyka-Nauka-Przemysł 62:48–53 (In Polish)
4. Cetin B, Aydilek AH, Guney Y (2010) Stabilization of recycled base material using high carbon fly ash. Resour Conserv Recycl 54:879–892
5. Sočo E, Kalembkiewicz J (2013) Adsorption of nickel(II) and copper(II) ions from aqueous solution by coal fly ash. J Environ Chem Eng 1:581–588
6. Ricou-Hoeffer P, Lecuyer I, Le Cloirec P (2001) Experimental design methodology applied to adsorption of metallic ions onto fly ash. Water Res 35:965–976
7. Sarbak Z, Kramer-Wachowiak M (2002) Porous structure of waste fly ashes and their chemical modifications. Powder Technol 123:53–58
8. Ahmaruzzaman M (2010) A review on the utilization of fly ash. Prog Energ Combust 36:327–363
9. Jiao F, Wijaya N, Zhang L, Ninomiya Y, Hocking R (2011) Synchrotron-based XANES speciation of chromium in the oxy-fuel fly ash collected from lab-scale drop-tube furnace. Environ Sci Technol 45:6640–6646
10. Seredin VV, Finkelman RB (2008) Metalliferous coals: A review of the main genetic and geochemical types. Int J Coal Geol 76:253–289
11. Mozgawa W, Jastrzębski W, Handke M (2006) Cation-terminated structural clusters as a model for the interpretation of zeolite vibrational spectra. J Mol Struct 792–793:163–169
12. Kantiranis N et al (2006) EPI-type zeolite synthesis from Greek sulphocalcic fly ashes promoted by H_2O_2 solutions. Fuel 85:360–366
13. Srivastava VC, Mall ID, Mishra IM (2007) Adsorption thermodynamics and isosteric heat of adsorption of toxic metal ions onto bagasse fly ash (BFA) and rice husk ash (RHA). Chem Eng J 132:267–278
14. Hsu T-C, Yu C-C, Yeh C-M (2008) Adsorption of Cu^{2+} from water using raw and modified coal fly ashes. Fuel 87:1355–1359
15. Kleinübing SJ, Guibal E, da Silva EA, da Silva MGC (2012) Copper and nickel competitive biosorption simulation from single and binary systems by Sargassum filipendula. Chem Eng J 184:16–22
16. Ho Y-S (2009) Kinetics of naphthalene adsorption on an activated carbon: comparison between aqueous and organic media. Chemosphere 76(4):433–438

17. Bedoui K, Bekri AI, Srasra E (2008) Removal of cadmium(II) from aqueous solution using pure smectite and Lewatite S 100: the effect of time and metal concentration. Desalination 223:269–273
18. Ferreiraa TR et al (2009) Cadmium(II) removal from aqueous solution using microporous titanosilicate ETS-4. Chem Eng J 147:173–179
19. Jiang M, Jin X, Lu X, Chen Z (2010) Adsorption of Pb(II), Cd(II), Ni(II) and Cu(II) onto natural kaolinite clay. Desalination 252:33–39
20. Rangel-Porras G, Garcia-Magno JB, Gonzalez-Munoz MP (2010) Lead and cadmium immobilization on calcitic limestone materials. Desalination 262:1–10
21. Cho H, Oh D, Kim K (2005) A study on removal characteristics of heavy metals from aqueous solution by fly ash. J Hazard Mater B127:187–195

Construction safety Risks Management and Construction Site

Marcela Spišáková(✉) and Tomáš Mandičák

Faculty of Civil Engineering, Technical University of Košice, 04001 Košice, Slovakia
marcela.spisakova@tuke.sk

Abstract. Architecture engineering and construction (AEC) represents one of the important industries with regard to the safety risks. Every construction site is a unique place. It is influenced by many factors that need to be monitored and managed by the construction safety management tools. Numerous incidents are reported daily in various sections of construction site. From very beginning of a construction implementation (f.e. excavation works) to final finishing works, construction sites are in proximity to various types of incidents. Different types of construction work require different types of precautionary measures. The identification, assessment and reduction of risks are the most important issues of safety in the workplace, to effectively address health and safety problems. The paper devotes to the analysis of safety risks at the construction site. An analytical knowledge of this issue is used when solving a case study. An object of the case study is an administrative building—Duett Business Residence in Košice, Slovakia. The aim of the case study is the processing of construction safety risks assessment through the point method assessment; and subsequently the proposal of measures to reduce potential safety risks during construction processes.

Keywords: Construction · Construction safety · Risk management · Point method assessment

1 Introduction

Architecture engineering and construction (AEC) is one of the most dangerous industries due to the occurrence of occupational accidents, often fatal accidents. This is confirmed by world and European statistics. According the Eurostat [1], which provided summary data within the European Union (EU), we know that:

- there were 3 355 fatal accidents at work in the EU during 2020, a decrease of 53 deaths compared with the year before,
- in 2020, more than a fifth of all fatal accidents at work (21.5%) in the EU took place within the construction sector,
- In 2020, 70% of all accidents at work in the EU caused wounds and superficial injuries, dislocations, sprains and strains, or concussions and internal injuries [1].

Moreover, AEC is characterized by the following features:

Z. Blikharskyy et al. (Eds.): CEE 2023, LNCE 438, pp. 440–447, 2024.
https://doi.org/10.1007/978-3-031-44955-0_44

- each building and each construction site are unique,
- more contractors are interested in construction project implementation,
- more contractors and subcontractors participate in construction process,
- construction process requires its own production place (construction site),
- construction process has stationary character—people, machines, products „follow" the construction,
- construction process requires many workers—in terms of number of workers, as well as professions,
- main construction works (earthworks, foundations, framework) are influenced by weather conditions,
- relatively long duration of construction—project implementation.

In view of this unfavorable fact, it is necessary to focus the attention of the construction management on risks analysis in construction projects. Construction project risks can occur in every phase of project including design, implementation, operation, renovation and demolition [2]. In general, the risk represents future unpredictable situation that results in a negative consequence [3] or involving exposure to danger. In other words, risks mean an exposure to loss or likelihood of loss multiplied by its probability of occurrence [4]. According ISO standards, risk can be understood as a combination of event´s consequence and probability of its occupancy [5]. Considering that, the risk can be expressed mathematically based the function according to Eq. (1):

$$R = f(P, C), \tag{1}$$

where R is risk, P is the possibility of a certain event occurring, and C is the impact or consequence of the event on the final effect of the particular project.

Risks of a construction projects mainly affect three basic construction parameters (price, time and quality) [6] among other also construction safety [7].

2 Safety Risks on Construction Site

Construction workers face numerous hazards every day, but some are however more frequent and have more serious consequences. Construction injury statistics are often unreliable due to injury underreporting. Most injuries which do not require immediate assistance of a medical professional and more than a few lost work days are seldom reported.

To structure the presentation of those more significant hazards, it was necessary to group them by type of construction workers who were exposed to the hazards, while the hazards to which all workers can be exposed to were grouped by injury type in Table 1. Classification of safety risks according to construction activities was processed on the basis of statistical data on accident rate at the world, European and national level [8–11]. Amongst those accidents with severe consequences, four types stand out by being responsible for more than half of all worker deaths in the construction industry. Those accidents and their percentages are: falls (33.5%), struck by object (11.1%), electrocutions (8.5%) and caught-in/between (5.5%) [8].

Table 1. Classification of safety risks according to construction activity.

Construction activity	Safety risk
Reinforcement works	Tripping and/or foot caught in the slab rebar Cutting and impaling body parts on the protruding bars Power tool accidents (angle grinder) Scratches, cuts, burns and other injuries when preparing or welding the bars for placement
Formwork works	Table saw injuries Hand tools injuries (hits, cuts) Formwork elements overturning and falling on the worker Accidents when dismantling formwork elements
Concrete works	Fall from the working platform when pouring concrete Struck by the concrete pump hose Fall and trip hazards on the placed rebar
Masonry works	Shotcreting accidents Hand tools injuries (hits, cuts, eye injuries)
Falls on the same level	Tripping hazard Slipping hazard Falls while carrying objects
Falls from height	Falls from roof and during high altitude assembly works Falls from mobile scaffolds Falls from fixed building scaffolds Falls through the scaffold Falls from unprotected building edges Falls through openings in the building structure Falls from ladders
Falls of objects from height	Fall of objects from scaffolds Fall of tools and materials during high altitude works Fall of objects carried by a crane Fall of objects from unprotected building edges Fall of objects through openings in the structure
Accidents during transit	Head and other body part injuries while walking through the construction site Struck by working machinery or equipment Injury from the carried object Hit by traffic
Other	Electrocution hazards Power tools hazards Caught in-between hazards Flying particles injuring worker's eyes Trench cave-ins Fall into a trench Fall of objects and machines into the trench

3 Materials and Methodology

The aim of paper is the processing of construction safety risks assessment through the point method assessment; and subsequently the proposal of measures to reduce potential safety risks during construction processes.

3.1 Material of Research

Construction safety risks assessment was processed through a case study. The object of the case study is the "Administrative building Duett" in Košice. Slovakia, which consists of two underground floors and two separate 10-story towers (see Fig. 1). The structural system is a reinforced concrete monolithic structure.

Fig. 1. Visualization of administrative building Duett.

According the possibilities of the construction process development, the construction site is characterized by: (i) limitation for establishing entrances and exits, (ii) lack of storage areas, (iii) a limited area for the development of the internal construction site communication system, (iv) a limited area for the location of service, production and social facilities of the construction site. The project of construction organization proposed measures to eliminate these limitations.

3.2 Methodology of Research

The methods for determining construction safety risks can be qualitative and quantitative [12]. Assessment of construction safety risks during the realization of "Administrative building Duett" was processed by point method assessment [12]. In the risk management methodology, risk is defined as the degree of adverse effects arising from a hazard, taking into account the probability of damage and side effects [5]. It can be expressed by Formula (2):

$$R = Q \times p, \tag{2}$$

where "Q" is probability (probability of an accident) and "p" is consequences (amount of the loss).

The combination of parity parameters and the negative effect determines the value of risk "R". The risk defined by the point method is determined on the basis of numerical risk assessment matrix (see Table 2).

Table 2. Numerical risk assessment matrix.

Probability "Q"	Consequence "p"			
	1	2	3	4
1	1	4	6	12
2	2	7	11	13
3	3	10	15	17
4	5	12	16	19
5	8	14	18	20

The probability value "Q" in case study vary from 1 to 5, where 1 means a very low probability, risk is almost excluded, accident is almost impossible; and 5 means a high probability, the risk is very frequent, accident is continuous. The consequences value "p" in case study vary from 1 to 4, where means negligible consequence; 4 means fatal consequence.

The result of using formula 2 can be summarized in Table 3, which describes the degree of risk of individual construction activities.

Table 3. Level of risk "R".

"R" value	Characteristics
1–3	Acceptable—system is safe, common procedures
4–11	Moderate—system is safe with training of attendants, tours, etc.
12–15	Unwanted—the application of protective measures
16–20	Unacceptable—the system is unacceptable: immediate application of corrective and protective measures

The biggest (unacceptable—value 20) risk is characterized by a very high frequency, dangerous consequence and a very high probability of occurrence of the event, The most favorable condition has value of 1.

4 Results and Discussion

The construction safety risk management is aimed at assessing risks and taking corrective measures. The level of risk can be in different degrees: from acceptable to unacceptable.

The source of safety risks occurrence during the construction of the Administrative building Duett were divided into 21 groups according to the construction activity. These activities included, for example: concrete work, masonry work, work at heights, demolition work, excavation, etc.

The degree of safety risk was determined for each source of danger through the point method of safety risk assessment; subsequently safety measures were identified. Construction safety risks assessment in case study was processed in tabular form (see Table 4). The values of probability (Q) and consequences (p) were determined by a group of 24 experts dealing with the management of safety risks at the construction site. They defined "Q" and "p" values for each safety risk. The resulting value was determined based on the arithmetic mean.

Table 4. Construction safety risks assessment for Administrative building Duett.

Source	Safety hazard	Risk	Risk level			Safety measure
			Q	p	R	
Concrete works	Formwork and removal, damaged joints Formwork, unsuitable release agent	Fall of formwork elements on the worker	2	3	11	Provide workers the shoes, clothes, gloves, helmets and ensure their correct use Comply with technological procedures during the assembly of formwork Ensure the transport of large formwork elements by a crane
Concrete works	Movement and work at height	Falling from a height when handling formwork, reinforcing and concreting	2	3	11	Provide workers the shoes, clothes, gloves, helmets and ensure their correct use Ensure safe access and workplaces, establish auxiliary work floors, ensure installation of handrails

Note Q—probability, p—consequences, R—risk

Table 4 provides an example of processing the construction safety risks assessment for Administrative building Duett for concrete works. Example shows that concrete work during the realization of a monolithic reinforced concrete building represents a moderate level of risk (R = 11), which is mainly associated with fall of formwork elements on the worker and falling from a height when handling formwork, reinforcing and concreting. The safety measures are also proposed, which are aimed at providing to workers the shoes, clothes, gloves, helmets and ensure their correct use, complying with technological procedures during the assembly of formwork, ensuring the transport of large formwork elements by a crane, and ensuring safe access and workplaces, establish auxiliary work floors, ensure installation of handrails.

A similar analysis and evaluation was carried out for 21 groups of construction activities, which represent the degree of safety risk. Processes that reach a moderate level of risk ($R = 7$) were identified:

- excavation work—unwanted interruption of underground lines,
- work with hand tools—flying of broken parts,
- mobile ladders—unwanted contact with parts of the ladder, unwanted movement of the ladders´ part,
- influences of weather and natural.

Other processes also reached a moderate risk level, but value $R = 11$.

Construction site management is mainly focus on management and monitoring of three basic parameters of construction—time, costs and quality [13, 14]. Document that deals with safety on the construction site is "Plan of occupational safety and health". According the long-term research (1996–2018) focused on processing the construction management documents emerged that the documentation "Plan of occupational safety and health" processing rate has an increasing trend [15]. This can contribute to the reduction of safety risks on the construction site.

5 Conclusion

The paper points to the need to solve construction safety risks on construction. Co Architecture engineering and construction is a dangerous industry in terms of the occurrence of fatal accidents. Reducing the risk of occupational accidents is possible through the analysis, assessment, monitoring and management of safety risks and the adoption of safety measures. Nowadays, we also know many sophisticated approaches that can be used. They are primarily based on information cloud modeling of buildings or using elements of virtual, mixed and augmented immersive reality.

Acknowledgments. Paper presents a partial research results of projects: project KEGA 009TUKE-4/2022 An interactive tool for designing a safe construction site in an immersive virtual reality; project VEGA 1/0336/22 Research on the effects of Lean Production/Lean Construction methods on increasing the efficiency of on-site and off-site construction technologies.

References

1. Accidents at work statistics. https://ec.europa.eu/eurostat/statistics-explained/index.php?title=Accidents_at_work_statistics#Analysis_by_activity, last accessed 2023/05/17

2. Marinho AJ, Couto JP (2020) Construction risk management in Portugal—identification of the tools/techniques and specific risks in the design and construction phases. In: 18th International probabilistic workshop. Springer, Cham, Switzerland, pp 237–253
3. Vivek A, Rao CHH (2022) Identification and analysing of risk factors affecting cost of construction projects. Material Today 60(3):1696–1701
4. Iqbal S, Choudhry RM, Holschemacher K, Ali A, Tamosaitiene J (2015) Risk management in construction projects. Technol Econ Dev Econ 21(1):65–78
5. Bahamid RA, Doh SI, Khoiry MA, Kassem MA, Al-Sharafi MA (2022) The current risk management practices and knowledge in the construction industry. Buildings 12(7):1016
6. Rose KH (2013) A Guide to the project management body of knowledge (PMBOK® Guide), 5th edn. Project Management Institute Inc, Newtown, PA, USA
7. Sousa VC, Almeida N, Dias L (2014) Risk-based management of occupational safety and health in the construction industry—part 1: background knowledge. Saf Sci 66(4):75–86
8. EHS Daily Advisor. https://ehsdailyadvisor.blr.com/2019/05/oshas-fatal-four-leading-causes-of-fatalities-in-the-workplace/, last accessed 2023/05/17
9. Construction statistics, Great Britain: 2021. https://www.ons.gov.uk/businessindustryandtrade/constructionindustry/articles/constructionstatistics/2021, last accessed 2023/05/17
10. Occupational Safety and Health Administration. https://www.osha.gov/stop-falls, last accessed 2023/05/17
11. Accidents at work—statistics on causes and circumstances. https://ec.europa.eu/eurostat/statistics-explained/index.php?title=Accidents_at_work_-_statistics_on_causes_and_circumstances#Cause_of_accident, last accessed 2023/05/17
12. Jayasudha K, Vidivelli B (2016) Analysis of major risks in construction projects. J Eng Appl Sci 11(11):6943–6950
13. Plebankiewicz E, Zima K, Wieczorek D (2021) Modelling of time, cost and risk of construction with using fuzzy logic. J Civ Eng Manag 27(6):412–426
14. Lesniak A, Janowiec F (2019) Risk assessment of additional works in railway construction investments using the Bayes network. Sustainability 11:5388
15. Mesaros P, Spisakova M, Mackova D (2018) Analysis of safety risks on the construction site. In: 13th International Scientific Conference on People, Buildings and Environment. Elsevier B.V, p 145566

Influence Analysis of the Most Common Defects and Damages on the Durability (Residual Resource) of RC Members: A Review

Roman Tytarenko[1](✉), Roman Khmil[1], Jacek Selejdak[2], and Yaroslav Blikharskyy[1]

[1] Lviv Polytechnic National University, Lviv 79013, Ukraine
roman.y.tytarenko@lpnu.ua
[2] Czestochowa University of Technology, 42-201 Czestochowa, Poland

Abstract. The world construction industry has been developing very fast in recent years. In turn, in accordance with modern economic trends, there is a need to optimize the design of new, as well as rationalize the use of existing building structures, in particular, reinforced concrete. Both new and existing RC structures are often subjected to different environmental and mechanical impacts, reducing their durability (residual resource). To extend the life cycle, and also for an objective assessment of the reliability and durability (residual resource) of RC structures (or structural elements), a reliable assessment of existing defects (damages) is necessary, as well as an analysis of their influence on further operation. In view of the above, all defects and damages of RC structural elements should be estimated, according to various criteria: formation reason and time, type, geometric dimensions, degradation degree, etc. The article considers the main types of defects (damages) of RC members, as well as the feasibility of further operation of such members with defects or damages. Furthermore, based on a detailed review of the literature on this problem, an influence analysis of the most common defects (damages) on the durability (residual resource) of RC structures (or structural elements) was carried out. Depending on the existing defects and damages (including corrosion of reinforced concrete), the classification of reasons for a reduction in durability (residual resource) of various RC members is also proposed.

Keywords: RC members · Defects · Damages · Corrosion · Durability · Residual resource

1 Introduction

The first projects of RC structures date back to the end of the 19th century—a time when, like a natural stone, this versatile and economical composite material was supposed "eternal" [1]. It is important to note here that since the middle of the 20th century, when many systems of roads, bridges, tunnels, and flyovers were built in European countries, terms like "durability" or "residual life" have not appeared in normative documents yet; obviously, this led to negative consequences in the operation of many structures that had to be repaired or, in worst cases, re-repaired [2].

Z. Blikharskyy et al. (Eds.): CEE 2023, LNCE 438, pp. 448–455, 2024.
https://doi.org/10.1007/978-3-031-44955-0_45

The durability design of new RC structures in Europe today is based on a prescriptive (normative) approach. According to [1], this means that if the structure resistance parameters limit values (concrete strength class, w/c ratio, crack width, cover depth, etc., which are necessary to achieve the required service life without significant repair work) are correctly used in construction, it will be possible to exclude most deterioration forms (including corrosion), that are due to incorrect design, construction practice, or material composition. Moreover, against the growth background of reconstruction volumes, especially in today's Ukraine, an important task is also the residual resource assessment of RC structures in operation [3–5].

In turn, ensuring the reliability of buildings and structures separate members (in particular, responsible ones) that are affected by destructive actions for a life cycle (which leads to a decrease in their reliability indexes) remains an important task; at the same time, lack of timely inspection, repair, or strengthening of structures (including RC) during operation leads to a progressive accumulation of defects and damages in them, and then to their failure [5].

In summary, the formulation of the durability (or residual resource) assessment problem, taking into account the influence of the most common damages (including their simulation), ultimately comes down to finding the stochastic function of member degradation, which, in most cases, is the subject of modern studies [4].

2 The Aim of Theoretical Research

Based on the above, it follows that the research of the operation of structures with damages is a very relevant task in our time; accordingly, the principal aims of this work are formulated, namely, the analysis of modern studies of the influence of damages and defects on the durability (or residual resource) of new and existing RC members, and to propose the direction for further research in this field.

3 Results and Discussion

Even though, as of today, quite a lot of works are dedicated to the development and review of new concepts for RC structures reliability (or non-failure) assessment, including those strengthened under load [6–8], studies of the reinforcement (or compressed zone of concrete) damages influence on the bearing capacity of RC bending members [9–11], analysis of the features of using BIM technologies in modern design [12], etc.—the transition to the durability assessment has not yet taken place. First of all, it is due to the complexity of mathematical apparatus and the lack of clear criteria in the selection of input statistical variables (for an objective assessment of durability) to the stochastic models [3, 4]. In turn, it determines that the issues of RC structures damage classification and, mostly, their impact (or impact prognosis) on durability, in our opinion, are not fully described both in various normative documents and research works; and the problem of regulatory support in this field (including the difficulties in combining the scientific basis of research into the process of degradation of RC structures with the requirements of design standards) is much more serious [4].

In Ukraine, for example, the only normative document that provides instructions on the inspection of the actual technical condition of different structures in operation is the Guidelines [13]. However, [13] lacks instructions for further quantitative assessment of the residual resource of investigated damaged elements; nevertheless, a detailed classification of RC structures damages (we summarized it in Table 1) depending on several criteria was presented in Javor's work [14].

Table 1. RC structures damage classification in accordance with [14].

No.	Criterion	Depending on	
1	Provenance	(1) Design errors	
		(2) Errors during assembly and construction operation	
		(3) Operational defects	
2	Damage degree	(1) Normal	
		(2) Satisfactory	
		(3) Unsuitable for normal operation	
3	Part of the building	(1) Foundations	
		(2) Over-ground structures	
		(3) Decoration elements	
		(4) Industrial components (insulation, compensators, etc.)	
4	Formation time	(1) During construction	
		(2) Within two years after construction completion	
		(3) Two (or more) years after construction completion	
5	Reason	Technological	(1) Acquired (during transportation, installation, or related operations) (2) Primary (violation of reinforcement placement design parameters, materials strength, prestress losses, etc.)
		Environmental	(1) Physical wear-in time (2) Aggressive environment (3) Alternating freeze-thaw cycles (4) Humidity, wetting (5) High temperature
		Extraordinary	(1) Natural phenomena (disasters, earthquakes) (2) Technological impacts (industrial accidents, resonant phenomena, etc.) (3) [a]Military impacts (missile strikes, etc.)

[a] This extraordinary reason, proposed by us, is very relevant for analysis in our time, especially for Ukraine, which is at war

Regarding the problem of ensuring durability: in Ukrainian Regulations [15, 16], for example, this issue is mentioned only in the context of the definition and general instructions regarding the minimum concrete class for new RC structures (depending on the environmental characteristics), etc.

In turn, the current profile normative documents regulating the reliability and durability of RC structures in the EU countries [17–19], and the USA [20], also do not provide clear methods for assessing the durability of members based on probabilistic calculation models (which would provide an opportunity to operate with any laws of input variables distribution). In accordance with [19], the expression $P(t)$—for estimating the probability of corrosion occurrence at a certain point in time due to carbonization—allows working only with a normal distribution.

All these above documents [17–20] consider durability based on the design and operation of RC structures experience—without monitoring and simulation of their actual work in specific environmental conditions. This leads to bad consequences: for example, the real resource of public structural elements is almost half as much as that given in [17].

From the point of view of the above-mentioned monitoring and simulation of the actual work of RC structures (in terms of determining their durability), in our opinion, qualitative studies have been done by Li [21] and Taffese [22]. In the first case [21], an assessment of the durability (using Monte Carlo simulation) of RC bridge members (taking into account the marine environment impact) with a service life of 120 years was carried out—the results show that all members have a low probability of failure ($P_f < 2\%$) after this time; furthermore, on the durability assessment basis, maintenance planning of structural elements (for normal and accidental working conditions) is also proposed in [21]. In the second case [22], a conceptual basis for autonomous durability monitoring and evaluation of RC structures in operation is formulated—which excludes the limitations of conventional methods (that are possible in the empirical models used there) by using the so-called "Internet of Things" (IoT).

In turn, Helland [23] talks about the failure probability, which leads to different design service lives of RC members (see Fig. 1) or various requirements to cover thickness for a design service life (for example, for de-passivation caused by carbonization).

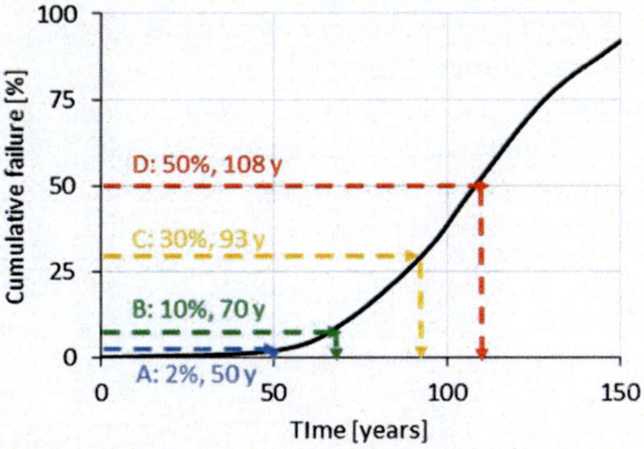

Fig. 1. Influence of the accepted failure probability on the prognosticated service life (in terms of de-passivation caused by carbonization): 50%—108 years; 30%—93 years; 10%—70 years; 2%—50 years [23].

A focal part of durability design is the acceptance criteria selection, namely, the limit states and the probability of exceeding these limit states; and, obviously, the greater the consequences (safety, economic, or social) of exceeding a specific limit state, the less the probability is accepted [1, 16, 17]. In accordance with Regulations [16], for example, the recommended stable values for the failure probability (P_f) are in the range from 1×10^{-4} to 1×10^{-6} (depending on the specific consequence for the 1st group of limit states). By the way, the works [24–26] are based on this fundamental "condition of limit state" definition principle.

The following studies [27–29] are dedicated to the relevant issues of identification, evaluation, and prognostication of RC structures damages (based on various approaches); in the works [30, 31], voluminous analytical reviews of methods for identification and assessment of damages were carried out—based on vibration [30] and acoustic emission [31] techniques in accordance. From a practical point of view, interesting is the research [32] in which the work simulation of damaged RC beams in the software complex "LIRA-SAPR" was carried out—the simulation results obtained there showed a good convergence of the residual bearing capacity compared with the full-scale experiment (the maximum difference was about 21%).

Finally, based on the above analysis of normative documentation [13, 15–20] and the latest studies [5, 9–11, 14, 27–32], we present below (see Table 2) our general recommendations regarding the assessment of defects and damages in RC bending members.

4 Conclusions

Based on a detailed analysis of the literature, it can be concluded that the problem of identification and assessment of defects and damages of RC structures (members) in operation is very complex, and it should be considered (in each case) with the account of different factors.

In turn, as of today, in the recent studies and profile standards of various countries, the impact issue of these defects and damages on the durability (or residual resource) of new and existing structures is not sufficiently developed—due to the complexity of mathematical apparatuses for the assessment of reliability, subjectivity when building stochastic models of the operation of members, etc.

In order to overcome the above-mentioned shortcomings, as well as the future unavoidable correction of normative documents in the durability field of RC structures, we propose in further studies to take into account:

(1) the action of a specific aggressive environment on the member operation in time;
(2) the approach to defects and damages identification and assessment outlined in the recommendations above, and additional evaluation of damages resulting from military impacts (for example, missile strikes);
(3) a probabilistic approach to durability assessment based on methods that would make it possible to work with various laws of input variables distribution (primarily, the Monte Carlo simulation);
(4) progress of approaches to the work simulation of damaged structures based on the finite element method (including BIM technology), and durability (residual resource) monitoring (including based on the IoT).

Table 2. Recommendations regarding the assessment of defects (damages) in RC beams.

No.	Type of defect or damage	Occurrence cause	Recovery measures
1	Normal cracks in the stretched zone	The bending moment action during overloading; reduction of the reinforcement diameter due to corrosion	Strengthening of normal sections; Protection against corrosion
2	Inclined cracks near supports	The action of transverse force and bending moment during overloading; insufficient area of transverse reinforcement	Strengthening of inclined sections; Protection against corrosion
3	Longitudinal cracks in the compressed zone	Crush of the compressed zone of concrete due to overloading	Strengthening of the compressed zone
4	Supportive cracks	Violation of the working reinforcement anchoring	Strengthening of supporting areas
5	Shrinkage cracks	Temperature-moisture and shrinkage deformations of concrete	Mashing of surface and injection of deep cracks
6	Cracks along the reinforcement, rust streams	Corrosion of reinforcement due to violation of the concrete cover, the action of an aggressive environment	Restoration of the concrete cover, protection of reinforcement from corrosion; beam strengthening
7	Chipping of the concrete cover	Mechanical damages during transportation and operation, reinforcement corrosion; fire influences	Restoration of destroyed areas of concrete; Beam strengthening
8	Unacceptable deflection	Overloading, insufficient prestressing, manufacturing defects	Unloading, strengthening

A comprehensive approach (with the account of the above factors) will significantly increase the reliability and objectivity of durability assessment models; and the problem of analysis and assessment of the influence of most common defects and damages on the durability (or residual resource) of structures (members), actually, requires future studies.

References

1. Geiker MR, Hendriks MAN, Elsener B (2023) Durability-based design: the European perspective. Sustain Resilient Infrastruct 8(2):169–184. https://doi.org/10.1080/23789689.2021.1951079

2. Geiker MR, Michel A, Stang H, Vikan H, Lepech MD (2019) Design and maintenance of concrete structures requires both engineering and sustainability limit states. In: Life-cycle analysis and assessment in civil engineering: towards an integrated vision—Proceedings of the 6th International Symposium on Life-Cycle Civil Engineering (IALCCE 2018). CRC Press and Balkema, pp 987–992
3. Tytarenko R, Khmil R, Selejdak J, Vashkevych R (2023) Probabilistic durability assessment of RC structures in operation: an analytical review of existing methods. Lecture Notes in Civil Engineering 290:408–415. https://doi.org/10.1007/978-3-031-14141-6_41
4. Tytarenko RY, Khmil RY (2022) To the problem of the residual resource assessment of reinforced concrete structures in operation. Resou-Saving Mat, Struct, Build Struct 41:339–345. https://doi.org/10.31713/budres.v0i41.38
5. Voskobiinyk OP (2010) Typological comparison of defects and damages of reinforced concrete, metal, and steel-reinforced concrete beam structures. Bull Lviv Polytechnic Nat Univ: Theory Building Pract 662:97–103
6. Khmil R, Tytarenko R, Blikharskyy Y, Vegera P (2018) Development of the procedure for the estimation of reliability of reinforced concrete beams, strengthened by building up the stretched reinforcing bars under load. Eastern-European J Enterprise Technol 5(95):32–42. https://doi.org/10.15587/1729-4061.2018.142750
7. Khmil RY, Tytarenko RY, Blikharskyy YZ, Vegera PI (2021) Improvement of the method of probability evaluation of the failure-free operation of reinforced concrete beams strengthened under load. In: IOP Conference Series: Materials Science and Engineering 1021(1):012014. https://doi.org/10.1088/1757-899X/1021/1/012014
8. Tytarenko R, Khmil R, Dankevych I (2021) Theoretical analysis of existing concepts to evaluate the non-failure of RC structures in operation. Theory Build Pract 3(2):1–6. https://doi.org/10.23939/jtbp2021.02.001
9. Blikharskyy Y, Selejdak J (2021) Influence of the percentage of reinforcement damage on the bearing capacity of RC beams. Construct Optimized Energy Poten (CoOEP) 10(1):145–150. https://doi.org/10.17512/bozpe.2021.1.15
10. Lobodanov M, Vegera P, Khmil R, Blikharskyy Z (2021) Influence of damages in the compressed zone on bearing capacity of reinforced concrete beams. Lecture Notes in Civil Engineering 100:260–267. https://doi.org/10.1007/978-3-030-57340-9_32
11. Vegera P, Vashkevych R, Blikharskyy Y, Khmil R (2021) Development methodology of determinating residual carrying capacity of reinforced concrete beams with damages tensile reinforcement which occurred during loading. Eastern-European J Enterprise Techn 4(112):6–17. https://doi.org/10.15587/1729-4061.2021.237954
12. Pawłowicz JA (2020) Computer-aided design in the construction industry—BIM technology as a modern design tool. Construct Optim Energy Potent (CoOEP) 9(2):89–96. https://doi.org/10.17512/bozpe.2020.2.10
13. National standard of Ukraine: Guidelines for inspection of buildings and facilities for identification and evaluation of their technical condition (DSTU-N B V.1.2-18:2016). SE "UkrNDNTs", Kyiv (2017)
14. Javor T (1991) Damage classification of concrete structures. The state of the art report of RILEM Technical Committee 104-DCC activity. Mat Struct 24:253–259. https://doi.org/10.1007/BF02472079
15. State construction regulations of Ukraine: Concrete and reinforced concrete structures. Basic principles (DBN V.2.6-98:2009). MinRegionConst of Ukraine, Kyiv (2011)
16. State construction regulations of Ukraine: General principles for reliability and constructive safety ensuring of buildings and civil engineering works (DBN V.1.2-14:2018). MinRegion of Ukraine, Kyiv (2018)
17. Standard E (2002) Eurocode—basis of structural design (EN 1990:2002). European Committee for Standardization, Brussels

18. ISO 2394:2015 (2015) International standard: General principles on reliability for structures, 4th edn. ISO, Geneva
19. ISO 13823:2008 (2008) International standard: General principles on the design of structures for durability, 1st edn. ISO, Geneva
20. ACI 318-19 (2019) An ACI standard: building code requirements for structural concrete. American Concrete Institute, Farmington Hills, MI
21. Li K, Wang P, Li Q, Fan Z (2016) Durability assessment of concrete structures in HZM sea link project for service life of 120 years. Mater Struct 49:3785–3800. https://doi.org/10.1617/s11527-015-0754-8
22. Taffese WZ, Nigussie E, Isoaho J (2019) Internet of things based durability monitoring and assessment of reinforced concrete structures. Procedia Computer Science 155:672–679. https://doi.org/10.1016/j.procs.2019.08.096
23. Helland S (2016) Performance-based service life design in the 2021 version of the European concrete standards—ambitions and challenges. In: International fib Symposium on Performance-Based Approaches for Concrete Structures (fib Symposium 2016). The International Federation for Structural Concrete, Cape Town, pp 32–41
24. Jitao Y, Liuzhuo C, Jun G, Ren X (2019) Structural durability and concept system of structural reliability. IOP Conference Series: Earth and Environmental Science 340(5):052035. https://doi.org/10.1088/1755-1315/304/5/052035
25. Nogueira CG, Leonel ED, Coda HB (2012) Reliability algorithms applied to reinforced concrete structures durability assessment. IBRACON Struct Mat J 5(4):440–450. https://doi.org/10.1590/S1983-41952012000400003
26. Lantukh-Liashchenko AI (2019) The resource probabilistic assessment of reinforced concrete members. Indust Construct Engineering Struct 3:13–17
27. Farhidzadeh A, Dehghan-Niri E, Moustafa A, Salamone S, Whittaker A (2013) Damage assessment of reinforced concrete structures using fractal analysis of residual crack patterns. Exp Mech 53(9):1607–1619. https://doi.org/10.1007/s11340-013-9769-7
28. Farhidzadeh A, Salamone S, Singla P (2013) A probabilistic approach for damage identification and crack mode classification in reinforced concrete structures. J Intell Mater Syst Struct 24(14):1722–1735. https://doi.org/10.1177/1045389X13484101
29. Karbassi A, Mohebi B, Rezaee S, Lestuzzi P (2014) Damage prediction for regular reinforced concrete buildings using the decision tree algorithm. Comput Struct 130:46–56. https://doi.org/10.1016/j.compstruc.2013.10.006
30. Fan W, Qiao P (2011) Vibration-based damage identification methods: a review and comparative study. Struct Health Monit 10(1):83–111. https://doi.org/10.1177/1475921710365419
31. Noorsuhada MN (2016) An overview on fatigue damage assessment of reinforced concrete structures with the aid of acoustic emission technique. Constr Build Mater 112:424–439. https://doi.org/10.1016/j.conbuildmat.2016.02.206
32. Klymenko YV, Antoniuk NR, Polianskyi KV (2019) Modeling the work of damaged reinforced concrete beams in the SC "LIRA-SAPR". Bull Odesa State Academy of Civil Engineering and Architecture 77, 58–65. https://doi.org/10.31650/2415-377X-2019-77-58-65

An Experimental Study of Beams Made of Autoclaved Aerated Concrete Masonry Units

Volodymyr Verba(✉) and Petro Kholod

Department of Building Constructions and Bridges, Lviv Polytechnic National University, Lviv 79013, Ukraine

volodymyr.b.verba@lpnu.ua

Abstract. The article represents experimental results obtained from four-point bending tests of beams, made of autoclaved aerated concrete (AAC) masonry units of D500 class, sawn and glued back together by special adhesive for AAC masonry. The aim was to develop the made-in-situ lintels which have not been sufficiently studied in the case of AAC block usage. The specimens in the current study include four beams, masonry units and specimens of the masonry. Two types of beams with a 1300 mm clear span were tested. In each beam, there was a reinforcement represented by two galvanized fully threaded M6 rods of 5.8 class. In the first two beams, steel bars were mounted externally as tie rods to provide the counteraction for a thrust on the supports. In the second pair of beams, steel bars were mounted inside the beams in predrilled holes. Steel bars in all beams were prestressed during making, which was achieved by a torque wrench. The obtained test data provide a basis for evaluating the accuracy of existing strength-design equations to predict the capacity of such beams. Test results show that the internal application of reinforcement at the bottom zone of the beams in combination with prestressing increases the load and deflection capacities. Both series of beams demonstrated, among other things, that the failure was accompanied by cracks in the tension zone and crushing of aerated concrete in the compressed zone in the pure bending zone as well as in the support zone due to shear.

Keywords: AAC masonry · Made-in-situ lintels · Four-point bending test · Autoclaved aerated concrete beams

1 Introduction

Traditional brick masonry construction is becoming outdated due to its low construction efficiency; thus, the use of more modern materials is a must. Masonry units made of autoclaved aerated concrete (AAC) are more modern, they have been introduced into civil engineering in Sweden in 1929 and then spread internationally. Nowadays mass production of AAC blocks is popular not only in Central and Eastern Europe, but also in China, India, the Middle East and other countries [1–3]. AAC masonry effectively reduces the dead load of the wall itself which leads to further savings in construction and maintenance costs [4]. AAC blocks are simpler and more economical to transport, they

Z. Blikharskyy et al. (Eds.): CEE 2023, LNCE 438, pp. 456–464, 2024.
https://doi.org/10.1007/978-3-031-44955-0_46

can make construction easier and faster. Blocks can also be handled by plain tools on the construction site with low wastage, which has led in the past decades to an increasing use of this material for non-structural applications, especially infill walls. Along with other advantages of AAC block masonry (such as environmental friendliness, fire resistance and good thermal and sound insulating properties), it has a list of disadvantages [5]. One of the major problems is the low compressive strength and even less tensile [4, 6–8], which limits the structural use of AAC masonry. To span the window openings or make a floor slab in AAC block construction, reinforced concrete is often used. However, reinforced concrete is much heavier than AAC blocks and has a high heat conductivity which can be problematic for energy-efficient AAC block construction.

Research on AAC masonry construction doesn't cover all known lintel types [9]. Most codes and manuals omit flat AAC block masonry lintels made directly on the construction site, while sister-type brick masonry beam is well-known and has been common masonry structural element since ancient times. Modern researches on reinforced brick masonry beams go back to 1837 [10]. It should be also remarked that the results obtained for common brick masonry are not fully applicable to AAC masonry because the latter belongs to "thin bed masonry" typology where units are glued together through layers with typical thicknesses ranging between 0.5 and 3 mm, which significantly affects masonry behavior. The tests showed an almost isotropic behavior of this masonry type [11].

Preliminary analysis showed the lack of dedicated code guidance or recommendations for the strength design of AAC masonry beams which leaves the design community to improvise, but it is commonly accepted that the strength design of reinforced masonry beams follows the same steps used for reinforced concrete beams [12–14]. This paper presents an experimental study on AAC block masonry beams where the beam is subjected to four-point loading. The influence of the position of reinforcement (external or internal) is investigated. Furthermore, the flexural response of both types of masonry beams is examined along with the description of cracking and failure modes.

2 Materials and Experimental Program

The experimental tests were carried out on two pairs of identical masonry fragments, which were prepared in a laboratory. We chose prefabricated autoclaved aerated concrete blocks from local manufacturer—AEROC—with density grade D500 and smooth edges. The dimensions of blocks were 100 × 200 × 610 mm. The physical characteristics of masonry units are summarized in Table 1.

One masonry unit was divided into two equal pieces and set as the supports. Other three blocks were sawn into four equal pieces and then glued back to form a compound beam out of twelve pieces (see Fig. 1). The thickness of the perpend and bed joints of the masonry was set to 3 mm. We used local pre-mixed cement-based AAC adhesive Ferozit. The mortar was received by mixing 25 kg of dry adhesive with approx. 6 liters of potable water. We prepared two series of beams for testing with the only difference between the series in the position of the reinforcement (Series 1—reinforcement placed externally on sides, Series 2—reinforcement is mounted inside the beam).

Two tested beams in each series were identical with a 1300 mm clear span. In each beam there were mounted two galvanized fully threaded M6 bars of 5.8 class with

Table 1. AAC block characteristics, declared by the manufacturer.

Location of manufacturing plant	Mean density in dry condition, kg·m^{-3}	Min. compressive strength, MPa	Min. frost resistance, cycles	Max. thermal conductivity in dry condition, W·(m·°C)$^{-1}$	Unit weight[a], kg
Obuhiv, UA	500	C 2.5	F100	0.12	7.22

[a] at 25% moisture content

Fig. 1. Test masonry beam with reinforcement placed externally (Series 1)

pretension. During the preparation of test samples of the 2nd series (see Fig. 2), we drilled two holes for reinforcement in each block. The reinforcement was positioned 3 cm up from the bottom edge of the beam in both series. The aim of reinforcement in the 1st series of beams was to act as tie rods to model the counteraction for a thrust on the supports, which otherwise is traditionally perceived by the wall masonry. To do so, we wrench the nuts with a controlled torque of 2 N·m. The required tension in steel bars in both series of beams was achieved by a torque wrench.

Fig. 2. Test masonry beam with reinforcement placed internally, instrumented (Series 2)

As anchors at the ends of the beams we used 4 mm thick sheet carbon steel plates. Based on research [15, 16], we chose a rectangular shape of anchor plate with effective dimensions 100 × 40 mm.

Along with the main experiments, there were several tests conducted, aiming to examine the mechanical characteristics of AAC and masonry. The exact quantity of specimens and the purpose of each test in this study are gathered in Table 2.

Table 2. Experimental program.

Type/shape of sample	Dimensions, mm	Quantity of samples	Test purpose
AAC masonry beam with external reinforcement	100 × 200 × 1833 (1300 mm clear span)	2	Bearing capacity of AAC masonry beam
AAC masonry beam with internal reinforcement	100 × 200 × 1833 (1300 mm clear span)	2	Bearing capacity of AAC masonry beam
Cubes	100 × 100 × 100	3	Compressive strength grade of AAC
AAC blockwork fragment	203 × 100 × 305	2	Characteristic compressive strength of masonry f_k

3 Testing Technique

The exposure of beams after manufacture before the start of the tests was 28 days, storage was carried out indoors at ambient temperature of 15 ± 2°C and at relative humidity of 50–60%.

The load was applied using the 12-tonne hydraulic jack and a distribution traverse made of timber plank to provide four-point bending. The loading was carried out in steps of 0.25–0.5 kN and was monitored by a spring dynamometer with a clock-type indicator. The delay at each stage was at least 5 min, during which readings were taken from the devices. Clock-type indicators with 0.01 mm resolution were placed on top of the supports and another indicator was placed in the middle of the span (with unlimited measuring range), measuring the vertical displacement of the relevant cross-sections of the beam.

Measurement of aerated concrete deformations was performed by precise indicators with 0.001 resolution installed on the base of 100 mm in the middle of the span. In addition, the same type indicators were installed at threaded bars with the help of special adapters to control their strain.

The formation of cracks was visually recorded, and the width of their opening was measured using a microscope with 24x magnification MPB-2 with 0.05 mm resolution.

Testing was conducted until we had reached the beam's failure.

4 Results and Discussion

The mean value of AAC compressive strength was 2.83 MPa with minimal value observed in experiments at 2.57 MPa (as determined on cubes with an edge length of 100 mm). Tests of the masonry units showed compressive strength $f_b = 2.6$ MPa.

The compressive strength of masonry, obtained from the experiment, was equal to $f_k = 2.06$ MPa. The failure mode for all materials specimens was common.

4.1 Results for the 1st Series of Beams (with External Reinforcement)

Beam No. 1. First cracks appeared after the 7th stage of loading (total load 2F = 1.75 kN). At the 9th stage, there were 4 cracks, two at the supports on the top edge and two in the pure bending zone on the bottom edge. The crack width at that moment was equal to 0.1 mm. In the 13th stage the biggest crack was 0.3 mm wide. At the 18th stage (total load 2F = 5.25 kN) the compressed zone near the left loading point split from the beam's body and was crushed which led to the beam destruction (Fig. 3a).

Beam No. 2. First cracking was heard during the 6th stage of loading. On the 9th stage it was clearly visible, that there was only one crack in the pure bending zone, and two parts of the beam were rotating around a plastic hinge at the midspan, causing the beam to tear off from the supports (Fig. 3b). At the 10th stage the split of the compressed zone was observed with crack width on the bottom edge equal to 3 mm. At the 14th stage the compressed zone was crushed (total load 2F = 3.5 kN) (Fig. 3c).

a b c

Fig. 3. Visualization of failures in the 1st series of beams: **a** failure of the compressed zone in the beam No. 1; **b** separation of the beam No. 2 from the support due to rotation around a plastic hinge; **c** crushing of the compressed zone in the beam No. 2.

4.2 Results for the 2nd Series of Beams (with Internal Reinforcement)

Beam No. 1. First cracks appeared at the 10th stage of loading near both of the supports (cracks on the left support due to negative bending moment and horizontal crack on the opposite side above the support). At the 12th stage cracks reached 0.3 mm width, at the later stages there appeared at least 5 vertical cracks in the pure bending zone. Final failure was accompanied by crushing of the compressed zone in the midspan and shear action at the supports (total load 2F = 6.75 kN) (Fig. 4a).

Beam No. 2. First cracks were heard on the 5–6th stages of loading. At the 9th stage the first vertical crack was noticed (0.1 mm wide) at the mid-span. In the late stages more cracks appeared throughout the beam length, both at the mid-span and at the supports. Final failure was caused by the combined action of bending moment and shear (total load 2F = 8.75 kN) (Fig. 4b).

a b

Fig. 4. Visualization of failures in the 2nd series of beams: **a** failure of the compressed zone in the beam No. 1; **b** failure of the beam No. 2 with the widest cracks in the area of maximum shear force, crushing of the compressed zone near the support.

Both beams in the 2nd series demonstrated, among other things, that failure occurred both in the pure bending zone with cracks in the tension zone and crushing of aerated concrete in the compressed zone as well as due to shear and damage to the supports.

4.3 Analysis of the Experimental Data

The load-deflection responses of both series of beams are shown in Fig. 5a. On this plot we also draw limiting lines that correspond to critical deflections at 1/300 and 1/500 of the effective span (taken between the middles of the support areas and equal to 1550 mm). It is clearly shown in Fig. 5a, that 2nd series of beams with internal reinforcement have significant improvement in load-carrying capacity and ductility.

We observed several different types of cracking patterns, where the lowest strength capacity in Series 1 Beam No. 2 corresponds to the emergence of only one crack in the pure bending zone. We believe that in real structural conditions with beams clamped in masonry and in case of usage of any internal reinforcement in beams this pattern is not likely to occur.

Figure 5b shows the load vs strain response of compressed AAC at the mid-span of the beams. On the plot, we marked the level of typical strain observed at maximum stresses for AAC ($\varepsilon = -0.002$ mm/mm, see [17]). We can also observe two jumps of strain on the plots there, which correspond to the beams where crushing of the compressed zone occurred under the base of tensometers and such sudden jumps we explain by the crushing of aerated concrete on the micro level there few stages of loading before failure.

Bends of the curvature of the graphs in Fig. 5 are explained by the sudden appearance of cracks, mainly on the supports, which changed the static scheme of the beams.

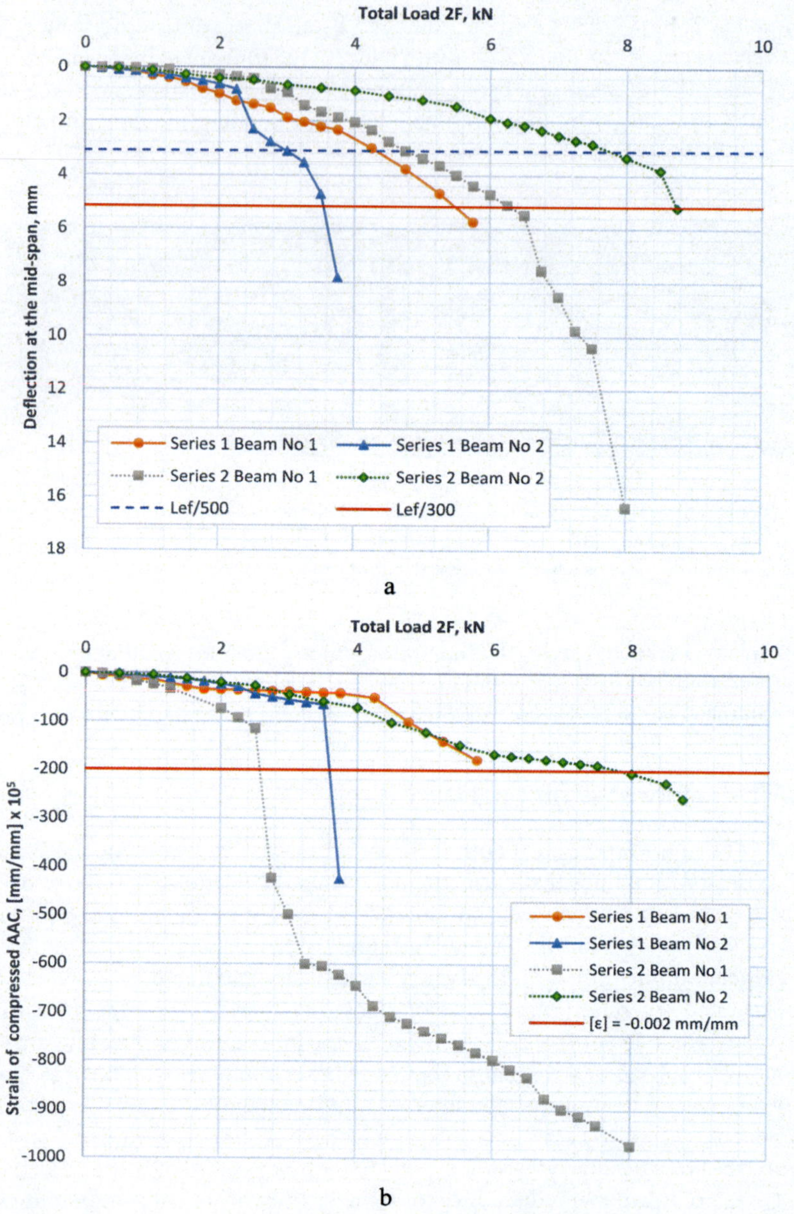

Fig. 5. Graphical representation of test results: **a** the load-deflection responses of beams; **b** the load vs strain response of compressed AAC at the mid-span of the beams.

5 Conclusions

The experimental results of masonry beams made of AAC blocks test were obtained in this study. In buildings with AAC masonry, we can design and use masonry beams or lintels made of AAC blocks without prefabrication. Beams with internal reinforcement seem to be more reliable in terms of flexural performance and load carrying capacity in comparison with masonry beams with external reinforcement on the sides. The following conclusions are drawn from this study:

- Conducted research showed that beams made on site from locally available D500 AAC blocks and having the same depth of cross-section as common blocks (200 mm) can bear the external load equivalent to q = 2.5 kN/meter of length per 10 cm of AAC beam width with typical for lintels clear span of 1300 mm with the most rigorous SLS requirements for the deflection at 1/500.
- We recommend using reinforcement bars beneath or inside of the compound beam instead of placing them on sides for the sake of security, because the observed failures in beams were caused by the crushing of concrete, therefore were brittle by nature.
- The general high variability of the results of the experiments, which was manifested by several different patterns of beam destruction and, accordingly, their different load-bearing capacity, is explained by the impossibility of manual manufacture of identical beams that consist of several parts. Obviously, this should be taken into account by increased reliability coefficients and requires additional research.

Acknowledgements. M.Eng. Alaa Abdalah and M.Eng. Fatima-Ezzahrae Oubuih are gratefully acknowledged for their contribution to the preparation and execution of experimental tests.

References

1. Van Boggelen W, History of autoclaved aerated concrete. Aircrete, https://www.aircrete.com/wp-content/uploads/2018/10/History-of-AAC.pdf, last accessed 2023/05/13
2. Hellers G, Schmidt R (2011) Autoclaved aerated concrete (AAC)—the story of a low-weight material. In: Boroński D (ed) 5th International Conference on Autoclaved Aerated Concrete "Securing a Sustainable Future". University of Technology and Life Sciences Press, Bydgoszcz, pp 63–70
3. Turski R, Rogala W (2022) Current situation and further development of AAC in Europe. Cement-Wapno-Beton = Cement lime concrete 27(3):154–165. https://doi.org/10.32047/cwb.2022.27.3.1
4. Rathi S, Khandve P (2016) Cost effectiveness of using AAC blocks for building construction in residential building and public buildings. IJRET 5(5):517–520
5. Kamal MA (2020) Analysis of autoclaved aerated concrete (AAC) blocks with reference to its potential and sustainability. J Build Mat Struct 7(1):76–86. https://doi.org/10.5281/zenodo.3950489
6. Raj A, Borsaikia AC, Dixit US (2019) Compressive and shear bond strengths of grooved AAC blocks and masonry. Mater Struct 52:116. https://doi.org/10.1617/s11527-019-1428-8
7. Raj A, Borsaikia AC, Dixit US (2020) Evaluation of mechanical properties of autoclaved aerated concrete (AAC) block and its masonry. J Inst Eng India Ser A 101:315–325. https://doi.org/10.1007/s40030-020-00437-5

8. Bhosale A, Zade NP, Davis R, Sarkar P (2019) Experimental investigation of autoclaved aerated concrete masonry. J Mat Civil Eng 31(7):04019109. https://doi.org/10.1061/(asce)mt.1943-5533.0002762
9. Technical Notes on Brick Construction. Reinforced Brick Masonry, https://www.gobrick.com/docs/default-source/read-research-documents/technicalnotes/17-reinforced-brick-masonry---introduction.pdf , last accessed 2023/05/13
10. Drobiec Ł, Jasiński R, Mazur W (2018) Analysis of AAC precast lintels embedded in walls different construction. Ce/Papers 2(4):367–376. https://doi.org/10.1002/cepa.830
11. Ferretti D, Michelini E, Gianpaolo Rosati G (2015) Mechanical characterization of autoclaved aerated concrete masonry subjected to in-plane loading: Experimental investigation and FE modeling. Const Build Mat 98:353–365
12. Loderer F, Trümer A (2019) Design of aerated concrete flat lintels = Bemessung von Porenbeton-Flachstürzen. Mauerwerk 23:414–420. https://doi.org/10.1002/dama.201900024
13. EN 1996-1-1 (2006) Eurocode 6, Design of masonry structures—part 1–1: General rules for reinforced and unreinforced masonry structures
14. Made Budiwati IA, Sudarsana IK (2019) Flexural tests of masonry beam with and without reinforced bar. MATEC Web Conf. 276:01018. https://doi.org/10.1051/matecconf/201927601018
15. Famulyak Y, Demchyna B, Sobczak-Piastka J (2019) The influence of size and form of plate anchors on their work at anchoring of reinforcement in the array of cellular concrete. AIP Conference Proceedings 2077:020017. https://doi.org/10.1063/1.5091878
16. Verba V, Kholod P (2020) Behaviour of square steel anchor plate in autoclaved aerated concrete block masonry. IOP Conf Series: Mat Sci Eng 737:012048. https://doi.org/10.1088/1757-899x/737/1/012048
17. Mollaei S, Babaei Ghazijahani R, Noroozinejad Farsangi E, Jahani D (2022) Investigation of behavior of MasonryWalls constructed with autoclaved aerated concrete blocks under blast loading. Appl Sci 12:8725. https://doi.org/10.3390/app12178725

Building Materials Circularity of Green Roofs

Silvia Vilcekova[1](✉), Eva Kridlova Burdova[1], Jana Budajova[1], and Peter Mesaros[2]

[1] Institute of Sustainable and Circular Construction, Faculty of Civil Engineering, Technical University of Kosice, Kosice, Slovakia
silvia.vilcekova@tuke.sk

[2] Institute of Technology, Economics and Management in Construction, Faculty of Civil Engineering, Technical University of Kosice, Kosice, Slovakia

Abstract. Green roofs are increasingly popular as a mitigation action against the reduction of green areas in Slovak cities. Therefore, it is essential to analyse and compare building materials circularity and environmental impact indicators during the whole life cycle of the roof assemblies. This study considers four design alternatives for green roofs. One Click LCA software was used to model the green roof design alternatives and determine the building circularity indicator which was used to represent the total materials circularity both in the use of materials for the project as well as end-of-life handling. This analysis can help identify areas where improvements can be made to improve the building circularity score of the green roofs. When assessing circularity, it is the mass proportion of materials that are either recycled or used as material, with 50% of materials either being recycled (with a loss of value, such as re-using concrete as aggregate) or used as energy (such as wood) or incineration of plastic products. The circularity of the compositions of the evaluated green roof structures is quantified as 18%, 15%, 14%, 17%, 57% and 71%. Compositions using wood and wood-based materials have a much higher circularity than concrete, steel and waterproofing.

Keywords: LCA · Building circularity · Building materials · Green roofs

1 Introduction

The linear system of consumption, production, and waste has led to harmful effects on the environment and the efficient use of materials. The transition to a circular economy can bring a range of benefits and mitigate these negative impacts [1]. Currently, the construction industry consumes a significant amount of natural resources and generates a substantial amount of waste. The rapid population growth and urbanization have further intensified these challenges. It's important to note that the industry has also been making efforts to address these issues [2]. The concept of a circular economy (CE) can indeed be an effective strategy for mitigating the consumption of energy and natural resources in the building and construction industry. One aspect of this strategy involves preserving and reusing historical buildings instead of demolishing them and constructing new ones [3].

Z. Blikharskyy et al. (Eds.): CEE 2023, LNCE 438, pp. 465–475, 2024.
https://doi.org/10.1007/978-3-031-44955-0_47

The circular economy paradigm can eliminate the problems with the traditional linear economy access in the construction industry [4]. It is known that it requires a closed-loop material cycle and seeks ways to apply circular principles [5]. Improving the circularity in the construction sector means ensuring greater efficiency of materials to avoid extraction of primary material. This effort could reduce the net negative environmental impacts associated mainly with construction materials. Assessing the building stock to identify elements that can be saved for reuse, repair, or repurposing is a crucial step in implementing circular practices in the construction industry [6]. Applying mentioned CE strategies in the construction industry is crucial for achieving carbon reduction targets set in the Paris Agreement and aligning with European policies. As already pointed out, the construction sector is a significant contributor to greenhouse gas emissions, thus adopting CE principles can help mitigate these emissions while promoting sustainable practices [7]. In December 2021, the Commission adopted a revised recommendation on the use of environmental footprint methods to help companies calculate their environmental performance based on reliable, verifiable and comparable information, and for other actors (e.g. public administrations, NGOs, business partners) to have access to such information. It will stimulate the industry to produce products that have better environmental properties [8]. European Green Deal aims to mobilize industry for clean and circular economy and emphasizes enabling buyers to make more sustainable decisions and reduce the risk of "greenwashing", reliable, comparable and verifiable information necessary [9]. While an operational economy strategy can be implemented in various sectors such as industry, waste, energy, buildings and transport, there is a life cycle assessment when optimizing new systems [10]. Stock and material flow analysis can be used to understand baselines through stockpiled materials and building material flow [1].

This study considers six design alternatives for green roofs. One Click LCA software was used to model the green roof design alternatives and quantify the building circularity indicator which represent the total materials circularity both in use of materials for the project as well as end of life handling. This analysis enables designers to analyse different scenarios, assess the impact of design choices and material selection, and identify strategies to improve circularity throughout the building life cycle. This also allows identifying which materials contribute most to circularity achievements and which may require further optimization. By understanding the circularity performance, targeted actions can be taken to enhance resource efficiency and waste reduction for specific materials.

2 Materials

Six green roof compositions were selected to quantify building circularity. The load-bearing structures of the four compositions are a reinforced concrete slab and two roof compositions are made of wood construction. The composition of the roof structures evaluated is presented in Table 1. The functional unit (FU) is 1m^2 and the estimated calculation period is 60 years.

3 Methods

Literature study, expert interviews and the expert panel showed that circularity, technically, consists of two components: (1) circular material usage, and (2) circular design. Other indicators such as material health, toxicity and material scarcity are preconditions and drivers, which are not influencing the buildings' level of circularity in any way. The Building Circularity Indicator (BCI) measures the extent to which the linear flows have been minimized and restorative flows maximized for its component systems, products, and materials and how long and intensively it is used compared to similar system-average

Table 1. Materials

	Thickness (m)	U – value (W/m^2 K)
Composition 1 (C1)	–	0.097
Melica ciliata	0.2500	
Volcanic lava substrate Vulkaterra Rasen 0/6	0.0500	
Hydrophilic mineral wool Isover INTENSE	0.0230	
Nop drainage foil Platon DE25	0.0024	
Geotextile geoNetex MB 300	0.0018	
Waterproofing foil resistant to overgrowth of roots FATRAFOL	0.0024	
Separation geotextile geoNetex MB 300	–	
Mechanical anchor Eljot H3	0.1000	
Thermal insulation XPS Styrodur 3000 CS SQ 200	0.1800	
Thermal insulation XPS Styrodur 3000 CS SQ 200	0.0100	
Adhesive for thermal insulation PUREN Dachkleber	0.0004	
Vapor barrier Isover Stopvap 90	0.2000	
Reinforced concrete ceiling slab in gradient 2°		
Composition 2 (C2)	–	0.099
Magnolia grandfolia	0.4500	
Volcanic lava substrate Vulkatree 0/16 mm	0.0016	
Filter geotextile geoNetex MB 200	0.0400	
Nop drainage foil Platon DE40	0.0024	
Geotextile geoNetex MB 300	0.0030	
Waterproofing foil resistant to overgrowth of roots ICOPAL	0.0024	
Separation geotextile geoNetex MB 300	–	
Mechanical anchor Eljot H3	0.1400	
Thermal insulation XPS Styrodur 3000 CS SQ 200	0.1800	
Thermal insulation XPS Styrodur 3000 CS SQ 200	0.0100	
Adhesive for thermal insulation PUREN Dachkleber	0.0003	
Vapor barrier JUTAFOL N 140 g/m^2 special	0.2000	
Reinforced concrete ceiling slab in gradient 2°		
Composition 3 (C3)	–	0.096
Red Mulberry	0.4000	
Intensive substrate Vulkaplus intenzív 0/16 mm	0.1000	
Hydrophilic mineral wool Isover INTENSE	0.0400	
Drainage foil DIADRAIN-40H	0.0020	
Geotextile PP FILTEK 300 g/m^2	0.0018	
Waterproofing foil resistant to overgrowth of roots Sikaplan	0.1800	
Thermal insulation from hardened polyurethane PUR/PIR foam	0.0100	
Adhesive for thermal insulation PUREN Dachkleber	0.0003	
Vapor barrier Mapeplan VB PE	0.2000	
Reinforced concrete ceiling slab in gradient 2°		

(*continued*)

Table 1. (*continued*)

	Thickness (m)	U – value (W/m^2 K)
Composition 4 (C4)	0.0350	0.099
Lawn carpet GREENDEK TR K 20	0.0500	
Intensive substrate GREENDEK lawn	0.2000	
Intensive substrate GREENDEK	0.0020	
Filter geotextile PP FILTEK 200 g/m^2	0.0200	
Nop drainage foil DEKDREN T20 GARDEN	0.0020	
Geotextile PP FILTEK 300 g/m^2	0.0053	
SBS modified asphalt, resistant to root growth ELASTEK 50	0.0040	
SBS modified asphalt GLASTEK 40 SPECIAL MINERAL	–	
Mechanical anchor Eljot H3	0.1500	
Thermal insulation EPS DEKPERIMETER SD 200	0.1800	
Thermal insulation EPS DEKPERIMETER SD 200	0.0100	
Adhesive for thermal insulation PUREN Dachkleber	0.0003	
Vapor barrier JUTAFOL N 140 g/m^2 special	0.2000	
Reinforced concrete ceiling slab in gradient 2°		
Composition 5 (C5)	0.020	0.096
Gypsum board 2 × 10 mm	0.140	
Wood fiber insulation	0.015	
OSB 3 board	0.280	
Cellulose wood fiber/BSH	0.035	
Wood fiber insulation	0.080	
Ventilated gap - 60 × 80	0.030	
OSB board 2 × 15 mm	–	
Waterproofing + geotextile	0.030	
Drainage layer-ceramist	–	
Filter fabric	0.080	
Substrate		
Composition 6 (C6)	0.170	0.095
Solid wood panel (cross-jointed wood with oak pins)	0.360	
Blown cellulose between I profiles SJ60s wood fiber insulation	0.015	
DHF wood fiber diffusion board	0.080	
Ventilated gap - 60 × 80, á = 625 mm	0.022	
OSB board 1 × 22 mm	–	
Insurance waterproofing	–	
Filter fabric	0.025	
Substrate		

products [11]. The Building Circularity tool plays a crucial role in tracking, quantifying, and optimizing the circularity of materials throughout the building life cycle. By providing a holistic picture and detailed breakdown per material type, this tool enables the calculation of the circularity percentage for a building or asset. This tool allows the tracking and measurement of the circularity of materials sourced and used during the building life cycle. It captures data on material inputs, such as recycled content, renewable resources, and sustainable sourcing, and assesses the efficiency of resource use and waste reduction strategies. This enables a quantitative assessment of the circularity to be achieved. It considers various factors such as material selection, design strategies, construction techniques, and end-of-life scenarios. This holistic perspective helps identify opportunities for improvement and informs decision-making at different stages of the building life cycle, and fosters a transition to more sustainable and resource-efficient building practices [12].

The BC calculation includes the construction material query, which is the part of the tool where material data, circularity score weighting factors are entered. The calculation period is where the life of the building is entered. When evaluating BC, the results are Materials Recovered and Materials Returned. Recycled, renewable or reused materials are very important in BC because they form the basis of the calculation. The percentage of recycled, renewable, or reused materials means the mass proportion of recycled, renewable, or reused materials in the product. This information does not affect the LCA results but is used to document the circularity of the material. Some products have a percentage of recycled, renewable, or reused materials defined with a default value that can be based on either the product or the product type. Waste defines the waste of materials on a construction site. The default values were chosen based on typical wastage and vary by construction process, building and design. Circularity Score Weighting Factors were entered for BC calculation.

Circularity Score Weighting Factors. Circularity Score Weighting Factors were entered for BC calculation. Different weighting factors of individual materials were chosen for the evaluation of the constructions.

Materials Recovered. Materials consist of virgin, renewable, recycled, reused materials. In the virgin materials item, new materials are evaluated if they are found in the project. New materials almost always have higher emissions compared to recycled, renewable or reused materials, so the multiplier is set to 0. Renewable materials can be regrown, so the default multiplier is set to 1. In the item, recovered recycled materials obtained for the project that have been recycled have lower emissions compared to newly manufactured materials, so the default multiplier is set to 1. If little or no processing is required, then they have almost no emissions, so the default multiplier is set to 1.

Materials Returned. Materials returned consist of materials that can be reused as materials, materials that can be recycled, materials that can be downcycled, materials that can be used as energy, materials that are disposed of (landfilled). In the case of materials that can be reused as materials, as well as reused materials recovered, if the materials can be reused for other projects, they reduce the emissions that are generated by that project, so the default multiplier is set to 1. Materials that can be recycled produce some emissions for the recycling process but tend to be as structurally chainable/usable as the original product, so the default multiplier is set to 1. Materials that can be downcycled end up being a product that is not as structurally strong as the original product and therefore can only be used to make a different product than the original. If we would want to create a product as strong as the original, we would need to add new materials and therefore the default multiplier is set to 0.5. Combustible materials produce emissions from the combustion process, but end up with recovered energy that can be used for various purposes (e.g. heating, electricity or manufacturing). Due to the generated emissions from combustion (and the possible release of carbon in organic products), the default multiplier is set to 0.5. In the case of materials that are disposed of (landfilled), landfilled materials can never fully decay or decompose, so the default multiplier for landfilled materials is set to 0.

Six green roof compositions were selected to quantify building circularity using One Click LCA software, license for Building Circularity.

4 Results and Discussion

Figures 1, 2, 3, 4, 5 and 6 present the results of buildings circularity for investigated green roof compositions. As we can see, the highest building circularity score is 71% for C6 and the lowest is 14% for C3. Compositions C1–C6 consist mainly of virgin materials (99.3%, 98.8%, 100%, 100%, 27.2% and 0% materials) and characterization factors are higher compared to the materials with recycled content. There are no renewable or reused materials in the compositions C1–C4. Compositions C5 and C6 comprise renewable materials (62.7% and 85.8%). Compositions C3 and C4 do not include materials with recycled content in product stage (A). On the other hand compositions C1, C2, C5 and C6 consist of materials with recycled content in proposition of 0.7%, 1.2%, 10.1% and 17.3% respectively. Compositions C1–C6 are characterized by no reused materials and recycling materials, therefore the values are 0. Except for composition C5, which has recycled materials with a value of 14.3%. Compositions C1–C4 have materials that can be downcycled (67.3%, 57.1%, 54.6% and 65.2%). Materials that are recycled but cannot be used alone because they are not strong enough and need to add other materials have a weight value of 0.5. Therefore, half the value of these materials is taken. Compositions C5 and C6 do not contain downcycled materials. Compositions C1–C6 have materials that can be used as energy with percentages of 2.7%, 1.9%, 2.1%, 2%, 53.9% and 78.7%, respectively. Materials that can be incinerated, produce emissions from incineration, but end up being regenerated (energy) also have a weight value of 0.5, so half the mass of these materials is taken into account. These compositions also have materials that are disposed of, landfilled (30%, 41%, 43%, 32.8%, 31.8% and 21.3%). Materials that are landfilled do not fully decompose, so they have a weight value of 0. The most recovered materials are wood (160%) and wood fiber insulation (166%) in C6 composition. The circularity is 83% for wood and 78% for wood fiber insulation (C6). In contrast, composition C3 does not have recovered materials. The best material in C3 is insulation, the circularity of which is 25%. Wood and wood-based materials can be recycled and used as energy.

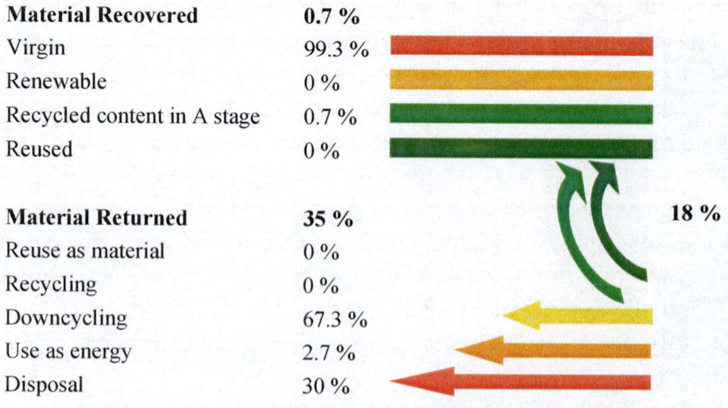

Fig. 1. Circularity of C1 composition

The composition of C1 mainly contains virgin materials up to 99.3%. In the category relating to the content of recycled material (in stage A), mineral wool makes up 0.68% of the total share and is taken to the landfill. In the downcycling category, reinforced concrete makes up 67.08%, of which 2% is reinforcement and the rest is concrete, which is used as a secondary raw material (aggregate) for foundations, for example. In the disposal category, rock wool makes up 0.906% and substrate with share of 29.35%, but it is not included in the BC calculation as stated above in the methodology.

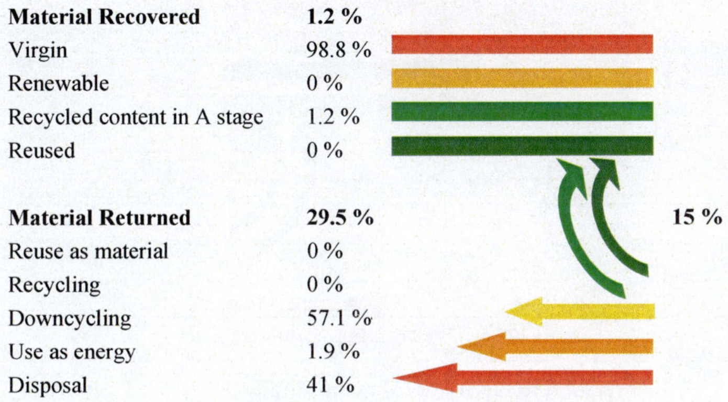

Fig. 2. Circularity of C2 composition.

Composition C2 mostly consists of virgin materials up to 98.84%. In the downcycling category, reinforced concrete makes up 57.24%, of which 2% is reinforcement and the rest is concrete, which is considered to be used as aggregates. In the energy category, in addition to recycled materials and materials that go to the landfill, there are those used for incineration and they make up 1.9%. In the disposal category, mineral wool makes up 1.55% and substrate 39.36% of the total share.

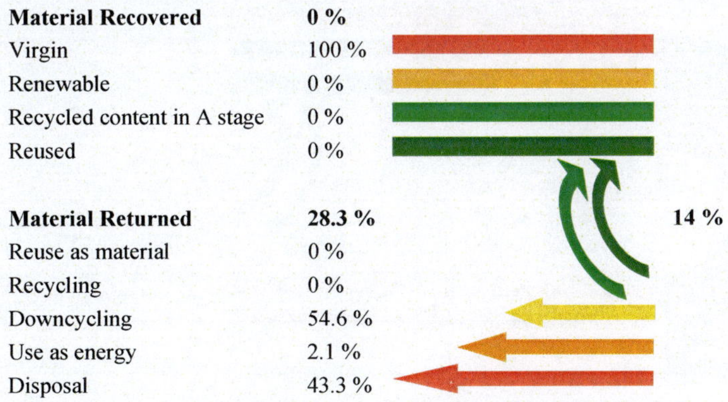

Fig. 3. Circularity of C3 composition.

The composition of C3 mainly contains virgin materials with percentage of 100%. In the downcycling category, reinforced concrete makes up 54.84%, of which 2% is reinforcement and the rest is concrete, which is used as a secondary raw material (aggregate). In the energy category, in addition to recycled materials and materials that go to the landfill, the others used for incineration are 2.07%. Regarding disposal, copolymer-bitumen roofing membrane makes up 0.815% and substrate with percentage of 42.27%.

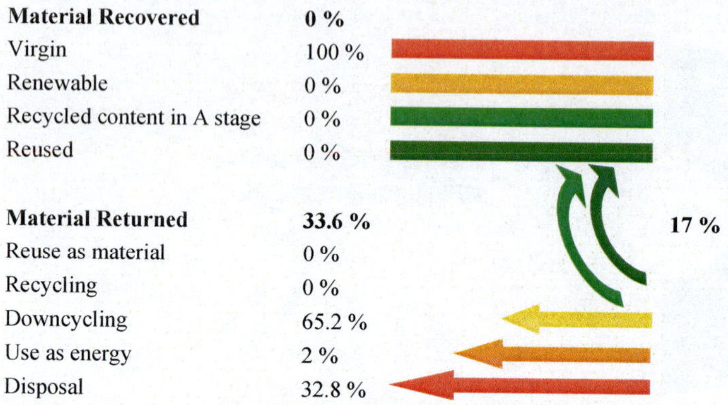

Fig. 4. Circularity of C4 composition.

In the composition of C4, there are mainly virgin materials up to 100%. In the downcycling category, reinforced concrete makes up 64.83%, of which 2% is reinforcement and the rest is concrete, which is used as a secondary raw material (aggregate). In the energy category, in addition to recycled materials and materials that go to the landfill, there are those used for incineration and they make up 2.0%. Concerning disposal, bitumen waterproofing film makes up 4.67% and substrate 28.5% of the total share.

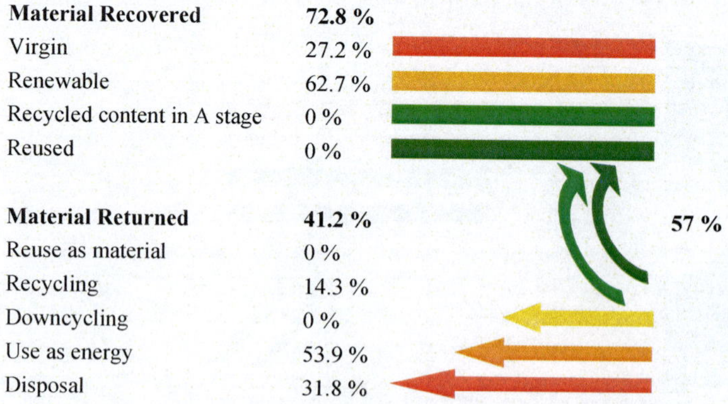

Fig. 5. Circularity of C5 composition.

Compared to the compositions above, C5 contains virgin materials in the percentage of 26.66%. In the renewable material category, materials make up 63.10%. In the category

Recycled content (in stage A), the material plasterboard, wood fiber insulation and organic insulation make up 10.41% of the total share. Gypsum plaster board is recycled and wood fiber insulation is incinerated. Relating to the recycling category, plasterboard makes up 14.58%. In the energy category, in addition to recycled materials and materials that go to the landfill, 53.66% is used for incineration. In the disposal category, wood fiber insulation makes up 31.75% of the total share.

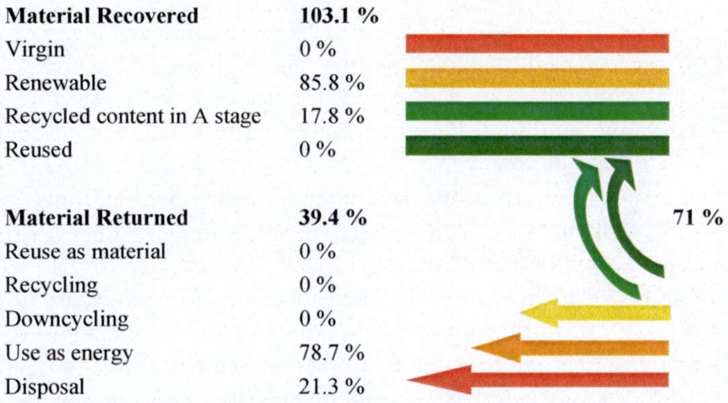

Fig. 6. Circularity of C6.

In the composition of C6, the virgin materials are in the negative values of −3.22%. In the renewable resources category, materials make up 85.75% and non-renewable materials 14.29%. In the category recycled content (in stage A), the material cellulose insulation and medium fiberboard (MDF) make up 17.07% of the total share. Cellulose insulation is landfilled and MDF is incinerated. In the energy category, in addition to recycled materials and materials that go to the landfill, 78.65% are used for incineration. In the disposal category, cellulose insulation and soil substrate make up 21.35% of the total share.

As mentioned above in the analysis related to compositions with load-bearing concrete structure, it should be noted that concrete at the end-of-life is considered to be used as an aggregate (secondary raw material) in foundations. The use of aggregates is addressed in a study [13] which analyzed concretes prepared using a new Eco-hybrid cement, fine recycled concrete aggregate, which was entirely derived from construction and demolition Waste and silica aerogel. Concerning the substrate, in our study the end-of-life is considered landfilling. Study [14] analysis four scenarios of trenches in terms of Life Cycle Costing and Life Cycle Assessment. Alternatives consisted of the classical solution; the classical solution with the reuse of soil; the control low-strength material and the eco-trench. According to the results eco-trench system, for which all material is reused, the environmental and economic impacts could be reduced by more than 80% and 50%, respectively. Study [1] identified as the highest mass among material types concrete, minerals, brick, and wood. There are stated that imperative for reducing the need for new building materials and waste is planning distinguishing materials based on components facilitates utilizing the accumulated materials at the end of life for new construction. Authors in study [15] pointed out that the material circularity potential presents a broad spectrum of pathways to incorporate circular-economy in material management. This research suggests the need for technical bodies to develop rules and standards for reuse of materials as well as a need for government leadership for promotion of circular-economy.

5 Conclusion

A circular building uses more recycled, renewable or reused resources and fewer original materials. Circular processes contributing to circularity can be grouped into 4 categories. Reduction by design, reducing the amount of material used, especially raw materials, should be applied as an overall guiding principle from the earliest stages of product and service design. From a user-to-user perspective, it is possible to refuse, reduce and reuse. From the user's point of view, repair, refurbish and renovation are possible. Reusage and recycling are possible from business to business perspective. Mining and energy recovery are processes used in some countries to manage waste, but the value of the material is consumed once and not returned to the cycle. Therefore, these processes are not further developed on this platform. In this study, 6 compositions were evaluated in terms of building circularity. Green roofs were included in the assessment, four compositions had a concrete supporting structure and two wooden. Circularity of the composition had values of 18%, 15%, 14%, 17%, 57% and 71%. Compositions using wood and wood-based materials have a much higher circularity than concrete, steel and waterproofing. The results presented in this study represent partial results and this research will continue with the aim of increasing the circularity scores of materials and buildings and thus reducing the amount of waste disposed in landfills. Thus, it should be emphasized that circularity in the construction industry means a reduction in the amount of waste generated mainly in manufacturing, construction as well as demolition. In conclusion, it should be noted that circularity in the construction means reduction of the amount of waste occurring during manufacturing as well as during the demolition. Incorporating circularity into the construction industry enables the creation of more sustainable and resilient buildings that contribute to a sustainable future and can also lead to cost savings and improved efficiency in the construction industry.

Acknowledgement. This study was supported by the Grant Agency of the Slovak Republic to support projects No. 1/0512/20 and project NFP313010T578.

References

1. Mohammadiziazi R, Bilec MM (2023) Quantifying and spatializing building material stock and renovation flow for circular economy. J Clean Prod 389:135765
2. Honarvar SMH, Golabchi M, Ledari B (2022) Building circularity as a measure of sustainability in the old and modern architecture: A case study of architecture development in the hot and dry climate. Energy Build 275:112469
3. Disli G, Ankaraligil B (2023) Circular economy in the heritage conservation sector: an analysis of circularity degree in existing buildings. Sustain Energy Technol Assess 56:103126

4. Khadim N, Agliata R, Marino A, Thaheem MJ (2022) Critical review of nano and micro-level building circularity indicators and frameworks. J Clean Prod 357:131859
5. Oluleye BI, Chan DWM, Antwi-Afari P, Olawumi TO (2023) Modeling the principal success factors for attaining systemic circularity in the building construction industry: an international survey of circular economy experts. Sustain Product Consump 37:268–283
6. Schaubroeck S, Dewil R, Allacker K (2022) Circularity of building stocks: modelling building joints and their disassembly in a 3D city model. Proc CIRP 105:712–720
7. Nußholz N, Cetin S, Eberhardt L, De Wolf C, Bocken N (2023) From circular strategies to actions: 65 European circular building cases and their decarbonisation potential. Res Conserv Recycl Adv 17:200130
8. Environmental Footprint Methods Recommendation, https://wayback.archive-it.org/org-1495/20221004164744mp_/https:/ec.europa.eu/environment/eussd/smgp/ef_methods.htm, last accessed 2023/05/18
9. Commission Recommendation of 16.12.2021 on the use of the Environmental Footprint methods to measure and communicate the life cycle environmental performance of products and organisations, 1–8 (2021)
10. Yang M et al (2023) Circular economy strategies for combating climate change and other environmental issues. Environ Chem Lett 21:55–80
11. de Vries BB, Chairman TU, Kunen ITT, BV BG (2016) Building circularity indicators. Doctoral dissertation, Eindhoven University of Technology
12. https://oneclicklca.zendesk.com/hc/en-us/articles/360014998199-Building-Circularity-Circular-Assessment
13. Ozçelikci E, Oskay A, Bayer IR, Sahmaran M (2023) Eco-hybrid cement-based building insulation materials as a circular economy solution to construction and demolition waste. Cement Concr Compos 141:105149
14. Josa I, Petit-Boix A, Casanovas-Rubio MM, Pujadas P, de la Fuente A (2023) Environmental and economic impacts of combining backfill materials for novel circular narrow trenches. J Environ Manage 341:118020
15. Mhatre P, Gedam VV, Unnikrishnan S (2021) Material circularity potential for construction materials—the case of transportation infrastructure in India. Resour Policy 74:102446

Air Distribution in a Room by The Swirled-Compact Compressed Air Jets

Orest Voznyak, Iryna Sukholova(✉), Oleksandr Dovbush, and Kyrylo Andreiko

Lviv Polytechnic National University, Lviv, Ukraine
iryna.y.sukholova@lpnu.ua

Abstract. The paper presents the research results for the characteristics of the forced air flows with increased level of turbulence within compressed room conditions. Flow boundaries for the direct and inverse air flows are determined. Analytical dependencies were found to determine air velocities in a direct and reverse air flow. The aim is to investigate the distribution patterns of air flows from a two-jet air distributor that develop in compressed conditions of the room and to design a mathematical model of swirled-compact air jets at compressed conditions of small-sized production premises. The mathematical model of air tidal by swirled-compact air jets at the room compressed conditions is improved. Graphs and analytical equations based on the experimental investigations results are presented. Characteristics of the swirled-compact air stream at its leakage at the compressed conditions of a premise are determined. The method of shear velocity for determining the velocity of a compressed air jet is substantiated.

Keywords: Air distribution · Air velocity · Swirled jet · Compact jet · Velocity attenuation

1 Introduction

As is known [4, 10–12], in the working area of the premises, standardized parameters of the indoor air, which is determined by the organization of air exchange, must be provided. So does energy saving [1, 6, 7, 13, 24, 26, 27, 30, 38]. The organization of air exchange largely depends on the nature of air currents and the development of ventilation jets in the room [31, 35]. There are many various designs of air distributors and schemes for tidal air jets both to the upper and to the working zone of the room by free [18] and compressed [33, 34] air streams.

It is suitable to use air distributors with a high intensity of attenuation of the air velocity and temperature of the tidal air jet [18, 29], especially for compressed [33, 34] one.

Until recently, the influence of the initial intensity of turbulence was not taken into account in the study of tidal air jets in ventilation [12] and heating [2] technology. It can be assumed that the initial intensity of turbulence affects the parameters of tidal air jets [31]. In real conditions, in ventilation networks, one can expect a significantly high

Z. Blikharskyy et al. (Eds.): CEE 2023, LNCE 438, pp. 476–483, 2024.
https://doi.org/10.1007/978-3-031-44955-0_48

intensity of turbulence on the approach to the tidal nozzles [35]. One of the methods of increasing the intensity of turbulence is the application of twisting the air jet [33, 34].

In this work, a tidal nozzle was chosen for research, which forms two air flows: a swirled air jet and a flat one, which develop in the limited space of a low-height room, that is, in compressed conditions. Leakage of the air jets in a limited space (compressed air jets) can occur according to a dead-end or flowing scheme. The research was carried out with a dead-end scheme, where the exhaust hole is located in the same plane as the tidal.

2 Analysis of Literary Data and Problem Statement

During room ventilation the normative air temperature and velocity in the serviced zone must be maintained [14, 15]. This is possible due to nozzle low attenuation coefficients of velocity m and temperature n [8]. Effective way turbulence increase is the swirled and compact air flows use. The dynamic indoor climate creation is also effective measure [36].

The graphs and analytical equations for the calculations of the velocities and temperatures of compressed air streams were considered and analyzed. It should be noted that it is advisable to substitute the corresponding attenuation coefficients m and n in the calculation formulas for air flows [37].

Especially noteworthy is the effect of air jet twisting. This provides more intensive turbulence of air flow and its velocity attenuation. The lowest attenuation coefficients of air streams m and n are needed exactly in the compressed conditions of production premises. In some works mathematical modeling [16, 17, 19–22, 32, 39] and numerical simulation [2, 3, 5] has been regarded. On the basis of the literature data analysis of the tidal air jets leakage it is summarized: the compression coefficients and twisting ones of the tidal air streams are needed; there is no suitable methodology for the calculation of the swirled and compact air streams in the room compressed conditions; so does an universal mathematical model of its leakage in the room compressed conditions.

3 The Purpose and Objectives of the Study

The purpose of this work is to investigate the distribution patterns of air flows from a two-jet air distributor (TJAD) that develop in compressed conditions of the room and to design a mathematical model of swirled-compact air jets at compressed conditions of small-sized production premises. To achieve this purpose, it is necessary to carry out the following investigation objectives:

- to analyze existing methodologies of the air streams calculation;
- to generalize and deepen the theory of the tidal air aerodynamics by swirled and compact air jets in a compressed conditions of the room;
- to perform an experimental research of the air distribution in the premise by swirled and compact air streams in compressed room conditions.

4 Methods, Materials and Research Results

Experimental studies were carried out on the installation shown in Fig. 1 under the following conditions, simplifications and assumptions:

- air jets are isothermal;
- the diameters of the TJAD were: $d = 150$ mm, $d = 200$ mm, $d = 250$ mm;
- the initial air jet velocity was within 5–10 m/s;
- air volume flow rate was approximately in the range of 320–1760 m^3/h;
- The velocity in the air jet flow was measured by a Testo-405 thermoelectric anemometer.
- Experimental points were fixed with a 10×10 cm coordinate grid.

Fig. 1. Photo of the experimental installation: a two-jet air distributor (TJAD) that forms a swirled and flat surface jet with twisting plates for an air stream swirling.

The results of the experiments proved the following:

1. The initial volume of air (the so-called transit flow), having reached the end of the room, turns around and moves towards the main flow.
2. The highest value of the velocity in the return flow was observed in the plane of the second critical section, in which the air flow in the return flow is the largest, and the area for the passage of the flow is the smallest.

3. In the first critical section, the air jet occupied about 25% of the cross-sectional area of the room, and in the second–40–45%.

The influence of limited space on the development of compressed air jets is practically convenient to take into account by entering the appropriate compression coefficients into the calculation formulas for the free air jets. Compression coefficients are the ratio of the air velocity, flow or amount of movement in a compressed air jet to the corresponding values under the conditions of a free air jet.

In this work, it is proposed to use the well-known "velocity shift" method [37] to find the compression coefficients of swirled and flat air jets. According to this method, the velocity of the compressed air jet v_c is determined as the difference between the velocity of the free air jet v and the so-called shear velocity v^*, which is constant in each cross-section of the flow. Then the axial velocity v_{xc} of the compressed jet should be determined by Eq. (1):

$$v_{xc} = v_x - \cdot v_x^*; \tag{1}$$

In the main section the compression coefficient k_c is determined from Eq. (2):

$$k_c = \frac{v_c}{v} = 1 - \frac{v^*}{v}; \tag{2}$$

The shear rate is determined from the equation of air flow in a compressed jet, according to which the resulting flow in any cross section of a dead-end jet is zero.

It is necessary to take into account the fact that in the initial section of the air jet, its properties depend on the initial compression, which is determined by the parameter $d/\sqrt{F}$ (F – is the cross-sectional area of the room in which the jet develops, m^2), and in the main section–from the initial compression of the relative distance from the air outlet $x/\sqrt{F}$, and attenuation coefficient m as its characteristics. Based on this, the compression ratio is obtained in the main section of the compressed swirled air jet. But it should be noted that the analytical method of determination is quite complex, so it is advisable to resort to experimental studies [40].

Figure 2 shows the results of experimental studies of the development of a swirled air jet under compressed conditions. In the course of the experiments, the boundaries of the air jets, the zones of the direct and reverse air flows have been determined.

The maximum velocity in the reverse flow is determined from the equation, in which the correction coefficient β is introduced, which takes into account the additional initial intensity of turbulence due to the twisting of the air jet Eq. (3):

$$v_{\max} = \beta v_0 \cdot \sqrt{\frac{F_0}{F}}; \tag{3}$$

where F_0 – nozzle opening area, m^2.

The coefficient β is determined by dependence Eq. (4):

$$\beta = \frac{2G}{4 - G^2}; \tag{4}$$

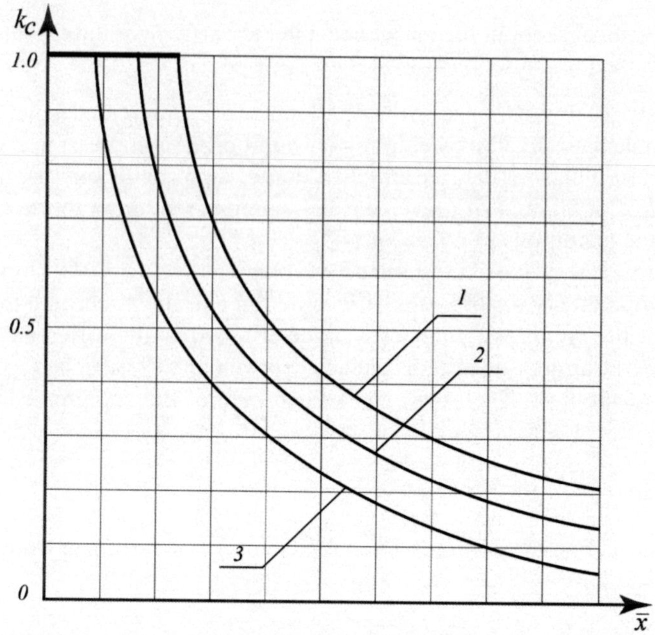

Fig. 2. Results of experiments on determination of compression coefficients by velocity: 1–d = 150 mm; 2–d = 200 mm; 3–d = 250 mm.

where G is the ratio of the maximum swirled velocity to the maximum axial velocity:

$$G = \frac{v_{sw\,\max}}{v_{x\,\max}}; \tag{5}$$

Experimental studies have shown that twisting affects the flow field, the expansion of the air jet, the processes of mixing and attenuation of the air jet velocity. All these characteristics are affected by the intensity of the flow twisting, which is characterized by the twisting parameter. This is the dimensionless ratio of the axial component of the momentum flow to the product of the axial component of the momentum flow and the equivalent radius of the tidal opening.

The adequacy confirmation of the theory of the experiment has been obtained.

5 Conclusions

1. The influence of the initial intensity of turbulence on the regularity of the air jets development, which flow from the two jet air distributor in the compressed conditions of the room, has been quantitatively established.
2. It was established that the initial section of the air jet decreases, and the axial velocity in the main section decreases more intensively with an increase in the diameter of the nozzle.
3. A two jet air distributor provides a high initial turbulence intensity and sufficient attenuation of the air velocity.

4. The generalized calculation dependencies for air distribution in a room by the compressed swirled and compact air streams are obtained.
5. The mathematical model of the compressed swirled and compact air streams is improved.

References

1. Adamski, M, Kiszkiel, P (2014) Condensation phenomena and frost problems in the air heat recuperators. MATEC Web Conf 18:01001
2. Basok B, Davydenko B, Farenuyk G, Goncharuk S (2014) Computational modeling of the temperature regime in a room with a two-panel radiator. J Eng Phys Thermophys 87(6):1433–1437
3. Basok B, Davydenko B, Isaev S, Goncharuk S, Kuzhel L (2016) Numerical modeling of heat transfer through a triple-pane window. J Eng Phys Thermophys 89(5):1277–1283
4. Bullova I, Kapalo P, Katunsky D (2021) Quantification of air rate change by selected methods in a typical apartment building. Buildings 11(4):1–20
5. Gorobets V, Trokhaniak V, Masiuk M, Spodyniuk N, Blesnyuk O, Marchishina Y (2021) CFD modeling of aerodynamic flow in a wind turbine with vertical rotational axis and wind flow concentrator. Agric Eng 64(2):159–166
6. Gulay B, Sukholova I, Dzeryn O, Shepitchak V (2020) Investigations of compact recuperators acoustic properties. In: Blikharskyy Z (ed) Proceedings of EcoComfort, vol 100, Lviv, Ukraine, September 16–18, Lecture Notes in Civil Engineering, pp 127–133
7. Gumen O, Spodyniuk N, Ulewicz M, Martyn Y (2017) Research of thermal processes in industrial premises with energy-saving technologies of heating. Diagnostyka 18(2):43–49
8. Hnativ R, Verbovskiy O (2019) Distribution of local velocities in a circular pipe with accelerating fluid flow. East Europ J Enterp Technol 2(7–98):58–63
9. Hulai B, Dovbush O, Piznak B, Kasynets M (2020) Studying equalization of the radial fans discharge flow. Lect Notes Civ Eng 47:119–126
10. Kabosova L, Katunsky D, Kmet S (2020) Wind-based parametric design in the changing climate. Appl Sci 10(23):8603
11. Kapalo P, Klymenko H, Zhelykh V, Adamski M (2020) Investigation of indoor air quality in the selected Ukraine classroom–case study. Lect Notes Civ Eng 47:168–173
12. Kapalo P, Sedláková A, Košicanová D, Voznyak O, Lojkovics J, Siroczki P (2014) Effect of ventilation on indoor environmental quality in buildings. In: The 9th international conference "environmental engineering". Vilnius, Lithuania. Selected Papers 2014, https://doi.org/10.3846/enviro.2014.265
13. Kapalo P, Spodyniuk N (2018) Effect of the variable air volume on energy consumption - Case study. IOP Conf Ser Mater Sci Eng 415(1):012027
14. Kapalo P, Sulewska M, Adamski M (2020) Examining the interdependence of the various parameters of indoor air. In: Blikharskyy Z (ed) Proceedings of EcoComfort 2020, vol 100, Lviv, Ukraine, September 16–18, 2020, Lecture Notes in Civil Engineering, pp 150–157
15. Kapalo P, Vilcekova S, Meciarova L, Domnita F, Adamski M (2020) Influence of indoor climate on employees in office buildings. A Case Study Sustain (Switzerland) 12(14):5569
16. Kharchenko Y, Blikharskyy Z, Vira V, Vasyliv B (2019) Study of structural changes in a nickel oxide containing anode material during reduction and oxidation at 600 °C. Springer Proc Phys 221:595–604
17. Klymenko H, Labay V, Yaroslav V, Gensetskyi M (2020) Criterial equation for the description of low-speed air distributor operation. Lect Notes Civ Eng 47:235–242

18. Korbut V, Voznyak O, Myroniuk K, Sukholova I, Kapalo P (2017) Examining a device for air distribution by the interaction of counter non-coaxial jets under alternating mode. East Europ J Enterp Technol 8(86):30–38. https://doi.org/10.15587/1729-4061.2017.96774
19. Krainskyi P, Vegera P, Khmil R, Blikharskyy Z (2019) Theoretical calculation method for crack resistance of jacketed RC columns. IOP Conf Ser Mater Sci Eng 708(1):012059
20. Labay V, Dovbush O, Yaroslav V, Klymenko H (2018) Mathematical modeling of a split-conditioner operation for evaluation of exergy efficiency of the R600A refrigerant application. Math Model Comput 5(2):169–177
21. Labay V, Savchenko O, Zhelykh V, Kozak K (2019) Mathematical modelling of the heating process in a vortex tube at the gas distribution stations. Math Model Comput 6(2):311–319
22. Labay V, Yaroslav V, Dovbush O, Tsizda A (2020) Mathematical modeling of an air split-conditioner heat pump operation for investigation its exergetic efficiency. Math Model Comput 7(1):169–178
23. Lee Y-K, Kim Y (2022) Analysis of indoor air pollutants and guidelines for space and physical activities in multi-purpose activity space of elementary schools. Energies 15(1):220
24. Lis P, Lis A, Janik M (2012) Aspects of the analytical heat consumption monitoring in local buildings' population. Rynek Energii 102(5):67–75
25. Lis A, Spodyniuk N (2019) The quality of the microclimate in educational buildings subjected to thermal modernization. E3S Web Conf 100:00048
26. Marushchak U, Sanytsky M, Pozniak O, Mazurak O (2019) Peculiarities of nanomodified portland systems structure formation. Chem Chem Technol 13(4):510–517
27. Novosad P, Pozniak O, Melnyk V, Braichenko S (2020) Porous thermal insulation materials on organic and mineral fillers. Lect Notes Civ Eng 47:354–360
28. Pietrucha T (2017) Ability to determine the quality of indoor air in classrooms without sensors. E3S Web Conf 17:00073
29. Spodyniuk N, Gulai B, Zhelykh V, Shapoval S (2019) Leveling of pressure flow of radial ventilator in mine ventilation system. Naukovyi Visnyk Natsionalnoho Hirnychoho Universytetu 6:80–86
30. Spodyniuk N, Lis A (2021) Research of temperature regime in the module for poultry growing. Lect Notes Civ Eng 100:451–458
31. Spodyniuk N, Voznyak O, Sukholova I, Dovbush O, Kasynets M, Datsko O (2021) Diagnosis of damage to the ventilation system. Diagnostyka 22(3):91–99
32. Tymchuk I, Malovanyy M, Shkvirko O, Zhuk V, Masikevych A, Synelnikov S (2020) Innovative creation technologies for the growth substrate based on the man-made waste - Perspective way for Ukraine to ensure biological reclamation of waste dumps and quarries. Int J Foresight Innov Policy 14(2–4):248–263
33. Voznyak O, Korbut V, Davydenko B, Sukholova I (2019) Air distribution efficiency in a room by a two-flow device. Adv Res Saving Technol Mater Civ Environ Eng 47:526–533. https://doi.org/10.1007/978-3-030-27011-7_67
34. Voznyak O, Savchenko O, Spodyniuk N, Sukholova I, Kasynets M, Dovbush O (2022) Improving of ventilation efficiency at air distribution by the swirled air jets. Pollack Periodica 17(1):123–127
35. Voznyak O, Spodyniuk N, Sukholova I, Savchenko O, Kasynets M, Datsko O (2022) Diagnosis of three types damages to the ventilation system. Diagnostyka 23(1):2022102
36. Voznyak O, Sukholova I, Myroniuk K (2015) Research of device for air distribution with swirl and spread air jets at variable mode. East Europ J Enterp Technol 6/7(78):15–23. https://doi.org/10.15587/1729-4061.2015.56235
37. Voznyak O, Spodyniuk N, Dovbush O, Savchenko O, Sukholova I, Kasynets M, Kirakevych I (2023) Research of the round convective air jet of the exhaust local ventilation. Pollack Periodica 18(1):60–65. https://doi.org/10.1556/606.2022.00691

38. Vranay F, Vranayova Z (2020) Influence of heat source choice on building energy certification process and CO_2 emissions. In: Blikharskyy Z, Koszelnik P, Mesaros P (eds) Proceedings of CEE 2019 on lecture notes in civil engineering vol 47, Lviv, Ukraine, September 11–13, 2019, pp 541–548
39. Zhuk V, Matlai I, Vovk L, Popadiuk I (2023) Analytical and experimental assessment of regulating volume of the stormwater storage tanks for rains of constant intensity. Lect Notes Civ Eng 290:459–469
40. Zhuk V, Verbovskyi O, Popadiuk I (2022) Experimental regulating parameters of bladder-type hydraulic accumulator. Int J Appl Mech Eng 27(1):232–243

Field Tests and Analysis of Flat Monolithic Reinforced Concrete Slabs

Yurii Vybranets[1(✉)], Svitlana Vikhot[1], and Sofiya Burchenya[2]

[1] Department of Building Production, Lviv Polytechnic National University, Lviv, Ukraine
yura.vybranets@gmail.com

[2] Department of Construction Technology and Organization, Lviv National Environmental University, Lviv, Ukraine

Abstract. Flat slabs in frame buildings are designed in the form of a smooth solid slab, with capitals (column heads), smooth or with capitals and contour beams around the perimeter of the building, with intercolumnar beams in one or two directions. When designing slabs, engineers are faced with the chaotic arrangement of technological openings, the variable pitch of columns, the non-linear work (need to reduce the modulus of elasticity of concrete, which affects the overall rigidity of structures). Analyzing the results of the reinforcement of similar flat slabs by different designers (the same scheme of work, approximately the same spans and loads), the engineers made a decision about field tests to establish their reliability and compliance with current standards. I would also like to note the lack of study of the work of flat slabs resting directly on the columns and the lack of recommendations regarding the location of technological openings in the regulatory literature. Sections of flat slabs in the zones of the largest spans (6.0–6.6 m) of three residential buildings in Lviv were taken for testing. Based on the results of field tests and instructions given in the technical and normative literature, recommendations were developed for the selection of constructive solutions and reinforcement of flat monolithic reinforced concrete slabs.

Keywords: Field tests · Modulus of elasticity · Flat slabs · Reinforcement · Deflections · Crack resistance · Strength

1 Introduction

More than 10 years ago, Ukraine took a course towards Europe not only politically, but also in the use of new materials to restore bearing capacity and increase durability [1, 2], new methods of calculating building structures made of common materials (metal, reinforced concrete) [3–10] and unpopular (glass) [11, 12]. A European approach is needed in the calculation of artificial structures [13–15] and in the reconstruction of buildings [16]. Calculations have become more complicated and now it is difficult to imagine a modern design engineer without skills in working with special computer programs. However, one should be very careful here – the computer always calculates correctly, but is the output data set correctly? There are discussions about this in the structure engineer's community and there are no unequivocal answers yet.

Z. Blikharskyy et al. (Eds.): CEE 2023, LNCE 438, pp. 484–497, 2024.
https://doi.org/10.1007/978-3-031-44955-0_49

The need to improve design solutions, insufficient study of the constructions of flat slabs, the constant development of engineering systems and modern architectural trends became the reason for the test. Another reason was the different results of reinforcement of similar flat slabs (thickness, load, spans). Engineers made a decision about field tests of monolithic reinforced concrete slabs on real objects in the city of Lviv to establish their reliability and compliance with current standards. The purpose of the study was to determine the stiffness characteristics (deflections), crack resistance of experimental structures, comparison of experimental results with theoretical data.

2 Methodology of Experiments

The fragments of three slabs - S-1, S-2, S-3 - were chosen for the tests. Cross-sections of maximum spans and largest calculated deflections are loaded. The concrete slabs have reached the design strength (performed at least 28 days before the test), free from partitions, formworks and other objects. Acts on hidden works certifying their reinforcement are drawn up on these slabs. Before the test, the actual strength of the concrete was established by a non-destructive method (using a sclerometer) and a geodetic survey was carried out to determine the existing deformations. Experimental tests were carried out in accordance with [17] on the construction site. According to the assigned tasks, the load on the structure was applied as close as possible to its real work. Experimental structures were loaded with pallets of building materials. Their size and weight depended on the material: gas block weight 830 kg; cement, weight 1500 kg; glue, weight 1200 kg. The load on the fragment of the slab was carried out in stages with a step of 20% to the control working value (Fig. 1). When the cracks develop intensively, the load is stopped. The readings of the measuring devices were taken 15 min after each level of load. There was a break of at least 30 min after the last reading.

During field tests, vertical deformation of the slab and cracks width were measured at each stage of loading.

Deflections of the structure were measured by deflection gauges with a division value of 0.01 mm and 0.1 mm. In addition, before and after the tests were measured with geodetic devices. The deflection gauges were installed on formwork racks that were not connected to the test plate (Figs. 2, 4 and 6). The wire of the deflection gauge was rigidly fixed to the test plate with dowels. Deflections of the structure were monitored in characteristic places of the slab (maximum theoretical deflections, middle part, column strip) according to the developed schemes. During the tests, crack formation and crack development were observed. The moment of crack formation was determined by visual inspection in characteristic places using a magnifying glass. When they appeared, the crack opening width was measured with an MPB-2 microscope. Measurements by devices, as well as an inspection of places for the appearance of cracks, were carried out after each load step and entered in the test log.

2.1 Program of Experimental Research

In the process of developing the test program, a numerical experiment was performed using the LIRA-SAPR software complex, which included two loading schemes for the

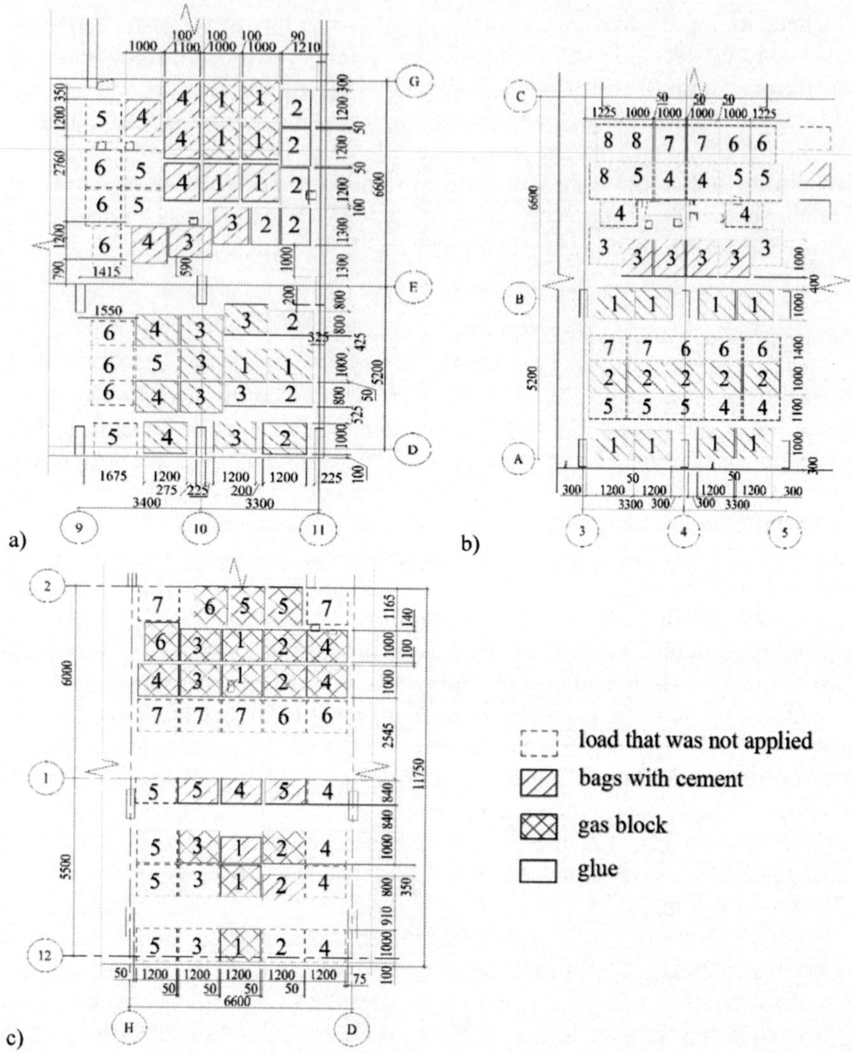

Fig. 1. Schemes of slabs loading (numbers indicate loading levels): a) S-1; b) S-2; c) S-3.

slab: the operational value of the load was applied over the entire area of the slab; the operational value of the load was applied to the fragment of the slab that was supposed to be tested.

At the same time, the values of the deflections in the second loading scheme were 6–10% smaller. Therefore, when determining the actual deflections when loading the overlap fragment, the value should be increased by 6–10%. Calculations are made both for the operational value of loads and for equivalent experimental loads. Uniformly distributed loads are assumed in the calculation. Only values of vertical loads were taken into account - seismic and wind effects were not applied.

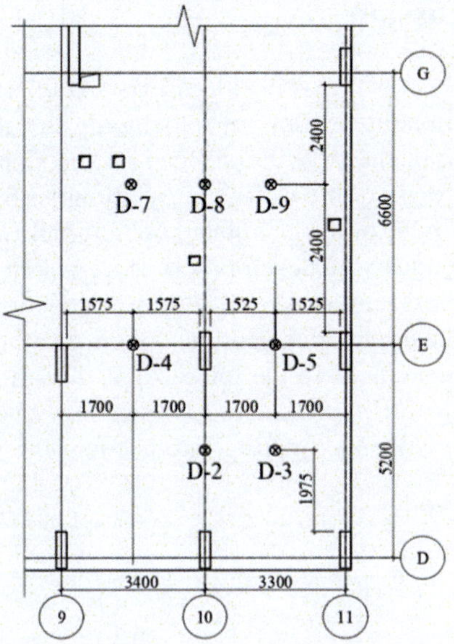

Fig. 2. Scheme of arrangement of deflection gauges S-1.

The background reinforcement of slabs S-1 and S-2 is made in steps of 200 mm in the lower zone with Ø10 A500S reinforcement, in the upper one - Ø8 A500S. Additional zones of the lower reinforcement are made of reinforcement Ø12 A500S in steps of 100 and 200 mm (S-1) and in steps of 200 mm (S-2), upper - Ø12 A500S in steps of 100 and 200 mm and Ø16 A500S in steps of 100 mm (S-1) and 200 mm (S-2). The background reinforcement of the P-3 is made in the lower and upper zones with Ø10 A500S reinforcement in step of 250 mm. Additional zones of the lower reinforcement are made with Ø10 A500S reinforcement in steps of 125, 250 and 500 mm, the upper - Ø10 A500S in steps of 125 and 250 mm and Ø16 A500S in steps of 125 mm. The length of the additional upper reinforcement above the columns in the span: for S-1 – 1.23 m, or 0.19 L (between axes E/1.1-G/1.1), for S-2 – 1.38 m or 0.21 L (between axes B-C), for S-3 - 1.0…1.2 m or (0.15…0.18) L. Additional reinforcement of the holes is made with 2 Ø16 A500S bars on each side in the upper and lower zones for S-1 and S-2, and 2 Ø10 A500S bars on each side in the upper, lower and corner (diagonal bars) zones for S-3. Welded compression frames with vertical reinforcement Ø10 A500S and horizontal reinforcement Ø8 A500S for S-1 and S-2 are arranged around all columns. Since the frames are made on the construction site by manual welding, using rebar of Ø8 is not allowed [18]. For S-3, compression reinforcement is made with bent rods Ø10 A500S on the ends of most of the columns. The total steel consumption per slab is: S-1 – 103.6 kg/m^3, S-2 – 91.6 kg/m^3, S-3 – 87.7 kg/m^3.

3 Results and Discussion

The S-1 slab was tested between axes 9–11 and D-G (Fig. 1a). The total applied load on this zone was approximately 430 kN (not including the slab's own weight), which is 5.21 kN/m^2. The value of the operational load on the slab according to the project (taking into account the weight of the walls and partitions) is 6.58 kN/m^2. Therefore, the load during the test was 79.2% of the calculated operational value. The test was stopped at the 4th stage of loading due to the development of a crack in the upper zone on the end of the diaphragm at the intersection of the G-9 axes (at the location of the hole) (Fig. 3). Cracks also appeared in the upper zone near the columns and barely noticeable (with an opening width of 0.05 mm) cracks in the lower zone. The maximum deflection during the tests was 6.9 mm, while the theoretical one is: 2.99 mm when calculated in the elastic stage and 6.23 mm when calculated taking into account the non-liner behavior of the materials (Table 1).

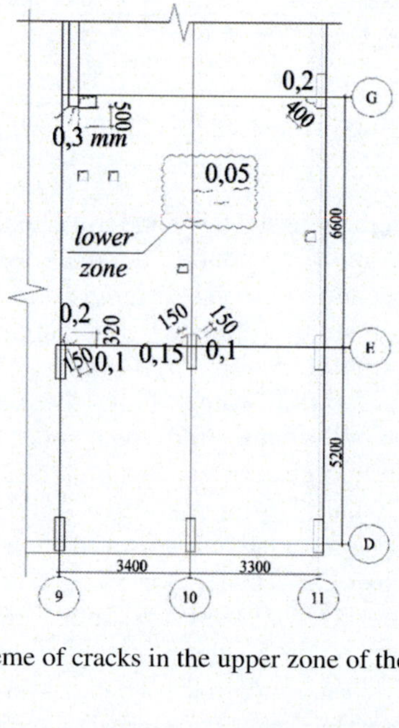

Fig. 3. The scheme of cracks in the upper zone of the slab S-1 (mm).

The S-2 slab was tested between axes 3–5 and A-C. The total applied load on this zone was approximately 26 kN (not including the slab's own weight).

Approximately 19.9 kN was applied in axes 3–5 and A-B, which is 5.0 kN/m^2. The value of the operational load on the slab according to the project (taking into account the weight of the walls and partitions) is 5.55 kN/m^2. Therefore, the load during the test was 90% of the calculated operational value. Approximately 6.1 kN was applied in axes 3–5 and B-C, which is 1.48 kN/m^2. The value of the operational load on the slab in this zone according to the project (taking into account the weight of the walls and partitions) is 6.16 kN/m^2. Therefore, the load during the test was 24% of the calculated operational value. The test was stopped at the 3rd stage of loading (only 24% of the

Table 1. The value of experimental and theoretical deflections of the slab S-1.

Stage	Experiment (mm)	Theoretical (mm)	
		Linear	Non-linear
D-2			
1	0.13	0.095	0.17
2	0.07	0.026	0.03
3	0.43	0.31	0.635
4	0.205	0.075	0.11
Total	0.835	0.506	0.945
D-3			
1	0.3	0.25	0.451
2	0.2	0.13	0.194
3	0.55	0.3	0.597
4	0.2	−0.01	−0.029
Total	1.25	0.67	1.213
D-4			
1	0.07	0.07	0.016
2	0.02	−0.015	−0.018
3	0.04	0.02	0.027
4	0.285	0.137	0.268
Total	0.415	0.212	0.293
D-5			
1	0.2	0.1	0.173
2	0.2	0.12	0.232
3	0.25	0.1	0.185
4	0.05	0	0
Total	0.7	0.32	0.59
D-7			
1	1.06	0.6	1.28
2	0.37	0.195	0.372
3	0.41	0.153	0.216
4	0.57	0.096	2.12
Total	2.41	1.044	3.988
D-8			

(*continued*)

Table 1. (*continued*)

1	2.2	1.13	2.448
2	1.1	0.5	0.975
3	0.9	0.25	0.344
4	2.7	1.11	2.464
Total	6.9	2.99	6.231
D-9			
1	2	1.17	2.675
2	1.5	0.85	1.865
3	1.1	0.22	0.315
4	2.1	0.67	1.4
Total	6.7	2.91	6.255

calculated operational value for this fragment) due to the appearance of cracks in the lower zone near the holes, which completely cut the strips between them (Figs. 4 and 5).

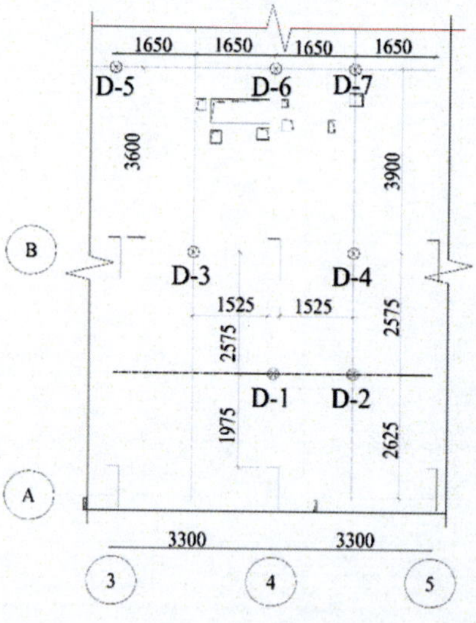

Fig. 4. Scheme of arrangement of deflection gauges S-2.

The maximum deflection during the tests was 0.61 mm, while the theoretical one is: 0.35 mm when calculated in the elastic stage and 0.71 mm when calculated taking into account the non-linear behavior of the materials (Table 2).

The S-3 slab was tested between the axes H-D and 2–12. The total applied load on this area was approximately 190 kN (not including the slab's own weight), which is 2.5 kN/m^2. The value of the operational load on the slab according to the project (taking

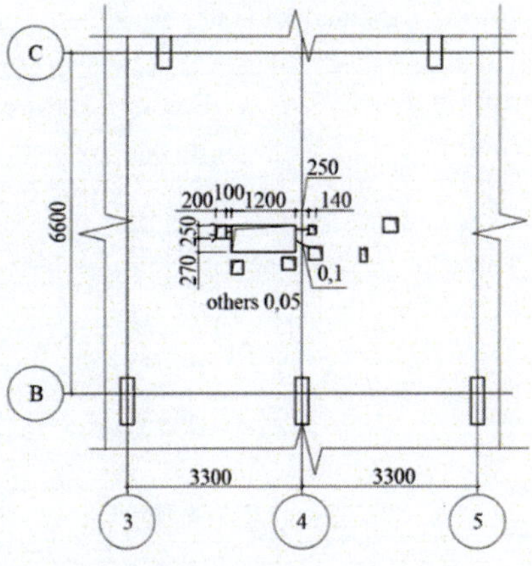

Fig. 5. The scheme of cracks in the lower zone of the slab S-2 (mm).

into account the mass of the walls and partitions) is 5.79 kN/m^2. Therefore, the load during the test was 43% of the calculated operational value. It should also be noted the presence of shrinkage cracks on the tested sample in the upper zone of the slab between axes 1–12 and H-D. The test was stopped at the 6th stage of loading due to intensive development of cracks in the upper zone along the face of the columns (Figs. 6 and 7).

The width of the crack was 0.1 mm, the length was 2300 mm, and the depth was 140 mm at the end of the slab. The maximum deflection during the tests was 3.12 mm, while the theoretical one is: 2.2 mm when calculated in the elastic stage and 3.24 mm when calculated taking into account the non-linear behavior of the materials (Table 3).

4 Conclusions

Tests stop in all slabs was caused by the development of cracks in the zones of maximum bending moments. In slabs S-1 and S-3 - in the upper zone (Figs. 2 and 4), in slab S-2 – in the lower zone (Fig. 3). The reason for the intensive development of cracks in the slab S-1 was an incorrectly located hole near the diaphragm. In the slab S-2, in our opinion, the reason was insufficient additional reinforcement around the holes. The cross-sectional area of the embedded (background and additional) reinforcement is 19.16 cm^2, and the cross-sectional area of the additional reinforcement around the holes is 8.04 cm^2, which is 42% of the removed reinforcement. It was in this place (along axis 4) where a crack appeared (Fig. 3) with the largest width (0.1 mm). In slab S-3, the length of the crack in the upper zone - approximately 2300 mm and the depth at the end of the slab - approximately 140 mm - was decisive for stopping the test, although the width of its was only 0.1 mm (Fig. 4). The possible reason for their early development, in our opinion, was shrinkage cracks caused by a violation of the slab concreting technology, as well as insufficient length of additional upper reinforcement above the columns in the span in the H and D axes. The total consumption of reinforcement for the tested slabs is within 87.7–103.6 kg/m^3.

Table 2. The value of experimental and theoretical deflections of the slab S-2.

Stage	Experiment (mm)	Theoretical (mm)	
		Linear	Non-linear
D-1			
1	0.11	0.051	0.07
2	0.43	0.316	0.653
3	0.065	−0.017	−0.016
Total	0.61	0.35	0.71
D-2			
1	0.14	0.069	0.1
2	0.317	0.303	0.6
3	0.093	−0.013	−0.01
Total	0.55	0.36	0.69
D-3			
1	0.15	0.083	0.125
2	0.05	0.007	0.03
3	0.14	0.104	0.123
Total	0.34	0.19	0.28
D-4			
1	0.108	0.087	0.125
2	0.04	0.015	0.04
3	0.17	0.081	0.107
Total	0.32	0.18	0.27
D-5			
1	0.02	0.005	0.004
2	0.01	−0.012	−0.012
3	0.27	0.165	0.187
Total	0.30	0.16	0.18
D-6			
1	0.05	0.013	0.016
2	-0.03	−0.019	−0.021
3	0.45	0.315	0.51
Total	0.47	0.31	0.51
D-7			

(continued)

Table 2. (*continued*)

1	0.01	0.012	0.014
2	−0.06	−0.017	−0.018
3	0.37	0.25	0.388
Total	0.32	0.25	0.38

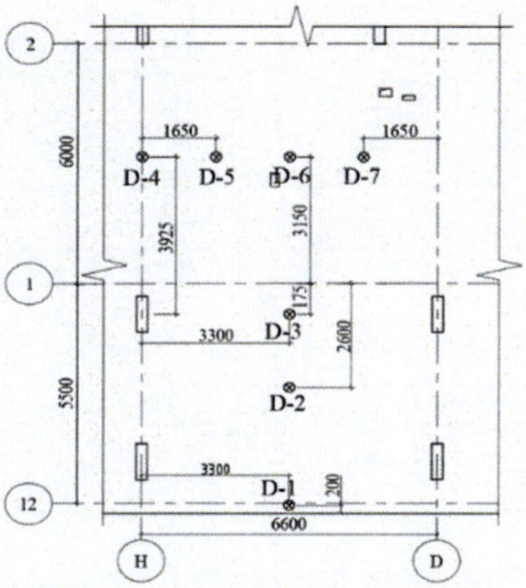

Fig. 6. Scheme of arrangement of deflection gauges S-3.

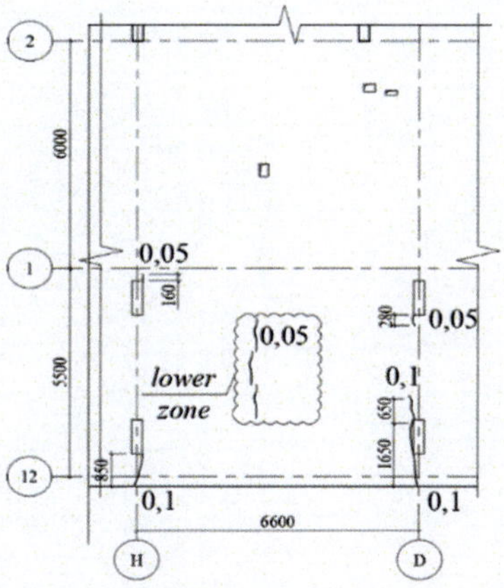

Fig. 7. The scheme of cracks in the upper zone of the slab S-3 (mm).

In all slabs during the test, deflections greater than the theoretical deflections calculated in the elastic stage were noticed (Tables 1, 2 and 3). The dependence of their increase at longer spans and higher stages of loading was also revealed. For the slab S-1 with spans up to 5.2 m, this ratio was 1.71 times on average and almost not depend of the load value. And with spans of 6.6 m, it was already 2.84 times on average, with an

Table 3. The value of experimental and theoretical deflections of the slab S-3.

Stage	Experiment, mm	Theoretical (mm)	
		Linear	Non-linear
D-1			
1	0.65	0.744	1.34
2	0.43	0.25	0.41
3	0.09	0.058	0.07
4	0.03	0.025	0.04
5	0.095	0.044	0.07
6	0.065	0	−0.002
Total	**1.36**	**1.12**	**1.93**
D-2			
1	0.5	0.658	1.19
2	0.35	0.377	0.56
3	0.25	0.17	0.2
4	0.2	0.174	0.26
5	0.3	0.25	0.38
6	0.17	0.006	0.006
Total	**1.77**	**1.64**	**2.60**
D-3			
1	0.49	0.475	0.85
2	0.37	0.311	0.46
3	0.31	0.207	0.265
4	0.455	0.367	0.55
5	0.545	0.443	0.66
6	0.12	0.033	0.035
Total	**2.29**	**1.84**	**2.82**
D-4			
1	0.14	0.148	0.21
2	0.08	0.081	0.1
3	0.22	0.228	0.33
4	0.285	0.249	0.37
5	0.215	0.106	0.14
6	0.21	0.156	0.19

(*continued*)

Table 3. (*continued*)

Total	**1.15**	**0.97**	**1.34**
D-5			
1	0.3	0.358	0.54
2	0.22	0.222	0.3
3	0.61	0.42	0.62
4	0.49	0.38	0.57
5	0.58	0.26	0.355
6	0.25	0.215	0.27
Total	**2.45**	**1.86**	**2.66**
D-6			
1	0.5	0.49	0.78
2	0.4	0.37	0.53
3	0.65	0.38	0.55
4	0.65	0.44	0.67
5	0.6	0.34	0.49
6	0.32	0.18	0.215
Total	**3.12**	**2.20**	**3.24**
D-7			
1	0.4	0.32	0.49
2	0.38	0.373	0.55
3	0.32	0.2	0.27
4	0.6	0.46	0.73
5	0.4	0.25	0.36
6	0.2	0.09	0.1
Total	**2.30**	**1.69**	**2.50**

average of 1.88 times on the 1st and 2nd stages of loading, and 3.8 on the 3rd and 4th times The decrease in the modulus of elasticity of concrete was $E_{cd} \leq (0.35\ldots0.75)\ E_c$.

According to the dynamics of the experimental deflections, it can be stated that they are significantly smaller than the limit values even when calculated with the lower value of the stiffness of the slab, which are determined taking into account the reinforcement, the formation of cracks and the development of plastic deformations in concrete and reinforcement [19].

References

1. Blikharskyy Z, Bobalo T, Kramarchuk A, Ilnytskyy B, Vashkevych R (2020) Bearing capacity of stone beam reinforced by GFRP. In: Proceedings of 2nd international scientific conference

on EcoComfort and current issues of civil engineering EcoComfort, vol 100, Lviv; Ukraine, 16–18 September 2020, pp 42–52. Lecture Notes in Civil Engineering. https://doi.org/10.1007/978-3-030-57340-9_6
2. Marushchak U, Sydor N, Braichenko S, Margal I, Soltysik R (2019) Modified fiber reinforced concrete for industrial floors. IOP Conf Ser Mater Sci Eng 708/1:012094. https://doi.org/10.1088/1757-899X/708/1/012094
3. ENV 1994-1-1 (1994) Eurocode 4. Design of composite steel and concrete structures. Part 1-1: General rules and rules for buildings: European prestandard, p 134
4. Ivanyk I, Vikhot S, Vybranets Y, Ivanyk Y (2018) Theoretical research into spatial work of a steel-reinforced concrete statically indeterminate combined structure. East Europ J Enterp Technol 5(7–95):13–22 (2018). https://doi.org/10.15587/1729-4061.2018.143023
5. Vybranets Y, Vikhot S (2023) Spatial calculation of metal truss structure in joint work with reinforced concrete slab. Lect Notes Civ Eng 290:450–458. https://doi.org/10.1007/978-3-031-14141-6_46
6. Vihot SI (2015) Strength and deformability metal composite structures based on sustainable design (candidate's dissertation). Available from ProQuest Dissertations and Theses database
7. Vybranets YY (2016) Strength and strain of combined metal systems, composite with concrete plate (candidate's dissertation). Available from ProQuest Dissertations and Theses database
8. Burchenya S, Sobczak-Piąstka J (2019) The research of bearing capacity and stress-strain behavior of bending complex steel concrete beams scientific session on applied mechanics X. In: Proceedings of the 10th international conference on applied mechanics C.2-7 physics and astronomy AIP conference proceedings, vol 2077, UNSP 020010. https://doi.org/10.1063/1.5091871
9. Burchenya S, Shmyh R (2016) Mathematical simulation of reinforced concrete beams with the external reinforcement in the form of steel cut and stretched sheet, Teka. Comm Motor Energ Agric 16(3):3–8
10. Burchenya S, Famulyak Y, Sobczak-Piastka J (2019) Modelling of work of cut and stretchy sheet in span beam structures with the mixed reinforcement. IOP Conf Ser Earth Environ Sci 362(1):012114. https://doi.org/10.1088/1755-1315/362/1/012114
11. Demchyna B, Surmai M, Tkach R (2019) The experimental study of glass multilayer columns using digital image correlation. Arch Mater Sci Eng 96(1):32–41. https://doi.org/10.5604/01.3001.0013.1990
12. Demchyna B, Surmai M, Tkach R, Hula V, Kozak R.: An analysis of using the method of two-dimensional digital image correlation in glass column research. Eastern-European Journal of Enterprise Technologies, 2020, 4(12–106), pp. 52–59. https://doi.org/10.15587/1729-4061.2020.209761
13. Kovalchuk V, Onyshchenko A, Fedorenko O, Harbel M, Parneta B, Voznyak O, Rybak R (2021) A comprehensive procedure for estimating the stressed strained state of a reinforced concrete bridge under the action of variable environmental temperatures. East Europ J Enterp Technol 2:23–30. https://doi.org/10.15587/1729-4061.2021.228960
14. Kovalchuk V, Sobolevska Y, Onyshchenko A, Bal O, Kravets I, Pentsak A, Voznyak O (2022) Investigating the influence of the diameter of a fiberglass pipe on the deformed state of railroad transportation structure «embankment-pipe». East Europ J Enterp Technol 2(7–116):35–43. https://doi.org/10.15587/1729-4061.2022.254573
15. Kovalchuk V, Koval M, Onyshchenko A, Kravets I, Bal O, Petrenko O, Milyanych A (2022) Determining the strained state of prefabricated metal corrugated structures of a tunnel overpass exposed to the dynamic loading from railroad rolling stock. East Europ J Enterp Technol 3(7–117):50–58. https://doi.org/10.15587/1729-4061.2022.259439
16. Kramarchuk A, Ilnytskyy B, Hladyshev D, Lytvyniak O (2020) Strengthening of the reinforced concrete tank of anaerobic purification plants with the manufacture of biogas, damaged as

a result of design and construction errors. In: IOP conference series: materials science and engineering, vol 1021: international scientific conference Energy efficiency in transport (EET 2020), 18–20 November, Kharkiv, Ukraine. Mode of access: https://doi.org/10.1088/1757-899X/1021/1/012017/pdf

17. DSTU B V.2.6-7-95 (1997) Precast concrete and reinforced concrete construction products. Test Methods
18. DSTU B V.2.6-169:2011 (2012) Connection of welded fittings and embedded products of reinforced concrete structures
19. DBN B V.2.6-98:2009 (2011) Concrete and reinforced concrete structures. Substantive provisions

Analysis of Building Structure Under Impact Load

Andrzej Wojnar[1](✉), Edyta Bernatowska[2], and Olena Chernieva[2]

[1] Rzeszow University of Technology, Ul. Poznanska 2, 35-959 Rzeszow, Poland
awojnar@prz.edu.pl

[2] Odessa State Academy of Civil Engineering and Architecture, Odesa, Ukraine

Abstract. The paper presents the analysis of progressive collapse in a building under impact load. The building with four-story frame structure and underground car park is taken into consideration. The car park structure consist of reinforced concrete columns. The above-ground part of the building consist of reinforced concrete slabs supported by steel columns. Two cases are considered: the impact of a car on a reinforced concrete columns of the car park, and the impact of the car on a steel columns on the ground floor. The computational analysis are carried out using two methods: equivalent static approach, and simplified dynamic approach according to EN 1991-1-7, Eurocode 1—Actions on structures—Part 1-7: General actions—Accidental actions, 2. Mitigation of the risk of progressive collapse in steel and composite building frames under exceptional events, D3-1: Design recommendations against progressive collapse in steel and steel-concrete buildings. In the equivalent static method the impact force is replaced with an equivalent static force, and internal forces in the structure are calculated using static analysis. In the dynamic approach, the structure is loaded with a dynamic force from the vehicle impact, and dynamic analysis is conducted using a linear-elastic material model. The calculation results include: values of internal forces, values of stress ratio, values of displacement at certain points (nodes) in the structure. Additionally, the paper presents the guidelines of designing buildings for the effects of vehicle impact included in EN 1991-1-7.

Keywords: Impact load · Metal structures · Concrete structures · Underground parking · Equivalent static method · Dynamic analysis

1 Introduction

Columns are critical elements of any structure. Their failure can lead to the catastrophic consequences of progressive failure. After the world shaking disasters as the 9/11 collapse of Twin Towers in New York City, the importance of the robustness design has been recognised. Robustness refers to the ability of a structure not to respond disproportionately to abnormal events or to initial damage. This response of a structure is characterised by a pronounced disproportion between a relatively minor abnormal event and the ensuing collapse of a major part or even the whole of a structure [1]. In structural design, procedures to design structures to resist conventional loads are well established

Z. Blikharskyy et al. (Eds.): CEE 2023, LNCE 438, pp. 498–509, 2024.
https://doi.org/10.1007/978-3-031-44955-0_50

but design for accidental loading conditions is increasingly requested by clients and occupants in modern engineering designs. The current draft revision of EN 1990 [2] for the forthcoming second generation of Eurocodes introduces Sect. 4.4 and Informative Annex E, which are solely dedicated to structural robustness. It provides strategies based on limiting the extent of damage, while the explicit design of structures for identified accidental action (e.g. impact, external and internal explosions), is covered within the scope of EN 1991 [3, 4]. In the latter case Eurocode 1991 [4] provides the next three designing strategies:

(a) designing the structure to have sufficient minimum robustness,
(b) preventing or reducing the action (e.g. protective measures),
(c) design structure to sustain the action.

This paper focuses on the analysis of progressive collapse in a building under impact load. The hazard coming from impact is typically associated with an incident involving vehicles. The road vehicle impact load is exceptionally complex, so it must be considered individually for every building structure. The consequences of the vehicle impact depend on the weight, speed, and direction (in relation with the building) of the vehicle. In conducted analyses two cases are considered: the impact of a car on reinforced concrete columns of the car park, and the impact of the car on a steel columns on the ground floor.

2 Vehicle Impact Load According to Eurocode

2.1 Eurocode 1: EN 1991-1-1

Most of the standards, replace the impact effects by the equivalent static impact load, which causes similar effects as the impact itself. These standards propose certain values of the impact load, or provide expressions for their calculation.

In Eurocode EN1991-1-1 [3] in Annex B, which is only informative and refers to vehicle barriers and parapets for car parks, we can find the expression for the characteristic horizontal force required to withstand the impact of a vehicle:

$$F = 0.5 \cdot m \cdot v^2 \Big/ (\delta_c + \delta_b), \tag{1}$$

where: m is the gross mass of the vehicle (in kg), v is the velocity of the vehicle (in m/s) normal to the barrier, δ_c is the deformation of the vehicle (in mm) and δ_b is the deformation of the barrier (in mm).

This force is normal to and uniformly distributed over any length of 1.5 m of a barrier for a car park. If the car park is designed for vehicles which do not exceed 2500 kg, Eurocode [3] specifies the values of the input parameters to obtain F: $m = 1500$ kg, $v = 4.5$ m/s, $\delta_c = 100$ mm. Additionally, if we assume a rigid barrier, for which δ_b may be given as zero, the characteristic force F appropriate to vehicles up to 2500 kg gross mass is taken as 150 kN. For vehicles with more than 2500 kg the same expression with the real mass of the vehicle can be used. The force determined according to Eq. (1) may be considered to act at bumper height. In the case of car parks intended for vehicles whose gross mass does not exceed 2500 kg this height may be taken as 375 mm above the floor level.

2.2 Eurocode 1: EN 1991-1-7

A slightly different approach is presented in the next part of Eurocode 1: EN-1991-1-7 [4].This code covers several different situations in which an impact action may occur and they are described in Chap. 4. The interaction between a colliding object (source of impact) and the impacted object (e.g., a column of a building) may be considered as "hard impact" (for structures for which the energy is mainly dissipated by the impacting body) or "soft impact" (for structures which are designed to absorb impact energy by elastic-plastic deformations of members). Depending on the consequences class of the structure, the Eurocode allows for two types of analyses:

- for structures in low and/or medium consequences class (up to CC2) [5], a static analysis such as the equivalent static approach is recommended. In this approach the impacted object is always considered rigid (hard impact), which leads to conservative results,
- for structures in high consequence class (CC3) [5], a dynamic analysis is required. This analysis can be a simplified or full dynamic. Hard impact or soft impact are both allowed.

Equivalent Static Approach
In this approach, the impact load is replaced by an equivalent static force F accounting for the effects of the load on the structure. This force has components parallel and perpendicular to the direction of travel. In the design process, these two components can be considered as independent, i.e., they should not be applied at the same time. Indicative equivalent static design forces due to vehicular impact on members supporting structures over or adjacent to roadways are given in Chap. 4 of Eurocode [4] and they depend on traffic category. The collision force from the impact of the car may be applied at $h = 0.50$ m above the level of carriageway. The area of application of this force should be 0.25 m height by 1.5 m width (or the member width, whichever is the smaller).

Simplified Dynamic Approach
In Annex C [4] a procedure to make a dynamic calculation of the problem is given. The assessment of the impact force F_{DYN} depends on the type of impact (hard impact or soft impact).

When hard impact is considered the structure is assumed as rigid and immovable and the colliding object deforms linearly during the impact phase. The dynamic interaction force is calculated from equation Eq. (2).

$$F_{\mathrm{DYN}} = v_r \sqrt{k \cdot m}, \tag{2}$$

where: v_r is the object velocity at impact, k is the equivalent elastic stiffness of the object (i.e. the ratio between force F and total deformation) and m is the mass of the colliding object.

The force due to impact may be considered as a rectangular pulse on the surface of the structure. If relevant, a non-zero rise time can be applied, see Fig. 1. The duration of the pulse follows from:

$$F_{\mathrm{DYN}}\,\Delta t = mv \quad \text{or} \quad \Delta t = \sqrt{m/k} \tag{3}$$

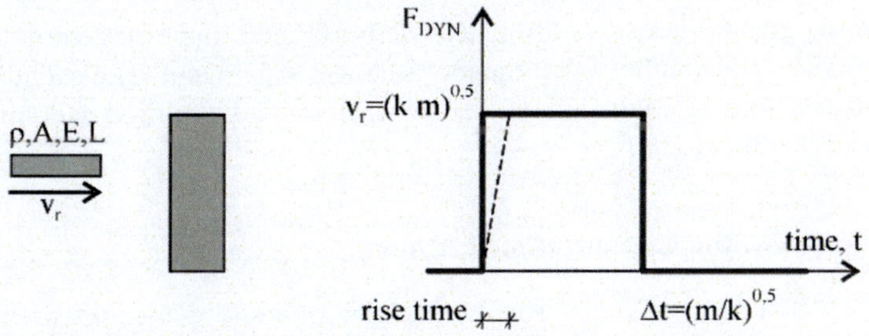

Fig. 1. Impact model [4] (F_{DYN}—dynamic force of interaction)

When the colliding object is modelled as an equivalent impacting object of uniform cross-section (see Fig. 1) equation Eqs. (4) and (5) should be used:

$$k = \mathrm{EA}/L; \tag{4}$$

$$m = \rho AL, \tag{5}$$

where: L is the length of the impacting object, A is the cross sectional area, E is the modulus of elasticity, ρ is the mass density of the impacting object.

When soft impact is considered the structure is designed to absorb the impact energy by plastic deformations. It is required that the structure ductility is enough to absorb the total kinetic energy 0.5 mvr^2 of the colliding object. Assuming a rigid plastic response of the structure, this requirement is satisfied by the condition of expression Eq. (6).

$$0.5m \cdot v_r^2 \leq F_0 \cdot y_0 \tag{6}$$

where: F_0 is the plastic strength of the structure, i.e. the limit value of the static force F; y_0 is its deformation capacity, i.e. the displacement of the point of impact that the structure can undergo.

Full Dynamic Approach

This is the most complex type of analysis of construction under impact load which takes into account separation of elements and possible collapse of structure after object collision. There are two possible ways to proceed: analysis where the action is explicitly modelled or the analysis where only the consequences of the impact are modeled, i.e., a column loss is simulated (which is called alternative load path analysis). The second approach can provide a good estimation of the structure robustness, without the complexity required to model impact actions, making it more attractive for designers. Detailed design guidelines for alternative load path analysis are described in the publication [6]. Examples of complex numerical analyses can be found in publications: [7–9].

3 Design Example

The following example presents a computational analysis of a shopping center structure. The analysis determined the internal forces values in the columns resulting from the impact of a passenger vehicle on a columns in the underground parking level or steel columns on the ground floor. Additionally, the horizontal displacement (deflection) of the column axis at the height of the impact force was calculated. The calculations were performed using both an equivalent static approach and a simplified dynamic approach according to EN 1991-1-7 [4].

3.1 Description of the Computational Model

The example discussed here concerns the analysis of a shopping center structure subjected to an exceptional load in the form of a vehicle impact. Static and dynamic calculations were carried out using the Finite Element Method and Autodesk Robot Structural Analysis software [10]. A three-dimensional computer model of the structure was created including beam and shell finite elements. The following geometric dimensions of the structure were adopted (see Fig. 2):

- the dimensions of the structure in plan: 30.00 × 30.00 m,
- height of underground parking: 3.00 m,
- height of above-ground storeys: 4.00 m,
- number of above-ground storeys: 4,
- spacing of columns: 6.00 m,
- spacing of steel frames: 6.00 m.

The thickness of the reinforced concrete walls of the underground parking was assumed to be 200 mm. The reinforced concrete columns of this storey were circular sections with diameter of 400 mm. The reinforced concrete slab above the underground parking was 250 mm thick, while floor slabs of intermediate storeys were 200 mm thick. The supporting structure of the roof was a 160 mm thick reinforced concrete slab. For all these elements, concrete C25/30 was assumed, while steel grade S355 was assumed for all steel elements in the structure.

The load-bearing capacity of cross-sections and the stability of steel elements in persistent design situation, were checked according to the interactive formulas included in the EN 1993-1-1 standard [11]. After optimization values of cross sections dimensions, the following types of profiles were obtained:

- steel columns of the first above-ground level—an H-beam cross-section: HD 360 × 162,
- steel columns of the remaining above-ground levels—an H-beam cross-section: HD 320 × 98,
- steel beams of intermediate floors—an I-beam cross-section: IPE 360,
- steel beams of the roof—an I-beam cross-section: IPE 270,
- vertical bracings—circular bars with a diameter of 24 mm.

The loads acting on the structure took into account: permanent loads, imposed loads (according to EN 1991-1-1 [3]), snow loads (according to EN 1991-1-3 [12]), wind loads (according to EN 1991-1-4 [13]) and accidental load from vehicle impact (according to EN 1991-1-7 [4]). Load combinations were created in accordance with the standard: EN 1990 [5] for persistent design situation, according to Eq. (7) and for exceptional

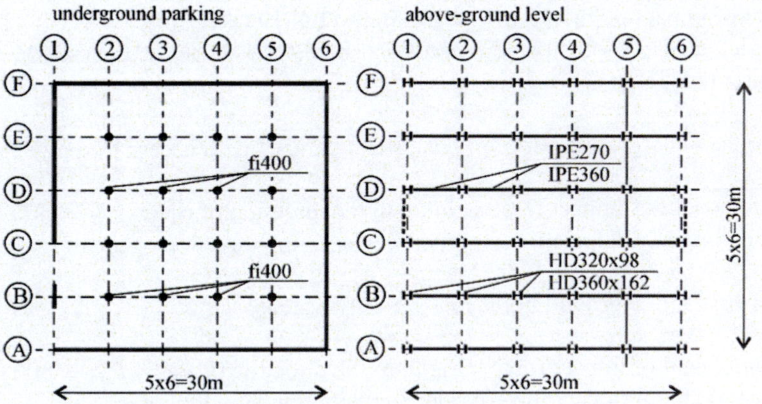

Fig. 2. The model geometry

situation, according to Eq. (8):

$$E_d = \max\begin{cases} \gamma_G \cdot G_k + \gamma_{Q,1} \cdot \psi_{0,1} \cdot Q_{k,1} + \sum_{i>1} \gamma_{Q,i} \cdot \psi_{0,i} \cdot Q_{k,i} \\ \xi \cdot \gamma_G \cdot G_k + \gamma_{Q,1} \cdot Q_{k,1} + \sum_{i>1} \gamma_{Q,i} \cdot \psi_{0,i} \cdot Q_{k,i} \end{cases} \quad (7)$$

$$E_d = G + A_d + \psi_{1,1} \cdot Q_{k,1} + \sum_{i>1} \psi_{2,i} \cdot Q_{k,i}, \quad (8)$$

where: G_k—characteristic value of a permanent action, $Q_{k,1}$—characteristic value of the leading variable action, $Q_{k,i}$—characteristic value of the accompanying variable action, A_d—design value of an accidental action, ξ, γ_G, $\gamma_{Q,1}$, $\gamma_{Q,i}$, $\psi_{0,1}$, $\psi_{0,i}$—factors according to EN 1990 [5].

The internal forces in the structure were determined as follows:

- by linear analysis—when checking the stability of the elements for the ultimate limit state for static analysis in persistent design situation,
- by linear analysis—in the case of using an equivalent static approach for an exceptional combination that takes into account vehicle impact,
- by dynamic analysis (time history analysis)—in the case of a dynamic approach that takes into account vehicle impact.

3.2 Equivalent Static Approach

The impact scenario includes perimeter columns (outermost) located along the traffic lanes for the underground parking and ground floor. The impact creates a force, whose components are parallel and perpendicular to the direction of travel. During the calculations, both components are considered independently (Fig. 3).

Assumptions for calculations:

- columns exposed to impact: C2, C3 (underground parking columns), C23, C24, C25 (ground floor columns)—see Fig. 3,

- height of impact point: 0.5 m—according to EN 1991-1-7 [4],
- the calculated equivalent static forces caused by the impact of a car on the structure are based on Table 4.1 of EN 1991-1-7 [4]. The following cases were considered:

  – for the underground parking: places and garages with access for passenger cars:

    component of impact force parallel to the direction of travel $F_{dx} = 50$ kN,
    component of impact force perpendicular to the direction of travel $F_{dy} = 25$ kN,

  – for first above-ground floor: roads in urban areas:

    component of impact force parallel to the direction of travel $F_{dx} = 500$ kN,
    component of impact force perpendicular to the direction of travel $F_{dy} = 250$ kN,

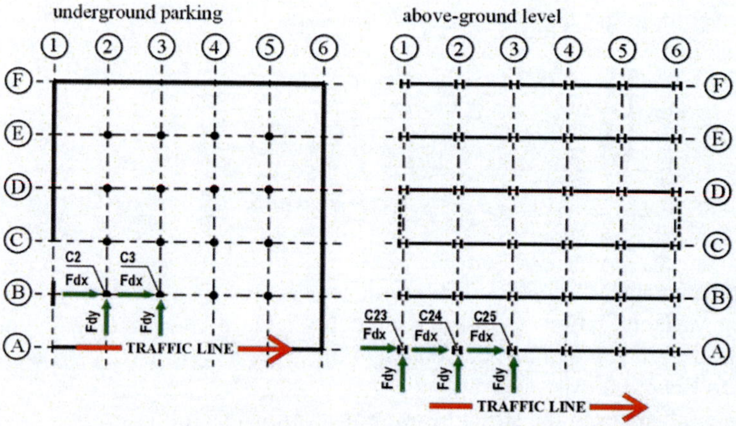

Fig. 3. The impact scenario—traffic lines and impact forces

In the computational model (Fig. 4), the force F_{dx} is directed along the X-axis of the global coordinate system, and the force F_{dy} is directed along the Y-axis of the global coordinate system).

The results of calculations include: values of internal forces in column, horizontal displacements of the column node where the impact force was applied, and stresses in column cross-section. They are presented in Table 1. The following column labels have been adopted for the Table: 1—column designation, 2—combination of loads, 3—value of axial force N, 4—value of bending moment M_Y relative to the y-y axis of the column's inertia, 5—value of bending moment M_Z relative to the z-z axis of the column's inertia, 6—stress values in the cross-section of the column, 7—ratio: load capacity of the cross section of the column (Fig. 5).

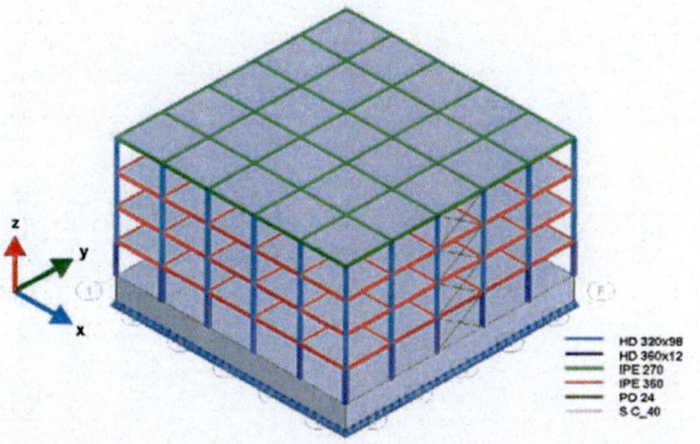

Fig. 4. The computational model of the building structure

Table 1. Results of calculations for Equivalent static approach

1	2	3	4	5	6	7
		N (kN)	M_Y (kNm)	M_Z (kNm)	σ (N/mm^2)	Ratio
	ULS	2415	0	31	24.2	0.78
C2	ACC1	1526	17	0	14.8	0.48
	ACC2	1526	0	8	13.4	0.43
	ULS	2418	0	42	25.9	0.84
C3	ACC1	1528	18	0	15.0	0.48
	ACC2	1528	0	8	13.4	0.43
	ULS	850	70	0	73.0	0.21
C23	ACC1	590	212	0	102.7	0.29
	ACC2	595	47	108	121.7	0.34
	ULS	1404	0	38	108.8	0.31
C24	ACC1	946	219	0	126.2	0.36
	ACC2	928	0	108	126.6	0.36
	ULS	1395	0	56	120.1	0.34
C25	ACC1	921	218	0	124.4	0.35
	ACC2	921	0	108	126.1	0.36

ULS—ultimate limit state according to EN 1990, ACC1—accidental design situations according to EN 1991-1-7, where F_{dx} is accidental (impact) load, ACC2—accidental design situations according to EN 1991-1-7, where F_{dy} is accidental (impact) load.

3.3 Simplified Dynamic Analysis

In a simplified dynamic approach, only two columns were chosen for further analysis. The selected columns were elements in which the stresses from bending and compression and ratio (calculated for ULS) reached maximum values, as shown in Table 1:

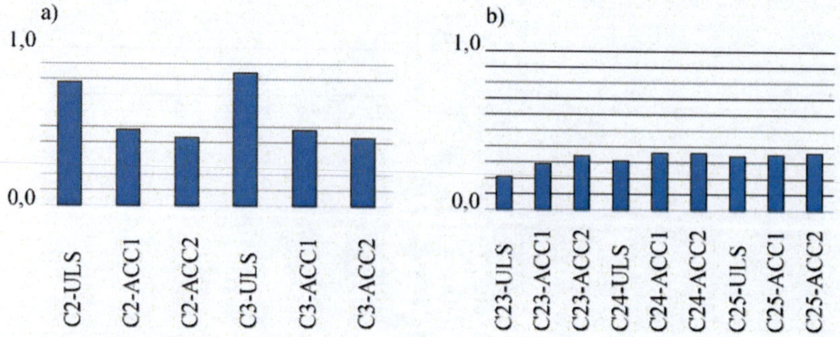

Fig. 5. The load capacity of the cross section of: **a** a concrete reinforcement columns of underground parking, **b** a steel columns of ground level

- column C3 of the underground parking,
- column C25 of the above-ground floor

Assumptions adopted for analysis (in accordance with point C.3 and Table C.1 of Eurocode EN 1991-1-7 [4]) were:

- type of impact—hard impact,
- dynamic effects were taken into account in the calculations, nonlinear material behavior was omitted,
- vehicle speed in the parking lot, $v_0 = 5$ km/h $= 1.39$ m/s,
- vehicle speed on the ground surface, $v_0 = 50$ km/h $= 13.89$ m/s,
- mass of the colliding object, $m = 1500$ kg,
- vehicle stiffness $k = 300$ kN/m $= 300000$ N/m.

The computational values of the dynamic force F_{DYN} caused by the vehicle impact were determined using the Eq. (2). And they are equal to:

- 30 kN for parking lot,
- 300 kN for above-ground floor.

In dynamic analysis, the force caused by impact is applied using a time function in the form of a rectangular impulse, with an instantaneous increase and a duration time equal to $t = 0.07$ s (calculated according to Eq. (3), Fig. 6).

Table 2 compares the values of internal forces and horizontal displacements of the columns resulting from equivalent static forces and dynamic forces. Load combinations were not considered. The analysis was conducted at the point of application of horizontal forces from impact. The following column designations have been adopted for the Table: 1—column identifier, 2—type of applied force: F_{dx}/F_{dy}—equivalent static force acting parallel/perpendicular to the direction of the traffic line, $F_{DYN,x}$—dynamic force acting parallel to the direction of the traffic line, 3—value of axial force N, 4—value of bending moment M_Y about the y-y inertial axis of the column, 5—value of bending moment M_Z about the z-z inertial axis of the column, 6—maximum value of the horizontal displacement of the column axis.

Values of axial force were equal zero, so they weren't presented in the table.

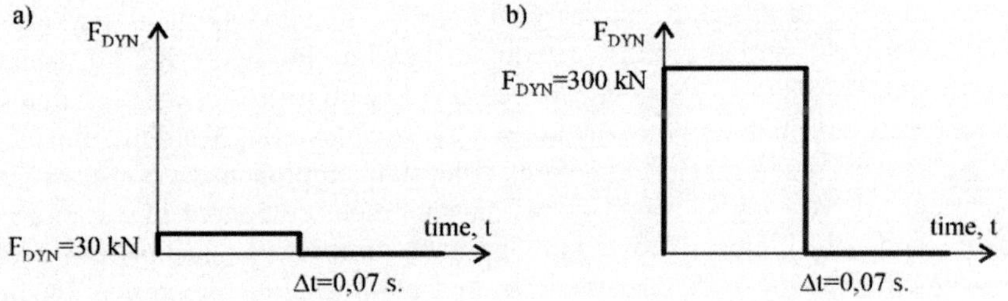

Fig. 6. The time-function of dynamic force, in case: **a** underground parking, **b** above-ground level

Table 2. Comparison of results of calculations

1	2	3	4	5
		M_Y (kNm)	M_Z (kNm)	u (mm)
	$\mathbf{F_{dx}}$	**6**	**0**	$\mathbf{u_x = 0.06}$
C3	F_{dy}	0	3	$u_y = 0.03$
	$\mathbf{F_{DYN,x}}$	**5**	**0**	$\mathbf{u_x = 0.06}$
	$\mathbf{F_{dx}}$	**218**	**0**	$\mathbf{u_x = 5.90}$
C25	F_{dy}	0	108	$u_y = 4.70$
	$\mathbf{F_{DYN,x}}$	**185**	**0**	$\mathbf{u_x = 2.20}$

4 Results Analysis and Conclusions

Based on the results presented above, the following conclusions have been drawn:

(1) In the case of reinforced concrete columns in an underground parking structure:

- in the considered scenario, the ultimate limit state governs the load-bearing capacity of the cross-section of the columns, with an utilization of 84%,
- accidental load combinations yield lower stress values due to reduction in the variable loads, as well as the small magnitude of the horizontal force resulting from impact and the corresponding bending moment it generates. The utilization of the cross-section's capacity ranges from 43 to 48%.

(2) In the case of above-ground level steel columns,

- in the considered scenario, the load-bearing capacity of the cross-sectional area of the columns is determined by the accidental load combination—utilizing 35–36% of the cross-sectional capacity,
- accidental load combinations yield higher stress values (despite reducing the variable loads values), due to the significant horizontal force resulting from impact and the corresponding bending moment it generates.

Comparing these two methods (the equivalent static approach and the simplified dynamic approach), it can be concluded:

- In the case of reinforced concrete columns in the underground parking, both approaches yield similar results in terms of bending moments and horizontal displacements. This is most likely due to the relatively high stiffness of the columns and the relatively low magnitude of the horizontal force resulting from the impact.
- For the above-ground level steel columns, the static approach is a conservative approach and yields a higher value of bending moment compared to the more accurate method—the simplified dynamic approach. Similar conclusions can be drawn regarding the horizontal displacements of the columns in the direction of the impact.
- It can be observed that multiplying the dynamic force F_{DYN} by the dynamic load factor (DLF) value of 1.40 as recommended in literature, yields force values close to those obtained in the equivalent static approach.
- The simplified dynamic approach takes into account both the impact force (directly related to the mass and velocity of the impacting object) and the duration of the impact impulse. This allows for a more accurate analysis of the behavior of the structure during and after the impact. However, this requires knowledge of the dynamic parameters of the structure (or structural elements) related to vibration damping. The behavior of the analyzed columns, in terms of changes in bending moment and horizontal displacement, are presented in Figs. 7 and 8 as an example.

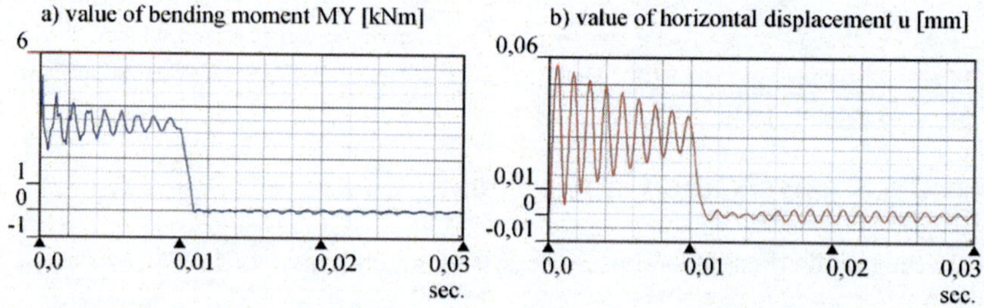

Fig. 7. Change in time of the bending moment and horizontal displacement values—underground parking column

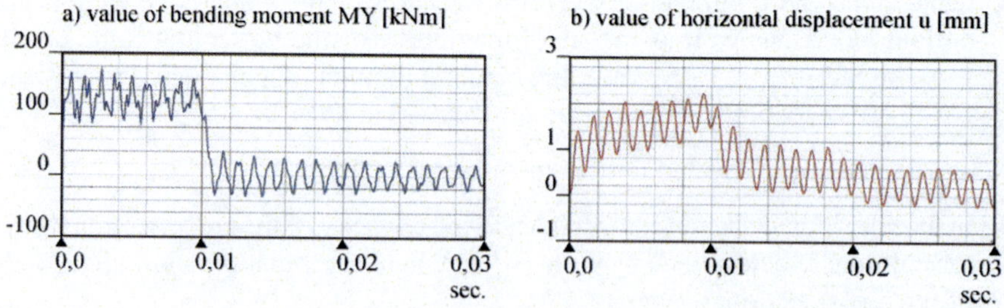

Fig. 8. Change in the time of the bending moment and horizontal displacement values—above-ground level

References

1. Starossek U, Haberland M (2012) Robustness of Structures. Int J Lifecycle Perform Eng 1:3–21
2. prEN 1990:2019 (2019) Eurocode 0—basis of structural and geotechnical design. European Committee for Standardisation
3. EN 1991–1-1 (2002) Eurocode 1: actions on structures—part 1–1: general actions—densities, self-weight, imposed loads for buildings. European Committee for Standardisation
4. EN 1991–1-7 (2006) Eurocode 1—actions on structures - part 1–7: general actions—accidental actions. European Committee for Standardisation
5. EN 1990 (2002) Eurocode 0—basis of structural design. European Committee for Standardisation
6. Demonceau J et al (2019) Design recommendations against progressive collapse in steel and steel-concrete buildings, 1st edn. European Convention for Constructional Steelwork
7. Ferrer B, Ivorra S, Segovia E, Irles R (2009) Impact load in parking steel column: code review and numerical approach. In: Papadrakakis M, Lagaros ND, Fragiadakis M (eds) Compdyn 2009 eccomas thematic conference on computational methods in structural dynamics and earthquake engineering, Rhodes
8. Al-Thairy H, Wang YC (2014) Behaviour and design of steel columns subjected to vehicle impact. Appl Mech Mater 566:193–198
9. Hyungoo K, Jinkoo K (2014) Progressive collapse of steel moment frames subjected to vehicle impact. J Perform Constr Facil 29(6):04014172
10. Autodesk Robot Structural Analysis Professional Version (2023)
11. EN 1993–1-1 (2006) Eurocode 3—design of steel structures—part 1–1: general rules and rules for buildings. European Committee for Standardisation
12. EN 1991–1-3 (2003) Eurocode 1: actions on structures—part 1–3: general actions—snow loads. European Committee for Standardisation
13. EN 1991–1-4 (2005) Eurocode 1: actions on structures—part 1–4: general actions—wind actions. European Committee for Standardisation

UV Radiation in Wastewater Disinfection

Justyna Zamorska(✉) and Izabela Kiełb-Sotkiewicz

Rzeszow University of Technology, Ul. Powstańców Warszawy 12, 35-959 Rzeszow, Poland
jzamor@prz.edu.pl

Abstract. Discharging treated municipal wastewater into rivers can pose a potential threat to the aquatic environment and human health. This danger results from the presence of numerous pathogenic microorganisms in the wastewater: viruses, bacteria, protists, fungi and intestinal parasite eggs. That is why it is so important to use highly efficient wastewater treatment methods. The average percentage of bacteria reduction achieved by purification processes is about 99%. However, despite such a high degree of reduction, the amount of indicator of fecal *coli* bacteria that enter rivers along with the runoff is significant and dangerous, which is why it is so important to use the third stage of wastewater treatment—disinfection. Many methods of wastewater disinfection are commonly used. The main advantage of physical processes, including ultraviolet radiation, is the absence of additional chemicals entering the environment and the avoidance of the formation of disinfection by-products. However, physical processes also have their drawbacks: their effectiveness largely depends on the sensitivity of the organism to radiation (including the phenomenon of photoreactivation), and the use of this method has a local effect and does not protect against re-contamination. The paper presents the results of the research on the effectiveness of UV radiation in the disinfection of treated wastewater, based on indicator bacteria, including: *coli* and *E. coli* bacteria, mesophilic and psychrophilic bacteria and spores. The research took into account different flow rates (radiation dose) and the effects of light and temperature on microorganism reconstruction. Traditional and modern methods of assessing the microbiological quality of wastewater were used, i.e. microbiological determinations by traditional culture methods, using luminometer (ATP) and flow cytometer (FCM).

Keywords: Municipal wastewater · Disinfection · UV radiation · ATP · FCM

1 Introduction

Discharging treated municipal wastewater into rivers can pose a potential threat to the aquatic environment and human health. The results show that not only raw wastewater but also treated wastewater can pose a serious microbiological threat to the quality of receiving waters. This is especially true when these waters are used for recreational purposes, fish farming, agricultural irrigation, and above all, for the production of drinking water. However, most wastewater treatment plants, even those using highly efficient treatment methods, are not equipped to effectively remove undesirable microorganisms

Z. Blikharskyy et al. (Eds.): CEE 2023, LNCE 438, pp. 510–520, 2024.
https://doi.org/10.1007/978-3-031-44955-0_51

of mainly fecal origin. Municipal wastewater, in addition to physical and chemical contaminants, contains a large number of pathogenic and opportunistic microorganisms, mainly of intestinal origin. The main sources of pathogenic microorganisms are urban wastewater (containing fecal matter and human urine) and industrial wastewater (e.g., from slaughterhouses), animal fecal matter flushed with rainwater from surfaces, and the fecal matter of rats inhabiting sewer systems. Sewage from hospitals also enters sewage systems. These are of a specific nature [1, 2]. Clinically relevant microorganisms, including drug-resistant and pathogenic bacteria, are found in greater numbers in hospital wastewater than in municipal wastewater. Including various strains of parasites and bacteria, the presence of hepatotropic viruses that cause hepatitis and HIV is common [3].

In Poland, there is no obligation to disinfect wastewater after the treatment process. The average percentage of bacteria reduction achieved in treatment processes is about 99%, which is very high. However, despite this high degree of reduction, the amount of fecal coliform indicator bacteria entering rivers with runoff is significant. This entails the fact of poor surface water quality and contributes to the steady deterioration of its sanitary condition [4].

The largest proportion of microorganisms in wastewater are viruses, of which there are about 140 types, and the number of their particles exceeds the number of bacterial cells five times. The most important viruses are adenovirus, calicivirus, hepatitis E virus, picornavirus (including enterovirus; poliovirus, hepatitis A virus), and rotavirus. Pathogenic bacteria of aquatic origin include: *Campylobacter jejuni, Campylobacter coli*, pathogenic strains of *Escherichia coli, Salmonella spp.* (including *S. typhi* and *S. paratyphi* species), *Shigella spp., Vibrio cholerae, Yersinia enterocolitica, Legionella spp., Pseudomonas aeruginosa, Aeromonas spp.* Parasite eggs (*Ascaris, Trichuris,* and *Toxocara*) can also be a problem [5, 6].

Many methods of disinfecting wastewater are commonly used. The most widely used method, especially in developing countries, is still chlorine, due to its cost-effectiveness and undemanding nature [7]. The main advantage of physical processes, including ultraviolet radiation, is the absence of the addition of various types of chemicals entering the environment and the avoidance of the formation of disinfection by-products. However, physical processes also have their disadvantages: their effectiveness is highly dependent on the sensitivity of the organism to radiation (including the phenomenon of photoreactivation), and the use of this method has a local effect and does not protect the water stream from recontamination [8, 9].

The paper presents the results of studies on the effectiveness of UV radiation in the hygienization of treated wastewater, based on indicator bacteria, including coliforms and E.coli, meso- and psychrophilic bacteria. The study took into account different flow rates (radiation dose) and the effect of light and temperature on the phenomenon of microorganism recovery. Traditional methodologies for assessing the microbiological quality of wastewater were used. A scientific novelty are determinations of microbiological quality of sewage using a luminometer (determination of ATP concentration) and a flow cytometer (FCM).

2 Methodology of the Research

2.1 Material and Its Initial Preparation

Sewage treated from the Municipal Wastewater Treatment Plant in Rzeszow and Tarnobrzeg was used for the study. Both treatment plants use the same treatment technology, but they differ in capacity. Sewage was collected from March to May, which resulted in a difference in their temperature (the temperature of the sewage ranged from 12 to 20°C. Tests should be carried out immediately after the sewage is delivered to the laboratory.

2.2 The Course of the Experiment

UV-C lamp BALLAST model EB-F6 with irradiance power of 4–6 W, 200–240 V, 50/60 Hz, 0.05 A was used for disinfection of wastewater. It is a lamp with a low-pressure radiator that emits radiation with a fixed specific wavelength (monochromatic radiation), which is usually 254 nm. Different exposure times were used, which is associated with different radiation doses: the flow of wastewater through the UV radiator.

- Dose of 70 $mWcm^{-2}$—exposure time: 30.7—referred to in the following as UV100,
- Dose of 89 $mWcm^{-2}$—exposure time 39.3 s—UV80,
- Dose of 118 $mWcm^{-2}$—exposure time 52.3 s—UV60,
- Dose of 177 $mWcm^{-2}$—exposure time 1 min 18 s—UV40.

After the disinfection process was carried out, the stability of the obtained disinfection effect was tested. Samples of wastewater (500 ml) were poured into sterile liter bottles and placed in incubators under the following conditions:

(1) in a refrigerator at 4°C without access to light,
(2) in an incubator at 22°C without light,
(3) in an incubator at 22°C with additional light exposure.

Analyses were performed on the day of wastewater disinfection and after 24, 48, 72 h, and 7 days.

2.3 Methodology of Microbiological Determinations

The effectiveness of the sewage disinfection processes was evaluated based on selected microbiological quality indicators (Table 1).

The tests were performed three times on wastewater, which varied significantly in turbidity—from a value of 1.54 to a value of 9.68 NTU. The results presented here are the average of these measurements.

Determinations of the total number of microorganisms present in the tested waters were performed using flow cytometry. Cells were counted using a Partec Cube 6 flow cytometer (Sysmex-Partec) equipped with a 488 nm blue laser, forward scattering detector (FSC), side scattering detector (SSC), and three fluorescence detectors (FL1–536 $\pm$ 20 nm, FL2–590 $\pm$ 25 nm, FL3N 615 nm).

Table 1. Indicators of microbiological quality of the studied sewage

1	Number of psychrophilic bacteria on the R2R agar media	Culture method by standard PN-EN ISO 6222:2004
2	Number of mesophilic bacteria on the R2R agar media	Culture method by standard PN-EN ISO 6222:2004
3	Number of coli and *Escherichia coli* bacteria	Culture method on chromogenic medium
4	FCM	Sysmex—Partec protocol for CytFlow Cube 6
5	Luminometry determination of ATP concentration*	PROMEGA protocol for 20/20 luminometer

* at work given as the RLU value—general and extracellular

A fluorescent dye, which is perfect for excitation by the blue line of the argon laser at 488 nm: SYBRGreen I nucleic acid stain (10,000 × diluted in DMSO0) [9] was used for the determinations. Due to the large number of loads in the tested sewage, the samples were diluted to a concentration of 10^{-1} ml, which allowed not to exceed the amount of 1,000 particles per second of the analyzed sewage flow.

3 Results

3.1 The Effectiveness of the Disinfection Process

Changes in the Number of Bacteria

The amount of coliform bacteria in the treated wastewater was on average 1740 cfu/100 ml. No *coli* and *E.coli* were found in the samples after UV disinfection.

Changes in ATP Values During the UV Disinfection Process

ATP is a good marker for living cells because its molecules are a product of cellular metabolism. With cell death, ATP synthesis is inhibited and existing molecules are broken down by the intracellular enzymes ATPase and phosphatases. During the disinfection processes, the bacteria are damaged, so the continuity of the cytoplasmic membrane is interrupted, ATP is released outside, thus the value of extracellular ATP increases [10].

ATP analysis was performed immediately after the disinfection process. In the treated wastewater, the value of total ATP was high, while extracellular ATP was very low. In contrast, after UV disinfection, the value of extracellular ATP increased rapidly, while the value of total ATP decreased. The lower the flow rate was (and therefore the concentration of UV radiation was higher), the value of extracellular ATP increased. This indicates that the disinfection process is proceeding correctly (Figs. 1, 2).

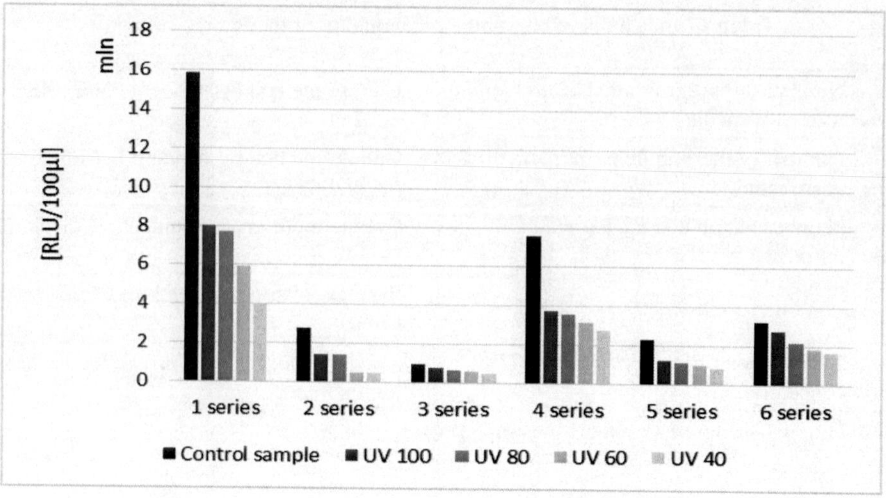

Fig. 1. RLU values for individual UV disinfection rates in subsequent series of tests

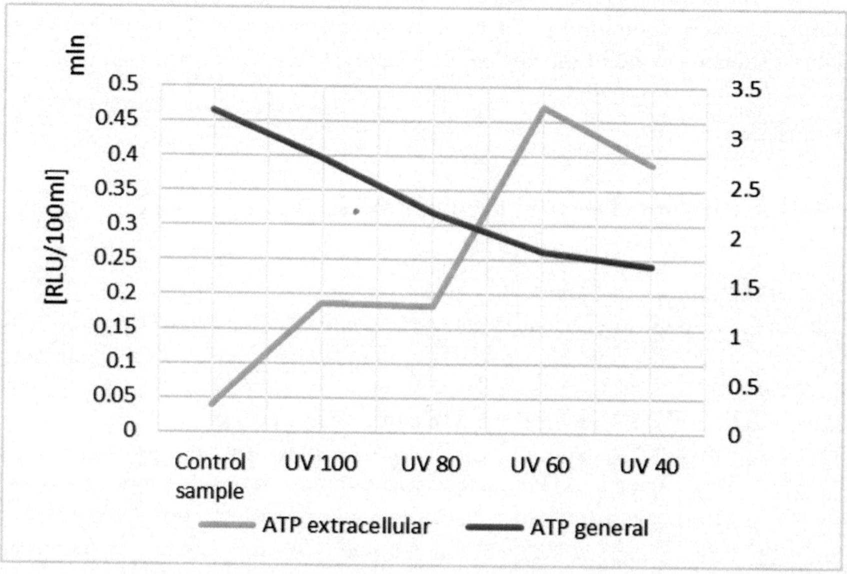

Fig. 2. Relationship between total ATP and extracellular ATP for individual UV disinfection rates

3.1.1 Changes in the Number of Particles During the UV Disinfection Process (FCM Analysis)

Based on the results of cytometric analysis, the effectiveness of UV disinfection was confirmed. In the control samples, a large population of live cells (HNA) was separated—349 454/100 µl, and a relatively small number of dead cells (LNA)—29 945/100 µl.

After passing through the UV irradiator at different flow rates, the number of dead bacteria increased significantly, while the number of live bacteria decreased (Fig. 3).

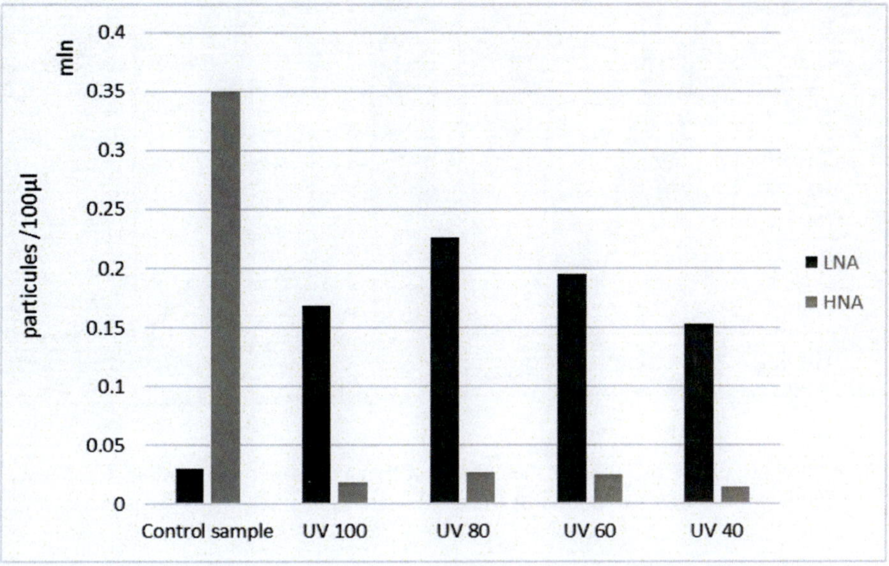

Fig. 3. The influence of UV radiation on the number of live cells [HNA] and dead cells [LNA]

Based on the results of cytometric analysis, the effectiveness of UV disinfection was confirmed. In the control samples, a large population of live cells (HNA)-quantity was separated, and a relatively small number of dead cells (LNA)-quantity). After passage through the UV light at different flow rates, the number of dead bacteria increased significantly, while the number of live bacteria decreased. (Fig. 4a, b).

3.2 Results of Studies on the Stability of Wastewater After the UV Disinfection Process

In this series of tests, stability tests were performed to determine how long the disinfection effect lasts, following the application of UV light. In the control samples, the lower temperature allowed the bacteria to survive longer in the wastewater (Fig. 5). Since these were solid culture conditions in all samples, an increase in bacterial counts was observed during storage. In the samples after UV40 disinfection (Fig. 6.), no microflora growth was observed after 24 h. However, at UV100, mesophilic bacteria were reactivated. The light caused faster and more intensive reactivation of bacterial cells (Fig. 7).

4 Discussion

Ultraviolet (UV) disinfection is increasingly used in wastewater treatment plants. In the disinfection process, UV radiation in the UV-B and UV-C range is the most effective. The most dangerous for living organisms—i.e. the most effective in disinfection—are

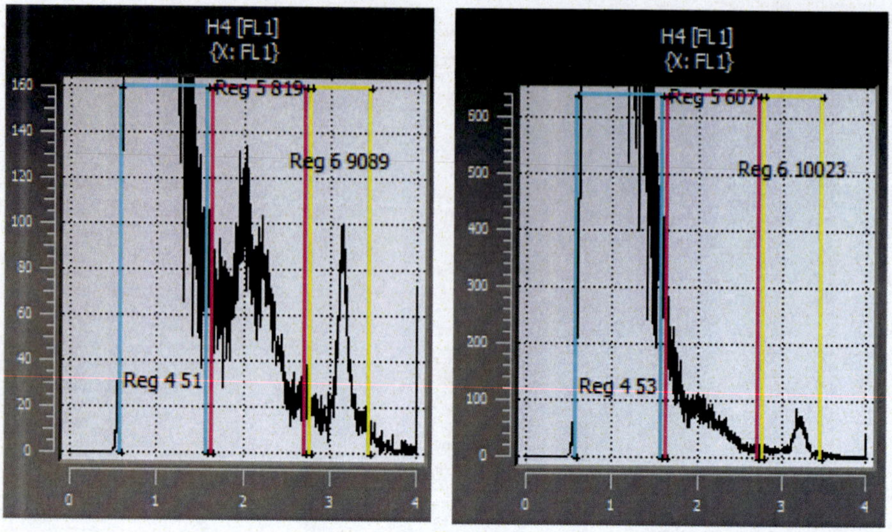

Fig. 4. **a** Influence of UV radiation on the number of living cells—control sample, **b** Influence of UV radiation on the number of living cells–UV100

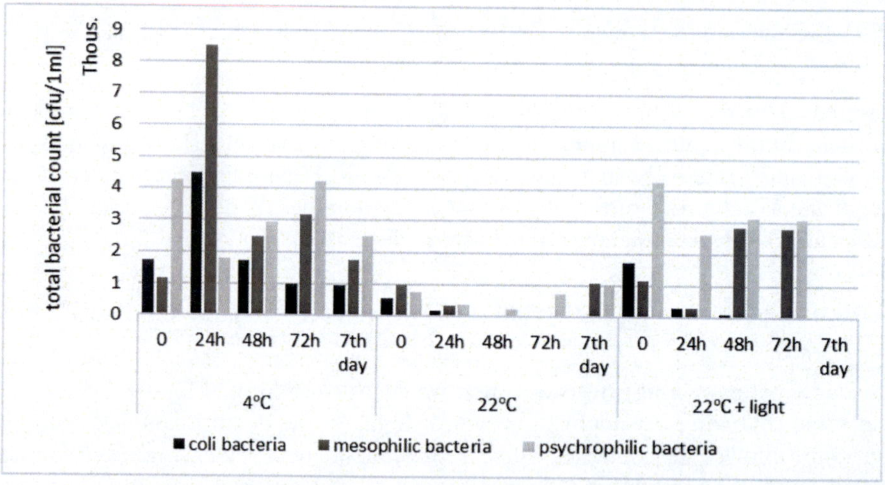

Fig. 5. The total number of bacteria on the control sample in the following hours of the experiment

UV-C waves [11]. This radiation penetrates the cell walls of microorganisms and permanently damages the DNA structure of microorganisms, resulting in their inability to reproduce [12]. Thanks to this, there is a complete elimination of pathogens dependent on the time and intensity of UV radiation. UV-C radiation causes changes in microbial nucleic acids, such as the breaking of the RNA or DNA chain, denaturation of proteins, the formation of pyrimidine dimers, which distort the strand and make it difficult to

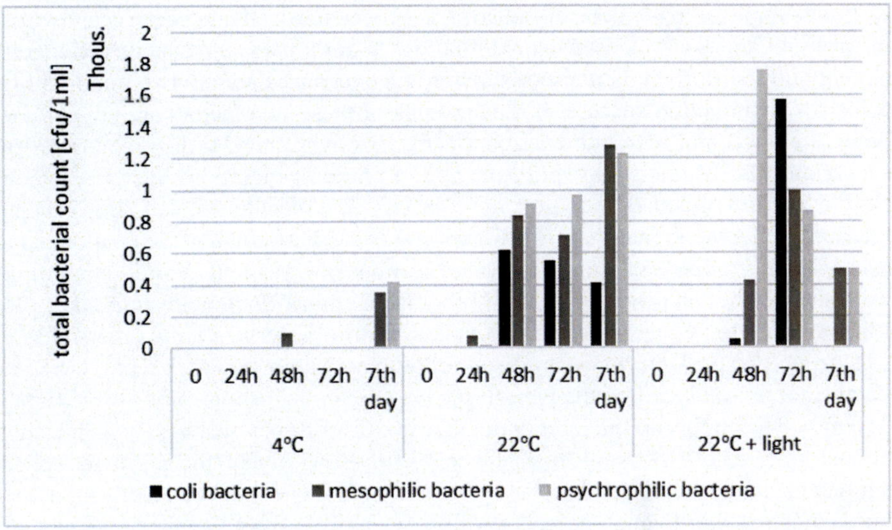

Fig. 6. The total number of bacteria on the UV40 sample in the following hours of the experiment

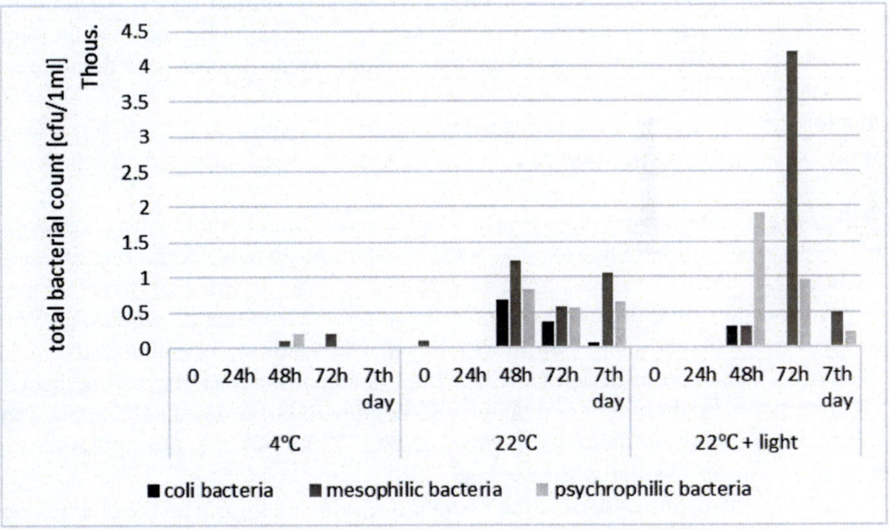

Fig. 7. The total number of bacteria on the UV100 sample in the following hours of the experiment

reconstitute [13]. The radiation also causes the formation of bonds between DNA and proteins, the formation of cross-links between two DNA strands, as well as cell-lethal dimers between thymine residues in the same DNA strand. The degree of sensitivity of microorganisms depends on several factors, among which can be distinguished the type, age, and size of microorganisms, their physiological state, environmental factors, and the form of the microorganism (vegetative forms are more sensitive than spore forms)

[14]. For example, a lower dose of radiation is needed to inactivate bacteria and viruses than for mold and yeast. Wastewater technology mainly uses UV-C lamps. Generally, the energy dose absorbed by microorganisms and exposure time are two important factors for UV disinfection success. If the amounts of these two factors are not sufficient enough, the microbial regrowth can recur [15]. However, the UV disinfection process has a serious disadvantage—the phenomenon of photoregeneration. Bacteria, to counteract the harmful effects of radiation, have developed a mechanism that allows them to repair damaged genetic material. A distinction is made between dark reparation, which consists of recognition, excision of damaged sections of DNA and incorporation of new ones in their place, and light reparation (photoactivation). Photoactivation takes place during exposure to visible light and occurs in coliform bacteria, among others [13].

There are two possible ways of repair—photoreactivation and dark repair [16]. Photoreactivation is catalyzed by the photolyase enzyme in the presence of visible light ($\lambda = 370$ nm). The energy of the light photons ($h\nu$) causes dimer cleavage, which leads to the transformation of DNA molecules. Once repaired, they regain the ability to replicate and infect the host in a manner similar to the state before UV-induced inactivation. Dark repair is much slower than photoreactivation.

Three mechanisms of dark repair have been proposed: repair by excision, recombination repair, and SOS error-prone repair [17]. The doses are maximized at a UV wavelength of $\lambda = 254$ nm. Excision is the predominant form of repair mechanism in the dark, in which repair is accomplished by physical incision (excision) of the dimer from the damaged DNA strand by an enzyme complex. What mechanisms dominate the cell reconstruction phenomenon is still unknown.

Photoreactivation-responsive organisms require 2.2 times more UV radiation to achieve 90% deactivation compared to organisms that do not undergo photoreactivation [18].

The phenomenon of photoregeneration will be very unfavorable when wastewater is discharged into a receiving body of water contaminated with organic compounds. It has been proven that residual BOD in treated and disinfected wastewater supports the regrowth of bacterial populations under favorable environmental conditions. When wastewater is discharged into clean waters, this phenomenon will be inhibited.

Somnath Basu observed a rapid increase in FC concentration and FC to total bacteria ratio between 24 and 48 h after UV disinfection compared to just after UV application. However, the growth rate gradually decreased after 72 to 144 h. Similar relationships were observed in the presented studies [19].

Both the intensity and dose of radiation have a significant impact on the effectiveness of removing microorganisms from disinfected wastewater. In practice, during the operation of the UV reactor, parameters such as flow, lamp condition, and radiation intensity are monitored [20]. In order to determine the effectiveness of the radiator, experimental studies are carried out using indicator organisms. MS2 bacteriophages, *Bacillus subtilis* spores, or *Cryptosporidium parvum* oocysts are most often used as indicator organisms. Coliform bacteria, which are sorbed in variable amounts by suspension particles, are not very precise.

The study highlighted rapid methods for assessing the microbiological quality of wastewater after the disinfection process, i.e. ATP concentration measurement and flow cytometry. Evaluation of microbiological quality performed by these methods seems to be a good alternative in the case of disinfection of wastewater, often of varying quality. In contrast, culture methods are of little use for wastewater disinfection. Too long a waiting time for the result would not allow a quick reaction in case of unsatisfactory quality of wastewater after the disinfection process.

5 Conclusions

Rapid microbiological techniques make it possible to access the microbiological quality of treated wastewater. The use of this assessment would allow to regulate the dose of radiation depending on the quality of disinfected wastewater. The high effectiveness of removing *coli* bacteria from wastewater with UV radiation in the range of tested flow rates was confirmed. However, the long-term effectiveness of the disinfection effects was found to be lacking, with the regrowth of coliforms and other microorganisms noted after some time. Bacterial reconstruction occurs between 24 and 48 h from the moment of disinfection with UV rays and this effect in conditions of permanent culture usually lasts up to 72 h. The light caused faster and more intensive reactivation of bacterial cells. The temperature has a significant impact on the rate of bacterial reconstruction—the smallest growth of bacteria occurred in after disinfection samples placed at 4°C.

References

1. Altin A, Altin S (2003) Characteristics and treatability of hospital (medical) wastewaters. Fresenius Environ Bull 12(9):1098–1108
2. Adanlokonon EAS (2018) At all: physicochemical and microbiological characterization of effluents from the "centre hospitalier universitaire de la mère et de l'enf and lagune (CHU-MEL)" discharged in the Cotonou lagoon in Benin. Int J Biol Chem Sci 12(4):1955–1964
3. Mackuľak T, Cverenkárová K, Staňová AV, Fehér M, Tamáš M, Škulcoá AB, Gál M, Naumowicz M, Špalková V, Bírošová L (2021) Hospital wastewater-source of specific micropollutants, antibiotic-resistant microorganisms, viruses, and their elimination. Antibiotics (Basel) 10(9):1070. https://doi.org/10.3390/antibiotics10091070. PMID: 34572652; PMCID: PMC8471966
4. Alemayehu TA, Weldetinsae A, Dinssa DA, Derra FA, Bedada TL, Asefa YB, Mengesha SD, Alemu ZA, Serte MG, Teklu KT, Woldegabriel MG, Kenea MA, van den Berg H, de Roda Husman AM (2020) Sanitary condition and its microbiological quality of improved water sources in the Southern Region of Ethiopia. Environ Monit Assess 192(5):319-1–319-9. https://doi.org/10.1007/s10661-020-08297-z
5. Chahal C, van den Akker B, Young F, Franco C, Blackbeard J, Monis P (2016) Pathogen and particle associations in wastewater: significance and implications for treatment and disinfection processes. Adv Appl Microbiol 97:63–119. https://doi.org/10.1016/bs.aambs.2016.08.001
6. Grudziński M, Pietruszka A, Sawicki W (2015) Anaerobic digestion in sanitization of pig slurry and biomass in agricultural biogas plant. J Microbiol Biotechnol Food Sci 5(2):524–526
7. Collivignarelli MC, Abbà A, Benigna I, Sorlini S, Torretta V (2017) Overview of the main disinfection processes for wastewater and drinking water treatment plants. Sustainability 10(1):86
8. Olańczuk-Nejman QB (2015) Dezynfekcja ścieków, Wyd. Seidel-Przywecki, Warszawa

9. Hammes FA, Egli T (2005) New method for assimilable organic carbon determination using flow-cytometric enumeration and a natural microbial consortium as inoculum. Environ Sci Technol 39(9):3289–3294. https://doi.org/10.1021/es048277c
10. Peperzak L (2023) The critical adenosine triphosphate (ATP) concentration in treated ballast water. Mar Pollut Bull 187:114506. https://doi.org/10.1016/j.marpolbul.2022.114506
11. Cutler TD, Zimmerman JJ (2011) Ultraviolet irradiation and the mechanisms underlying its inactivation of infectious agents. Anim Health Res Rev 12(1):15–23
12. Gromiec M, Gromiec T (2013) Dezynfekcja wody i ścieków za pomocą promieniowania UV. Wodociągi i Kanalizacja 7–8:44–47
13. Hallmich C, Gehr R (2010) Effect of pre- and post-UV disinfection conditions on photoreactivation of fecal coliforms in wastewater effluents. Water Res 44:2885–2893. https://doi.org/10.1016/j.watres.2010.02.003
14. Słoboda M, Włodyka-Bergier A (2015) Analiza możliwości zastosowania zaawansowanych metod utleniania do dezynfekcji ścieków komunalnych. Logistyka 4(CD3):9773–9779
15. Khodadadi M, Khorami-Pour S, Askari H et al (2022) Investigating disinfection efficiency and regrowth control of microorganisms in urban sewage effluent using ultraviolet radiation, ozone, gamma-ray, and dielectric barrier discharge plasma. Water Air Soil Pollut 233:492. https://doi.org/10.1007/s11270-022-05965-3
16. Salcedo I, Andrade JA, Quiroga JM, Nebot E (2007) Photoreactivation and dark repair in UV-treated microorganisms: effect of temperature. Appl Environ Microbiol 73(5):1594–1600. https://doi.org/10.1128/AEM.02145-06
17. Livneh Z, Cohen-Fix O, Skaliter R, Tamar ET (1993) Replication of damaged DNA and the molecular mechanism of ultraviolet light mutagenesis. Crit Rev Biochem Mol Biol 28(6):465–513. https://doi.org/10.3109/10409239309085136
18. Solecka M (2011) Zastosowanie promieniowania ultrafioletowego do dezynfekcji ścieków miejskich. Ekologia i Technika 9(3):115–214
19. Basu S, Page J, Wei IW (2007) UV disinfection of treated wastewater effluent: influence of color, reactivation and regrowth of coliform bacteria. In: Environmental engineer: applied research and practice, vol 4. American Academy of Environmental Engineers and Scientists, pp 1–8
20. Artichowicz W, Luczkiewicz A, Sawicki JM (2020) Analysis of the radiation dose in UV-disinfection flow reactors. Water 12:231. https://doi.org/10.3390/w12010231

Microbiological Stability of Post-fermentation Leachate

Monika Zdeb[1](✉), Justyna Zamorska[1], and Adam Masłoń[2]

[1] Department of Water Purification and Protection, Faculty of Civil, Environmental Engineering and Architecture, Rzeszów University of Technology, Rzeszów, Poland
mzdeb@prz.edu.pl

[2] Department of Environmental and Chemistry Engineering, Faculty of Civil, Environmental Engineering and Architecture, Rzeszów University of Technology, Rzeszów, Poland

Abstract. Wastewater is an environment very rich in a number of microorganisms, including many pathogens. The conducted research aimed to determine the microbiological quality of the post-fermentation leachates from municipal treatment plants and the parameters of their hygienization. Thermal treatment of liquid from mechanical dewatering of sewage sludge from sewage treatment plants results in complete hygienization, which leads to a fully sanitary-safe liquid fertilizer preparation. The preparation obtained does not contain pathogenic bacteria and intestinal parasite eggs. Compared to other natural fertilizers or waste fertilizing products, the use of a liquid form greatly facilitates its penetration into the soil, which does not pose a threat to the health and life of animals and people. Microbiological stability of leachate after hygienization was check and to indicate whether the hygienized leachate can be stored (e.g., in sales packages). The tests performed showed very rapid regrowth of the bacterial microflora of both raw and hygienized leachates. The number of microorganisms after 20 days of storing hygienized leachate at temperatures of 15 and 22°C increases significantly. In the context of using such leachates as fertilizer, it is suggested to use them immediately after the hygienization process. No coliform bacteria were detected in the leachate stored for 20 days, but the total number of microorganisms increased so much that it may pose a sanitary threat.

Keywords: Post-fermentation leachates · Hygienization · Fertilizer product

1 Introduction

Fertilization is a basic agricultural procedure that significantly affects plant productivity and yield quality. To supplement the frequently occurring deficiency of one or more of the nutrients in a given soil or alternatively to increase the yield of green matter, mineral fertilizers are used. The nutrient content of the soil is limited. Most of the minerals present in the soil come from dead plant parts, but these are increasingly being exported from where they are grown. The result of this activity is soil depletion, and in order to inhibit this process, the soil must be replenished with nutrients—using fertilizers [1, 2]. For proper growth, plants need mineral nutrients such as nitrogen, phosphorus,

Z. Blikharskyy et al. (Eds.): CEE 2023, LNCE 438, pp. 521–532, 2024.
https://doi.org/10.1007/978-3-031-44955-0_52

potassium, as well as secondary nutrients, which include sulfur, magnesium and calcium. Micronutrients are also essential [3]. Currently, a huge problem for the agricultural sector is to ensure that there is enough water. This problem affects many EU countries, including Poland. Recent legal regulations indicate the possibility of using non-potable water for agricultural purposes (Proposal for a Regulation of the European Parliament and of the Council on Minimum Requirements for Water Reuse. COM (2018) 337 final, 2018/0169 (COD), Brussels 2018) [4].

Regular use of fertilizers is essential to maintain high yields, which entails increasing financial outlays. One of the alternatives is to the use of mineral fertilizers that are available on sale, is the use of wastewater treatment plant wastes that have fertilizer and irrigation properties [5, 6]. Also one of these could be the use of leachate from the anaerobic treatment of sewage sludge or post-fermentation pulp to produce a product that improves soil properties [7–9].

The Regulation of the Minister of Agriculture and Rural Development of 18 June 2008 on the implementation of certain provisions of the Act on Fertilizers and Fertilization regulates the maximum values of pollutants contained in fertilizers [10]. The presence of heavy metals in fertilizers is limited, and the presence of *Salmonella* bacteria and eggs of intestinal parasites is unacceptable.

The literature lacks information on the microbiological quality of post-fermentation effluents. Theoretically, a properly conducted methane fermentation process preceded by hygienization or sterilization of waste should result in the elimination of pathogenic microorganisms [11].

Despite the awareness of the microbiological risk associated with the use of this type of waste as a fertilizer, there is no information in the literature on the microbiological quality of digestion leachate. Theoretically, a properly conducted methane fermentation process preceded by hygienization or sterilization of waste should result in the elimination of pathogenic microorganisms [12]. When thinking about the use of digestion leachate and its introduction into the environment, a very important issue is its sanitary condition. As a product formed from wastewater, digestate can contain many microorganisms. The largest proportion of microorganisms in wastewater are viruses, of which there are about 140 types, and the number of their particles exceeds the number of bacterial cells five times. The most important viruses are adenovirus, calicivirus, hepatitis E virus, picornavirus (including enterovirus; poliovirus, hepatitis A virus), and rotavirus. Pathogenic bacteria of aquatic origin include: *Campylobacter jejuni, Campylobacter coli,* pathogenic strains of *Escherichia coli, Salmonella spp.* (including *S. typhi* and *S. paratyphi* species), *Shigella spp., Vibrio cholerae, Yersinia enterocolitica, Legionella spp., Pseudomonas aeruginosa, Aeromonas spp.* Parasite eggs (*Ascaris, Trichuris,* and *Toxocara*) can also be a problem [13, 14].

The aim of the research was to determine the microbiological quality of post-fermentation effluents from municipal sewage treatment plants and their microbiological stability. The novelty is the possibility of using this waste for agricultural purposes.

2 Methods

2.1 Material and Its Preliminary Preparation

Mesophilic digestion leachates obtained from the Municipal Wastewater Treatment Plant in Rzeszow were used for the study. Sample preparation consisted of percolating the leachate obtained from the wastewater treatment plant through a sterile filter with a mesh diameter of 0.5 mm.

2.2 The Course of the Experiment

- Evaluation of the effectiveness of the hygienization process

  The pretreated digestion leachate was hygienized using temperature and/or ultrasound. For each test variant, the volume of hygienized leachates was 300 ml. For each of the two research variants, microbiological determinations were performed three times.

- First variant

  Studies of the effect of temperature on the hygienization process of post-fermentation leachate were conducted using a temperature magnetic stirrer. The leachate was heated to temperatures of 60, 70 and 80°C, and then maintained for 15 and 30 min. During heating, the leachates were continuously stirred at a speed of 30 rpm. Determinations to check the effectiveness of hygienization were tested just after the temperature was reached and after 15 and 30 min of temperature maintenance.

- Second variant

  An ultrasonic generator was used to study the effect of sonication on the sanitary condition of post-fermentation leachate. The leachates were sonicated for 10, 15, and 20 min at an ultrasonic intensity of 20 kHz and an amplitude of 30 μm. After each sonication variant, the post-fermentation leachates were also heated to 70 and 80°C, which was maintained for 15 min. During heating, the leachates were continuously mixed at 30 rpm. Determinations to check the effectiveness of leachate hygienization were tested after the specified sonication times, just after the temperature was reached, and after it was maintained for 15 min.

- Study of biological stability of leachate after hygienization process

  The pretreated post-fermentation leachate was hygienized using temperature and sonication. For each test variant, the volume of hygienized leachates was 900 ml. They were heated at 60 and 80°C for 30 min and sonication for 20 min.

The leachates after the hygienization processes were incubated to test microbiological stability in four variants:

- post-fermentation leachates hygienized for 30 min at a temperature of 80°C (H),
- post-fermentation leachates sonicated for 20 min (S),
- post-fermentation leachates sonicated for 20 min and hygienized for 30 min at 80 (S + H),
- raw post-fermentation leachates (not hygienized) (C).

Hygienized leachate of 0.2 L volume was stored in sterile glass bottles placed in a laboratory incubator in two temperature options: 22 and 15°C. Tests to control their sanitary condition were conducted immediately after hygienization, and on the 1st, 2nd, 5th and 20th day after hygienization.

2.3 Methodology of Microbiological Determinations

The effectiveness of the post-fermentation leachate hygienization processes was evaluated based on selected microbiological quality indicators (Table 1).

Table 1. Indicators of microbiological quality of the studied post-fermentation leachate

No	Parameter/indicator	Method/procedure
1	Total number of psychrophilic bacteria on A-agar	Culture method by standard PN-EN ISO 6222:2004
2	Total number of mezophilic bacteria on A-agar	Culture method by standard PN-EN ISO 6222:2004
3	Number of coli bacteria and *Escherichia coli*	Culture method on chromogenic medium
4	Presence of *Salmonella sp.*	Culture method and biochemical tests by standard PN-EN ISO 6579-1:2017-04
5	The number of eggs of intestinal parasites	Detection and quantification—by standard PN-Z-19005:2018-10
6	Luminometry determination of ATP concentration*	PROMEGA protocol for 20/20 luminometer

* at work given as the RLU value—general and extracellular

3 Results and Discussion

3.1 Quality of Post-fermentation Leachates After the Hygienization Process

The microbial counts in the tested leachates are significantly lower than in raw wastewater, and comparable to the counts reported for wastewater after the biological treatment process [15]. In the raw post-fermentation leachate, *Salmonella* bacteria were found, and the absence of live eggs of intestinal parasites has been reported.

The post-fermentation leachate filtration process will not reduce microbial counts. The process of heating the leachate regardless of the temperature used resulted in a reduction in the number of mesophilic and psychrophilic bacteria. The absence of *coliforms* and *Escherichia coli* and *Salmonella* bacteria was also observed. The sonication process alone did not affect the reduction of mesophilic bacteria With the time of sonication, the effect of reducing *coliforms* and *E. coli* was obtained (Figs. 1, 2). The increase in the number of mesophilic bacteria determined by the culture method after the sonication process can be explained by the fact that ultrasound causes the flocculent suspension of microorganisms to disintegrate. In the culture method, better colony growth from single cells is evident [16].

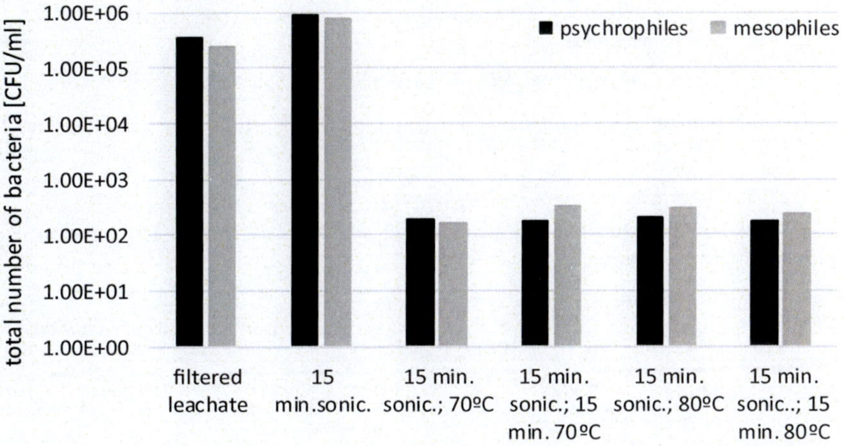

Fig. 1. The total number of bacteria in post-fermentation leachate after hygienization at 60°C and sonication

3.2 Microbiological Stability of Hygienized Post-fermentation Leachate

Studies of thermal hygienization of digestion leachates indicate an increase in the total number of psychrophilic and mesophilic bacteria and RLU values after this process. *Coliform* bacteria were not noted in the hygienized post-fermentation leachate. At 60°C, there is a resurgence of coliform bacteria after just the first day of leachate storage. No *coliform* growth was observed in post-fermentation leachates after hygienized at 80°C.

Regrowth of *coliform* and *Escherichia coli* populations was observed on day 1 after heating the leachate at 60°C for 30 min. When higher temperatures (80°C) were applied, no *coliforms* and *E. coli* were detected until day 5 (Table 2).

The results of the total number of both mesophilic and psychrophilic bacteria and the number of RLUs indicate that the temperature-hygienized leachates are not biologically stable. Microflora regrowth was noted after the first day of heating, and on day 5 the total number of bacteria is many times higher than on the day of hygienization (Figs. 3, 4).

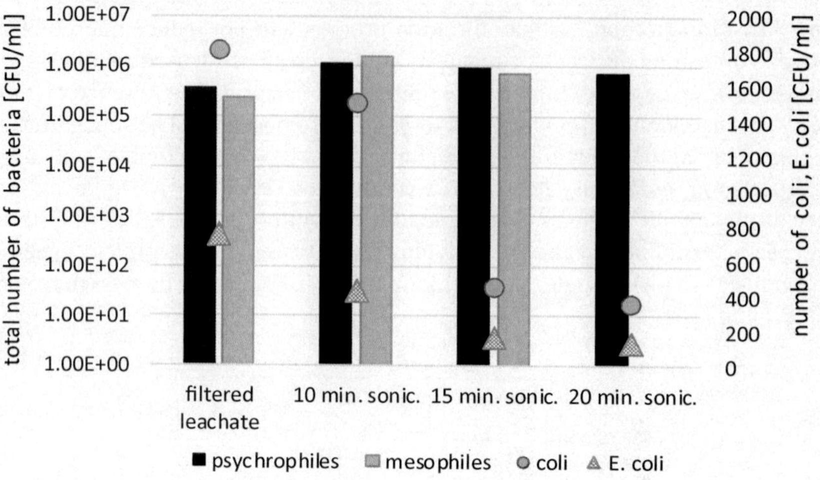

Fig. 2. The total number of bacteria in post-fermentation leachate after hygienization at 80°C and sonication

Table 2. Changes in the number of coli bacteria and *Escherichia coli* in post-fermentation leachate after temperature hygienization

Test sample	Number of *coli* bacteria (CFU/ml)	Number of E*scherichia coli* (CFU/ml)
raw post-fermentation leachate (without hygienization)	1.800	600
60°C, day 0 immediately after hygienization	0	0
hig. 60°C; 1 day after hygienization	1	0
hig. 60°C; 5day after hygienization	300	10
hig. 80°C; 0 day after hygienization	0	0
hig..80°C; 1 day after hygienization	0	0
hig. 80°C; 5 day after hygienization	0	0

On the basis of the obtained test results, appropriate parameters of hygienization of post-fermentation leachates were selected and the experiment was repeated to check their stability.

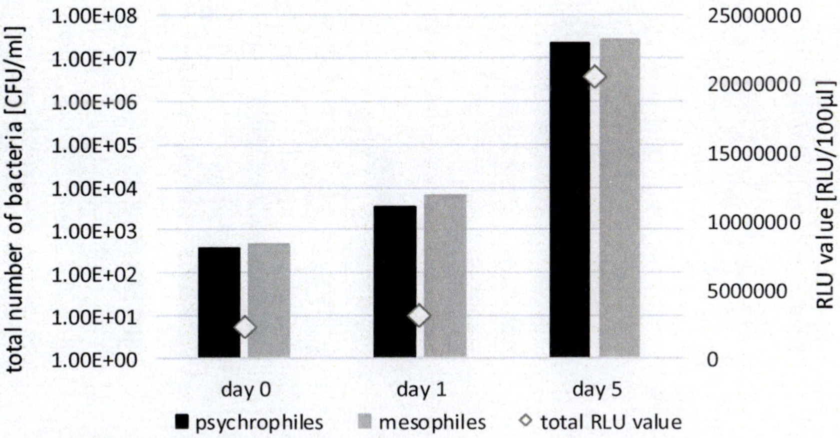

Fig. 3. Changes in the total number of bacteria and RLU values in post-fermentation leachate after thermal hygienization at 60°C

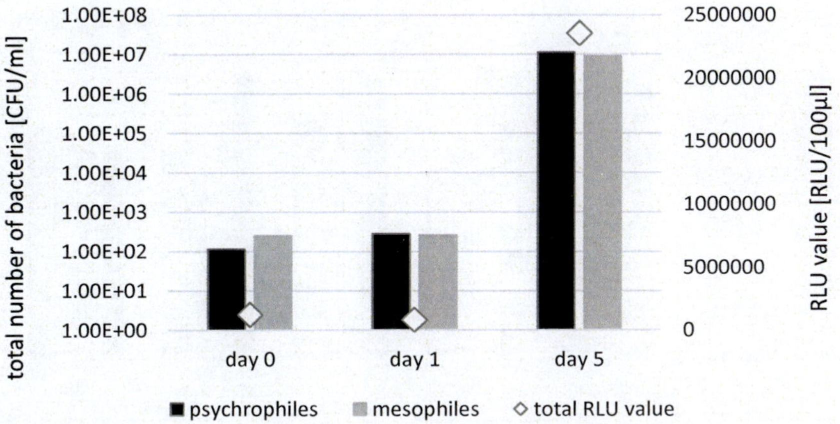

Fig. 4. Changes in the total number of bacteria and RLU values in post-fermentation leachate after thermal hygienization at 80°C

The hygienized effluents were not microbiologically stable. Already after the first day of incubation, a very rapid increase in the number of mesophilic and psychrophilic bacteria was observed—both in the samples of post-fermentation effluents incubated at 15 and 22°C. It was also confirmed that sonication does not fulfill the role of hygienizing the tested material (Figs. 5, 6, 7, 8).

A positive aspect of the research results obtained is the fact that the coliform bacteria did not regrowth until the 20th day of storage both at 22 and 15°C (Tables 3, 4).

The biological instability of digestion leachates is due to the presence of biogenic compounds in them, mainly inorganic nitrogen, phosphorus, and organic carbon. In addition, the fertilizer prepared in this way lacks factors that limit the growth of microflora.

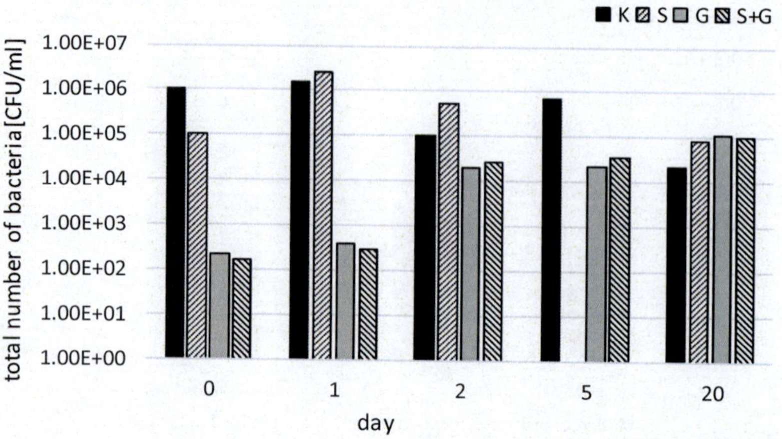

Fig. 5. Changes in the total number of psychrophilic bacteria in fermentation effluents after hygienization stored at 15°C

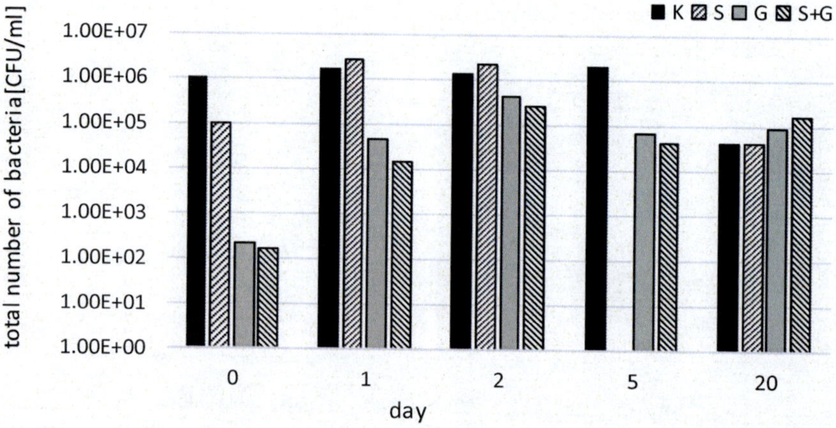

Fig. 6. Changes in the total number of psychrophilic bacteria in fermentation effluents after hygienization stored at 22°C

The potential use of hygienized leachate is associated with sanitary risks [17]. The literature lacks information on the microbiological quality of post-fermentation effluents. However, there are studies indicating a high content of organic compounds and biogenic substances [18, 19]. Their presence determines the fertilizing value, but at the same time is the basis for the regrowth of microflora in such material. Many attempts are made to use waste materials in various fields of economy, especially in agriculture and forestry. However, the basis for the use of materials such as sewage sludge or post-fermentation leachates or effluents from landfills is sanitary safety.

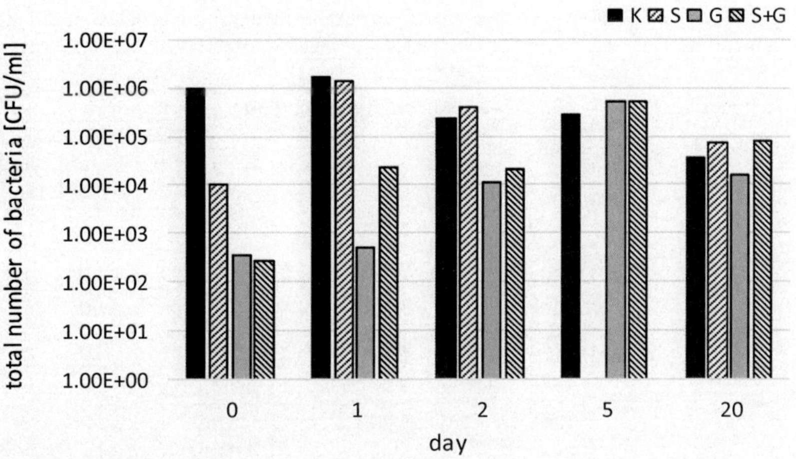

Fig. 7. Changes in the total number of mezophilic bacteria in fermentation effluents after hygienization stored at 15°C

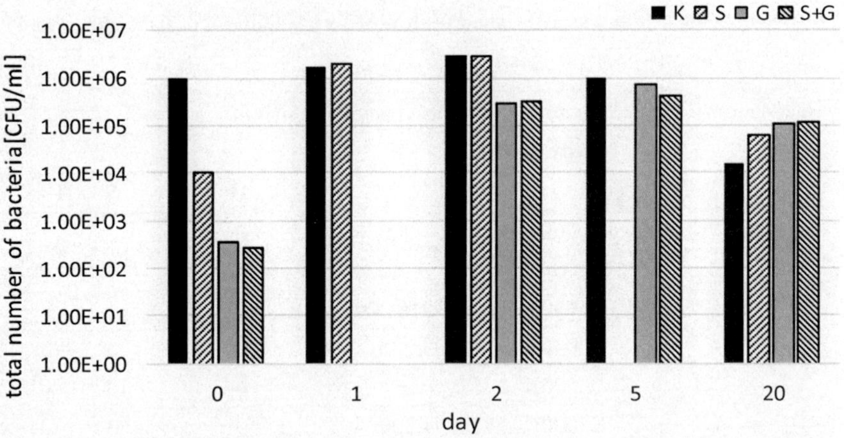

Fig. 8. Changes in the total number of mezophilic bacteria in fermentation effluents after hygienization stored at 22°C

Table 3. Changes the number of coli bacteria in post-fermentation leachates stored at 15Â°C after hygienization

Day of incubation	Number of coliform bacteria (CFU/ml)			
	Study sample			
	C	S	H	S + H
0	7.000	1.400	0	0
1	1.800	6.000	0	0
2	3.000	300.000	0	0
5	5.000	4.000	0	0
7	1.600	n.b	0	0
9	1.500	3.200	0	0
14	1.200	2.000	0	0
20	800	1.100	0	0

Table 4. Changes the number of coli bacteria in post-fermentation leachates stored at 22Â°C after hygienization

Day of incubation	Number of coliform bacteria (CFU/ml)			
	Study sample			
	C	S	H	S + H
0	7.000	1.400	0	0
1	9.000	290.000	0	0
2	3.000	210.000	0	0
5	7.000	5.000	0	0
7	18.000	n.b	0	0
9	12.000	4.000	0	0
14	9.000	3.900	0	0
20	8.000	3.200	0	0

4 Conclusion

The use of post-fermentation leachates as fertilizer is a good alternative and solution to the problem of this waste in treatment plants. The leachate, after anaerobic stabilisation contain valuable elements such as phosphorus, nitrogen, and potassium that can improve soil properties and therefore be used to produce fertilizers. When considering the use of post-fermentation leachates, their sanitary condition is very important. A properly conducted hygienization process will allow the use of this waste material as a fertilizer in agriculture. The thermal hygienization allows to obtain a product that is sanitary safe (no *coli* bacteria, *Salmonella* and intestinal parasite eggs). Heating the post-fermentation leachates to a temperature of 80°C and maintaining it for 30 min results in the hygienization of the tested material. These properties persist for 20 days and the material should be used within this period. However, under these conditions, a large number of both psychrophilic and mesophilic bacteria still remain in the post-ferentation leachates. This creates a potential threat to rebuild microflora.

References

1. Pilarska A, Wolna-Maruwka A, Piechota T, Pilarski K, Szymańska M, Wolicka D (2015) Preliminary studies of post-fermentation pulp from a biogas plant nad its compost as potential organic fertilizer. Nauka Przyroda Technol 9(2):1–13. https://doi.org/10.17306/J.NPT.2015.2.19
2. Wójcik P, Dyśko J, Kaniszeski S, Kowalczyk W, Nowak J (2014) Sustainable fertilization of horticultural plants (in Polish)
3. Frąc M, Matyka M, Rozbicki J, Tryjanowski P (2022) Regenerative agriculture—the concept of sustainable food production and agroecosystems improvement. Regenerative agriculture for agroecosystems. Nauka, 4:155–164. https://doi.org/10.24425/nauka.2022.142927. Retrieved from www.journals.pan.pl.
4. Proposal for a regulation of the European parliament and of the council on minimum requirements for water reuse (2018) COM/2018/337 final—2018/0169 (COD), Brussels
5. Czarnota J, Masłoń A (2020) Organic and organic-mineral fertilizers based on municipal sewage sludge (in Polish). Gaz Woda i Technika Sanitarna 4:14–18. https://doi.org/10.15199/17.2020.4.3
6. Chew KW, Chia SR, Yen H-W, Nomanbhay S, Ho Y-C, Show PL (2019) Transformation of biomass waste into sustainable organic fertilizers. Sustainability 11(8): 2266-1–2266-19. https://doi.org/10.3390/su11082266
7. Zamrisham NAF et al (2023) State of the art in anaerobic treatment of landfill leachate: a review on integrated system, additive substances, and machine learning application. Water 15:1303. https://doi.org/10.3390/w15071303
8. Montusiewicz A, Lebiocka M (2011) Co-digestion of intermediate landfill leachate and sewage sludge as a method of leachate utiliztion. Biores Technol 102(3):2563–2571. https://doi.org/10.1016/j.biortech.2010.11.105
9. Łagocka A, Kamiński M, Cholewiński M, Pospolita W (2016) Health and Environmental benefits of utilization of post-fermentation pulp from agricultural biogas plants as a natural fertilizer. Kosmos Probl Nauk Biologicznych 65(4):601–607
10. Regulation of the Minister of Agriculture and Rural Development (2008) Regulation of the minister of agriculture and rural development on the implementation of certain provisions of the Act on fertilizers and fertilization. Journal Laws vol 119, item 765
11. Kovačić Đ, Lončarić Z, Jović J, Samac D, Popović B, Tišma M (2022) Digestate management and processing practices: a review. Appl Sci 12:9216. https://doi.org/10.3390/app12189216
12. Sahlström L, Aspan A, Bagge E, Danielsson-Tham M-L, Albihn A (2004) Bacterial pathogen incidences in sludge from Swedish sewage treatment plants. Water Resourses 38(8):1989–1994
13. Grudziński M, Pietruszka A, Sawicki W (2015) Anaerobic digestion in sanitization of pig slurry and biomass in agricultural biogas plant. J Microbiol Biotechnol Food Sci 5(2):524–526. https://doi.org/10.15414/jmbfs.2015.4.6.534-526

14. Chahal C, van den Akker B, Young F, Franco C, Blackbeard J, Monis P (2016) Pathogen and particle associations in wastewater: significance and implications for treatment and disinfection processes. Adv Appl Microbiol 97:63–119. https://doi.org/10.1016/bs.aambs.2016.08.001
15. Łagód G, Sobczuk H, Suchorab Z (2006) Application of saprobiontic microorganisms community analisys in the calibration of the model description of sewage self—purification in sewer system. Ecol Chem Eng 13:265–275
16. Markl D, Zeitler JA (2017) A review of disintegration mechanism and measuremant techniques. Pharm Res 34:890–917. https://doi.org/10.1007/s11095-017-2129-z
17. Wolska M, Mołczan M (2015) Assessment of the stability of water introduced into the water supply network (in Polish). Ochrona środowiska 37:51–56
18. Urbanowska A, Kotas P, Kabsch-Korbutowicz M (2019) Characteristics and methods of post-fermentation mass management in biogas plants (in Polish), vol 1, p 41
19. Kowalczyk-Juśko A, Szymańska M (2015) Digestate as fertilizer for agriculture. In: Foundation for the development of polish agriculture (in Polish). Retrieved from https://ksow.pl/files/Bazy/Biblioteka/files/Poferment_nawozem_dla_rolnictwa_01.pdf. ISBN 978–83–937363–6–2

Frequency Analysis of Hydrological Groundwater Drought in the Eastern Part of Slovakia Using Standardized Streamflow Index

Martina Zeleňáková[1](✉), Tatiana Soľáková[1], Hany F. Abd-Elhamid[1,2], and Pavol Purcz[1]

[1] Department of Environmental Engineering, Faculty of Civil Engineering, Technical University of Kosice, 040 01 Košice, Slovakia
martina.zelenakova@tuke.sk

[2] Department of Water and Water Structures Engineering, Faculty of Engineering, Zagazig University, Zagazig 44519, Egypt

Abstract. Hydrological drought is a recurring extreme hydrological phenomenon that negatively affects surface water levels and groundwater levels in basin. The monitored territory, i.e. the vicinity of Vychodoslovenská lowland ranks among the most vulnerable to the phenomenon of hydrological drought. This paper deals with the one-dimensional frequency analysis of historical severe drought events in groundwater in four sub-catchments namely: Dunajec and Poprad, Bodva, Bodrog, and Hornád using Standardized Streamflow Index (SSI) computed in 24-months' time scales for identification of this phenomenon for the reference period from 1972 to 2019. The analysis was done at seven water meter stations situated in the Eastern part of Slovakia. Then, a severe hydrological drought event was identified if the index value was less than −1.5. At the beginning of 2009, the most serious and longest groundwater drought event was recorded in the Hornad sub-basin at Kysak station, which lasted 28 months. The results showed that the year 1994 was a significant year in which severe hydrological drought was recorded at five water meter stations. Predicting the natural phenomenon with the help of inter-arrival time helps in prevention of drought, as well as in introduction of regional preventive measures ensuring the minimization of negative impacts on people and their activities and the environment. The shortest drought return time is recorded in Kysak namely 2.5 years, and the longest drought return time 11.9 year is recorded in Humenne.

Keywords: Frequency analysis · Hydrological drought · Standardized streamflow index (SSI) · Lack of groundwater levels · Risk of drought · Slovakia

1 Introduction

Water resources in territory of eastern Slovakia are unevenly distributed in time and space and are hydraulically connected with groundwater. The elementary cause of hydrological drought is a persistent deficit of rainfall, the intensity of which is further supported by

Z. Blikharskyy et al. (Eds.): CEE 2023, LNCE 438, pp. 533–542, 2024.
https://doi.org/10.1007/978-3-031-44955-0_53

other factors such as: air humidity, wind speed, sunlight, terrain slope, soil type and its hydrolimits. Due to the mutual interaction of the mentioned factors, the meteorological drought deepens and participates in the occurrence of the hydrological drought, which leads to direct and indirect negative consequences. A direct negative consequence is decreasing surface water and groundwater levels and decreasing the yield from springs. Indirect consequences include the negative impact on fauna and flora, as well as human activities [1].

Hydrological drought is rarely occurring phenomenon with temporary negative and severe deviation from average value of stream flow (or average value of level of groundwater) during a specific time in an area [2]. The most serious hydrological droughts in the 21st century were in Slovakia and in Europe especially in 2003, 2011–2012, 2015 [3]. While 2012 is classified as extremely dry, this year up to 97% of stream stations from the total territory of Slovakia did not reach the average annual flow [4]. Quantification of hydrological drought can be done using the theory of run or using indexes [5]. In Slovakia, the first method is more preferred. Drought indexes are especially suitable for spatial and temporal comparisons different basins. Among the drought indexes: Standardized Runoff Index by Shukla and Wood (2008) and Streamflow drought index by Nalbantis and Tsakiris (2009) [6, 7].

The Climate Atlas of Slovakia study from 2015 confirmed the changing climate in some Slovak basins. Adapting to climate change should lead to management or eliminating its negative effects (e.g. extreme droughts) through preventive, operational and organizational measures. Preventive measures ensure the retention of water in the territory and include for example: drainage canals, irrigation systems, melioration canals, sluices, dams, planting of riparian vegetation, restoration of wetlands and others [4]. Monitoring and predicting the development of drought are classified as operational measures [4]. In the territory of eastern Slovakia, groundwater drought monitoring is underway using probes and springs. Hydrological drought monitoring of groundwater levels has been carried out by the Slovak Hydrometeorological Institute since 2017. Currently, 27 probes and 4 springs are active in the given area. Monitoring outputs (i.e. groundwater level) are freely available on the website: www.shmu.sk.

Drought index SSI is an indicator that use streamflow series to determine the beginning and end of drought as well as quantifying hydrological drought into several classes as shown in Table 1. The Standardized Streamflow Index (SSI) was created by Nalbantis (2008) [8] and Vincente-Serrano et al. (2012) [9] for the detection of hydrological drought and identification its characteristics such as duration, severity, intensity and inter-arrival time.

In this study long time scales of the index SSI-24 is used to estimate the lack of groundwater at seven water meter stations located in Eastern part of Slovakia. Quantitative assessment of severe groundwater drought using the SSI-24 index is basis for evaluating the current and spatial distribution of water resources and for predicting future development using the average inter-arrival time. Our findings are beneficial for regional groundwater drought risk management.

Table 1. Classification of hydrological drought using SSI [8].

SSI intervals	SSI classes	Probability events (%)
$SSI \geq 0$	No-drought	50
$0 > SSI \geq -1$	Mild drought	34.1
$-1 > SSI \geq -1.5$	Moderate drought	9.2
$-1.5 > SSI \geq -2$	Severe drought	4.4
$SSI < -2$	Extreme drought	2.3

2 Material and Methods

2.1 Study Area

The monitored area represents eastern part of Slovakia with total basin areas of 14,495 km^2, of which agricultural land occupies 47.13%. Forests and semi-natural areas occupy 46.76% and artificial areas represent 5.42% and the last 0.69% belong to wet areas and waters.

The monitored area is shown in Fig. 1. It is classified as a moderately sensitive (sub-catchment: Poprad and Dunajec, Bodrog, Hornád) to very sensitive sub-catchment (Bodva) to the occurrence of hydrological drought [4].

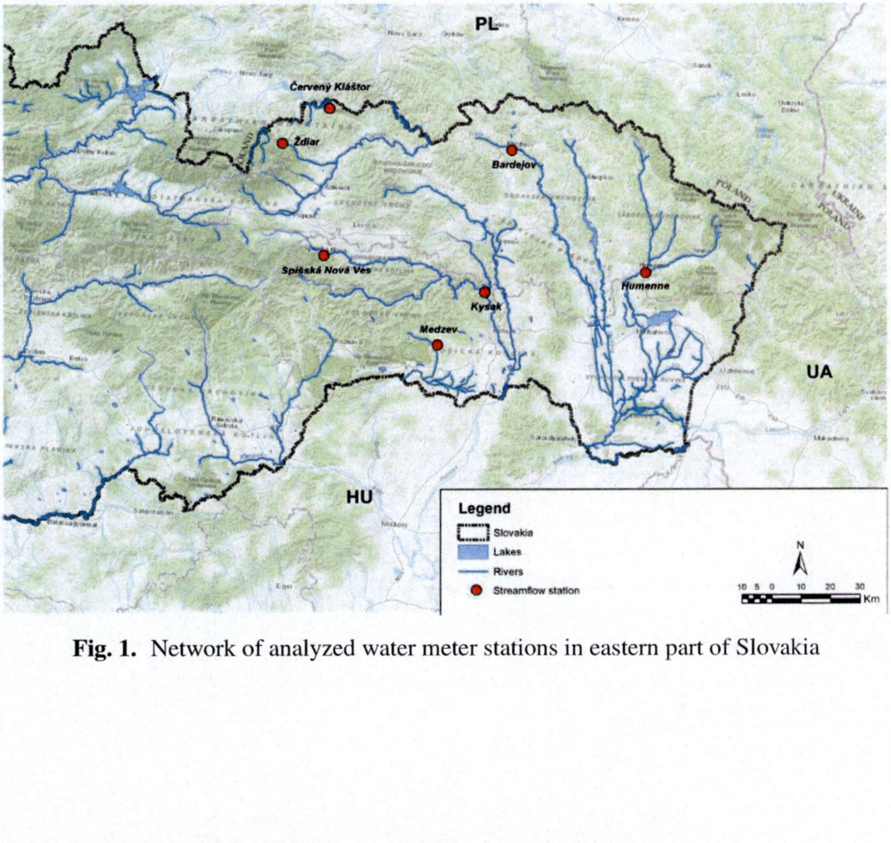

Fig. 1. Network of analyzed water meter stations in eastern part of Slovakia

The current study deals with seven stations: **Červený Kláštor** (49° 23′ 30″N, 20° 24′ 19″E), Ždiar (49° 16′ 18″N, 20° 15′ 44″E), Bardejov (49° 18′ 56″N, 21° 12′ 44″E), Humenné (48° 56′ 09″N, 21° 54′ 24″E), Kysak (48° 51′ 30″N, 21° 13′ 00″E), Spišská Nová Ves (48° 56′ 41″N, 20° 33′ 42″E) and Medzev (48° 42′ 1″N, 20° 53′ 37″E). Average monthly flow values were provided by Slovak Hydrometeorological Institute of Košice for a period of year 1972–2019 for a stations **Červený Kláštor**, Ždiar, Bardejov, Humenné, but Kysak station has a time period from 1973 to 2019 and Spišská Nová Ves station has a time period from 1974 to 2019.

2.2 Methodology

Even before the actual calculation of SSI index, it is necessary to work with monthly stream flow data by aggregating into 24-months time scale using so called "moving window". We further subject the data thus prepared to the statistical Kolmogorov-Smirnov test at a significance level of 5%. Next, we determine the value of the statistic D, on the basis of which we select the most suitable theoretical probability density function for each month separately for an approximately 47-year time period (1973–2019). The best-fitting theoretical probability density function is selected if D converges to 0, which means that the probability density function best describes the empirical 47-year stream flow time series. The selection of the best-fitting probability distribution function is done in the EasyFitt program. The EasyFitt program offers several probability density functions as we can see in Table 2 and based on the maximum likehood method, it is possible to determine their parameters [10]. If we selected the best-fitting probability density function, we will calculate its cumulative distribution function. The final step is transformation of the cumulative distribution function field of the formula Eq. (1) into a standard distribution which actually represents the SSI values.

$$SSI = \Phi^{-1}(F(x)) \tag{1}$$

The severe hydrological drought begins when the SSI value is less than −1.5. The duration of this event represents the time interval during which the SSI value is constantly less than −1.5. The severity of drought is a sum of SSI values during the duration of drought. Inter-arrival time is a time between the start of the first drought event and the start of the next drought event [12].

3 Results and discussion

The main task of this work was the identification of a severe hydrological drought of groundwater and the predict the severe groundwater drought in the future in the evaluated area as well as identification of sensitivity of this area to hydrological drought. The identification of a severe drought event in groundwater is done using SSI index. The results of temporal drought analysis at the seven water meter stations are presented in Figs. 2, 3, 4, 5, 6, 7, 8 and 9 and Table 3. The historical temporal analysis of drought divided into decades is shown in Fig. 2.

Table 2. Chosen probability density function and the mathematical formula of its cumulative distribution [11].

Probability density function	Cumulative distribution function
Gen. extreme value	$F_{(x)} = \exp(-(1 + \mathrm{k}z)^{-\frac{1}{\mathrm{k}}})$ $z = \frac{x-\mu}{\sigma}$
Weibull 3P	$F_{(x)} = 1 - \exp(-(\frac{x-\gamma}{\beta})^{\alpha})$
Gamma 3P	$F_{(x)} = \frac{\Gamma_{\frac{(x-\gamma)}{\beta}}(\alpha)}{\Gamma(\alpha)}$
Gen. Gamma	$F_{(x)} = \frac{\Gamma_{(\frac{x}{\beta})^{k}}(\alpha)}{\Gamma(\alpha)}$
Burr	$F_{(x)} = 1 - (1 + (\frac{x}{\beta})^{\alpha})^{-\mathrm{k}}$
Lognormal 3P Pearson 6 Normal	$F_{(x)} = \left(\frac{ln(x-\gamma)-\mu}{\sigma}\right)$ $F_{(x)} = \mathrm{I}_{\mathrm{x}/(\mathrm{x}+\beta)}(\alpha_1, \alpha_2)$ $\boldsymbol{F}_{(\boldsymbol{x})} = \left(\frac{(x-\mu)}{\sigma}\right)$

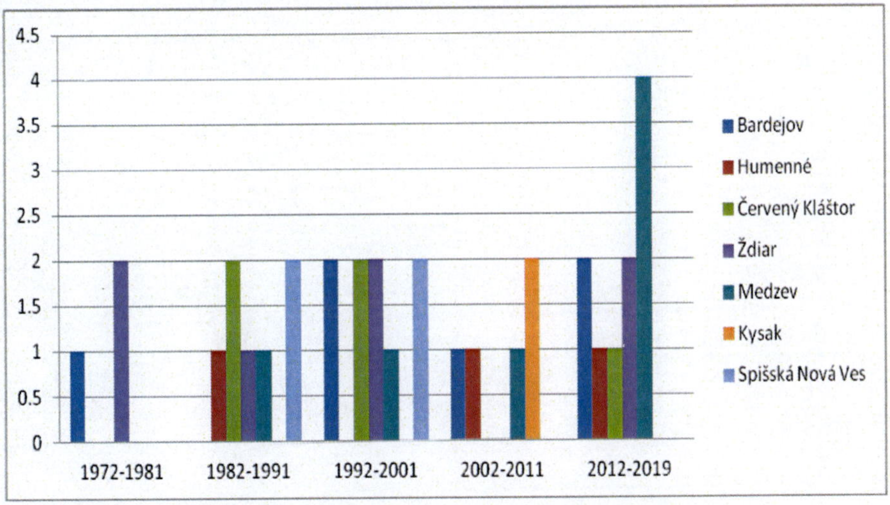

Fig. 2. Frequency analysis of the severe groundwater drought event by SSI-24.

It can be highlighted that the last incomplete decade 2012–2019 was characterized by an excessive occurrence of episodes of severe groundwater drought. In the last decade the Bodva basin faced of four severe hydrological droughts.

The total number of groundwater drought episodes in eastern Slovakia during the period 1972–2019 was 34 with a total duration of 276 days and cumulative severity of −517.9 and a cumulative intensity of −59.2. The evaluation within individual basins can be read from Table 3.

Table 3. Characterization of severe groundwater drought by SSI-24

Sub-basin	Water meter station	No of Events	Duration (months)	Cumulative severity	Cumulative intensity	Inter-arrival time (year)
Bodrog	Bardejov	6	49	−88.7	−10.5	8.2
	Humenné	3	38	−76.6	−5.5	11.9
Dunajec a Poprad	Červený Kláštor	5	42	−5.8	−8.8	6.5
	Ždiar	7	42	−74.1	−11.8	6.9
Hornad	Kysak	2	31	−69.3	−4.0	2.5
	Spišská Nová Ves	4	36	−62.7	−6.7	10.9
Bodva	Medzev	7	38	−70.7	−11.9	4.9

Figures 3 and 4 indicate the temporal analysis of hydrological drought in sub-basin Bodrog. The total number of severe hydrological drought is 9 with duration of 87 months in the years: 1972–73, 1991, 1994, 1995, 2003–04, 2013, 2015–17. The longest 25-month severe groundwater drought was recorded at Humenne station in April 2015. A severe hydrological drought can be expected in Bodrog sub-basin for an average of 8.2 to 11.9 years.

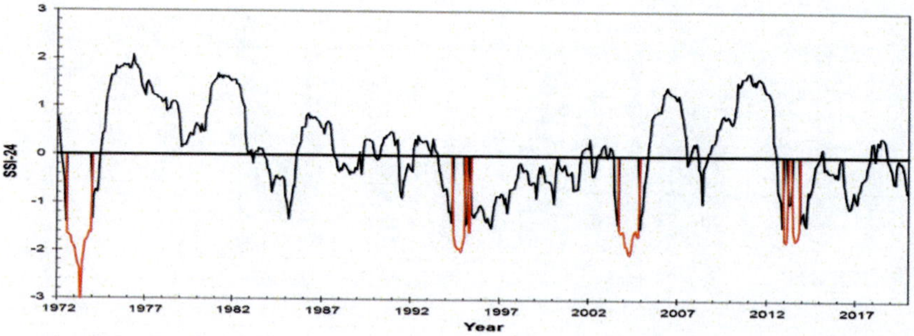

Fig. 3. Histogram of SSI-24 for Bardejov station in the period (1972–2019).

Figures 5 and 6 represent the temporal analysis of hydrological drought in sub-basin Dunajec and Poprad. The total number of severe hydrological drought is 12 with duration of 84 months in the years: 1972–73, 1987–88, 1993–95, 1994–95, 1996 2012–13. The longest 22-month severe groundwater drought was recorded at **Červený Kláštor** station

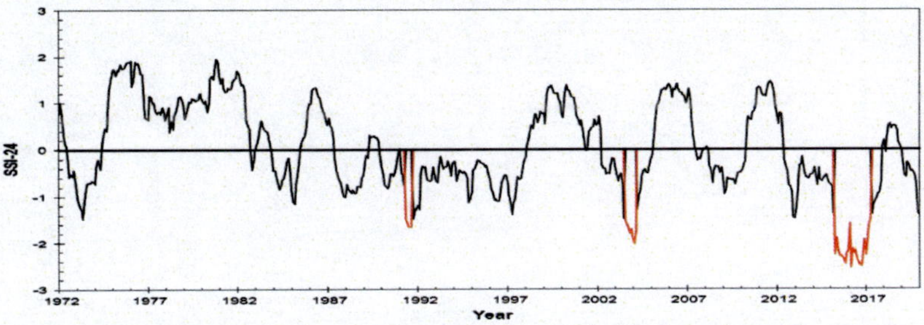

Fig. 4. Histogram of SSI-24 for Humenne station in the period (1972–2019).

in August 1993. A severe hydrological drought can be expected in Dunajec and Poprad sub-basin for an average of 6.5 to 6.9 years.

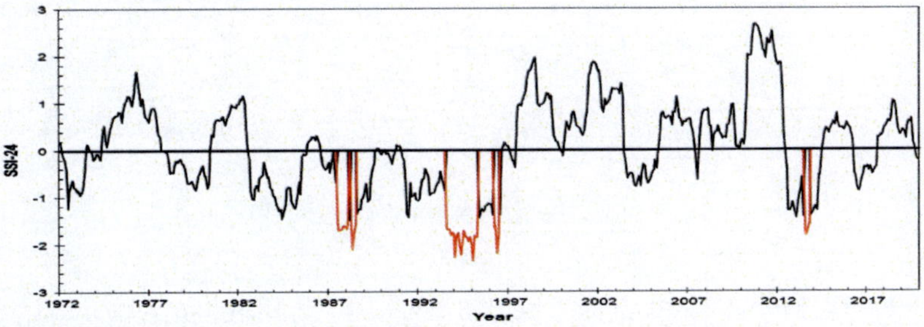

Fig. 5. Histogram of SSI-24 for Červeny Klaštor station in the period (1972–2019).

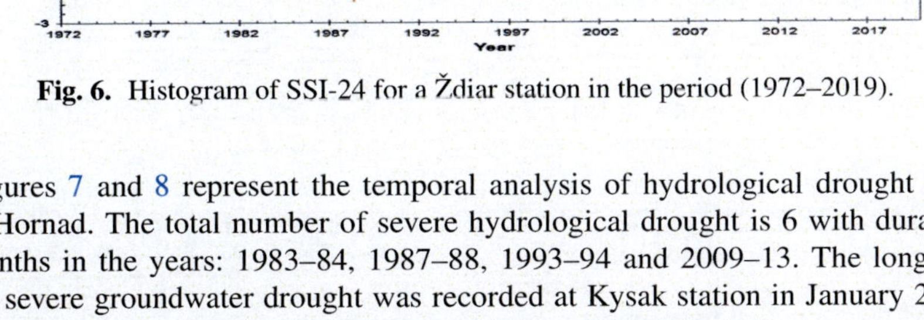

Fig. 6. Histogram of SSI-24 for a Ždiar station in the period (1972–2019).

Figures 7 and 8 represent the temporal analysis of hydrological drought in sub-basin Hornad. The total number of severe hydrological drought is 6 with duration of 67 months in the years: 1983–84, 1987–88, 1993–94 and 2009–13. The longest 28-month severe groundwater drought was recorded at Kysak station in January 2009. A severe hydrological drought can be expected in Hornad sub-basin for an average of 2.5 to 10.9 years.

Figure 9 represents the temporal analysis of hydrological drought in sub-basin Bodva. The total number of severe hydrological drought is 7 with duration of 38 months in the

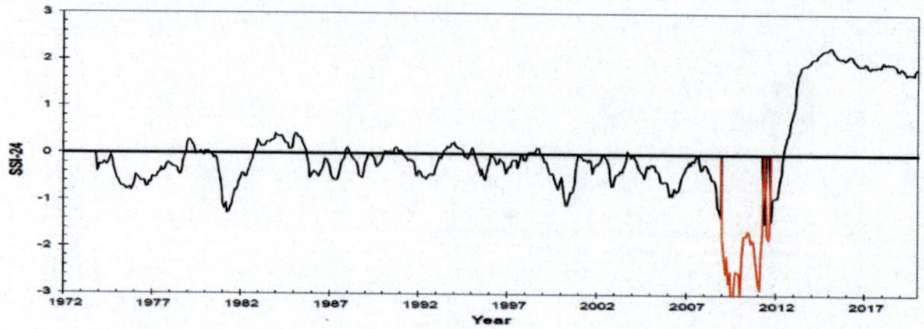

Fig. 7. Histogram of SSI-24 for a Kysak station in the period (1973–2019).

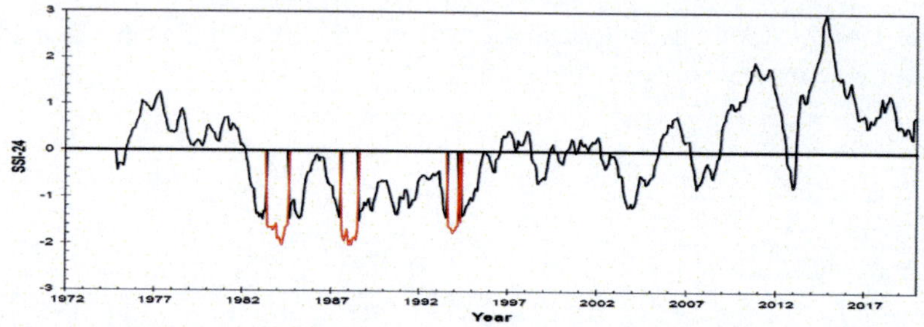

Fig. 8. Histogram of SSI-24 for a Spišská Nová Ves station in the period (1974–2019).

years: 1990–91, 1993–94 and 2003–04, 2019. The longest 14-month severe groundwater drought was recorded in March 2003. A severe hydrological drought can be expected in Bodva sub-basin for an average of 4.9 years.

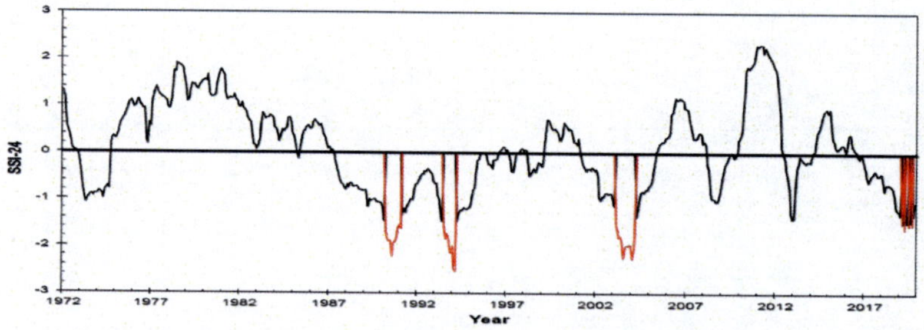

Fig. 9. Histogram of SSI-24 for a Medzev station in the period (1972–2019).

The smallest inter-arrival time of groundwater drought was recorded at Kysak station. We can consider this result to be less significant because the water levels are influenced by human activity (ie the dam regulates the flow) and it is not possible to clearly determine when this phenomenon will appear again. Then the smaller inter-arrival time of severe groundwater drought was identified at the Medzev station, where the severe hydrological drought can be expected in 2023. We can expect a serious hydrological drought in the Dunajec and Poprad basin in 2025 and in Bodrog basin in 2028 to 2029.

4 Conclusions

The increase in the temperature of the earth 's surface or change in precipitation pattern on the Eurasian continent affects water resources of groundwater bodies and their use for various industries. Groundwater drought is not only a consequence of atmospheric processes, but also of hydrological processes and local characteristics of the basin. The current study investigated the temporal occurrence of hydrological drought at the local level. For this purpose, a standardized streamflow index (SSI) in a 24-month time scale was calculated for 7 water meter stations in the territory of eastern Slovakia over a period of 47 years from 1973 to 2019. In a sub-basin Hornad, the hydrological drought became less frequent, less intense and longer. In the water measure station Kysak, it is not possible to clearly determine the return of the hydrological drought. For other basins the return of hydrological drought applies, which is written in previous section. Potentially the most affected by the severe hydrological drought are the areas around the water meter stations Medzev and Ždiar, where droughts are relatively shorter and more frequent. Knowing the temporal dynamics of occurrence of hydrological drought on a local basis is an important step in the process of monitoring and management of water resources in the basin.

Acknowledgments. This work was supported by the Slovak Research and Development Agency under the Contract no. APVV-20-0281, and a project funded by the Ministry of Education of the Slovak Republic VEGA1/0308/20 "Mitigation of hydrological hazards, floods, and droughts by exploring extreme hydroclimatic phenomena in river basins".

References

1. SHMU. www.vuvh.sk/rsv2/download/02_Dokumenty/10_Podpone_dokumenty_metodiky/SUCHO_2011beata.pdf. Last accessed 1 Mar 2023
2. COM: Commission of the European Communities Addressing the challenge of water scarcity and drought in European Union (2007) Communication from the Commission to the European Parliament and Council. COM (2007) 414 final, Brussels, 18 July 2007
3. Fendeková M, Poórová J, Slivová V (eds) (2018) Hydrologic drought in Slovakia and forecast of its development. Comenius University, Faculty of Science, Bratislava
4. MZP SR. www.minzp.sk/files/sekcia-vod/hodnota-je-voda/h2odnota-je-voda-akcny-plan-riesenie-dosledkov-sucha-nedostatku-vody.pdf. Last accessed 10 Mar 2023
5. Yevjevich VM (1967) An objective approach to definitions and investigations of continental hydrologic drought. In: Hydrology papers no. 23, Colorado State University
6. Shukla S, Wood A (2008) Use of a standardized runoff index for characterizing hydrologic drought. Geophys Res Lett 35:L02405
7. Nalbantis I, Tsakiris G (2009) Assessment of hydrological drought revisited. Water Resour Manage 23(5):881–897
8. Nalbantis I (2008) Evalution of hydrological drought index. European Water 23(24):67–77
9. Vincente-Serrano SM, López_Moreno JI, Begueria S, Lorenzo-Lacruz J, Azorin-Molina C, Morán-Tejeda E (2012) Accurate computation of a streamflow drought index. J Hydrol Eng 17(2):229–358
10. Kottegoda N, Rosso R (1997) Statistics, probability and reliability of civil and environmental engineers. McGraw-Hill, New York

11. Zeleňáková M, Soľáková T, Purcz P, Abd-Elhamid HF (2023) Analysis of hydrological drought in the Eastern part of Slovakia using standardized precipitation index. In: Sherif M, Singh VP, Sefelnasr A, Abrar M (eds) Water resources management and sustainability. Water Science and Technology Library 121:119–128 (2023)
12. Madadgar S, Moradkhani H (2013) Drought analysis under climate change using copula. J Hydrol Eng 18(7):746–759

Green Zones at TUKE Campus

Martina Zeleňáková[1] (✉), Rastislav Fijko[1], Viktor Karľa[1], Igor Hančovský[1], Marcel Behun[1], Ervín Lumnitzer[1], and Hany F. Abd-Elhamid[1,2]

[1] Technical University of Kosice, Košice 04001, Slovakia
martina.zelenakova@tuke.sk
[2] Zagazig University, Zagazig 44519, Egypt

Abstract. This paper deals with the creations of green zones at the Technical University in Košice (TUKE) campus and counts on the construction of a retention pond subsidized by rainwater from the reconstructed flat roof with park arrangements and vegetation walls both in the interiors of the entrance areas to the rectorate and on the facade of the adjacent civil protection warehouses. The proposed solutions aim to restore its use at a new, higher level, primarily for the representative purposes of the university and its workplaces with the creation of green vegetation areas as part of raising awareness of climate change. The concept of a green university presented in paper is designed to increase and strengthen the water retention capacity. Retained rainwater can rehydrate agricultural and urbanized landscapes. Rainwater that is currently draining uselessly should seep into the soil as a matter of priority, thus restoring and strengthening groundwater supplies. Recharged aquifers give the vegetation enough moisture to allow evaporation, which pumps heat from the superheated surface to the higher colder layers of the atmosphere.

Keywords: Rainwater management · Green roof · Green wall · Retention basin · TUKE campus

1 Introduction

Due to the increasing number of requirements for the construction of the territory for various needs, the natural green area is being degraded and used. Green elements artificially created on already existing buildings can be a possibility of mitigating the negative impacts of the development of the territory [1–4]. The main users of green campus are students. The students are a foundation to create and increase more sustainable green campuses.

The purpose of this paper is introduction of retention pond design. The site for the construction of the retention pond is located in the campus of the Technical University in Košice (TUKE), in its central part. The object of the pond and the collection system is designed as a set of construction objects that together create a functional unit serving to retain water in the country.

Z. Blikharskyy et al. (Eds.): CEE 2023, LNCE 438, pp. 543–550, 2024.
https://doi.org/10.1007/978-3-031-44955-0_54

2 Description and Design of the System

The construction is designed as a system of a pond and an infiltration shaft with a rain garden, into which the collected rainwater from the surrounding buildings is diverted through a system of pipes and shafts and is then controlled into the infiltration basin. Construction object of TUKE retention pond consists of the following partial objects as shown in Figs. 1 and 2:

- Retention pond
- Rainwater connection from object L9A
- Rainwater connection from the object to KCN 2731/148
- Retention tank
- Suspension shaft
- Pier
- Water connection
- Electrical connection and lighting;

In addition to two operating units:

- Gas station ČS1 at L9A
- TUKE pond technology shaft L9A.

Fig. 1. Layout of the system compoonents

2.1 Retention Pond

The newly proposed area is planned as a space for rest and relaxation for students, teachers and passers-by who are just passing through the university campus, or it will serve as an area for university representation and create a new space for the rector's building in the direction from the parking lot and the inner campus. The entire design is conceptually designed so that it complements the environment as best as possible from the point of view of architectural function and from the point of view of a functional solution, it offers the possibility to manage rainwater on the university property, or it may also have an educational character. The proposal envisages the construction of a

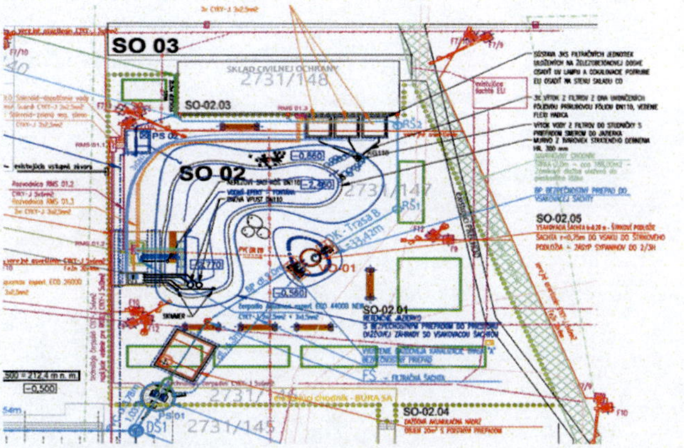

Fig. 2. Detailed description of the system compoonents

retention pond with the placement of a smaller number of medium-sized fish throughout the year, which will create a suitable and clean environment for life. The immediate surroundings of the pond will be supplemented with suitable aquatic plants as shown in Fig. 3. The overall impact is completed by a pier located above the water surface, placed on separate foundation strips (see the statics section of the pier). Underneath the pier area, there is a pond technology in a concrete shaft (machine room) and a prefabricated pump shaft for storing the pumps needed for water circulation.

The designed parameters of the proposed pond are as following:

- The total volume of the pond, including the shallows, is 187.2 m^3
- The volume of the pond watered for height 1.6–2.0 m is 163.0 m^3
- The permanently irrigated area for height 1.6–2.0 m is 90.43 m^2
- The permanently watered total area, including shallows is 171.8 m^2

Floor plan of the pond: the shape of the pond is designed in such a way that it is as suitable as possible integrated into the surrounding environment. In plan, it is a bent lenticular shape with a bend in the middle part from 90°. In its vicinity, several areas have been created for rest and relaxation, but also for a direct view of the water and the surroundings (pier).

Water effect located on the pier, oase Waterfall XL 90 is proposed—a stainless steel waterfall in the number of 5 pcs. For the needs of the design, it is necessary to connect from the back, possibly the bottom. The designed element ensures an even water pattern and a precisely flowing waterfall. The proposed water element can be supplemented with ambient lighting of the waterfall with insulation through the passage for lighting to highlight the contour of the water element.

Bench: a concrete bench placed on a concrete base is proposed. The bench consists of agate wood, dried, without surface treatment. The total number of proposed benches is 5 on the ground and 1 needs to be modified and installed in the area of the pier.

Pedestrian path: the proposed pedestrian path is planned across the campus of the Technical University.

Fig. 3. Retention pond

2.2 Rainwater Connection from Main Building

Rainwater from the roof of the building is currently drained by gravity, through the roof rain gutters and individual internal waste stone pipes with a passage through the base plate in the basement of the building into the horizontal distribution of the rain sewer and continues as a combined distribution with the sewage sewer to the city sewer. It is proposed to disconnect four internal drains and relieve the capacity of the existing sewage system.

Calculated flow of rainwater amount (Eqs. 1 and 2):

$$Q_d = q \times S_s \times \psi_s; \tag{1}$$

$$Q_d = 178 \times (0.0629 + 0.0354) \times 0.9) = 15.75\, l_s^{-1}. \tag{2}$$

Retention tank design.

Tank volume (Eq. 3):

$$Q_d \times 900s = 15.75 \times 900 = 14,175 \text{ liters.} \tag{3}$$

2.3 Rainwater Connection from the Object to Shaft

Construction object of rainwater connection from the object to shaft, solves the drainage of rainwater from the roof of the existing building of the civil protection warehouse in the area of interest on the plots indicated in the basic characteristics of the building, so that they are controlled to be discharged through the pipe into the infiltration shaft.

Calculated flow of rainwater amount (Eqs. 4 and 5):

$$Q_d = q \times S_s \times \psi_s, \tag{4}$$

$$Q_d = 178 \times (0.0071) \times 0.9) = 1.137\, l_s^{-1}. \tag{5}$$

Minimum volume of infiltration shaft required (Eq. 6):

$$Q_d \times 900s = 1.137 \times 900 = 1024 \text{ liters}. \tag{6}$$

2.4 Retention Tank

Construction object of retention tank solves the management of rainwater drained from the roof of the existing building of the main TU building in the area of interest on the plots indicated in the basic characteristics of the building so that they are controlled by a pipe through a filter shaft into an infiltration shaft.

2.5 Infiltration Shaft

Construction object of infiltration shaft serves to retain water from the proposed green roof of the main TU building and from the roof of the nearby building of the civil protection warehouse. Drained water will be drained into the gravel layer of the subsoil determined according to the geological composition.

The new approach to dealing with rainwater is based on the recommendations of the European Union. It is preferable to let rainwater infiltrate into the subsoil in the same area where it falls. The retention infiltration garden will mainly serve for the drainage of larger reduced areas. This type of device for absorbing and retaining rainwater is always designed as an open tank. It serves to capture the entire volume of rainwater from the surface runoff from individual objects. The supplied volume of rainwater soaks into the bottom and slopes of the garden and then into the drain. A safety overflow must be designed on the retention rain tanks. It is recommended to design a fortification around the place of the concentrated inflow into the reservoir.

The building is designed as an attractive depressed place in the garden of the campus of TUKE. The water should be retained in it for a maximum of 48 h.

Calculated flow of amount of rainwater (Eqs. 7 and 8):

$$Q_d = q \times S_s \times \psi_s; \tag{7}$$

$$Q_d = 178 \times (0.0071 + 0.0629 + 0.0354) \times 0.9) = 16.88\, l_s^{-1}. \tag{8}$$

Required minimum volume of infiltration shaft: the net volume of the shaft is proposed to be removed at time t = 15 min, the volume of 50% of all supplied water, while the total volume of water at t = 15 min is 15,192 m^3. The remaining volume of water will be retained in an underground retention tank. The proposed rain garden with an accumulation volume of 13.8 m^3 located around the infiltration shaft will also help to retain water immediately.

The whole concept of rain gardens is designed in such a way that it positively affects the natural and overall operation of individual facilities and does not disturb them in any way. It is possible to present the functionality and technical solution of individual elements on individual surfaces, thus pointing out the importance of handling water and its cycle.

For the needs of the proposed solution, the total area of the rain garden was proposed to be 35 m^2. Pipes from three directions are brought into the garden area, which are equipped with a frog valve of the required size at the point of discharge to the surface.

In case of increased intensity of precipitation, the garden is equipped with an overflow shaft, which drains the excess water into the subsoil. The surroundings of the garden are completed with plants.

2.6 Pier

The floor of the building consists of a wooden floor made of agave wood of minimum strength class D30. The thickness of the floor is 50 mm. The wooden floor must be properly fixed to the steel beams. A heavy asphalt waterproofing strip is inserted between the floor and the steel beams.

2.7 Water Connection

The water connection is led from the connection point to the entrance to the water meter shaft and is used for hygienic and social purposes.

Calculation of water needs:

Average daily water requirement (Eq. (9)):

$$Q_p = n.q = 435\, l/d. \tag{9}$$

Maximum daily water requirement (Eq. (10)):

$$Q_m = Qp.kd = 870\, l/d. \tag{10}$$

Maximum hourly water requirement (Eq. (11)):

$$Q_h = 1/24.Qp.kd.kh = 36.25\, l/h; \tag{11}$$

Annual water consumption—estimated:

Q_{rok} = 158.7 m^3/year—according to Decree 397/2003.

2.8 Electrical connection and lighting

It is a set of street lamps put in both vertical and horizontal directions oriented on several levels as shown in Fig. 4. These are prefabricated steel columns of a specific manufacturer.

The mentioned measures for smart green campus TUKE will enhance its useability, apparence and contribute to mitigate impacts of climate changes.

Fig.4. Apperance of lighting

3 Conclusions

Rainwater management in Slovakia faces many challenges today, as the approach used in Slovakia a few years ago and for that reason it is necessary to adapt to ongoing changes. The increase in the temperature of the earth's surface and changes in precipitation pattern in the continent affects surface and groundwater water resources. Hydrological risks—floods and droughts are not only a consequence of atmospheric processes, but also of hydrological processes and local characteristics of the basin and finally. The concept of a green university is designed to increase and strengthen the water retention capacity. Retained rainwater can rehydrate agricultural and urbanized landscapes. Rainwater that is currently draining uselessly should seep into the soil as a matter of priority, thus restoring and strengthening groundwater supplies. Recharged aquifers give the vegetation enough moisture to allow evaporation, which pumps heat from the superheated surface to the higher colder layers of the atmosphere.

Acknowledgments. This work was supported by the Slovak Research and Development Agency under the Contract no. APVV-20-0281, and a project funded by the Ministry of Education of the Slovak Republic VEGA1/0308/20 "Mitigation of hydrological hazards, floods, and droughts by exploring extreme hydroclimatic phenomena in river basins".

This work has been supported by project of the Ministry of Education of the Slovak Republic ID 20210812131460210: Creation of green zones at TUKE and raising awareness of climate change.

References

1. Fendeková M, Poórová J, Slivová V (eds) (2018) Hydrologic drought in Slovakia and forecast of its development. Comenius University, Faculty of Science, Bratislava

2. Fachrudin HT (2020) Green campus concept based on architect perspective. IOP Conf Ser Mater Sci Eng 801:012028
3. Zeleňáková M, Purcz P, Hlavatá H, Blišťan P (2015) Climate change in urban versus rural areas. Procedia Eng 119:1171–1180
4. VanWoert ND, Rowe DB, Andresen JA, Rugh CL, Fernandez RT, Xiao L (2015) Green roof stormwater retention. J Environ Qual 34(3):1036–1044
5. Fijko R et al (2023) Documentation for building permit for retention pond, 3–20

Safety of the Water Supply System from the System Operator Perspective

Jakub Żywiec(✉), Barbara Tchórzewska-Cieślak, and Janusz Rak

Department of Water Supply and Sewerage Systems, Faculty of Civil, Environmental Engineering and Architecture, Rzeszow University of Technology, Al. Powstancow Warszawy 6, 35-959 Rzeszow, Poland
j.zywiec@prz.edu.pl

Abstract. The collective water supply system (CWSS) belongs to the critical infrastructure systems, being at the same time an anthropotechnical system, i.e. a system consisting of mutually interacting man—technical object—environment. Ensuring an appropriate level of reliability and safety of water supply to consumers is one of the current problems of the water supply sector. Specialist literature indicates that 70–90% of failures of technical systems are related to the impact of the human factor. In the case of the CWSS, it is estimated that man is responsible for 75% of the occurring failures. The influence of the human factor in the assessment of the reliability of the CWSS cannot be omitted. The paper presents the role of the operator in the CWSS, indicates the types of human errors and their causes. A review of the human reliability assessment (HRA) methods was made and the perspective of using the HRA methods in the analysis of the reliability and safety of the CWSS was presented.

Keywords: Water supply · Safety · Reliability · Human error

1 Introduction

One of the most common causes of failure, apart from technical or environmental reasons, is also the direct or indirect impact of the human factor [1]. It can affect the system at the design stage, at the stage of building or finally during its operation, creating conditions for failure [1]. Statistics show that human activities contribute to 60–90% of failures in various technical systems [1]. The data available in the literature indicate that the human factor had an impact on up to 75% of failures in collective water supply systems (CWSS) [2]. The impact of the human factor on the functioning of the technical system may be internal, including authorized activities (staff), or external in relation to acts of vandalism, terrorism or its incorrect use (actions of third parties) [3]. A man in the anthropotechnical system may turn out to be the weakest link. For this reason, in the early 1960s, research on the human reliability began, to determine the probability of the technical staff performing the task correctly in a given time period during the operation of the system. From the very beginning, there was a multidisciplinary approach to the human reliability assessment (HRA), which included the sciences of reliability, ergonomics and

Z. Blikharskyy et al. (Eds.): CEE 2023, LNCE 438, pp. 551–561, 2024.
https://doi.org/10.1007/978-3-031-44955-0_55

psychology [1]. More than 70 different HRA methods have been presented over the last 60 years [4]. Methods of assessing operator reliability have been widely used in many fields, including construction [5], energy [6], aviation [7], space [8], sea transport [9], land transport [10] or water supply [2].

The CWSS is an example of an anthropotechnical system, i.e. the system of man—technical object—environment [3]. Due to the task of the CWSS, which is to provide consumers with water in the appropriate quality and quantity, under the required pressure, at any time and at an acceptable price, the impact of the human factor in assessing the reliability of the CWSS cannot be omitted [11].

The paper presents the role of the operator in the collective water supply system, indicates the classification of human errors and their causes. The classification of HRA methods and the perspective of their use in the analysis of reliability and safety of the functioning of collective water supply systems were also presented.

2 The Role of the Operator in the Operation of the CWSS

Water and sewage companies should create vacancy for:

- water treatment plant (WTP) operator—responsible for managing and controlling the operation of the water treatment plant,
- sewage treatment plant (STP) operator—responsible for managing and controlling the operation of the sewage treatment plant,
- central operator—responsible for monitoring and coordinating activities ensuring the maintenance of the proper condition of the water supply and sewage infrastructure.

The task of the CWSS operator is to maintain an appropriate level of reliability and safety of the functioning of the system and its interdependent subsystems by controlling devices and objects remotely or manually. The CWSS consists of the following subsystems: WIsb—water intake subsystem, WTsb—water treatment subsystem, WPsb—water pumping subsystem, WSsb—water storage subsystem, WDsb—water distribution subsystem [3].

The reliability of the CWSS is understood as its ability to perform its functions in accordance with the requirements in terms of operation and safety, i.e. to provide consumers with water in the right amount, of the right quality, under the right pressure, at an acceptable price at any time [3]. The appropriate amount of water is understood as the amount resulting from the consumer's current demand for water, and in the event of a crisis situation, the supply of the minimum amount of water, i.e. 7.5 $dm^3/I \cdot d$ [12]. The quality of the supplied water must not pose a threat to the life or health of consumers and must meet the conditions contained in the Regulation of the Minister of Health on the quality of water intended for human consumption [13]. Appropriate pressure of the supplied water should ensure the comfort of using sanitary utensils and enable the use of household appliances supplied with tap water [3]. The unit price for a cubic meter of tap water should remain at a level acceptable to consumers, taking into account the operating costs necessary to ensure reliable operation of the system [3]. The operation of the CWSS operator should ensure the fulfillment of these conditions at any time [3].

Currently, the basic tool of the operator's work are ICT systems that enable remote management of individual elements of the subsystems. The Supervisory Control And

Data Acquisition (SCADA) system is commonly used, it is an IT system, superior over individual systems controlling subsequent processes of the technological line. SCADA allows the operator to supervise the technological processes. The main function of the system is to collect current data from the facilities (including measurements of: flow, pressure, dose of reagents, water level in tanks, physical and chemical parameters of water), their visualization and presentation to the operator in a clear graphical way. In addition, SCADA enables control of technological processes, alarming in the event of a failure and data archiving [14]. Operator's tasks in individual subsystems are presented below:

- water intake subsystem (WIsb): for surface water intakes, it is recommended to observe the lowest water levels during the year, observe flood waters, bottom and rubble movement, remove coastal vegetation, protect water against blooms and control the quality of intaken water [14]. For underground water intakes, it is recommended to control the quality of water, monitor the efficiency of the well, observe the water level in the well, remove sediments formed on the well filter, and, if necessary, renovate and remove sand form the well [14].
- water treatment subsystem (WTsb): Due to the large variety of water treatment technologies and devices dedicated to them, there are many operational activities aimed at maintaining the facilities in an efficient state. The most commonly used surface water treatment processes include: coagulation, filtration, and disinfection. Water coagulation is a process involving the removal of colloids and hard-settling particles from water, which are responsible for the turbidity and color of water. It occurs as a result of adding chemicals to the treated water, the so-called coagulants and substances supporting the coagulation process—weights and flocculants. Operational tasks in the coagulation system include: inspections and maintenance of devices and equipment of technological facilities, adjustment of reagent doses to water quality parameters, and maintenance of sludge removal systems in an efficient state [15]. The doses of reagents are usually controlled automatically, depending on the flow or turbidity of the water [15]. The water filtration process usually occurs after the coagulation or aeration process on the filters. In this process, as a result of water filtration through a bed filled with porous material, suspended impurities are removed. Operating activities in filter maintenance include: rinsing the filters, replacing the filter bed filling, checking the technical condition of the filter drainage, washings troughs, pipelines and filter accessories [15]. The operator supervises the operation of the filters based on parameters such as: filtration speed, length of the filtration cycle, value of hydraulic pressure losses on the filter and filtered water turbidity. Filter operation control consists in maintaining a constant filtration speed with increasing pressure losses on the filter, determining the end of the filtration cycle, and then carrying out the filter rinsing procedure and putting it back into operation [15]. The purpose of the disinfection process is to neutralize living and spore forms of pathogenic organisms and to protect water against secondary pollution in the water supply network [15]. Due to the use of dangerous chemical substances, the operation of devices intended for water disinfection and the preparation of disinfectants must be carried out with particular care in compliance with applicable health and safety requirements [15]. Most often, automatic control systems are used, where the dose of the disinfectant

depends on the instantaneous flow rate [15]. It is possible to manually control the process (e.g. changing the disinfectant dose, switching on/off) by the operator from the computer station in the control room and locally from the level of technological facilities,

- water pumping subsystem (WPsb): the operation of pump units is a complex task and requires strict adherence to the required procedures regarding, among others, starting or stopping the pump unit [14]. Periodically, it is required to test the efficiency of the pump unit's operation, which allows you to assess the operating conditions and the degree of wear of the pump, and, if necessary, qualify it for repair. Among the causes of improper operation of the pump, manifested, for example, by excessive energy consumption or noise, the following can be distinguished: wear or faulty construction of pump components, leaks in the suction line, air in the discharge pipeline, incorrect adjustment of the pump operation [14],
- water storage subsystem (WSsb): during the operation of tanks, it is recommended to monitor the water level in the tank, control water quality, tank condition (leakage control) and valve chamber, as well as tank maintenance (washing, disinfection) [14],
- water distribution subsystem (WDsb): proper operation of water supply network consists in maintaining the pipes and their fittings in the proper technical condition by conducting systematic inspections and maintenance works, quick removal of failures, constant analysis of the operating conditions of the water supply network in terms of the quantity and quality of water and pressure values, as well as planning and carrying out renovation works [14].

3 Taxonomy of Human Errors and Their Causes

One of the features of a human being as an element of the anthropotechnical system is fallacy [16]. It is expressed in making incorrect operator decisions and false assessment of the environment and technical condition of the system [16]. Operator errors can have different causes and consequences, so their identification is crucial when looking for opportunities to reduce their number. There are many taxonomies of operator errors [17]. The most popular of them was presented by Reason in 1990 [17]. It was used e.g. while investigating the accident at the Chernobyl nuclear power plant. This division assumes that the operator's errors result from error in the functioning of human cognitive functions, i.e. inattention, memory problems and lack of knowledge. There are three categories of errors:

- slip—execution error, the operator, as a result of inattention, does not perform the planned activities correctly,
- lapse—execution error, the operator does not perform the planned activities correctly as a result of memory problems,
- mistake—planning error, the activities planned to be performed by the operator are inappropriate for the correct execution of the task.

Another division of operator error types is by operator intent [16]:

- intentional errors—the operator knowingly performed an action that was inappropriate for the situation, which in his opinion was the right thing to do at the time or was a better course of action; the mistake is an intended error,

- unintentional errors—the operator unknowingly performed an action inappropriate for a given situation, which caused that he did not achieve the intended results; unintentional errors include a slip and a lapse.

The operator can make a error in doing or not doing the action at the right time, which makes them stand out as [11]:

- commission—the operator performs an action that should not be performed,
- omission—the operator does not perform an action that should be performed.

Incorrect assessment of the situation by the operator may result in errors related to the hypothesis put forward by the operator, among which it stands out as [16]:

- type I errors—result from the adoption of a false hypothesis, which results in the operator performing actions that he should not have performed. Type I errors are commission errors. With regard to the reliability of facilities, this error consists in adopting a hypothesis about the facility's inoperability, which results in financial losses resulting from the failure to use the efficient facility, e.g. stopping the process of water production at the WTP,
- type II errors—result from the rejection of a true hypothesis, which results in the operator's failure to perform the actions that he should perform. Type II errors are omission errors. With regard to the reliability of objects, this error consists in rejecting the hypothesis of the object's inoperability and continuing operation, posing a threat to people and the environment.

Figure 1 shows the classification of operator errors.

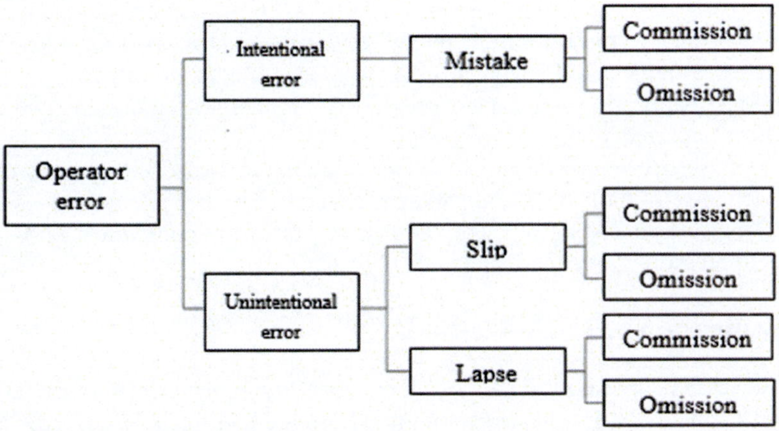

Fig. 1. Types of operator errors (based on [17])

The operator during work is characterized by the following states of efficiency [16]:

- state of efficiency—proper functioning in terms of psychophysical,

- state of potential inefficiency—a temporary limit state in which it is possible to lose psychophysical fitness,
- state of inefficiency—the operator is completely incapable of acting in terms of psychophysical.

Many factors influence the operator's loss of efficiency [16, 18]. The most important of them are:

- unfavorable environmental conditions (lighting, vibrations, noise, microclimate, exposure to hazardous substances),
- improper ergonomics of the operator's station—lack of resources (tools, materials, documentation) to perform the task,
- improper organization of work (complexity of tasks, lack of team cooperation, difficulties in communication and information flow),
- poor psychophysical condition of the operator:
  - fatigue—physical or mental, related to the circadian rhythm of sleep and wakefulness,
  - distraction—results from confusion caused by too much information,
  - stress—occurs when the operator feels psychological discomfort regarding the conditions or requirements related to the implementation of the task in a situation where they exceed his capabilities,
- low competence of the operator related to the lack of knowledge and experience, resulting in a lack of understanding of the purpose of the task and the rules of its implementation,
- low motivation of operators manifesting itself with:
  - routine—caused by repetitive activities, tedious work, results from many years of practice with the simultaneous loss of awareness of existing threats,
  - pressure—appearing as a result of pressure from superiors, lack of the right amount of time, incorrect definition of tasks,
  - lack of assertiveness—meaning the lack of ability to refuse to perform a task due to lack of self-confidence, complexes and fear,
  - negligence—consisting in incorrect assessment of the consequences of actions due to pressure or inexperience.

4 Classification of HRA Methods

With the development of research on operator reliability in technical systems, about 70 different HRA methods [1]. The most important HRA methods include: TESEO (Tecnica Empirica Stima Errori Operatori); THERP (Technique for Human Error Rate Prediction); HEART (Human Error Assessment and Reduction Technique); CREAM (Cognitive Reliability and Error Analysis Methods); ATHEANA (A Technique for Human Event Analysis); SPAR-H (Simplified Plant Analysis Risk Human Reliability Assessment); NARA (Nuclear Action Reliability Assessment); SHERPA (Simulator for Human Error Probability Analysis).

Among the HRA methods, quantitative methods, qualitative methods and mixed methods can be distinguished [1]. An alternative division assumes the existence of three generations of HRA methods. It takes into account the year of creation of the method and differences in the methodology [1]. It is often assumed that the first generation methods (1970–1990) focus on the quantitative assessment of operator reliability, the second generation methods (1990–2005) take into account the qualitative aspect, while the third generation methods (2005–present) combine assumptions of the two previous generations [1].

The characteristics of individual groups of methods are presented below:

- Quantitative methods in assessing the operator's reliability: the assessment of the operator's reliability using quantitative methods focuses on determining the numerical value of HEP, omitting the analysis of the reasons for the operator's actions [1]. Quantitative methods are often criticized due to the lack of consideration of the impact of contextual, environmental or organizational factors on operator reliability and the subjectivity of expert assessment [1]. The results obtained using quantitative methods often assume overestimated values [1]. The example of quantitative methods are TESEO, THERP and HEART.
- Qualitative methods in assessing operator reliability: the qualitative aspect in assessing operator reliability allows for examining the impact of contextual factors accompanying the assessed situation [1]. Qualitative methods make it possible to take into account issues related to the psychophysical state of the operator during the task [1]. Analyzes using qualitative methods allow the identification of potential situations in which an operator error may occur. They can also be used for retrospective analysis [1]. Examples of qualitative methods are CREAM, ATHEANA and SPAR-H.
- Mixed methods in the assessment of operator reliability: due to the limitations of the first and second generation HRA methods, the third generation is currently being developed [9]. There are two main directions of development of new methods of assessing operator reliability. One of them consists in the use of known methods and their modifications, e.g. with the use of Bayesian reasoning, the use of fuzzy logic theory or the construction of decision models. The second direction assumes the use of computer simulations. As a result, third-generation HRA methods reduce the limitations of older methods, allowing for a more realistic operator reliability analysis. Examples of mixed methods are NARA, SHERPA and Fuzzy-Bayes CREAM.

5 The Perspective of Using the HRA in the Reliability and Safety Assessment of CWSS

Due to the fact that the CWSS is an anthropotechnical system, and the role of the operator in ensuring an appropriate level of reliability and safety of the CWSS operation, the use of HRA results in the process of assessing the reliability or safety of the CWSS seems necessary. HRA analyzes provide us with information on the probability of making an error by the operator, so we can prevent undesirable situations related to the operator's actions by introducing appropriate corrective actions at the stage of design, builidng or operation of the system. In order to directly use the results of HRA analyses, existing methods of system reliability assessment can be adapted, taking into account the influence of the operator on the technical object. The basic value describing reliability is the reliability index, which determines the probability that the system will be operational in a given period of time [30]. It can be determined from formula [19]:

$$K_g = \frac{T_p}{T_p + T_n}, [-] \tag{1}$$

where:

T_P—mean time between failure [d],

T_n—mean repair time [d].

Among the most common reliability assessment methods are the one-parameter method, the two-parameter method and the state review method [19]. The one-parameter method is about determining the reliability index for the entire system based on the knowledge of this index for individual components of the system. The calculations are made on the basis of the reliability scheme of the tested system, determining the reliability index for each structure [19]. In the case of the two-parameter method, the reliability scheme of the system should be similarly used and calculations of the remaining reliability indicators should be performed on the basis of the known two reliability indicators of all components of the tested system [19]. The method of reviewing the system states is applied assuming the independence of the elements present in the system. It consists in distinguishing subsets of the state of efficiency and failure in the system, the sum of which creates a full set of events. This is a laborious method, because, 2^L combinations should be considered here, where L is the number of elements in the system that affect the reliability of the system as a whole [19].

The result of the HRA analysis in the form of the human error probability value (HEP) can be used to determine the operator's reliability index, as the probability that the operator will perform his work, being in a state of efficiency, without making an error in given environmental conditions and a specified time period. It can be determined by formula:

$$K_{g\ operator} = 1 - HEP \tag{2}$$

where:

HEP—human error probability.

Having the operator's reliability index, we can include it as an element of the system and use it in the reliability analysis using the one-parameter, two-parameter method or the state review method. The operator and the technical object form an anthropotechnical pair with a serial structure [16], where the damage to the entire system occurs as a result of damage to the technical object or operator's error. Figure 2 shows the serial structure for the operator and the technical object.

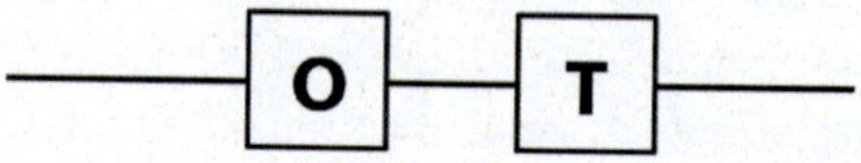

Fig. 2. Serial structure of O (operator) and T (technical object) (based on [16])

The introduction of elements corresponding to a human in the analysis of the reliability of technical systems allows for a better representation of reality. Due to the fact that the operator and the object form a serial structure, the value of the reliability index for the anthropotechnical pair can be determined from the formula [19]:

$$K_g = \prod_{i=1}^{n} K_{gi}, [-] \tag{3}$$

where:

K_{gi}—reliability index of the i-th element of the serial structure.

For an example filter object whose reliability index according to manufacturer's data is 0.98, after taking into account the influence of the operator, the reliability index varies in the range of 0.098 to 0.9702 depending on the operator's reliability index. Table 1 shows an example of a calculation.

Table 1. Impact of operator reliability on filter reliability

Reliability index		Kg operator									
		0.1	*0.2*	*0.3*	*0.4*	*0.5*	*0.6*	*0.7*	*0.8*	*0.9*	*0.99*
$K_{g\ filter}$	0.98	0.098	0.196	0.294	0.392	0.49	0.588	0.686	0.784	0.882	0.970

As observed, the influence of the operator on the reliability of the system operation cannot be omitted, and according to the authors, it should be included as a permanent element of the reliability analysis. The use of the results of the HRA analysis in the CWSS reliability assessment may also be the basis for the development of new CWSS safety assessment methods. After estimating the value of the effects (losses) resulting from the operator's actions, it will be possible to determine the value of the risk associated with the operation of the operator—r_{op}, in accordance with the formula [3]:

$$r_{op} = P \cdot C \tag{4}$$

where:

P—parameter determining the probability of undesirable events related to the operator's actions,

C—parameter determining the value of losses.

The process of analyzing the reliability and safety of the CWSS, taking into account the operator's actions, will reduce the number of undesirable events related to the operator's activities within the CWSS and will allow to make appropriate, rational operational decisions in water supply companies.

6 Conclusions

Collective water supply systems are an example of an anthropotechnical system [11]. The CWSS operator as an element of the human-technical object-environment system has a significant impact on the level of reliability and safety of the system's operation.

The work presents in detail the role of the operator in the CWSS, detailing the tasks in individual subsystems, i.e. the water intake subsystem, water treatment subsystem, water pumping subsystem, water storage subsystem and water distribution subsystem. The classification of human errors and their causes is shown. A review of methods for assessing the reliability of the HRA operator is presented. The paper also indicates the prospect of using the HRA analysis in the assessment of CWSS reliability and safety based on the operator's reliability index and the risk parameter related to the operator's operation. Further research on the impact of the human factor on the functioning of the CWSS is crucial in order to maintain the reliability and security of water supplies at the required level.

The approach allowing to identify and limit the negative impact of the operator on the functioning of the CWSS refers to the recommendations of Directive (EU) 2020/2184 of the European Parliament and of the Council of December 16, 2020 on the quality of water intended for human consumption [20], which emphasizes safety of water supply by introducing a risk-based approach to threats to reduce adverse events within the CWSS.

References

1. French S, Bedford T, Pollard SJT, Soane E (2011) Human reliability analysis: a critique and review for managers. Saf Sci 49(6):753–763
2. Tang Y, Wu S, Miao X, Pollard SJT, Hrudey SE (2013) Resilience to evolving drinking water contamination risks: a human error prevention perspective. J Clean Prod 57:228–237
3. Rak J (2004) Istota ryzyka w funkcjonowaniu systemu zaopatrzenia w wodę. Oficyna Wydawnicza PRz, Rzeszów
4. Tao J, Qiu D, Yang F, Duan Z (2020) A bibliometric analysis of human reliability research. J Clean Prod 260:121041
5. Tu J, Lin W, Lin Y (2015) A Bayes-SLIM based methodology for human reliability analysis of lifting operations. Int J Ind Ergon 45:48–54
6. Tavakoli M, Nafar M (2021) Modification of the FFTA method for calculating and analyzing the human reliability of maintenance groups in power transmission grids. Int J Syst Assur Eng Manag 12
7. Guo Y, Sun Y (2019) Human reliability quantification in flight through a simplified CREAM method. Adv Human Factors Transp, 762–773
8. Wang X, Zhang W, Pan X, Liu T (2019) Method for crew human reliability analysis based on CREAM in advances in human error, reliability, resilience and performance. Springer, pp 225–234
9. Zhou Q, Wong YD, Loh HS, Yuen KF (2018) A fuzzy and Bayesian network CREAM model for human reliability analysis—the case of tanker shipping. Saf Sci 105:149–157
10. Nowakowski W, Ciszewski T, Łukasik Z (2018) Metody oceny wpływu czynnika ludzkiego na bezpieczeństwo w transporcie. Autobusy: technika, eksploatacja, systemy transportowe 19(6)
11. Rak J, Tchórzewska-Cieślak B, Żywiec J (2019) Czynnik niezawodności człowieka w systemach zaopatrzenia w wodę. Instal 3
12. Rak J, Piegdoń I, Homa K (2013) Kryzysowe zaopatrzenie Rzeszowa w wodę. J KONBiN 1(25)
13. Rozporządzenie Ministra Zdrowia z dnia 7 grudnia 2017 r w sprawie jakości wody przeznaczonej do spożycia przez ludzi (Dz.U. 2017 poz. 2294)
14. Knapik K, Bajer J (2010) Wodociągi. Wydawnictwo Politechniki Krakowskiej, Kraków

15. Kowal A, Świderska-Bróż M (2007) Oczyszczanie wody. Podstawy teoretyczne i technologiczne, procesy i urządzenia. Wydawnictwo Naukowe PWN, Warszawa
16. Smalko Z, Nowakowski T, Tubis A (2020) Zarys niezawodnościowej teorii zagrożeń. Oficyna Wydawnicza Politechniki Wrocławskiej, Wrocław
17. Reason J (2000) Human error: models and management. BMJ 320(7237):768
18. Fruggiero F, Fera M, Lambiase A, Di Pasquale V (2019) Linking human factors to assess human reliability in human-friendly robotics. Springer, Switzerland
19. Kwietniewski M, Roman M, Kłoss-Trębaczkiewicz H (1993) Niezawodność wodociągów i kanalizacji. Arkady, Warszawa
20. Dyrektywa Parlamentu Europejskiego i Rady (UE) 2020/2184 z dnia 16 grudnia 2020 r. w sprawie jakości wody przeznaczonej do spożycia przez ludzi

Author Index

Z. Blikharskyy et al. (Eds.): CEE 2023, LNCE 438, pp. 563–565, 2024.
https://doi.org/10.1007/978-3-031-44955-0

Z